Advances in Soil Science

GLOBAL CLIMATE CHANGE AND TROPICAL ECOSYSTEMS

Advances in Soil Science

Series Editor: *B. A. Stewart*

Published Titles

Interacting Processes in Soil Science
R. J. Wagenet, P. Baveye, and B. A. Stewart
Soil Biology: Effects on Soil Quality
J. L. Hatfield and B. A. Stewart
Crops Residue Management
J. L. Hatfield and B. A. Stewart
Soil Processes and Water Quality
R. Lal and B. A. Stewart
Subsoil Management Techniques
N. S. Jayawardane and B. A. Stewart
**Soil Management: Experimental Basis for Sustainability
and Environmental Quality**
R. Lal and B. A. Stewart
Soil Management and Greenhouse Effect
R. Lal, J. M. Kimble, E. Levine, and B. A. Stewart
Soils and Global Change
R. Lal, J. M. Kimble, E. Levine, and B. A. Stewart
Soil Structure: Its Development and Function
B. A. Stewart and K. H. Hartge
Structure and Organic Matter Storage in Agricultural Soils
M. R. Carter and B. A. Stewart
Methods for Assessment of Soil Degradation
R. Lal, W. H. Blum, C. Valentine, and B. A. Stewart
Management of Carbon Sequestration in Soil
R. Lal, J. M. Kimble, R. F. Follett, and B. A. Stewart
Soil Processes and the Carbon Cycle
R. Lal, J. M. Kimble, R. F. Follett, and B. A. Stewart
Global Climate Change and Pedogenic Carbonates
R. Lal, J. M. Kimble, and B. A. Stewart
Global Climate Change and Tropical Ecosystems
R. Lal, J. M. Kimble, and B. A. Stewart
Global Climate Change: Cold Regions Ecosystems
R. Lal, J. M. Kimble, and B. A. Stewart
Assessment Methods for Soil Carbon
R. Lal, J. M. Kimble, R. F. Follett, and B. A. Stewart

Advances in Soil Science

GLOBAL CLIMATE CHANGE AND TROPICAL ECOSYSTEMS

Edited by

R. Lal
J.M. Kimble
B.A. Stewart

CRC Press
Taylor & Francis Group
Boca Raton London New York

CRC Press is an imprint of the
Taylor & Francis Group, an **informa** business

CRC Press
Taylor & Francis Group
6000 Broken Sound Parkway NW, Suite 300
Boca Raton, FL 33487-2742

First issued in paperback 2019

© 2000 by Taylor & Francis Group, LLC
CRC Press is an imprint of Taylor & Francis Group, an Informa business

No claim to original U.S. Government works

ISBN-13: 978-1-56670-485-4 (hbk)
ISBN-13: 978-0-367-39908-5 (pbk)

Visit the Taylor & Francis Web site at
http://www.taylorandfrancis.com

and the CRC Press Web site at
http://www.crcpress.com

ABOUT THE EDITORS

Dr. R. Lal is a professor of soil science in the School of Natural Resources at The Ohio State University. Prior to joining Ohio State in 1987, he served as a soil scientist for 18 years at the International Institute of Tropical Agriculture, Ibadan, Nigeria. Professor Lal is a fellow of the Soil Science Society of America, American Society of Agronomy, Third World Academy of Sciences, American Association for the Advancement of Sciences, Soil and Water Conservation Society, and the Indian Academy of Agricultural Sciences. He is a recipient of the International Soil Science Award, the Soil Science Applied Research Award of the Soil Science Society of America, the International Agronomy Award of the American Society of Agronomy, the International Agronomy Award of the American Society of Agronomy, and the Hugh Hammond Bennett Award of the Soil and Water Conservation Society. He is past president of the World Association of the Soil and Water Conservation Society and the International Soil Tillage Research Organization.

Dr. John Kimble is a Research Soil Scientist at the USDA Natural Resources Conservation Service National Soil Survey Laboratory in Lincoln, Nebraska. Dr. Kimble manages the Global Change project of the Natural Resources Conservation Service, and has worked more than 15 years with the U.S. Agency for International Development projects dealing with soils-related problems in more than 40 developing countries. He is a member of the American Society of Agronomy, the Soil Science Society of America, the International Soil Science Society, and the International Humic Substances Society.

Dr. Hari Eswaran is the national leader for World Soil Resources at the U.S. Department of Agriculture, Natural Resources Conservation Service. His major work has been on properties, genesis and classification of soils of the tropics. More recently, his focus is on land degradation and desertification and has used GIS and models to evaluate global soil resources.

Dr. B.A. Stewart is Distinguished Professor of Agriculture, and Director of the Dryland Agriculture Institute at West Texas A&M University. Prior to joining West Texas A&M University in 1993, he was Director of the USDA Conservation and Production Research Laboratory, Bushland, Texas. Dr. Stewart is a past president of the Soil Science Society of America, and was a member of the 1990-93 Committee on Long Range Soil and Water Policy, National Research Council, National Academy of Sciences. He is a Fellow of the Soil Science Society of America, American Society of Agronomy, Soil and Water Conservation Society, a recipient of the USDA Superior Service Award, and a recipient of the Hugh Hammond Bennett Award of the Soil and Water Conservation Society.

PREFACE

Tropical ecosystems, regions between tropics of Cancer and Capricorn, play important roles in global processes, economic issues, and political concerns. Among numerous processes, tropical ecosystems affect the global carbon cycle. Two predominant tropical ecosystems include tropical rainforest (TRF) and tropical savanna (TS). The TRF ecosystems are mostly concentrated around $10°$ north and south of the equator, where precipitation exceeds evapotranspiration for more than 9 months in a year. These regions, also termed the humid tropics, support tropical forest vegetation. Tropical regions where precipitation exceeds evapotranspiration for 6 to 8 months a year support semi-deciduous TRF ecosystems. The TS ecosystems normally exist in regions where precipitation exceeds evapotranspiration for 4 to 6 months a year, or where soil-related constraints (e.g., sub-soil acidity, Al toxicity or P deficiency) limit growth of herbaceous vegetation and woody perennials. The TS ecosystems may be dominated by trees, shrubs or grasses depending on rainfall regime, water balance, and soil characteristics. In their natural state, tropical ecosystems support a large quantity of above- and below-ground biomass, and constitute a major part of the terrestrial C pool.

Conversion of natural to agricultural and forestry ecosystems can disturb the ecologic balance, disrupt the C cycle, deplete soil and biotic C pools, and lead to emission of C (as CO_2 and CH_4) and N (as N_2O, and NO_x) into the atmosphere. These gases (CO_2, CH_4 and N_2O) being radiatively-active (greenhouse gases) can influence the global climate. Numerous experiments from around the world have demonstrated emissions of greenhouse gases from conversion of tropical ecosystems to agricultural, pastoral and silvicultural land uses. The magnitude of emission, however, depends on the method of removing the natural vegetation cover (e.g., mechanical, chemical, manual or burning), land use (e.g., arable, pastoral or plantation), soil management (e.g., tillage methods, residue management, fertilizer use, irrigation, etc.), and cropping/farming systems (e.g., crop rotations and crop combinations, agropastoral systems, agroforestry systems, etc.). An important factor affecting the C balance is the land use and farming system. Traditional shifting cultivation and related bush fallow or resource-based systems have a very different impact on the carbon balance than science-based systems involving judicious/discriminate use of external input.

Despite the importance of tropical ecosystems and their impact on the global C cycle, there is a lack of literature on systematic assessment of the C pool and fluxes in undisturbed state versus anthropogenically-perturbed ecosystems. Therefore, a workshop was organized in Belem, Brazil to highlight the significance of tropical ecosystems in global C cycle.

The workshop entitled "Carbon Pools and Dynamics in Tropical Ecosystems" was held in Belem, from 1-5 December, 1997. About 30 participants were drawn from 10 countries including Australia, Brazil, Colombia, Holland, Kenya, Germany, United Kingdom, Venezuela and the United States. The objectives of the workshop were to: (i) assess C pool in soils and biomass of tropical ecosystems, (ii) evaluate the magnitude of C flux from natural and managed ecosystems, (iii) determine the impact of anthropogenic activities, land use and land cover, and management on C pools and fluxes, (iv) evaluate carbon dynamics in tropical ecosystems in relation to soil quality and agricultural productivity, and (v) identify methodological and modeling potentials and constraints to determine C pools and fluxes at different scales.

This workshop, second in a series of three regional workshops organized for specific ecoregions (e.g., arid, tropics, cold), emerged as a recommendation of an international symposium "Carbon Sequestration in Soils" held in Columbus, OH, in July, 1996. The workshop was jointly organized by The Ohio State University, Columbus, OH (USA), National Soil Survey Center of USDA-NRCS at Lincoln, NE (USA) and EMBRAPA-CPATU, Belem, Para (Brazil). The financial support for the workshop was provided by USDA-NRCS, Ohio State University and EMBRAPA.

This volume is based on some of the papers presented at the workshop. Additional papers were sought from scientists with long-standing experience in the region. All papers published in this volume were reviewed by the editorial committee and appropriately revised and edited. A total of 25 papers are logically arranged in seven separate but interrelated sections. These papers represent data on C pool and fluxes from case studies in 12 countries of tropical regions. The first section entitled

"Carbon pool in tropical ecosystems" contains four papers dealing with characteristics of tropical ecosystems and magnitude and attributes of soil and biotic C pools. In addition to an overview paper, three other papers deal with C pool in soils of Brazil, Australia and the Pacific, and India.

Land use impacts on C pool and dynamics are discussed in three separate sections dealing with Africa, tropical America, and Asia and the Pacific. The section dealing with soils of tropical Africa contains four chapters. One of the four chapters collates the available information on slash-and-burn practices in the humid tropics, the second deals with crop residue and fertility management effects on C pool in soils of semi-arid west Africa, and the two chapters from western Nigeria are based on case studies dealing with the C sequestration potential through restoration of degraded soils.

The third section deals with land use in tropical America and comprises six chapters, four of which are from different regions of Brazil, one from the savanna region of Colombia, and one is an overview paper dealing with C dynamics in the semi-arid regions in general.

The fourth section dealing with Asia and the Pacific contains two manuscripts, one each from India and Australia. Both of these manuscripts assess the impact of land use and soil management on C pool. The fifth section deals with terminology, generic concepts and basic processes affecting soil carbon pool and dynamics. The sixth section, dealing with monitoring and prediction, contains six chapters addressing the issue of C assessment in laboratory and extrapolation to different scales. Critical issues of soil sampling, sample preparation, analytical techniques, and scaling procedures from point to landscape and regional scales are discussed in this section. The seventh section contains one concluding chapter that summarizes the discussion of all sections, identifies knowledge gaps and prioritizes research and development issues.

The organization of the symposium and the publication of this volume was made possible by cooperation and funding from USDA-NRCS, The Ohio State University and EMBRAPA-CPATU. Special thanks are due to Dr. Adilson Serrao and his colleagues at EMBRAPA-CPATU for their generous hospitality and logistic support. The field tour was successfully organized by Adilson Serrao and scientists from the EMBRAPA-CPATU. We also thank the staff of the Inter-American Institute of Global Change and Dr. Carlos Cerri of the University of São Paulo at Piricicaba. The editors thank all authors for their outstanding efforts to document and present in a timely fashion their information on the current understanding of soil processes and the carbon cycle. Their efforts have enhanced the overall understanding of pedospheric processes in tropical ecosystems and how to better use soils as a sink for carbon while also managing soils to minimize pedosphere contributions of carbon dioxide and other greenhouse gases to the atmosphere. These efforts have advanced the frontiers of soil science and improved the understanding of tropical ecosystems into the broader scientific arena of linking soils to the global carbon cycle, soil productivity and environmental quality in these ecologically sensitive regions. Thanks are also due to the staff of the Sleeping Bear Press and CRC for their efforts in publishing this information on time to make it available to the overall scientific community.

We especially thank Lynn Everett for her support in organizing the conference and for handling the flow of chapters to and from the authors throughout the review process. Her tireless efforts, good humor, and good nature are greatly appreciated. We also offer special thanks to Brenda Swank for her help in preparing this material and for her assistance in all aspects of the workshop and preparation of the manuscript. Maria Lemon of "The Editors Inc." helped in editing several chapters. The efforts of many others were also very important in getting this relevant and important scientific information out in a timely manner.

The Editorial Committee

CONTRIBUTORS

R.L. Ahuja, CCS Haryanna Agricultural University, Hissar 125 004, India.

J. Alegre, ICRAF-Peru, Lima, Peru.

D. Arrouays, Institut National de la Recherche Agronomique, Unite de Science du Sol SESCPF, 45160 Ardon, France.

A. Bationo, IFDC/ICRISAT, B.P. 12404 Niamey, Niger.

J. Bauhus, Division of Forestry, CSIRO, P.O. Box 4008, Canberra City, Australian Capital Territory 2604, Australia.

M. Bernoux, Centro de Energia Nuclear na Agricultura, Universidade de São Paulo, Avenida Cenenario 303, CP 96, CEP 13400-970 Piracicaba, São Paulo, Brazil.

T. Bhattacharyya, National Bureau of Soil Survey and Land Use Planning, Amravati Road, Nagpur 440 010, India.

C.L. Bielders, ICRISAT, B.P. 12404 Niamey, Niger.

O. Brunini, Instituto Aronomics Campinas, São Paulo, Brazil.

O.M.R. Cabral, EMBRAPA Meio Ambiente, Jaguariuna, São Paulo, Brazil.

J.O. Carter, Department of Natural Resources, Resource Sciences Centre, Brisbane, Queensland, Australia.

C. Castilla, Apto 103, Cali, Colombia.

C.C. Cerri, Centro de Energia Nuclear na Agricultura, Universidade de São Paulo, Avenida Centenario 303, CP 96, CEP 13400-970, Piracicaba, São Paulo, Brazil.

H.H. Cheng, Department of Soil Science, Water and Climate, University of Minnesota, 1991 Upper Buford Circle, St. Paul, MN 55108-6028, U.S.

K. Coleman, Soil Science Department, IACR-Rothamsted, Harpenden, Herts. AL5 2JQ, U.K.

D.G. Cordeiro, EMBRAPA-Acre, P.O. Box 381, Rio Branco, Acre, Brazil.

M. Cravo, EMPRAPA-CPAA, Manaus, Amazonas, Brazil.

A.D. Culf, IH, Wallingford, OX108BB, U.K.

R.C. Dalal, Department of Natural Resources, Resource Sciences Centre, Brisbane, Queensland, Australia.

H.R. Da Rocha, Departmento de Ciencias Atmosfericas / IAG, Universidade de São Paulo, Rua do Matao 1226 - Cid Universitaria, São Paulo, O5508-900, Brazil.

M.A.F. DaSilva Dias, Departmento de Ciencias Atmosfericas/ IAG, Universidade de São Paulo, São Paulo, Brazil.

E.A. Davidson, The Woods Hole Research Center, P.O. Box 296, Woods Hole, MA 02543, U.S.

M. Denich, ZEF, University of Bonn, Walter-Flex-Str. 3, 53113 Bonn, Germany.

J.A. Elbers, Winard Staring Centre/ DLO, Wageringen, The Netherlands.

P. Falloon, Soil Science Department, IACR-Rothamsted, Harpenden, Herts. AL5 2JQ, U.K.

P.M. Fearnside, INPA, CP 478, CEP 69011-970 Manaus, Amazonas, Brazil.

B.J. Feigl, Centro de Energia Nuclear na Agricultura, Universidade de São Paulo, Avenida Cenenario 303, CP 96, CEP 13400-970 Piracicaba, São Paulo, Brazil.

S. Fifer, NASA/Goddard Spaceflight Center, Code 923, Biospheric Sciences Branch, Greenbelt, MD 20771, U.S.

G. Fisch, CTA/IAE-ACA, São Jose dos Campos-São Paulo, 12228-908, Brazil.

H.C. Freitas, Departmento de Ciencias Atmosfericas/ IAG, Universidade de São Paulo, São Paulo, Brazil.

K. Hairiah, Brawijaya University, Malang 65145, Indonesia.

J. Ingram, Soil Science Department, IACR-Rothamsted, Harpenden, Herts. AL5 2JQ, U.K.

D. Jenkinson, Soil Science Department, IACR-Rothamsted, Harpenden, Herts, AL5 2JQ, U.K.

M. Kanashiro, EMPRAPA Amazonia Oriental, Belem, Pará, Brazil.

P.K. Khanna, Division of Forest Resources, CSIRO, P.O. Box 4008, Canberra City, Australian Capital Territory 2604, Australia.

J.M. Kimble, National Soil Survey Center, USDA-NRCS, Federal Building, Room 152, 10 Centennial, Mall North, Lincoln, NE 68508-3866, U.S.

D. Kimes, NASA/Goddard Spaceflight Center, Code 923, Biospheric Sciences Branch, Greenbelt, MD 20771, U.S.

J. Kotto-Same, IRAD, P.O. Box 2067, Yaounde, Cameroon.

R. Lal, The Ohio State University, School of Natural Resources, 2021 Coffey Road, Columbus, OH 43210, U.S.

E. Levine, NASA/Goddard Spaceflight Center, Code 923, Biospheric Sciences Branch, Greenbelt, MD 20771, U.S.

M.A. Ligo, EMBRAPA Meio Ambiente, Jaguariuna, São Paulo, Brazil.

M.C.M. Macedo, EMBRAPA-CNPGC, CP 154, Campo Grande, Mato Grosso do Sul, CEP 79002-970, Brazil.

Y. Malhi, IERM, University of Edinburgh, Edinburgh, EH9 3JU, U.K.

M.C. Manna, Indian Institute of Soil Science, Nabibagh, Berasia Road, Bhopal 462038, India.

A.U. Mokwunye, United Nations University, University of Ghana, Legon, Accra, Ghana.

A. Moukam, IRAD, P.O. Box 2067, Yaounde, Cameroon.

C. Neill, Marine Biological Laboratory, The Ecosystems Center, Woods Hole, MA 02543, U.S.

R. Nelson, NASA/Goddard Spaceflight Center, Code 923, Biospheric Sciences Branch, Greenbelt, MD 20771, U.S.

A.D. Nobre, INPA, CP 478, CEP 69011-970 Manaus, 69083-000, Amazonas, Brazil.

C.A. Nobre, CPTEC, INPE, Cachoeira Paulista, CEP 12630-000, São Paulo, Brazil.

D.K. Pal, National Bureau of Soil Survey and Land Use Planning, Amravati Road, Nagpur 440 010, India.

C.A. Palm, Tropical Soil Biology and Fertile Programme, P.O. Box 30592, Nairobi, Kenya.

L. Pásztor, GIS Laboratory, RISSAC, Herman OTTO Utca 15, Budapest 1022, Hungary.

M.C. Piccolo, Centro de Energia Nuclear na Agricultura, Universidade de São Paulo, Avenida Cenenario 303, CP 96, CEP 13400-970 Piracicaba, São Paulo, Brazil.

D.V.S. Resck, EMBRAPA-CPAC, Km 18 Br 020 CP 08223, Planaltina, Distrito Federal, CEP 73301-970, Brazil.

A. Riese, INIA, Pucallpa, Peru.

C.A. Robinson, Division of Agriculture, West Texas A&M University, WTAMU Box 60278, Canyon, TX 79016, U.S.

V. Rodrigues, EMBRAPA-Rondônia, P.O. Box 406, Porto Velho, Rondônia, Brazil.

G.B. Singh, Indian Council of Agricultural Research, Krishi Bhavan, New Delhi, India.

J. Smith, Soil Science Department, IACR-Rothamsted, Harpenden, Herts. AL5 2JQ, U.K.

P. Smith, Soil Science Department, IACR-Rothamsted, Harpenden, Herts. AL5 2JQ, U.K.

P. Snowdon, Division of Forestry, P.O. Box 4008, Canberra City, Australian Capital Territory 2604, Australia.

W. Sombroek, Edafologo, Projeto PPG7 - SPRN/AAP, Instituto de Prote, Ambiental de Amazonas, Rua Recife 3280, Parque 10, 69050-030, Manaus, Amazonas, Brazil.

B.A. Stewart, Dryland Agriculture Institute, Division of Agriculture, West Texas A&M University, WTAMU Box 60278, Canyon, TX 79016, U.S.

A. Swarup, Indian Institute of Soil Science, Nabibagh, Berasia Road, Bhopal 462038, India.

J. Szabó, GIS laboratory, RISSAC, Herman OTTO Utca 15, Budapest 1022, Hungary.

M. van Noordwijk, ICRAF-Indonesia, P.O. Box 161, Bagor 16001, Indonesia.

S. Vadivelu, National Bureau of Soil Survey and Land Use Planning, Amravati Road, Nagpur 440 010, India.

C.A. Vasconcellos, EMBRAPA-CNPMS, Rodovia MG 424 Km 65 CP 151, Sete Lagoas, MG, CEP 35701-970, Brazil.

M. Velayutham, National Bureau of Soil Survey and Land Use Planning, Amravati Road, Shankar Nagar P.O., Nagpur 440 010, India.

L. Vilela, EMBRAPA-CPAC, Km 18 Br 020 CP 08223, Planaltina, Distrito Federal, CEP 73301-970, Brazil.

P.L.G. Vlek, Institute of Agriculture in the Tropics, University of Goettingen, Goettingen, Germany.

C. Von Randow, Departmento de Ciencias Atmosfericas/ IAG, Universidade de São Paulo, São Paulo, Brazil.

S.P. Wani, ICRISAT, Patancheru 5023424, Hyderabad, Andhra Pradesh, India.

P.L. Woomer, c/o Dr. Nancy Karanja, Department of Soil Science, University of Nairobi, P.O. Box 30197, Nairobi, Kenya.

CONTENTS

Section I. CARBON POOL IN TROPICAL ECOSYSTEMS

Chapter 1. Tropical Ecosystems and the Global C Cycle 3
R. Lal and J.M. Kimble

Chapter 2. Carbon Stocks in Soils of the Brazilian Amazon 33
C.C. Cerri, M. Bernoux, D. Arrouays, B.J. Feigl and M.C. Piccolo

Chapter 3. Carbon Pools in Forest Ecosystems of Australasia and Oceania 51
P.K. Khanna, P. Snowdon and J. Bauhus

Chapter 4. Organic Carbon Stock in Soils of India 71
M. Velayutham, D.K. Pal and T. Bhattacharyya

Section II. LAND USE AND CARBON POOL IN SOILS OF AFRICA

Chapter 5. Slash-and-Burn Effects on Carbon Stocks in the Humid Tropics 99
P.L. Woomer, C.A. Palm, J. Alegre, C. Castilla, D.G. Cordeiro, K. Hairiah, J. Kotto-Same,
A. Moukam, A. Riese, V. Rodrigues and M. van Noordwijk

Chapter 6. Crop Residue and Fertilizer Management to Improve Soil Organic Carbon
Content, Soil Quality and Productivity in the Desert Margins of West Africa 117
A. Bationo, S.P. Wani, C.L. Bielders, P.L.G. Vlek and A.U. Mokwunye

Chapter 7. Restorative Effects of *Mucuna Utilis* on Soil Organic C Pool of a Severely
Degraded Alfisol in Western Nigeria 147
R. Lal

Chapter 8. Land Use and Cropping System Effects on Restoring Soil Carbon Pool of
Degraded Alfisols in Western Nigeria 157
R. Lal

Section III. LAND USE AND CARBON POOL IN SOILS OF TROPICAL AMERICA

Chapter 9. Impact of Conversion of Brazilian Cerrados to Cropland and Pastureland
on Soil Carbon Pool and Dynamics 169
D.V.S. Resck, C.A. Vasconcellos, L. Vilela and M.C.M. Macedo

Chapter 10. Soil Carbon Accumulation or Loss Following Deforestation for Pasture in the Brazilian Amazon. 197
C. Neill and E.A. Davidson

Chapter 11. The Potential and Dynamics of Carbon Sequestration in Traditional and Modified Fallow Systems of the Eastern Amazon Region, Brazil 213
M. Denich, M. Kanashiro and P.L.G. Vlek

Chapter 12. Greenhouse Gas Emissions from Land-Use Change in Brazil's Amazon Region 231
P.M. Fearnside

Chapter 13. Land Use Impact on Carbon Dynamics in Soils of the Arid and Semiarid Tropics 251
B.A. Stewart and C.A. Robinson

Section IV. LAND USE AND CARBON POOL IN SOILS OF ASIA AND THE PACIFIC

Chapter 14. Impact of Land Use and Management Practices on Organic Carbon Dynamics in Soils of India 261
A. Swarup, M.C. Manna and G.B. Singh

Chapter 15. Soil Organic Matter Dynamics and Carbon Sequestration in Australian Tropical Soils 283
R.C. Dalal and J.O. Carter

Section V. BASIC SOIL PROCESSES AND CARBON DYNAMICS

Chapter 16. Soil Aggregation and C Sequestration 317
R. Lal

Section VI. MONITORING AND PREDICTION

Chapter 17. Methods of Analysis for Soil Carbon: An Overview 333
H.H. Cheng and J. Kimble

Chapter 18. Modeling Soil Carbon Dynamics in Tropical Ecosystems 341
P. Smith, P. Falloon, K. Coleman, J. Smith, M.C. Piccolo, C. Cerri, M. Bernoux, D. Jenkinson, J. Ingram, J. Szabó and L. Pásztor

Chapter 19. Evaluating Tropical Soil Properties with Pedon Data, Satellite Imagery, and Neural Networks 365
E. Levine, D. Kimes, S. Fifer and R. Nelson

Chapter 20. Geographic Assessment of Carbon Stored in Amazonian Terrestrial Ecosystems and Their Soils in Particular 375
W.G. Sombroek, P.M. Fearnside and M. Cravo

Chapter 21. Carbon Dioxide Measurements in the Nocturnal Boundary Layer over Amazonian Tropical Forest 391
G. Fisch, A.D. Culf, Y. Malhi, C.A. Nobre and A.D. Nobre

Chapter 22. Atmospheric CO_2 Fluxes and Soil Respiration Measurements over Sugarcane in Southeast Brazil 405
H.R. Da Rocha, O.M.R. Cabral, M.A.F. Da Silva Dias, M.A. Ligo, J.A. Elbers, H.C. Freitas, C.Von Randow and O. Brunini

Section VIII. RESEARCH AND DEVELOPMENT PRIORITIES

Chapter 23. What Do We Know and What Needs to Be Known and Implemented for C Sequestration in Tropical Ecosystems 417
R. Lal and J.M. Kimble

Index 433

Section I.

CARBON POOL IN TROPICAL ECOSYSTEMS

CHAPTER 1

Tropical Ecosystems and the Global C Cycle

R. Lal and J.M. Kimble

I. Introduction

The rapid increase in atmospheric concentration of CO_2 and other greenhouse gases (CH_4, N_2O, NO_x) since about 1850 has raised numerous questions of global significance. Several are important: What is the role of tropical ecosystems as sources or sinks for atmospheric CO_2, and how does the potential change in climate alter ecological processes and basic functions of such ecosystems? How do anthropogenic perturbations of tropical ecosystems affect atmospheric CO_2 concentration, and how would these ecosystems function under raised levels of atmospheric CO_2? What is the role of soils of tropical ecosystems in the global C cycle, and what soil management options can exploit the full potential of these soils as major sinks for atmospheric CO_2?

By virtue of their size and the biomass contained in them, tropical ecosystems are important to the global C cycle. Total land area of tropical ecoregions is about 4.9×10^9 ha. On the basis of rainfall regime, tropical regions can be grouped into different categories. Regions receiving >1000 mm of rainfall per year are termed the humid tropics (Plates 1, 2, 3), those with 500 to 1000 mm as semi-arid tropics (Plates 4, 5, 6), and those with < 500 mm as arid or dryland tropics (Plate 7) (Figure 1). Of the total land area, the humid tropics comprise 1.18×10^9 ha (24%), semi-arid 2.4×10^9 ha (49%), and arid tropics 1.32×10^9 ha (27%). The humid tropics have a generally favorable soil moisture regime throughout the year with a growing season from 9 to 12 months. However, short dry spells may cause drought stress to shallow-rooted annuals or seasonal crops during the growing season. The growing season in semi-arid tropics may range from 5 to 9 months, and the rainfall distribution may be monomodal or bimodal. Drought stress may be frequent even during the rainy season, and supplemental irrigation can enhance crop growth and yield. Most arid or dry tropics have a short rainy season and the growing season varies from 3 to 5 months. Supplemental irrigation can prolong the growing season and increase crop production.

On the basis of soil moisture regime, the balance between rainfall and evapotranspiration as regulated by soil properties, the tropics can be classified into different zones (Tables 1 and 2). The moist savanna zone occupies an area of 12.6%, dry humid 9.6%, typic humid 12.9% and per humid 5.9% of the total tropics.

Predominant soils of the humid tropics are Oxisols, Ultisols and Alfisols, which together cover 920 $\times 10^6$ ha or 78.3% of the humid tropical land (Table 3). These three soil orders also constitute large areas of the semi-arid tropics (1.54×10^9 ha or 64.1%). In addition, other soils of the semi-arid tropics comprise Psamments and Lithosols (272×10^6 ha or 11.3%), Inceptisols (192×10^6 ha or 36.6%), Aridisols (582×10^6 ha or 44.2%), Vertisols and Andisols. For the tropics as a whole, most predominant soils are Oxisols, Ultisols and Alfisols comprising 51% of the land area, Psamments and Lithosols covering 17% of the land area, and Aridisols covering 14% of the land area.

Plate 1. Canopy structure of the tropical rainforest in the Amazon (Dr. B.A. Stewart is in foreground).

Plate 2. Understorey of the rainforest ecosystem.

Plate 3. Soil surface under TRF covered with leaf litter.

Plate 4a. Tropical savanna vegetation with (a): large density of trees.

Plate 4b. Tropical savanna vegetation with (b): less density of trees.

Plate 5. A tropical grass savanna.

Plate 6. Treeless savanna vegetation with perennial shrubs.

Plate 7. Scrub vegetation of the arid tropics.

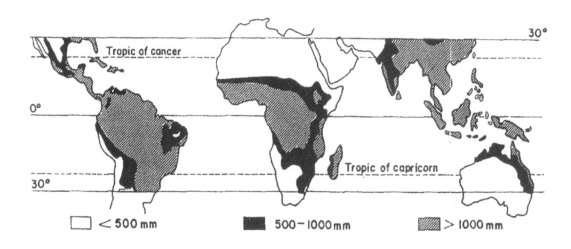

Figure 1. Annual rainfall regimes of tropical and subtropical regions of the world.
(Adapted from Sanchez and Isbell, 1978.)

Table 1. Area occupied by different kinds of soil moisture regimes. This is an estimate of the potential forest area within the tropics; total area of tropics (considered as the zone between the Tropics of Cancer and Capricorn) is 49.6 km^2

Soil moisture regimes	Area (10^6 km^2)	Percent of tropics
Moist savanna	6.2	12.6
Dry humid	4.7	9.6
Typic humid	6.3	12.9
Per humid	2.9	5.9
Others	29.6	59.0
Total	49.7	100.0

Table 2. Estimates of different types of TRF for 1985 and 2000

Biome	1985 area (10^6 km^2)	2000 area (10^6 km^2)
Equatorial TRF (humid)	4.4	4.0
Tropical seasonal (subhumid)	6.0	5.5
Moist/dry forest (semi-arid)	6.0	5.5
Total	16.4	15.0

(Modified from Bruenig, 1996.)

Table 3. Distribution of major soils of the tropics

| Soil group | Moisture regimes | | | | Percent of tropics |
	Humid	Semi-arid	Dry and arid	Total	
Oxisols, Ultisols, and Alfisols	920	1540	51	2511	51
Psamments and Lithic group	80	272	482	834	17
Aridisols	0	103	582	685	14
Aqueps and Fluvents	146	192	28	366	8
Vertisols, Mollisols	24	174	93	291	6
Andepts, Hopepts	5	122	70	207	4
Total area	1175	2403	1316	4896	100
% of tropics	24	49	27	100	

(Modified from Sanchez, 1976 and VanWambeke, 1992.)

II. Tropical Ecosystems

Combination of the soil and rainfall regime governs the existence of principal biomes or ecoregions of the tropics. Two principal biomes of the tropics are tropical rainforest and tropical savannas.

A. Tropical Rainforest

The tropical rainforest (TRF) ecosystems consist of approximately one-sixth of the deciduous forests of the world and occur between 6 to 10° north and south of the equator in South America, Africa, Southeast Asia, and the Pacific. Primeval TRF covered more than 90% of the biome's land surface before the advent of humankind. Bruenig (1996) estimated that cumulative area of all TRF was 5.5 x 10^6 km^2 in 1965, 5.0 in 1975, and 4.4 in 1985. The area of TRF is projected to be 4.0 x 10^6 km^2 in 2000 and 3.0 x 10^6 km^2 in 2050. Bruenig (1996) estimated temporal changes in TRF (million ha) in tropical America from 1850 to 1985 as follows:

	1850	1985
Evergreen	226	212
Seasonal	616	445
Open	380	211
Total	1222	868

The distribution of land area of the TRF in three continents and the rate of deforestation are shown in Tables 4 and 5. The current area under TRF is estimated at 1.76 billion ha, with a rate of deforestation at 15.4 million ha yr^{-1} (Faminow, 1998). The types of vegetation in three ecoregions of Asia, Africa and tropical America are quite different from one another, highly variable or diverse (Jordan, 1983; Almeda and Pringle, 1988), and are characterized with a large quantity of above- and below-ground biomass (Plates 1-3). The annual addition of biomass in TRF (e.g., litter, branches and dead roots) in the mature forest is about 5 Mg ha^{-1}yr^{-1} compared with about 1 Mg ha^{-1} yr^{-1} in temperate forest. The range of organic matter returned to the soil is 3 to 15 Mg ha^{-1} yr^{-1} in TRF compared with 1 to 8 Mg ha^{-1} yr^{-1} in temperate forests.

Table 4. Estimates of forest cover and deforestation in the humid tropics (1000 ha)

Continent	Forest 1980	Forest 1990	Annual deforestation 1981–1990	Rate of change 1981–1990 (% yr^{-1})
Africa	289,700	241,800	4,800	-1.7
Latin America	825,900	753,000	7,300	-0.9
Asia	334,500	287,500	4,700	-1.4
Total	1,450,100	1,282,300	16,800	-1.2

(Modified from NRC, 1993.)

Table 5. Tropical rainforests and the rate of deforestation

Region	Area (10^6 ha)	Rate of deforestation (10^6 ha yr^{-1})
Central Africa	204.10	1.08
Tropical southern Africa	100.46	0.84
West Africa	55.60	0.53
South Asia	63.90	0.50
Southeast Asia	210.60	2.80
Mexico	48.60	0.59
Central America	19.50	0.36
Brazil	561.10	3.42
Andean region and Paraguay	241.80	2.25
Total	1,505.70	12.37

(Modified from WRI, 1993; Southgate, 1998; and Faminow, 1998; Faminow estimated total TRF at 1756.3 M ha and the rate of deforestation at 15.4 M ha yr^{-1}.)

Therefore, tropical forests may add about 5 times more organic matter to the soil than temperate forests.

B. Tropical Savannas

The tropical savannas (TS) cover a large land area especially in South America, Australia and Africa (Misra, 1979; Figure 1). With an annual rainfall of 500 to 1000 mm, the vegetation of tropical savannas is very diverse (Plates 4-6). The range of organic matter added to the soil (litter, branches and dead roots) is 0.5 to 1.5 Mg ha yr^{-1} in tropical savanna compared with 1.5 Mg ha yr^{-1} in temperate climate prairies (Singh and Joshi, 1979). Fresh organic matter additions to the soil are similar in tropical and temperate savanna ecosystems. Cerrados of Brazil occupy more than 170 x 10^6 ha, corresponding to about 20% of the land area of that country. Predominant soils of the Brazilian Cerrados are Oxisols (94.5 x 10^6 ha or 56% of the Cerrado land area), Entisols (34.3 x 10^6 ha or 20%), Ultisols (19.1 x 10^6 ha or 11%), Alfisols (7.0 x 10^6 ha or 4%) and Lithosols (15.1 x 10^6 ha or 9%) (Kornelius et al. 1978).

Table 6. Annual additions (b) and conversion rate (m) of fresh organic matter and annual additions (a) and decomposition rate (k) and equilibrium levels C of topsoil organic matter in temperate and tropical ecoregions

Parameter	Forests		Savannas	
	Topsoil	Temperate	Tropical	Temperate
(a) Fresh organic matter				
1. Annual additions (b)	3.8–6.0	0.8–1.7	0.4–1.4	1.4
2. Conversion rate (m)	47–51	47–52	43–50	37
(b) SOC content				
1. Annual additions (a)	2.0–2.9	0.4–0.9	0.2–0.7	0.5
2. Decomposition rates (k)	2.0–5.2	0.4–1.0	1.2–1.3	0.4
3. Equilibrium levels (c)				
(i) Mg ha^{-1} yr^{-1}	55–400	87	16–55	134
(ii) %	1.2–9.0	2.0	0.4–1.2	3.0

(Modified from Grabriels and Michiels, 1991; Greenland and Nye, 1959.)

III. Soil Organic Carbon (SOC) Content in Tropical Ecosystems

The general principle of SOC dynamic, conversion from organic matter to humus and decomposition rate, is similar in tropical vs. temperate ecoregions. Greenland and Nye (1959) proposed the following general relationship describing synthesis and decomposition under steady state conditions (Equations 1 and 2).

$$C = b*m/k \qquad \text{Eq. 1}$$
$$a = b*m = C*k \qquad \text{Eq. 2}$$

where C = the amount (Mg ha^{-1}) or % of SOC stored in the soil at equilibrium
  b = the annual amount of fresh organic matter added to the soil (Mg ha^{-1} yr^{-1})
  m = the conversion rate of fresh organic matter into SOC (humification coefficient, %)
  a = the annual addition of SOC, and
  k = the annual decomposition rate of SOC (mineralization coefficient, %)

The data in Table 6 show the range of coefficients used in Equations 1 and 2 for temperate and tropical forests and savanna ecosystems under natural or undisturbed conditions. The conversion rate (m) of fresh organic matter into SOC is about 30 to 50% for both ecoregions in tropics and temperate climates. However, the annual decomposition rate (k) can be about 2 to 5 times more in the tropics than in the temperate climate. The k factor is about 5 times more in TRF than temperate forest, and about 3 times more in tropical than temperate savannas. Jenkinson and Ayanaba (1977) observed that rate of decomposition of organic matter was 4 times more at Ibadan, Nigeria than at Rothamsted, U.K.

There are many factors that affect the magnitude of the decomposition constant k (Table 7). The decomposition is increased by practices that exacerbate soil disturbance, e.g., mechanized clearing, biomass burning, plowing, low-input agriculture. The k factor is increased by soil degradative processes, e.g., physical degradation, soil erosion, nutrient depletion, etc. In contrast, decomposition

Table 7. Factors affecting the magnitude of decomposition constant k in TRF ecosystem following clearing and arable land use

Factors that increase k	Factors that decrease k
1. Increase in maximum soil temperature, aeration, and denitrification	1. Decrease in maximum soil temperature, aeration, and denitrification
2. Mechanized deforestation, burning	2. Manual land clearing, growing cover crops
3. Plowing, row cropping, monoculture	3. No-till, mixed cropping
4. Low-input agriculture	4. Science-based agriculture
5. Accelerated erosion	5. Erosion control measures
6. Nutrient depletion and soil degradation	6. Soil restoration and soil fertility enhancement

Table 8. Effect of management systems on k for some soils of the west African Savanna

Country	Cropping system	Clay + silt (%)	k (%)	Duration (yrs)
Burkina Faso	Sorghum	12	1.5–2.6	10
Cameroon	Cotton-cereal	17	2.5–3.2	5
Ivory Coast	Cotton-cereal	–	0.4–0.6	3
Senegal	Millet-groundnut	3	4.3–7.0	5
Senegal	Cereal-legume	11	3.2–5.2	17
Chad	Cotton-cereal	11	0.5–2.8	20
Togo	Cotton-cereal	10	1.1–2.4	20

(Adapted from Pieri, 1991.)

is decreased by land use and soil/crop management practices that reduce soil disturbance, e.g., manual land clearing, growing cover crop, no-till and residue mulching, soil management with science-based input, restoration of degraded soils (Table 8).

These analyses show that with 3 to 5 times higher rate of addition of organic matter and 5 times more rate of decomposition in TRF than temperate forests, the equilibrium level of SOC content in undisturbed soils is similar in both cases. Sanchez et al. (1982) observed that SOC content in soils of the tropics are similar to those of temperate climates. However, cultivation decreases SOC content more rapidly in tropical than in temperate climates.

IV. Factors Affecting SOC Pool in Soils of the Tropics

The SOC pool under undisturbed soils of TRF and TS ecosystems may be larger and highly variable depending on rainfall regime and soil conditions. The SOC pool in undisturbed soils may range from 15 to 17 kg m^{-2} in TRF of Indonesia (Figures 2 and 3), 18 to 29 kg m^{-2} in subtropical rainforest of Brazil (Figure 4) and 10 to 12 kg m^{-3} in semi-deciduous rainforest of Nigeria (Figure 5). Under favorable conditions, the SOC pool in soils of the TS may be more than that of the TRF. The data in

SOC Pool (kg/m²)

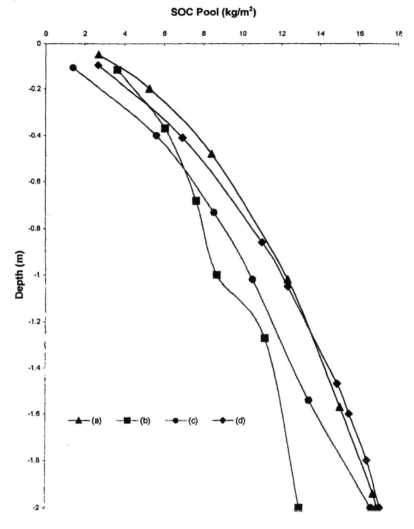

Figure 2. Land use effects on SOC pool in some soils of TRF in Sumatra, Indonesia (a) Forest, (b) Cultivated, (c) Recently Cleared, (d) Rubber Plantation. (Unpublished data of J.M. Kimble from NRCS database.)

Figure 4 shows the SOC pool under undisturbed soils as high as 28.6 kg m⁻². Anthropogenic perturbations leading to deforestation (Plates 8,9,10) and conversion to agricultural land uses (Plates 11 and 12) may have a strong effect on SOC content. Principal activities that affect SOC content and its quality are deforestation, cropping duration, tillage methods and soil/crop management.

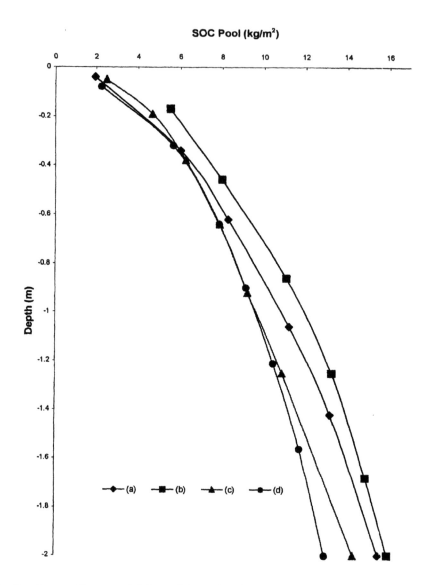

Figure 3. Land use effects on SOC pool in soils of TRF in Sumatra, Indonesia (a) Forest, (b) Recently cleared (c) Cultivated, (d) Cultivated. (Unpublished data of J.M. Kimble from NRCS database.)

A. Deforestation

Removal of TRF vegetation leads to soil and ecological degradation (Goodland and Irwin, 1975). Deforestation decreases SOC content (Lal, 1995), and the magnitude of decrease may be more in mechanized (Plate 8) than in manual (Plates 9 and 10) systems of deforestation. Lal and Cummings (1979) observed that soon after deforestation, the SOC content in the surface soil horizon was lower in plots cleared with bulldozers than in those cleared with manual clearing methods. Biomass burning (Plate 9), following manual clearing, had no drastic impact on SOC content. In the upper Amazon

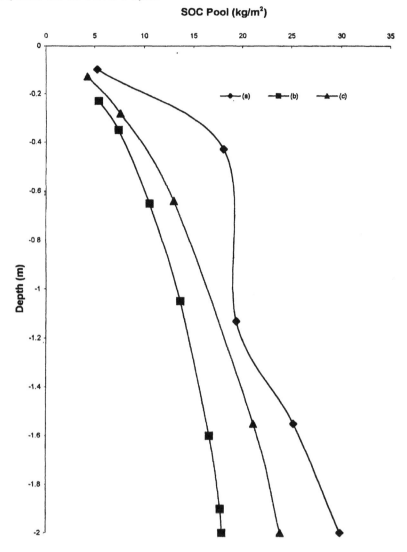

SOC Pool (kg/m²)

Figure 4. Land use effects on SOC pool of some soils of sub-tropical rainforest regions of São Paulo, Brazil (a) Forest, (b) Cultivated, (c) Cultivated. (Unpublished data of J.M. Kimble from NRCS database.)

Basin, Alegre and Cassel (1986) also observed a more drastic decrease in SOC content of mechanized-cleared than manually-cleared (slash-and-burn) plots. The data in Figures 2 and 3 show that deforestation of TRF in Sumatra, Indonesia decreased SOC pool by 2 to 3 kg m^{-2}. Similarly, the losses of SOC pool by deforestation in subtropical Brazil were 5 to 10 kg m^{-2} (Figure 4).

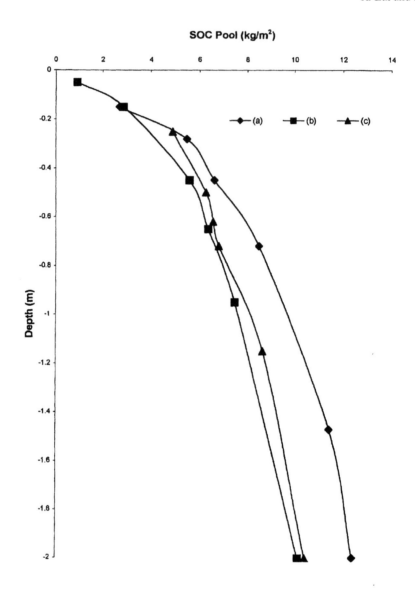

Figure 5. The SOC pool in some upland soils under semi-deciduous rainforest in southwestern Nigeria (a) Iwo soil series, (b) Egbeda soil series, (c) Ibadan soil series. (Recalculated from Moormann et al., 1975.)

B. Cultivation

Conversion of tropical rainforest to agricultural land use can drastically alter the soil C pool. Principal agricultural and forestry systems are pastures, grain crops (Plate 11) and plantation crops (Plate 12). Cropping usually leads to decline in SOC content, especially in management systems that accentuate soil erosion or lead to depletion of soil fertility. Brams (1972) observed decrease in SOC content by cultivation to almost 1m depth in both forest and savanna soils. The rate of decrease depends on the antecedent SOC content, soil properties, climate and the management. In Casamance,

Senegal, Siband (1974) observed that SOC content in a savanna soil managed with traditional farming systems decreased from an initial level of about 2.8% to about 0.7% in 180 years. The rate of decrease followed an exponential function. In a study on soils from Mali and Burkina Faso, Dutartre et al. (1993) observed a strong correlation between aggregation and SOC content. In western Nigeria, Lal (1995) observed a rapid decline in SOC content following deforestation and continuous cultivation. Conversion of TS in Brazil to arable land also resulted in reduction in the SOC pool by 0.5 to 3 kg m^{-2} (Figure 6). In contrast to the Brazilian TS, the data in Figure 7 show drastic reduction in SOC pools in cultivated soils of Mali, with extremely low levels of SOC pool of only 5 to 6 kg m^{-2}. Compared to some coarse-textured Alfisols, heavy-textured Vertisols may have high SOC content and thus high inherent soil fertility (Figure 8).

C. Cropping System

Because of the differences in input of organic matter (roots and shoot biomass), and in soil cover and micro-climate beneath the canopy, cropping systems can have a strong effect on SOC content. In the Para State, eastern Amazonia region of Brazil, Andreux et al. (1994) observed that SOC content in arable land was extremely sensitive to soil erosion because of its close association with fine mineral fraction which is easily eroded. Similar observations were made in Nigeria by Lal (1976) and in Cameroon by Angue-Abane (1988). Experiments conducted on eroded Alfisols by showed that use of cover crops within a rotation cycle increased SOC content. Deep rooted cover crops and pastures can add a considerable quantity of root biomass to the sub-soil horizon (Chopart, 1980; Lamotte and Bourliene, 1978). Experiments conducted at CIAT by Fisher et al. (1998) have shown a considerable C sequestration potential of improved pastures and tropical grasses. In Brazil, Serrao et al. (1978) also demonstrated pastures improve SOC content (Table 9). The data in Figure 6 from TS region of Brazil show lower levels of SOC pool under pasture than under cropland. High stocking rate and poor management can lead to drastic decline in SOC pool and soil compaction even under pasture (Plate 13). In contrast, data in Figure 9 from Malawi show relatively higher SOC pool under pasture than under cropland.

There are numerous forest silvicultural and agrosilvicultural options for sustainable management (Dawkins and Philip, 1998). Use of contour hedgerows of *Leucaena leucocephala* and *Gliricidia sepium* maintained SOC content at a higher level than without them (Lal, 1989a; 1995). Positive effects of agroforestry systems on SOC content are due to the beneficial impacts on soil and water conservation, and additions of organic material through prunings and root mass. Use of crop residue mulch also maintains SOC content at a higher level than without mulch. Similar to the benefits of agroforestry techniques, positive effects of mulching on SOC content are also due to soil and water conservation, lower average and maximum soil temperatures under mulch than in unmulched soil surface, return of biomass to the soil, increase in soil biodiversity, and strengthening of the nutrient cycling mechanisms.

D. Conservation Tillage

Reduction or elimination of plowing, soil turnover, has beneficial effects on SOC content especially in soils of the humid and sub-humid tropics (Lal, 1989b). Several long-term experiments conducted on Alfisols in western Nigeria showed that SOC content was maintained at a higher level in no-till than plow-till methods of seedbed preparation (Lal, 1989b; 1997). Higher level of SOC in no-till, especially in the surface about 15 cm layer, is due to soil and water conservation, return of biomass to the soil, favorable soil moisture and temperature regimes, and increase in soil biodiversity.

Plate 8a. Deforestation by mechanical means: (a) tree pusher.

Plate 8b. Deforestation by mechanical means: (b) shear blade used to cut tree.

Plate 8c. Deforestation by mechanical means: (c) shrub remover.

Plate 9a. Biomass burning for land clearing.

Plate 9b. Biomass burning for land clearing.

Plate 9c. Biomass burning for land clearing.

Plate 10. Deforestation by manual clearing for shifting cultivation.

Plate 11a. Land conversion to grain crops: (a) corn.

Plate 11b. Land conversion to grain crops: (b) soybean.

Plate 12a. Land conversion to plantation: (a) papaya.

Plate 12b. Land conversion to plantation: (b) passion fruits, etc.

Plate 12c. Land conversion to plantation: (c) guarana.

Plate 12d. Land conversion to plantation: (d) coconut.

Plate 12e. Land conversion to plantation: (e) coffee.

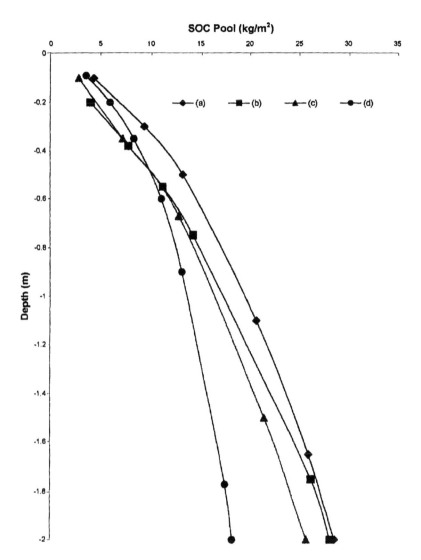

Figure 6. The SOC pool in soil of tropical savanna regions of Berazillia, Brazil (a) Native savanna, (b) Cultivated, (c) Cultivated, and (d) Pasture. (Unpublished data of J.M. Kimble from NRCS database.)

E. Soil Fertility Management

Nutrient depletion due to traditional low input farming systems has adverse impact on SOC content. Maintenance of soil fertility through judicious use of fertilizers and organic manures is likely to maintain SOC content at a higher level than in systems based on low-input strategy. Conversion of C in crop residue and other organic material applied to the soil into humus requires nutrients. The wider the C:N ratio of the organic material applied, the more is the need for applying N as fertilizers to convert biomass into humus. In addition, applications of P and S may also be needed to sequester C in the biomass into humus (Himes, 1998).

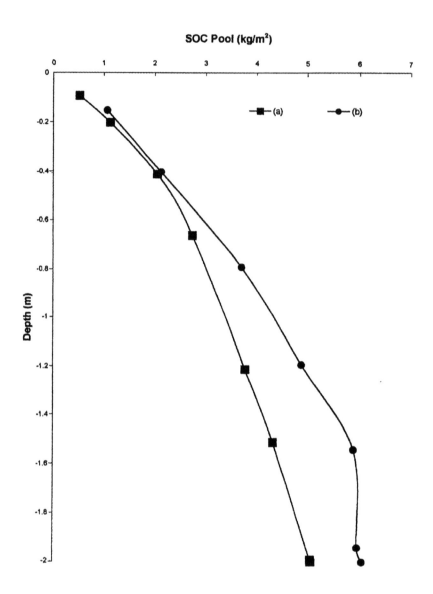

Figure 7. The SOC pool in two cultivated Alfisols of the savanna region of Mali, West Africa: (a) Site A, (b) Site B. (Unpublished data of J.M. Kimble from NRCS database.)

V. Soil Restoration and C Sequestration

Restoration of degraded soils is an important strategy of increasing SOC content and sequestering C within the terrestrial ecosystems. Degraded soils are characterized by low SOC content, and have a high potential to sequester C through adoption of appropriate soil restorative measures. Relevant soil restorative measures include those that facilitate establishment of any vegetative cover that adds a large quantity of biomass into the soil, e.g., growing cover crops, establishing trees, etc. The rate of change of SOC content under soil restorative measures can be computed using the following equation proposed by Jenny (1949) and others:

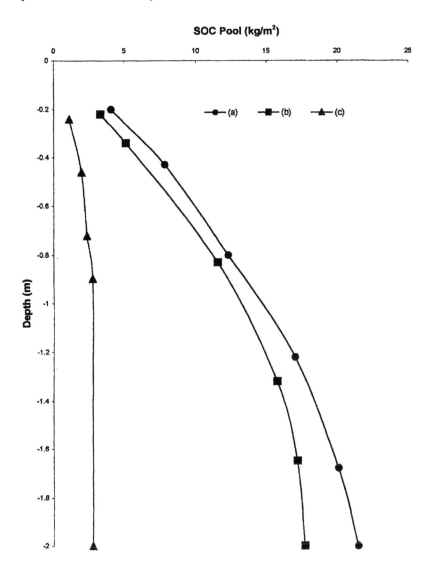

SOC Pool (kg/m²)

Figure 8. The SOC pool in two Vertisols and one Alfisol in the savanna region of Zimbabwe, South Africa: (a) Vertisol I, (b) Vertisol II, (c) Alfisol. (Unpublished data of J.M. Kimble from NRCS database.)

$$\frac{dC}{dt} = -KC + A \qquad \text{Eq. 3}$$

where C is the stock of SOC content at a time t, A is the amount of SOC added or synthesized during the time under consideration, and K is the decomposition constant. Because the term A is difficult to

Table 9. Effect of pasture establishment on SOM content of 0-20 cm layer for three soils in Brazil

Duration of pasture establishment (yrs)	Dark Red Latosol, Northern Mato Grosso (Oxisol)	Yellow Latosol, Para (Oxisol)	Red-Yellow Podzolic soil, Para (Ultisol)
Texture	Loamy	Very clayey	Loamy
Forest	1.95	2.79	1.17
Burnt forest	1.31	–	1.04
1	0.99	2.04	1.04 (just planted)
2	1.07	–	1.32
3	–	3.09	–
4	1.39	2.20	1.20
5	0.98	1.90	0.93
6	0.98	1.90	0.93
7	1.07	1.77	1.34
8	1.20	1.69	1.08
9	1.30	2.34	1.19
10	0.98	–	0.93
11	1.00	3.37	–
12	–	–	–
13	–	2.80	–

(Adapted from Serrao et al., 1978.)

estimate, the decomposition constant K is usually substituted by k which represents the total annual loss of SOC content and not only just due to decomposition of humus. Substituting K for k leads to Equation 4.

$$dC/dt = -kc \qquad \text{Eq. 4}$$

Integration of Equation 4 provides the relationship to estimate SOC content at the time t (Eq. 5).

$$C_t = C_o(1-k)^t \qquad \text{Eq. 5}$$

where C_o refers to the antecedent or initial SOC content, and C_t the SOC content after time t. The decomposition constant k can be computed by rearrangement of Equation 5 into Equation 6.

$$k(\%) = (1-e^{(\log Co - \log Ct)/t}) \times 100 \qquad \text{Eq. 6}$$

The constant k(%) thus computed is a measure of the annual rate of SOC sequestration in the soil through adoption of appropriate restorative measures. Pieri (1991) computed k for different soil and crop management systems for some soils of the semi-arid regions of West Africa. The data in Table 6 show that the magnitude of k depends on cropping system, soil management, and clay + silt content.

Restoration of degraded soils has a tremendous potential for C sequestration, improving soil quality and increasing productivity. Simple restorative measures with potential for C sequestration may include adoption of science-based agricultural techniques, e.g., use of fertilizers and organic

Plate 13. Soil compaction under pasture in TRF region of Brazil.

manures, mulch farming, conservation tillage, agroforestry measures, cover crops and mixed farming systems, and establishment of vegetative cover on denuded soil surfaces. Restoration of 1 x 10⁹ ha of soils in the tropics has a potential to sequester C at the rate of 1.5 Pg yr⁻¹ (increasing SOC content at 0.01 % yr⁻¹ to 1m depth for a mean bulk density of 1.5 Mg m⁻³). This vast potential can be realized through judicious management of natural resources.

VI. Conclusions

The TRF and tropical savannas are important biomes of the tropics. Under undisturbed/natural conditions, the SOC content of soils of the tropics is similar to those of temperate regions. Anthropogenic perturbations, however, cause rapid decline in SOC content in soils of the tropics. With conversion to an arable land use, the SOC content can decrease more rapidly in the tropics because of the higher rate of decomposition (about 4 times). The decomposition constant k is generally greater with mechanized and burning vs. manual and no burning methods of deforestation, plow-till vs. no-till system of seedbed preparation, monoculture vs. rotations based on cover crops and mixed cropping systems, tradition vs. systems based on judicious use of external inputs. Therefore, important farming/cropping systems to enhance SOC content in soils of the tropics are plantation crops, improved and properly stocked pastures, and improved cropping systems managed with science-based inputs, e.g., no-till farming, agroforestry, judicious use of fertilizers and inputs, etc. Restoration of degraded soils in the tropics has a vast potential to increase SOC content, enhance soil

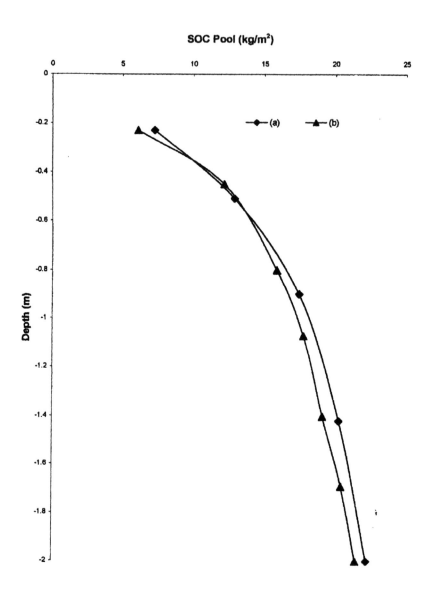

Figure 9. The SOC pool in savanna soil of Malawi, South Africa: (a) Native pasture, (b) Cultivated land. (Unpublished data of J.M. Kimble from NRCS database.)

quality, improve biomass productivity and sequester C in soil and the terrestrial ecosystems. The C sequestration potential of soil restorative measures in the tropics is as much as 1.5 Pg yr^{-1} or about 50% of the total global warming potential due to the current increase in concentrations of radiatively-active gases in the atmosphere. This vast potential can be realized through judicious management of natural resources. Soil degradation is a severe problem in the tropics, and it leads to a rapid decrease in SOC content. Restoration of degraded soils has a potential for C sequestration and greenhouse effect mitigation. Judicious management of tropical ecosystems may be an important strategy for global environmental regulation.

References

Alegre, J.C. and D.K. Cassel. 1986. Effect of land clearing methods and post-clearing management on aggregate stability and organic carbon content of a soil in the humid tropics. *Soil Sci.* 142: 289-295.

Almeda, F. and C.M. Pringle (eds.) 1988. *Tropical Rainforests: Diversity and Conservation.* California Academy of Sciences, San Francisco, CA.

Andreux, F. Ph. Dutartre, B. Guillet, T. Chone and T. Desjardins. 1994. The status of soil organic matter in selected fragile ecosystems. p. 389-403. In: N. Senesi and T.M. Miano (eds.), *Humic Substances in the Global Environment and Implications on Human Health.* Elsevier Science B.V., Amsterdam.

Angue-Abane, M. 1988. Biodynamique des humus et cycles biogeochimiques des éléments dans des sites forestiers et des cultivés en cacaoyers du centre-sud Cameround. Doctoral Thesis, University of Nancy, France.

Brams, E.A. 1972. Cation exchange as related to the management of tropical soils. Unpublished, Prairie View, TX, U.S.

Bruenig, E.F. 1996. *Conservation and Management of Tropical Rainforests: an Integrated Approach to Sustainability.* CAB International, Wallingford, U.K. 339 pp.

Chopart, J.L. 1980. Etude an champ des systèms racinaires des principles cultures pluviales an Senegal Carachide-milsorgho-riz pluvial. Thèse Doct. Prod. Veg. Qual. Prod. Inst. Natl. Polytech Toulouse.

Council of Environment Quality. 1982. Global 2000 Report to the President, Department of State, Washington, D.C.

Dawkins, H.C. and M.S. Philip. 1998. *Tropical Moist Forest Silviculture and Management: A History of Success and Failure.* CAB International, Wallingford, U.K. 359 pp.

Dutartre, Ph., F. Bartli, F. Andreux, J.M. Portal and A. Auge. 1993. Influence of contents and nature of organic matter on the structure of some sandy soils from west Africa. *Geoderma* 56: 459-478.

Faminow, M.D. 1998. *Cattle, Deforestation and Development in the Amazon: an Economic, Agronomic and Environmental Perspective.* CAB International, Wallingford, U.K. 253 pp.

Fisher, M.J., R.J. Thomas and I.M. Rao. 1998. Management of tropical pastures in acid-savannas of South America for carbon sequestration in the soil. p. 405-420. In: R. Lal, J.M. Kimble, R.F. Follett and B.A. Stewart (eds.), *Management of Carbon Sequestration in Soil.* CRC Press, Boca Raton, FL.

Gabriels, D. and P. Michiels. 1991. Soil organic matter and water erosion processes. p. 141-152. In: W.S. Wilson (ed.), *Advances in Soil Organic matter Research: The Impact on Agriculture and the Environment,* The Royal Society of Chemistry, London.

Goodland, R.J.A. and H.S. Irwin. 1975. *Amazon Jungle: Green Hell to Red Desert.* Elsevier, Amsterdam. 155 pp.

Greenland, D.J. and P.H. Nye. 1959. Increases in C and N contents of tropical soils under natural fallows. *J. Soil Sci.* 9:284-299.

Himes, F.L. 1998. Nitrogen, sulfur and phosphorus and the sequestering of carbon. p. 315-320. In: R. Lal, J.M. Kimble, R.F. Follett and B.A. Stewart (eds.), *Soil Processes and the Carbon Cycle.* CRC Press, Boca Raton, FL.

Jenkinson, D.S. and A. Ayanaba. 1977. Decomposition of carbon-14 labelled plant material under tropical conditions. *Soil Sci. Soc. Am. J.* 41:912-915.

Jenny, H. 1949. Comparative study of decomposition rates of organic matter in temperate and tropical regions. *Soil Sci.* 68:63-69.

Jordan, C.F. 1983. Productivity of tropical rainforest ecosystems and the implications for their use as future wood and energy sources. p. 117-135. In: F.B. Golley (ed.), *Tropical Rain Forest Ecosystems: Structure and Function.* Elsevier, Amsterdam.

Kornelius, E., M.G. Saueressig and W.J. Goedert. 1978. Pastures establishment and management in the Cerrado of Brazil. p. 1-147. In: P.A. Sanchez and L.E. Tergas (eds.), Pasture Production in Acid Soils of the Tropics. Beef Program, CIAT, Cali, Colombia.

Lal, R. 1976. Soil erosion problems on Alfisols in western Nigeria and their control. IITA Monograph 1, Ibadan, Nigeria. 208 pp.

Lal, R. 1989a. Agroforestry systems and soil surface management of a tropical Alfisol. III. Soil Chemical Properties. *Agroforestry Syst.* 8:113-132.

Lal, R. 1989b. Conservation tillage for sustainable agriculture. *Adv. Agron.* 42:85-197.

Lal, R. 1995. Sustainable management of soil resources in the humid tropics. United Nations University Press, Tokyo, Japan. 146 pp.

Lal, R. 1997. Long-term tillage and maize monoculture effects on a tropical Alfisol in western Nigeria. II. Soil chemical properties. *Soil & Tillage Res.* 42:161-174.

Lal, R. and D.J. Cummings. 1979. Clearing a tropical forest: effects on soil and microclimate. *Field Crops Res.* 2:91-107.

Lamotte, M. and F. Bourliene. *Problème d´Ecologie: Structure et Fonctionnement des Ecosystèmes Terrestres.* Masson, Paris. 345 pp.

Misra, K.C. 1979. Introduction: tropical grasslands. p. 189-195. In: R.T. Coupland (ed.), *Grassland Ecosystems of the World: Analysis of Grasslands and Their Uses.* Cambridge University Press, Cambridge.

Moormann, F.R., R. Lal and A.S.R. Juo. 1975. Soils of IITA. Technical Bulletin 3, IITA, Ibadan, Nigeria.

NRC. 1993. Sustainable agriculture and the environment in the humid tropics. National Research Council, Washington, D.C. 702 pp.

Pieri, C.M.M.G. 1991. *Fertility of Soils: a Future for Farming in the West African Savannah.* Springer-Verlag, Berlin.

Sanchez, P.A. 1976. *Properties and Management of Soils in the Tropics.* John Wiley & Sons, New York.

Sanchez, P.A. and R.F. Isbell. 1978. A comparison of the soils of tropical Latin America and tropical Australia. In: Pasture Production in Acid Soils of the Tropics, Proceedings of a seminar held at CIAT, Cali, Colombia, 17-21 April, 1978.

Sanchez, P., M.P. Gichuru and L.B. Katz. 1982. Organic matter in major soils of the tropical and temperate regions. Trans. 12th Int'l Cong. Soil Sci. (New Delhi) 1:99-114.

Serrao, E.A.S., I.C. Falesi, J.B. de Veige and J.F.T. Neto. 1978. Productivity of cultivated pastures on low fertility soils in the Amazon of Brazil. p. 1-195. In: P.A. Sanchez and L.E. Tergas (eds.), *Pasture Production in Acid Soils of the Tropics*, Beef Program, CIAT, Cali, Colombia.

Siband, P. 1974. Evolution des caractères et de la fertilité d´un sol rouge de Casamance. *L´Agronomie Tropicale* 29:1228-1248.

Singh, J.S. and M.C. Joshi. 1979. Primary production in tropical grasslands. p. 197-218. In: R.T. Coupland (ed.), *Grassland Ecosystems of the World: Analysis of Grasslands and Their Uses.* Cambridge University Press, Cambridge.

Southgate, D. 1998. *Tropical Forest Conservation: an Economic Assessment of the Alternatives in Latin America.* Oxford University Press, New York. 175 pp.

Van Wambeke, A. 1990. *Soils of the Tropics: Properties and Appraisal.* McGraw-Hill, New York. 343 pp.

World Resources Institute. 1996. World Resources 1996-97. Washington, D.C.

CHAPTER 2

Carbon Stocks in Soils of the Brazilian Amazon

C.C. Cerri, M. Bernoux, D. Arrouays, B.J. Feigl and M.C. Piccolo

I. Introduction

Soil organic carbon represents a major pool of carbon within the biosphere. Recent concerns about rising levels of atmospheric CO_2 have directed attention to carbon stocks in soils of the world (Post et al., 1982; Eswaran et al., 1993; Sombroek et al., 1993; Batjes, 1996) and to their role as both a source and sink of carbon. Over short periods of time, changes in vegetation and in land use patterns have a marked effect on topsoil carbon storage. One of the main issues of concern relating to the effect of land use changes on soil carbon balance is the effect of the conversion of tropical forest to agriculture and grazing (e.g., Detwiler, 1986; Houghton et al., 1991; Lugo and Brown, 1993). Such conversion affected 200 million hectares between 1980 and 1995 (FAO, 1997).

Tropical soils represent at least 32% of the total mass of organic C stored in the soils of the world (Eswaran et al., 1993). Among tropical ecosystems, the Amazonian forest is known to play a major role in C sequestration and release (Cerri et al., 1994). However, global estimates of C storage in this ecosystem are few (Moraes et al., 1995). These carbon pools are difficult to estimate because of the limited availability of reliable, complete and uniform data for soils (C concentration and bulk density down to a sufficient depth) as pointed out by Schlesinger (1986) and Batjes (1996). Other principal reasons why C pools are difficult to estimate are the still limited knowledge of the extent of soil types (Sombroek et al., 1993; Batjes, 1996), the high spatial variability of soil C even within one soil map unit, and the confounding effects of the factors controlling the soil organic carbon cycle (Melillo et al., 1984; Pastor and Post, 1986; Parton et al., 1987).

The objective of this chapter is to compare different techniques in order to obtain reliable estimates of the carbon stocks in the Brazilian Amazon basin. Our goal is to provide more accurate estimates of soil organic carbon under primary uncut 'old growth' forest in order to better assess the relative impact of human disturbance caused by clear-cutting and agricultural or grazing use.

II. Material

The nested zones used for the study are represented in Figure 1. The continental scale is the whole Amazon basin. The regional scale is defined as the Brazilian part of a 6° square, from 60 to 66° West longitude and 8 to 14° South latitude, over the Rondônia state. The local study is focused on an 26,000 ha cattle ranch named "Nova Vida" (10°10'05" S and 62°49'27" W) located between the cities of Ariquemes and Jaru in Rondônia, along the road BR-364.

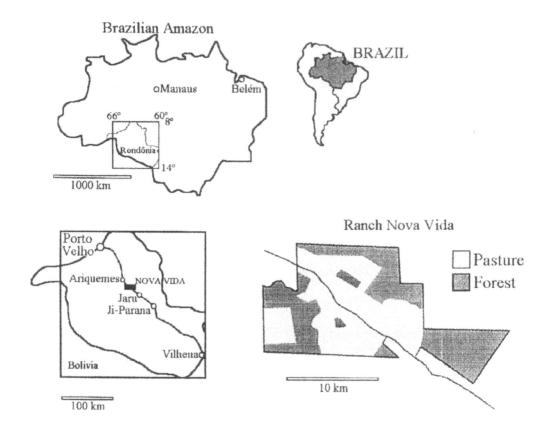

Figure 1. Spatial scales covered in this study.

A. Continental Scale

According to the considerations developed by Jacomine and Camargo (1996), two main soil divisions of the Brazilian soil classification, Latossolos (Oxisols) and Podzólicos (Ultisols and Alfisols) cover nearly 75% of the total area (Figure 2) of the legal Amazon basin (Rodrigues, 1996). The remainder is distributed among 13 soil divisions, only two of which are more than 5% of the Amazon basin: Plintossolos (Inceptisols, Oxisols and Alfisols) and Gleissolos (Entisols and Inceptisols) representing 7.4 and 5.3%, respectively.

The Brazilian Latossolos correspond to well drained Oxisols in the Soil Taxonomy (Soil Survey Staff, 1990), and the FAO-Unesco soil map legend identifies them as Ferralsols. The Podzólicos belong to the Alfisols (when eutrofic) and to the Ultisols (when dystrofic) orders of the Soil Taxonomy, and most of them fall into the Acrisols, Nitosols and Lixisols (Luvisols) of the FAO-Unesco map legends (Van Wambeke, 1991). Most Alfisols, Ultisols and Oxisols belong to low-activity clay soils. Ultisols usually occupy younger geomorphic surfaces than Oxisols with which they are often associated in landscapes (Moraes et al., 1996). These soils are thick mineral soils with thickness often > 2 m.

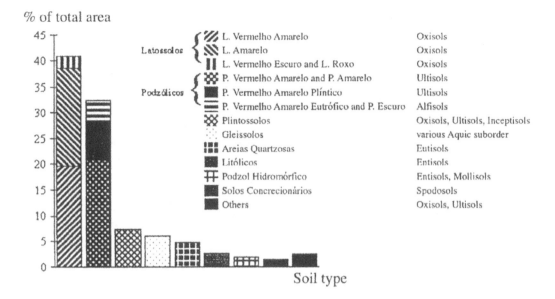

Figure 2. Relative distribution of the main soil types covering the Amazon basin.

B. Regional Scale

At the regional scale the main soil types are the same (Figure 3), and their relative extents are similar. The Podzólicos and the Latossolos are the dominant types covering respectively 40 and 34% of the area. The Plintossolos (Inceptisols, Oxisols and Alfisols) and the Gleissolos (Entisols and Inceptisols) still represent a noticeable extent, accounting for 5 and 3% of the area. The Litólicos (Entisols and Mollisols) and the Areais quartzosas (Entisols) show extents representing 7.5 and 5% of the total area. In total, these 6 soil types cover 95% of the area. The main difference between the continental and the regional levels concerns the extent of the Podzois Hidromórficos which represent less than 0.5% in Rondônia and around 2% at the continental scale.

C. Local Scale

Geomorphologically the ranch area can be divided into two landscape patterns: (1) a sedimentary landscape of convex rolling hills with flat tops at about 200 m, and (2) rocky hills with more irregular diameters and shapes because of a complex incision by a temporary river network. The elevations range between 130 to 170 m and the higher hills exhibit an Inselberg feature. Most hill flat tops have a thin lateritic-like gravel mantle which probably results from the dismantling of a sedimentary rock composed of fine ferruginous sandstone occurring as a horizontal laminar layer (Moraes et al., 1995).

The Latossolos and Podzólicos soil types are dominant in the ranch, represented by a red-yellow podzolic latosol (Prado, 1996) classified as Kandiudult in the U.S. Soil Taxonomy (Soil Survey Staff,

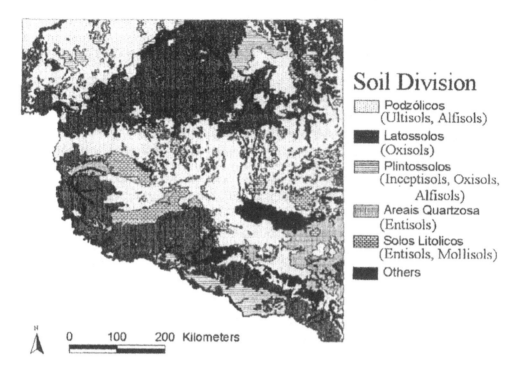

Figure 3. Simplified soil map of the regional study zone.

1990), and a red-yellow podzolic soil (Prado, 1996) or Ultisol (Paleudult, Tropudult). The red-yellow podzolic latosol is dominant in the sedimentary landscape pattern, occurring on both the top position and the upper part of the slopes of the convex low hills. It is present in association with the red-yellow podzolic soil in the rocky hills pattern. More details about the soils can be found in Moraes et al. (1996).

III. Methods

A. Continental Scale

The original database built and utilized by Moraes et al. (1995) was used at the continental scale. Briefly, the database included soil profile information for 1162 soil pits collected across the Amazon basin. The carbon concentration reported (Ministério das Minas e Energia, 1973-1982) was determined using the Walkley–Black method (Walkley and Black, 1934). C stocks (expressed in kilograms per square meter) by horizon were calculated as the result of the product of bulk density, C concentration and thickness of the horizon. Moraes et al. (1995) used as estimates of the soil bulk densities means of directly reported information for 474 soil horizons. They gave estimates of C stocks in the 0–0.2 m and the 0–1.0 m layers. The carbon stock for the Brazilian Amazon basin was calculated for each soil type by multiplying the mean C stocks by the area of each soil type from a digitized soil map at scale 1:5,000,000 (Empresa Brasileira de pesquisa Agropecuaria, 1981). In this study, using strictly the same methodology as the one described above, the stocks for the 0–0.3 m soil layer were also calculated.

B. Regional Scale

A specific soil database was elaborated from physico-chemical results of soil horizons sampled during the end of the 1970s and the beginning of the 1980s, and published as reports (Ministério das Minas e Energia, 1978; 1979; Rodrigues, 1980; Empresa Brasileira de Pesquisa Agronomica, 1983).

In these reports, carbon concentrations (Walkley–Black method) are expressed in g/100g of fine earth (< 2 mm fraction) by horizon. For estimating the carbon storage, a first correction was applied considering that the soil fraction > 2 mm is carbon free. Soil bulk density (BD) was estimated using multiple regressions from other available parameters (clay content, organic carbon (OC) and pH for example). These regressions were established from data on soil horizons spread over the whole Amazon basin (Bernoux et al., 1998) The following equations were used to estimate BD for the two main soil types:

(1) for the Latossolos:

$$BD = 1.419 - 0.0037 \text{ Clay}\% - 0.061 \text{ OC}\% \text{ for A horizons}$$
$$BD = 1.392 - 0.0044 \text{ Clay}\% \text{ for B horizons}$$

(2) for the Podzólicos:

$$BD = 1.133 - 0.041 \text{ OC}\% + 0.0026 \text{ Sand}\% \text{ for A horizons}$$
$$BD = 1.718 - 0.0056 \text{ Clay}\% - 0.068 \text{ pH}_{water} \text{ for B horizons}$$

Reported results of soil analyses for 3016 soil horizons, corresponding to 796 soil profiles, were stored into the regional database. But, only 782 profiles were georeferenced, of which only 639 (2534 soil horizons) contained carbon concentration data and were used in this study.

The regional carbon stocks were evaluated for the 0–0.3 and 0–1.0 m layers using two different approaches. The first approach is based on carbon stocks averaged by soil type multiplied by the extent of these soil types. It corresponds to the methodology used by Moraes et al. (1995). Two digitized soils maps were considered for this purpose; one was the 1:5,000,000 soil map already used at the continental level (Empresa Brasileira de pesquisa Agropecuaria, 1981). The second map was obtained by digitization of the 1:1,000,000 soil maps published in Volumes 16 and 19 of the Radambrasil soil survey report (Ministério das Minas e Energia, 1978; 1979) covering the regional study zone.

The second approach is based on interpolation of carbon stocks using geostatistics. Geostatistical treatments were carried out using the theory first presented by Matheron (1965) and applied to pedometrics by Burgess and Webster (1980a, b). The experimental semivariogram for separation vector h is calculated using to the following formula:

$$\gamma(h) = \frac{1}{2N(h)} \sum_{i=1}^{i=N(h)} \left[v(x_i) - v(x_i + h) \right]^2 \qquad \text{Eq. 1}$$

where $v(x_i)$ is the value of carbon density at spatial location x_i and N(h) the number of pairs with separate vector h.

The geostatistical analyses were run using GS+ software on PC (Gamma Design Software, 1991). Omnidirectional semivariograms were calculated using a 50km step (lag distance) to a maximum of

600 km, but using the population after removal of the extremes (values beyond three times the interquartile range from the quartiles)

Two types of validation were conducted: an internal validation known as "Jackknife" (Tukey, 1958; Cressie, 1993) and an external validation with preserved data (Bourennane et al., 1996). The cross-validation technique ("Jackknife") consists of testing the validity of the semivariogram model by kriging at each sampled location using all other neighboring samples and then comparing the estimates with the real values. A mean error (ME) close to zero indicates no systematic bias, and a root mean square error (RMSE) close to unity shows a good fit of the semivariogram model and its parameters to the data set. But this validation is nothing more than a validation of the fitted semivariogram to the original data; it does not validate the reliability of the prediction methods for external data. Before each variographic analysis, 10% of the initial data were kept out (one of ten values of the data after geographical classification by increasing longitude and latitude locations), and the geostatistical analyses were conducted with the remaining data (90% of the information). The 10% of values preserved were used to calculate the mean error (ME2) and the root mean square error (RMSE2). The ME2 should be close to zero for unbiased methods, and the RMSE2 should be small for an unbiased and precise prediction. After cross and external validations, maps of estimated values were obtained by block kriging using square blocks of 1km².

C. Local Scale

A 36 km² area under native forest, belonging to the rocky hills pattern describe above, (see Figure 1) was systematically sampled following a pseudo-regular 500 meters grid. The sampling was performed by using cylinders (diameter 0.15 m, height 0.10 m). The 0–0.1 m, 0.1–0.2 m and 0.2–0.3 m layers were collected for BD and total OC determination. The number of sampling sites was 121. Total OC was analyzed by dry combustion on a Carmograph 12A analyzer.

IV. Results and Discussion

A. Continental Scale

The first estimate of 47 Pg of carbon contained in the upper 1 m of soil calculated by Moraes et al. (1995) had been determined using the average carbon content for each soil type multiplied by the total area of each soil type from a digitized 1:5,000,000 map of the Brazilian Amazon. Table 1 reports the average values used by the authors for calculation, together with the associated standard deviation (SD), standard error (SE) and the median.

If the SD is used as a mean for evaluating the accuracy of estimates, the error associated would be 11.6 Pg of carbon, i. e., nearly 24.5% of the mean value. The use of the averages is subject to caution, even more when the number of samples is reduced. As the sample size for most of the soil types is less than 30, outlying values may have a marked influence on the mean. The calculation with the median values being less sensitive to extremes values, leads to a lower global estimate of 41 Pg. Of this amount stored in the upper meter of soil, 23.4 Pg C (using medians) were stored in the top 0.3 m. Using means instead of medians would lead to an increase by more than 16% of this value, reaching 27.2 Pg C. The associated error using SD is 4.5 Pg C. The upper 0.3 m contains 57% of the 0–1.0 m carbon stock indicating an important potential for marked changes in Amazon soils stocks as a result of conversion of the forests for agriculture or ranching.

Table 1. Variability in organic carbon content for the soil units of the Brazilian classification[a]

Brazilian classification	U.S. soil taxonomy classification[b]	Symbol	n	Mean	Median	SD	SE
						kg C m^{-2}	
Areias Quartzosas	Psamments	AQ	43	9.43	8.38	4.23	0.62
Areias Quartzosas Hidromórficas	Psamments	AQH	11	9.39	8.27	5.44	1.41
Brunizens Avermelhados	Chernozems	BA	8	15.57	13.22	8.05	2.55
Cambissolos Tropicais distróficos	Inceptisols	Ctd	34	7.75	7.06	3.12	0.44
Cambissolos Tropicais eutróficos	Inceptisols	Cte	14	6.84	5.74	4.06	0.96
Gleissolos distróficos	Aquic suborders	Gd	46	12.24	10.33	9.38	1.27
Gleissolos eutróficos	Aquic suborders	Ge	11	7.2	6.81	2.67	0.74
Plintossolos Hidromórficos	Oxisols, Ultisols	LH	43	7.77	6.29	4.58	0.65
Latossolos Amarelos	Oxisols	LA	87	8.49	7.84	3.55	0.37
Latossolos Roxos	Oxisols	LR	3	21.65	10.4	19.68	11.36
Latossolos Vermelho Amarelos	Oxisols	LVA	155	10.51	9.42	4.88	0.39
Latossolos Vermelho Escuros	Oxisols	LVE	35	9.3	6.96	4.98	0.83
Planossolos	Alfisols, Ultisols, Mollisols	P	8	9.47	7.34	7.41	2.34
Podzois	Spodosols	Po	4	11.42	10.02	4.7	2.35
Podzois Hidromórficos	Spodosols	PH	13	18.53	8.84	25.45	6.17
Podzólicos Vermelho Amarelo	Ultisols	PVA	315	9.5	8.35	4.92	0.26
Podzólicos Vermelho Amarelos eutróficos Ta[c]	Alfisols	PVAa	29	7.58	7.68	2.27	0.38
Podzólicos Vermelho Amarelos eutróficos Tb[c]	Alfisols	PVAb	49	8.73	7.8	3.95	0.49

Table 1. continued –

Brazilian classification	U.S. soil taxonomy classification[b]	Symbol	n	Mean	Median	SD	SE
					—kg C m^{-2}—		
Podzólicos Vermelho Amarelos plínticos	Ultisols	PVAp	25	9.41	8.5	4.54	0.91
Solonetz-Solodizados/Solonchak	Aridisols	S	1	2.32			
Solos Aluviais eutróficos	Entisols (Fluvents)	Sae	11	7.24	6.19	2.41	0.7
Solos Aluvias distróficos	Entisols (Fluvents)	Sad	11	6.77	7.03	2.69	0.7
Solos Concrecionários	not classified	SC	10	13.69	11.92	5.25	1.36
Solos Litólicos distróficos	Lithic subgroups	Sld	31	8.62	6.72	5.89	1.06
Solos Litólicos eutróficos	Lithic subgroups	Sle	5	11.6	5.04	9.37	4.19
Terras Roxas distróficas	Ultisols	Trd	6	15.04	15.76	4.7	1.66
Terras Roxas eutróficas	Alfisols	TRe	9	11.97	9.74	6.62	1.77
Vertissolos Eutróico	Vertisols	V	1	11.81			

[a]n is the number of samples per soil unit, SD the standard deviation and SE the standard error.

[b]Soils were mapped in the Brazilian classification and only the higher taxa can be correlated accurately with the U.S. Soil Taxonomy (Beinroth, 1975).

[c]Ta = cation exchange capacity $\geq$ 24 cmol kg^{-1} clay; Tb = cation exchange capacity < 24 cmol kg^{-1} clay.

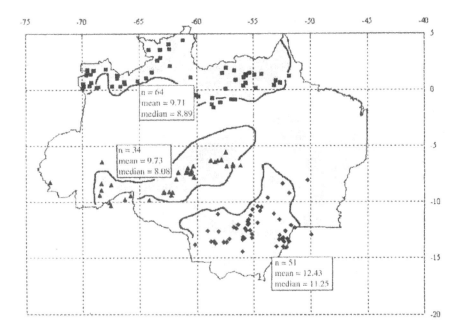

Figure 4. Variation in means and medians according to the geographical localization, case of the Latossolos Vermelho Amarelo.

It must be stressed that these continental estimates cannot be completely accurate. Their reliability is reduced by numerous factors, the principals being (1) the limited accuracy of the soil maps, (2) the limited availability of data for these soils, (3) the spatial variability in carbon content, bulk density and stoniness within one soil map unit, (4) the non-homogeneity of the analytical methods used, (5) the lack of information concerning the land-use changes. The reliability of the above estimates is reduced mainly by the lack of information concerning the soil fraction > 2 mm which can locally be large and is generally carbon free. It appears therefore that this estimate of 41 Pg C would probably be lower taking into account this effect of the soil fraction > 2 mm. But the strongest criticism that can be made is related to the way the calculation is made.

This calculation is based on average or median by soil type multiplied by the total area of each type disregarding the geographical disparity that may exist due to different vegetation, climate, or geology for example. To consider this point, the soil classified as Latossolos Vermelho Amarelo (n= 155, mean =10.51 and median =9.42) in the database was selected (Figure 4). For the 149 pits having geographic coordinates, the mean C content is 10.60 and the median 9.67. Although no significant correlation between the stock and the longitude or latitude of the pits was observed, a geographic discrepancy still appears. Visually, three groups can be considered according to the corresponding soil unit distribution in the basin, in order to apply a local mean or median to each cluster of soil unit. The southern one shows a mean and a median significantly higher than the others. Variations in some controlling factors of soil OC might explain these higher values. The southern part of the Amazon basin has a more marked dry season and its forests are less dense.

Table 2. Classical statistics concerning the soil carbon pools in the upper 0.3 m (P0.3) and in the upper meter of the soil (P1.0)[a]

Variable	Number	Average	Median	Minimum	Maximum	Standard deviation
			kg C m^{-2}			
P0.3	474	3.895	3.555	0.319	24.317	2.546
P1.0	324	7.296	6.294	0.605	41.618	4.505

[a]See Table 1 for symbol meaning, except GH standing for Gleissolos Húmicos and GNH for Gleissolos Não Húmicos.

B. Regional Scale

1. Soil Map Approach

Soil carbon pools in the upper 30 centimeters (Variable P0.3) and in the upper meter of the soil (Variable P1.0) were calculated when possible for each soil profile. Table 2 reports classical statistics for these variables. It was possible to calculate P0.3 in nearly 75 % of the original 639 profiles, but P1.0 was calculated for only 50% of the profiles. This is due to the fact that numerous profiles have reported carbon results for only two horizons: the topsoil horizon (commonly 0–x m, where x is ranging from 0.1 to 0.3 m) and a deeper horizon.

The higher values correspond to soil profiles classified as "Gley Humico" (Humic Gley) or "Latossolo Roxo" (Dark Red Oxisols) which present single point occurrences only.

A first approximation of the regional stock could be calculated multiplying the values of the mean or of the median by the total soil extent. Using the mean values would lead to a regional stock of 2400 +/- 1500 Tg stored in the first meter, with 1300 +/- 850 Tg in the first 0.3 m. Using the median, these values are reduced to 2100 Tg in the first meter with 1120 Tg contained in the upper 0.3 m. In order to refine these values, elementary statistics where calculated by taxonomic unit (Table 3).

The regional stock is obtained multiplying the carbon content of each taxonomic unit by its area. This calculation leads to similar results using either the soil map at 1:5,000,000 scale (Embrapa) or the soil map at 1:1,000,000. Regional stocks (0–1.0 m) ranged between 2100 Tg using median values, and 2300 Tg using means, and exhibit an associated error (based on SD) of 900Tg. Using the map from Embrapa (1:5,000,000) instead of the Radam (1:1,000,000) leads to an increase of only 25 Tg. The carbon content for the 0–0.3 m layer represents 1135 Tg (using medians) with an associated error of 540 Tg. It is striking that the results both for 0–0.3 m and 0–1.0 m layers are very similar using (1) the median value of all P0.3 and P1.0 values or (2) the medians of P0.3 and P1.0 segregated by soil type. But in the second calculation the associated errors are reduced to less than 50% (43% for the 0–1.0 m layer and 48% for the 0–0.3 m layer) of the amount, whereas they reached 71% (0–1.0 m layer) and 65% (0–0.3 m) in the first calculation .

When using the continental database (Moraes et al., 1995), the regional stocks are overestimated, reaching 2700 Tg using the 1:5,000,000 map or 2675 Tg using the 1:1,000,000 soil map. This overestimation of about 600 Tg is mainly attributable to the non-consideration of the soil fraction > 2 mm in the Moraes' database. If Rondônia is representative of the Amazon Basin, C stock for the whole basin would be around 31.5 Pg in the first meter of soil, with 54% concentrated in the first 0.3 m. This last estimate is only two thirds of the first estimate made by Moraes et al., (1995).

Table 3. Carbon stored in the first 0.3 m (P0.3) and in the first meter (P1.0) by taxonomic unit

Unit[a]	P0.3					P1.0					
	n	Mean	Min.	Max.	SD	Median	n	Mean	Min.	Max.	SD	Median
AQ	18	4.12	2.19	9.2	1.88	4.05	17	8.96	4.68	19.13	4.29	6.88
AQH	7	3.01	0.32	5.83	2.3	1.88	4	6.02	0.61	13.88	5.78	4.79
BA	8	7.53	4.7	13.31	2.83	7.12	6	11.14	6.61	19.59	4.78	10.37
Ctd	36	3.64	1.04	6.23	1.35	3.75	14	6.5	2.3	13.21	2.71	5.51
Cte	6	4.91	2.87	11.88	3.45	3.57	5	10.61	5.19	23.04	7.63	6.23
GH	6	16.33	7.59	24.31	7.08	15.59	4	23.32	9.57	33.79	10.25	24.96
GPH	12	4.67	2.84	7.16	1.35	4.36	9	7.99	3.71	10.86	2.54	7.77
LA	27	3.65	1.97	6.38	1.03	3.53	26	7.3	4.94	10.71	1.48	7.32
LR	4	6.46	3.49	12.49	4.1	4.93	3	19.29	7.12	41.62	19.36	9.14
LVA	53	3.78	1.93	9.57	1.35	3.61	34	7.49	4.4	19.46	3.03	6.97
LVE	13	3.73	2.31	5.56	1.13	3.49	10	6.9	3.78	11.52	2.37	6.65
PH	6	4.96	0.76	12.54	5.52	2.02	4	3.18	1.34	7.62	2.98	1.88
P	19	3.41	1.01	9.57	2.08	3.16	12	7.06	1.8	27.69	6.77	5.78
PVA	202	3.41	0.61	13.51	1.95	2.96	154	6.4	1.44	23.12	3.14	5.84
PVE	9	3.09	1.11	6.23	1.42	2.75	4	5.9	4.16	9.96	2.73	4.74
SA	10	2.7	1.07	5.26	1.08	2.54	9	5.57	3.45	11.66	2.45	5.13
SL	22	3.95	0.97	7.45	1.87	4.08	1	6.73				6.73
TR	16	4.89	1.99	8.96	2.07	4.34	8	9.52	4.57	15.76	3.8	8.44

[a]See Table 1 for symbol meaning, except GH standing for Gleissolos Húmicos and GNH for Gleissolos Não Húmicos.

Therefore, a reasonable level for the continental estimate of C store in the soil upper meter of the Brazilian Amazon would be between 30 Pg to 45 Pg.

2. Geostatistical Approach

a. Regional Scale

Experimental semivariogram for the variables P0.3 and P1.0 were all best modeled (i.e., more satisfactory results of the validations) using a nugget effect plus a spherical model according to the function:

$$\gamma(h) = c_o + c_1\left\{ \frac{3h}{2a} - \left(\frac{h}{a}\right)^3 \right\} \text{ for } 0 < h \leq a$$

$$\gamma(h) = c_o + c_1 \text{ for } h > a \qquad \text{Eq. 2}$$

$$\gamma(0) = 0$$

where a is the range of the model, c_0 the nugget and c_1 the sill values.

C.C. Cerri, M. Bernoux, D. Arrouays, B.J. Feigl and M.C. Piccolo

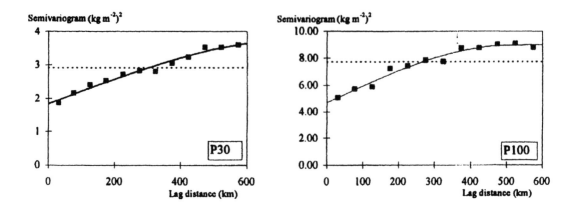

Figure 5. Experimental semivariograms and fitted models for variables P0.3 and P1.0

Table 4. Results of the validation procedures for the carbon stocks in the first 0.3 m (P0.3) and first top meter (P1.0) at regional scale

Variable	Cross validation (Jackknife)		External validation	
	ME	RSME	ME2	RMSE2
		kg m^{-2}		
P0.3	-0.007	1.078	0.197	1.520
P1.0	-0.015	1.059	0.065	2.668

The experimental semivariogram for P0.3 (Figure 5) was fitted with a nugget effect of 1.84 (kg m^{-2})^2 plus a spherical model using a sill value of 1.93 (kg m^{-2})^2 and range of 772 km. This result shows that carbon densities in the first 0.3 m exhibit a spatial structure at the regional scale. However the high nugget effect reveals that about half of the spatial variability may appear within distances inferior to a few tens of kilometers. Results of the fitting of the experimental semivariogram for P1.0 show the same pattern with a nugget effect of 4.70 (kg C m^{-2})^2 and a sill value of 4.26 (kg C m^{-2})^2, but the range is reduced to 513 km.

Results of the validation procedures reported in Table 4 show that the models of the semivariograms are well adapted to the data used to calculate the experimental semivariograms (cross-validation) and to external data (external validation).

Maps of estimated carbon densities and of the associated kriged standard deviation (KSD) by block kriging are shown in Figure 6. Each map is made up of 334,107 1km^2 blocks. For the carbon stocks in the upper 0.3 m the kriged values ranged from 2.09 to 6.00 kgm^{-2}, and KSD from 0.36 to 0.76 kg C m^{-2}, their average values being respectively 3.58 kg m^{-2} (mean) and 0.51 kg m^{-2} (KSD). The mean of the kriged values is close to the median of the population of the carbon densities for the first 0.3 m without segregation of soil type. This results in a regional soil carbon content of 1195 Tg, very close to the one previously calculated with the median, but now the associated error is reduced considerably to 170 Tg, i.e., only 14.2% of the amount. Concerning the upper meter estimates, kriged values ranged from 4.27 kg C m^{-2} to 10.40 kg C m^{-2} with a mean value of 6.665 kg C m^{-2}, resulting in a regional

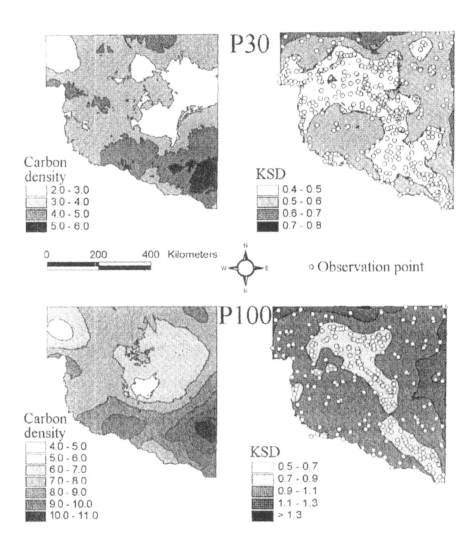

Figure 6. Carbon stocks down to 1m depth obtained by block kriging of values (P0.3 and P1.0) and maps of associated errors.

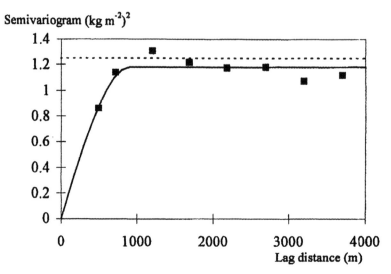

Figure 7. Experimental semivariogram and model for the carbon stored in the first 0.3 m at the local scale (Ranch Nova Vida).

stock of 2220 Tg indicating an intermediary level between the previous estimates. The associated KSD ranged from 0.54 to 1.31 kg C m^{-2}, resulting in a CV ranging from 7.7% to 24.2% with a mean of 13.5% giving an associated error of 295 Tg.

The geostatistical approach gives regional estimates very similar to those derived from the classical approach based on a soil map, but the advantage of the geostatistical method is that it leads to a much lower global associated error and provides a map of the punctual kriged error.

b. Local Scale

Soil carbon stocks in the first 30 centimeters of the 121 sampling sites varied from 1.34 to 7.34 kg C m^{-2}, the mean and the median being respectively 3.09 and 2.85 kg C m^{-2} with a SD of 1.12 kg C m^{-2}. These values are close to the values encountered at the regional scale for the Podzólicos Vermelho Amarelos, and they are also in agreement with the previous amount of 3.30 kg C m^{-2} to 0.3 m calculated from a few sampled pits in the same area (Moraes et al., 1995). The experimental semivariogram was best modeled using a spherical model without nugget effect (Figure 7). The model parameters were a sill value of 1.18 (kg m^{-2})^2 and a range of 899 meters. Results of the cross validation procedure showed a ME of -0.001 kg C m^{-2} and a RMSE of 1.09 kg C m^{-2}. It should be noticed that this experimental semivariogram is in agreement with the regional semivariogram as the sill value of the first one is inferior to the nugget effect of the second one, indicating the presence of different scales of spatial variation in the nested structure.

Kriged carbon densities values and associated KSD resulting from a block kriging (square block of 0.25 ha) are mapped in Figure 8. The 14,534 kriged values ranged from 1.41 to 6.50 kg C m^{-2}, with a mean value of 3.07 kg C m^{-2} closed to the mean of the input data. KSD values ranged from 0.17 to 1.16 kg C m^{-2}. The mean KSD value is 0.706 kg C m^{-2}, but this mean is influenced by a few high values. Only 6.5% of the area have KSD values superior or equal to 1.0 kg C m^{-2}, these values are located in zones of extrapolation rather than interpolation (border effect) and zones of lacking information. The total carbon stocked in this forest sector is equal to 111.7 Gg (1Gg = 10^9 g) with an associated error of 23% (Figure 8).

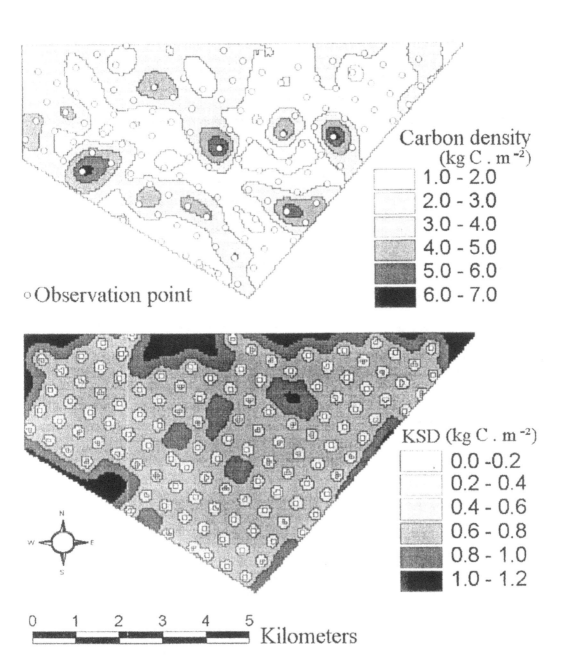

Figure 8. Carbon stock for the upper 0.3 m obtained by block kriging and map of associated errors. (Ranch Nova Vida).

If the carbon stored in this corresponding sector of forest using the regional map of kriged valu
is used, the corresponding stock is 114.7 Gg C. This little difference tends to validate the use of t
results from the regional scale study at a local scale.

V. Conclusion

The C stored in the upper soil horizons represents the pool most sensitive to changes if the nati
forest is used for agriculture or converted to pasture for ranching. Because the original data we
sampled from sites with native vegetation in the absence of significant disturbance, the estima
represent a valuable baseline for evaluating the original states and assessing the effects of land u
changes on soil C stocks in the Amazon Basin at various scales.

The comparison of different techniques gave quite similar overall estimates of the carbon sto(
at the local, regional and continental levels. However, the use of geostatistics on georeferenced d;
provided more accurate estimates (i.e., the associated error is reduced) of soil organic carbon und
primary uncut 'old growth' forest.

Moreover, the geostatistical approach appears to be an essential tool to analyze georeferenc
information and may contribute in the understanding of spatial variability at various scales. Howev
published soil surveys generally lack georeferenced information, and therefore the ability to appl
geostatistical approach may be limited.

The recent increasing development of remote sensing techniques results in a sharp increase
available georeferenced information. The combination of data derived from soil surveys w
geographical information systems, remote sensing products and geostatistical approach might b
useful research direction in order to provide more accurate overall estimates of C stored in soils. F
instance, estimates might be improved by using an external drift in the kriging of soil parame
(Bourennane et al., 1996) or by incorporating variability into soil map unit delineation's (Rogow;
and Wolf, 1994). Further work will lead to the development of new knowledge in this area.

Acknowledgments

Research support was provided by the Fundação de Amparo à Pesquisa do Estado de São Pau
(FAPESP) with contracts 94/6046-0 and 95/1451-6, and by the Fundação Coordenação
Aperçoamento de Pessoal de Nível Superior (CAPES-MEC) with grant number 2129/95.

References

Batjes, N.H. 1996. Total carbon and nitrogen in the soils of the world. *Eur. J. Soil Sci.* 47:151-1€
Beinroth, F.H. 1975. Relationships between U.S. soil taxonomy, the Brazilian soil classificati
    system, and FAO/UNESCO soil units. p. 92-108. In: E. Bornemisza and A. Alvarado (eds.), S
    Management in Tropical America, North Carolina State University, Raleigh.
Bernoux, M., D. Arrouays, M. Volkoff, C.C. Cerri and C. Jolivet. 1998. Bulk densities of Brazili
    Amazon soils related to other soil properties. *Soil Sci. Soc. Am. J.* 62:743-749.
Bourennane, H., D. King, P. Chéry and A. Bruand. 1996. Improving the kriging of a soil variat
    using slope gradient as external drift. *Eur. J. Soil Sci.* 47:476-483.

Burgess, T. M. and R. Webster. 1980a. Optimal interpolation and isarithmic mapping of soil properties. I. The semi-variogram and punctual kriging. *J. Soil Sci.* 31:315-331.

Burgess, T. M. and R. Webster. 1980b. Optimal interpolation and isarithmic mapping of soil properties, II, Block kriging. *J. Soil Sci.* 31:333-341.

Cerri, C.C., M. Bernoux and G.J. Blair. 1994. Carbon pools and fluxes in Brazilian natural and agricultural systems and the implication for the global CO_2 balance. *Trans. 15th Int. Cong. of Soil Sci.,* Acapulco, Mexico. Vol 5a: 399-406.

Cressie, N.A.C. 1993. *Statistics for Spatial Data,* (Revised ed.). Wiley Interscience, New York. 900 pp.

Detwiler, R.P. 1986. Land use change and the global carbon cycle: the role of tropical soils. *Biogeochemistry* 2:67-93.

Empresa Brasileira de pesquisa Agropecuaria. 1981. Mapa de Solos do Brasil, escala 1:5,000,000.

Empresa Brasileira de Pesquisa Agronomica. 1983. Levantamento de reconhecimento de média intensidade dos solos e avaliação da aptidão agrícola das terras do Estado de Rondônia. Contrato Embrapa/SNLCS-Governo do Estado de Rondônia. Rio de Janeiro, Brazil. 895 pp.

Eswaran, H., E. van Den Berg and P. Reich. 1993. Organic carbon in soils of the world. *Soil Sci. Soc. Am. J.* 57:192-194.

FAO. 1997. State of the World's Forests. Food and Agriculture Organization of the United Nations. 200 pp.

Gamma Design Software. 1991. GS+: Geostatistics for the agronomic and biological sciences. Version 1.1. Gamma Design Sofware, Plainwell, MI.

Houghton, R.A., D.L. Skole and D.S. Lefkowitz. 1991. Changes in the landscape of Latin America between 1850 and 1985 II. Net release of CO2 to the atmosphere. *For. Ecol. Manage.* 38:173-179.

Jacomine, P.K.T. and M.N. Camargo. 1996. Classificação pedológica nacional em vigor. p. 675-689. In: V.H. Alvarez, L.E.F. Fontes and M.P.F. Fontes (eds.), *Solo nos Grandes Domínios Morfoclimáticos do Brasil e o Desenvolvimento Sustentado.* SBCS-UFV, Viçosa-MG, Brazil.

Lugo, A.E. and S. Brown. 1993. Management of tropical soils as sinks or souces of atmospheric carbon. *Plant Soil* 149:27-41.

Matheron, G. 1965. *Les Variables Régionalisées et Leur Estimation.* Masson, Paris. 306 pp.

Melillo, J.M., R.J. Naiman, J.D. Aber and A.E. Linkins. 1984. Factors controlling mass loss and nitrogen dynamics of plant litter decaying in northern streams. *Bull. Mar. Sci.* 35:341-356.

Ministério das Minas e Energia. 1973-1982. Projeto RADAMBRASIL, programa de integração nacional. Levantamento de recursos naturais. Vol. 1-27. MME/DNPM, Rio de Janeiro, Brazil.

Ministério das Minas e Energia. 1978. Projeto RADAMBRASIL, programa de integração nacional. Levantamento de recursos naturais. Vol. 16, Folha SC-20 "Porto Velho", 663 p., MME/DNPM, Rio de Janeiro, Brazil.

Ministério das Minas e Energia. 1979. Projeto RADAMBRASIL, programa de integração nacional. Levantamento de recursos naturais. Vol. 19, Folha SD-20 "Guaporé", 368 p., MME/DNPM, Rio de Janeiro, Brazil.

Moraes, J.L., C.C. Cerri, J.M. Melillo, D. Kicklighter, C. Neill, D.L. Skole and P.A. Steudler. 1995. Soil carbon stocks of the Brazilian Amazon basin. *Soil Sci. Soc. Am. J.* 59:244-247.

Moraes, J.F.L., C.C. Cerri, B. Volkoff and M. Bernoux. 1996. Soil properties under Amazon forest and changes due to pasture installation in Rondônia, Brazil. *Geoderma.* 70:63-81.

Parton, W.J., D.S. Schimel, C.V. Cole and D.S. Ojima. 1987. Analysis of factors controlling soil organic matter levels in Great Plains. *Soil Sci. Soc. Am. J.* 51:1173-1179.

Pastor, J. and W.M. Post. 1986. Influence of climate, soil moisture and succession on forest carbon and nitrogen cycles. *Biogeochemistry* 2:3-27.

Post, W.M., W.R. Emmanuel, P.J. Zinke and A.G. Stangenberger. 1982. Soil carbon pools and world life zones. *Nature* 298:156-159.

Prado, H. 1996. *Manual de Clasificação de Solos do Brasil*. 3rd ed., FUNEP, Jaboticabal, Brazil, 194 pp.

Rodrigues, T.A. 1980. Estudo expedito de solos do Território Federal de Rondônia para fins de classificação, correlação e legenda preliminar. Contrato EMBRAPA/SNLCS-Governo do Território Federal de Rondônia. SNLCS, Boletim Técnico n° 73, Rio de Janeiro, Brazil.

Rodrigues, T.E. 1996. Solos da Amazônia. p 19-60. In: V.H. Alvarez, L.E.F. Fontes and M.P. F. Fontes (eds.), O Solo nos Grandes Domínios Morfoclimáticos do Brasil e o Desenvolvimento Sustentado. SBCS-UFV, Viçosa-MG, Brazil..

Rogowski, A.S. and J.K. Wolf. 1994. Incorporating variability into soil map unit delineations. *Soil Sci. Soc. Am. J.* 58:163-174.

Schlesinger, W.H. 1986. Changes in soil carbon storage and associated properties with disturbance and recovery. p. 194-200. In: J.R. Trabalka and D.E. Reichle (eds.), *The Changing Carbon Cycle, a Global Analysis*. Springer-Verlag, New York.

Soil Survey Staff, 1990. Keys to Soil Taxonomy, 4th ed. *SMSS Tech. Monogr.*, 6. SMSS, Blacksburg, VA, 422 pp.

Sombroek, W.G., F.O. Nachtergaele and A. Hebel. 1993. Amounts, dynamics and sequestrations of carbon in tropical and subtropical soils. *Ambio* 22:417-426.

Tukey, J.W. 1958. Bias and confidence in not-quite large samples (Abst.). *Annals of Mathematical Statistics* 29:614.

Van Wambeke, A. 1991. *Soils of the Tropics: Properties and Appraisal*. McGraw-Hill, New York. 343 pp.

Walkley, A. and I.A. Black. 1934. An examination of the Degtjareff method for determining soil organic matter and a proposed modification of the chromic titration method. *Soil Sci.* 37:29-38.

CHAPTER 3

Carbon Pools in Forest Ecosystems of Australasia and Oceania

P.K. Khanna, P. Snowdon and J. Bauhus

I. Introduction

In the 200 years since European settlement, over 1 million km² of forest and woodlands within the intensive land use zone of Australia has been cleared or thinned and converted into pasture, crop-land, or plantations (Graetz et al., 1995). An international comparison of land clearing estimates for the period 1981—1990 showed that Australia ranked as the eighth largest land clearer in the world, the only developed nation among the top 15 countries listed (Department of the Environment, Sports and Territories, 1995). The estimated land clearing in Australia for 1990 was 650,000 ha, which is more than half the area that was cleared in the Brazilian Amazon for that year. In some areas of the country land clearing is still continuing. For example, a nationwide survey showed that in 1994 farmers in Australia expressed an intention to clear 3.3 million ha of forests and woodlands during the next five years (Wilson et al., 1995). Eighty eight percent of this proposed area was in Queensland. Most of the land intended to be cleared was located in the pastoral and wheat-sheep zone of the drier inland, and would therefore carry low open forest or woodlands. Primary forests in New Zealand have experienced a similar change while in more recent times large changes in land use have begun in the Pacific (FAO, 1995).

Any major change in land use must result in significant changes in the C pools of the aboveground and belowground ecosystem components. Together with fossil fuels, release of C from aboveground and belowground biomass due to burning or decomposition constitute the most important C fluxes entering the atmospheric pool. The total amount of C stored in forests and woodlands (both above-ground and belowground) has declined over time, primarily due to clearing for agriculture. Changes in vegetation type or land management practices can also lead to changes in the equilibrium levels of soil organic matter (SOM). Decline in soil organic matter is considered to be an important factor contributing to deleterious changes in soil properties and to processes which may decrease the productive potential of a site. A decrease in SOM may decrease the mineralization and cycling of C, N, P and S; the retention and buffering of cations and anions; the soil pH; the soil structure and water retention; the root development; and the nutrient uptake. In recognition of this important role, change in soil organic matter status has been included as one of the indicators that could be used to assess the effects of forest management on ecological sustainability of forests (Commonwealth of Australia, 1997).

This chapter describes carbon pools in different forest and woodlands in Australasia and the Oceania (Pacific) region. The quantities of C in different components of forest and woodland ecosystems in Australia are presented. A balance sheet approach is used to describe changes in C pools

Table 1. Area under forest vegetation in Australia

Forest type	Tall (>30 m)	Medium (11–30 m)	Low (2–10 m)	Total
		(million ha)		
Closed forest	1.64	1.96	1.03	4.63
Open forest	5.48	28.09	1.41	34.98
Mallee forest		0.23	3.95	4.18
Woodland	1.07	76.08	27.32	104.47
Mallee woodland		0.87	6.71	7.58
Plantations		1.04[a]		1.04[a]

Closed forest >80% canopy cover, open forest 50–80% canopy cover, woodlands 20–50% canopy cover, mallee = low growing eucalypts with multiple stems.
(Adapted from Commonwealth of Australia, 1997; ABARE[a], 1997.)

and fluxes for four management scenarios with commonly employed practices: (a) where old-growth or mature forest or woodland is harvested and converted into regrowth forests, plantation or pasture, (b) where regrowth is harvested and regenerated, (c) where plantation is harvested and replanted for a subsequent rotation, and (d) where pasture land is converted into plantations for wood production.

II. Assessment of Land Under Forests

Land and forest management is primarily the responsibility of State and Territory Governments in Australia. In cooperation with these agencies the Commonwealth Government has been responsible for developing a national approach to forest management, including the National Forest Policy Statement (Commonwealth of Australia, 1992). Regional Forest Agreements have been or are being developed which allocate the land among different forest users by considering the broad conservation and industry goals. These agreements are based on following defined criteria for the conservation and sustainable management of Australian forests as given by the 'Montreal Process'. These initiatives have led to a re-evaluation of the use of Australia's forest estate as described in the National Forest Inventory and the National Plantation Inventory (National Forest Inventory, 1997).

The current definition of native forest in Australia includes areas dominated by trees usually with a single stem with a potential mature height exceeding two meters and with a potential overstory crown cover of 20%. Under this definition about 72% of Australian forests are publicly owned with about 14% managed by State forest authorities for multiple use including wood production. Based on this estimate total native forest amounts to about 156 million hectares or 20 % of Australia's land area (Table 1). Earlier estimates have been made on more narrow definitions of forest and woodlands. Thus, the Resource Assessment Commission (Commonwealth of Australia, 1992) estimated that the area under forests was about 43 million ha in 1990. In the FAO (1995) document which provided global synthesis of forest resources, 39.8 million ha were estimated as forest land with an additional 105.8 million ha as woodlands giving a total of 145.6 million ha. Plantations established for wood production in Australia cover an area of about 1.04 million ha but there are plans to increase this to 3 million ha by 2020.

About 29%, or 7.9 million ha (FAO estimates are different, Table 2), of New Zealand is occupied by forest (Zwartz, 1997). Indigenous forest amounts to 6.4 million ha but only 23% of this is available

Table 2. Land area, total forest area, wood volume and aboveground biomass of forests in the Pacific region, Australia and New Zealand[a]

Country	Land area	Forest + other wooded land	Forest	Plantation	Wood volume		Aboveground biomass	
		(million ha)			$m^3\,ha^{-1}$	total, million m^3	$Mg\,ha^{-1}$	Total Tg
Papua New Guinea	45.28	42.11	36.03	0.030	168	6063	191	6890
Solomon Islands	2.80	2.46	2.41	0.016	115	275	201	481
New Caledonia	1.83	1.29	0.71	0.009	113	79	197	138
Fiji	1.83	0.86	0.85	0.078	118	91	206	160
Vanuatu	1.22	0.81	0.81	0.007	123	99	216	173
Others	1.07	0.36	0.14	0.009		16		28
Total Pacific Islands	54.03	47.89	40.95	0.149	162	6623	193	7870
Australia	754.40	145.53	39.84	1.04[b]	83	3306	61	2430
New Zealand	26.78	7.47	7.47	1.48[c]	53	396	28	209
Grand total	835.21	200.99	88.26	2.67		10325		10509

[a]These estimates may vary from those given by other authorities elsewhere in the text; [b] ABARE, 1997; [c]Zwartz, 1997. (Adapted from FAO, 1995.)

for commercial utilization. There are 1.5 million ha of plantations of which 91% is *Pinus radiata* and 3% hardwood, predominantly eucalypts. In 1989, total plantation (1.24 million ha) carbon storage above and belowground (but excluding soil organic matter) was estimated to be 88 Tg C (Hollinger et al., 1993).

Most of the land area in the Pacific islands is under forest (75%) or woodland (14 %) with only a very small area under plantations (Table 2). Plantations for wood production are increasing at the expense of the primary forest. Of all the Pacific countries, Papua New Guinea (PNG) has proportionally the largest area under forest, and it has been estimated that 114 million m^3 of wood was harvested from 4.6 million ha of PNG forests during 50 years to 1990 (Saulei, 1997).

During the period from 1980 to 1990 in the Pacific region, a very significant area under forests was lost to other forms of land use (FAO, 1995). According to the estimates given by FAO (1995), the natural forest area of the Pacific region decreased annually by 131,000 ha, and if an annual increase in plantations of 8,100 ha is considered, a net annual loss of 123,000 ha occurred for 1980 to 1990. Most of it, 83,000 ha annually, was lost in Papua New Guinea. Considering a biomass density of 191 Mg ha^{-1} (FAO, 1995), the total C lost in the aboveground biomass during 1980 to 1990 in the Pacific region was approximately 120 Tg. It is very hard to make estimates of how much of the removed C has already been converted to atmospheric CO_2. Few estimates are available for changes in soil C.

III. Carbon Density and Carbon Pools in the Aboveground Components

The average volume of wood or growing stock of global forests is estimated to be 114 m^3 ha^{-1} which is almost the same in the developing (103 to 125 m^3 ha^{-1}) as in the developed (92 to 129 m^3 ha^{-1}) countries (FAO, 1995). This is usually estimated by applying a formula to the overbark diameter of the trees at breast height. In order to calculate the aboveground biomass one needs to know the density of wood and the biomass of all other vegetation components (branches, twigs, leaves, etc). Sometimes a relationship between tree diameter and biomass is sought and used to calculate total biomass values. Carbon is usually calculated from biomass data by assuming a conversion factor of 0.5 (the exact value lies between 0.4 and 0.5).

The aboveground biomass in different forest types is expected to vary significantly depending upon species composition and their growth rates. Climatic parameters, age of trees, nutrition and other growth parameters are important ancillary factors. But in order to develop biomass figures for forests at the national level an average value of biomass for a given type of forest is used (ideally it should represent mean age of the forest based on weighted mean area of the forest). The uncertainty in such an average value can be great and the biomass data obtained for a given forest may have a large error. Limitations of the biomass data given in Table 3 are evident. For example, most exploitable primary and regrowth forests in Australia are included in the open forest type and the values given in Table 3 are so close to each other (200 to 279 Mg ha^{-1}) that they defy general experience in the field. The Commonwealth of Australia has very recently decided to improve upon these figures by field sampling and the use of appropriate mensurational techniques. It may take several years before more reliable data become available. In the meantime the values given in Table 3 must suffice. Highest quantities of aboveground carbon in Australia are stored in the woodlands of medium height (Table 3) primarily because of the vast areas under these ecosystems (Table 1). These woodlands are also those most subject to clearing for agricultural purposes and thus have the potential to contribute in a significant way to carbon dioxide emissions.

Average C density data for aboveground biomass in forests of the Pacific region are shown in Table 2. These values are comparable to those for woodland or the lower end of open forest type in Australia (Table 3), which are expected to be low for most of the forests in the Pacific region. C

Table 3. Mean aboveground biomass and total carbon in the aboveground components of various forest ecosystems of Australia

Forest type	Mean aboveground biomass density (Mg ha^{-1})			Total C in aboveground biomass in Australia (Tg)			
	Tall	Medium	Low	Tall	Medium	Low	Total
Closed forest	450	356	300	369	349	155	873
Open forest	279	272	200	764	3821	142	4727
Mallee forest	–	39	39	–	5	77	82
Woodland	200	150	100	107	5706	1366	7179
Mallee woodland	–	39	39	–	17	131	148
Plantations	–	244	–	–	127	–	127
Total							13136

(Adapted from Commonwealth of Australia, 1997.)

density in particular forest types can exceed these values. Thus, Enright (1979) estimated an aboveground biomass value of 286 Mg ha^{-1} in a rainforest near Bulolo (PNG) which was similar to 295 Mg ha^{-1} in a lower montane rain forest reported by Edwards and Grubb (1977). The total biomass measured by Edwards and Grubb (1977) was 350 Mg ha^{-1} which included 40 Mg ha^{-1} of roots and 15 Mg ha^{-1} of under-story plants. A reasonable aboveground biomass value for such closed forests would therefore lie between 300 and 400 Mg ha^{-1}. However, the reliability of the biomass data for Pacific region can be considered to be quite low and is likely to remain so without significant injection of funds.

IV. Pools of C in the Soil

Carbon in the soil can be divided into the following three components.

- Soil organic matter – Nonliving organic matter excluding root material, charcoal and inorganic C. Different fractions of SOM have varying turnover rates.
- Root C - living and relatively undecomposed dead fine or woody components of roots which are usually not included in the analysis of soil C.
- Charcoal and inorganic C.

In addition, the amount of C in the litter layer can form a significant proportion (2 to 11% in two forests shown in Table 4) of the total C in the soil. In many studies on soil C this fraction is either completely ignored or inadequately measured. Amounts of C present in coarse roots are very difficult to measure directly and are usually assessed by using allometric relationships with stem diameters. The commonly employed method of collection and analysis of soils captures most of the C associated with organic matter, but only varying proportions of that present as roots and charcoal in soils. Depending upon site conditions, the underestimation of soil C can range from 8 to more than 50% if the contribution by litter layer, coarse roots and part of fine roots is not included (Table 4). In soils low in organic C (Table 4) coarse roots can account for the same amount of C as soil organic matter.

Carbon in charcoal is considered to be sequestered for a very long time. The amount of charcoal, also called black carbon, will depend upon intensity and frequency of fire; the higher the intensity and

Table 4. Amount of C in various components of a mature *Eucalyptus pauciflora* stand and a 20-year old *Pinus radiata* plantation near Canberra, Australia; percent values for various components refer to the respective total amounts in the aboveground or belowground fractions

Component	E. pauciflora		P. radiata	
	Mg C ha⁻¹	%	Mg C ha⁻¹	%
Aboveground				
Leaves and twigs	5.8	4.9	9.3	7.7
Branches	25.7	21.8	18.9	15.7
Stem wood + bark	79.2	67.3	92.0	76.5
Understorey	7.0	5.9		
Belowground				
Litter layer	8.5	2.2	6.8	11.2
Soil 0-60 cm	347.5	90.7	25.4	41.9
Fine roots <5 mm	3.3	0.9	2.6	4.3
Coarse roots	23.9	6.2	25.9	42.7

(Adapted from the data from Keith et al., 1997, and Ryan et al., 1996.)

frequency, the higher will be the amount of charcoal in a soil. Skjemstad et al. (1990) observed that 30% of C in four surface soils occurred as charcoal. The quantity of charcoal deeper in the soil is expected to be much lower. For the surface soil the history of site management (frequency and intensity of burning residues) will be an important factor. Kuhlbusch and Crutzen (1995) have estimated the global black carbon formation to be 50 to 270 Tg yr⁻¹. Despite fire being an integral part of the disturbance ecology of most Australian forest types, charcoal may only be a small fraction of the soil C pool in these forests.

There are considerable areas of Australian soils which contain carbonate in the regolith (Hubble et al., 1983). Calcareous soils occur extensively in southern Australia particularly in semi-arid environments where leaching is not a dominant factor. The native vegetation is commonly mallee or desert scrub. Calcareous red earths and red-brown earths (Alfisols) which contain free carbonate in the profile are widely spread and often support woodland. Calcareous dune soils can occur on coastal margins.

V. Sources and Sinks of Carbon in Forest Ecosystems

Dixon et al. (1994) concluded that the world's forests are presently a net source of atmospheric C. Forests in the high and mid latitudes presently accumulate 0.74 ± 0.1 Pg (10^{15} g) C per year, whereas clearing and change in land use of forests in low latitudes (largely tropical forests) release approximately 1.65 ± 0.4 Pg C. The combined net annual contribution of forests to atmospheric C is 0.9 ± 0.4 Pg. The latter is equivalent to about 16% of the carbon emissions from the burning of fossil fuels by e.g., industry and transport sectors, estimated to be 5.5 Pg yr⁻¹.

Out of a total of 165 Tg of CO_2 (equivalent) emitted annually in Australia, 107 Tg is attributed to land clearing (Table 5). Burning of vegetation is still an important practice to clear land and this can contribute to almost half of the total emission, the remainder is attributable to reduction in soil carbon. The net annual emissions of C related to land clearing and forestry practices in Australia are estimated to be in the order of 57 Tg of CO_2 equivalent (Table 5).

Table 5. Estimates of emissions and sinks of CO_2 equivalent due to land clearing and forestry practices in 1995 in Australia

Emissions	1995	Sinks	1995
	(Tg of CO_2-equivalent)		(Tg of CO_2-equivalent)
Clearing of vegetation		Regrowth	23
(A) Burning	47.1	Forestry growth	79.7
(B) Decay of debris	6.7	Other	6.5
(C) Soil including roots	52.8		
Forest harvesting	58.6		
Total emissions	165.2	Total sinks	108.2
		Net emissions	57

(Adapted from NGGIC, 1997.)

There has been a significant reduction in the estimates of emission from land clearing activities in Australia since 1990 (from 206.2 Tg in 1990 to 165.2 Tg per year in 1995). This was primarily related to a reduction in land clearing activities and the associated burning of residues and to decomposition of belowground debris and soil organic matter. Due to decreases in the area of clearing primary forest for establishing regrowth, there has been a decrease in the sink of C in the regrowth. Forest harvesting practices have also changed (value added practices) leading to a higher degree of wood utilization and thus the removal of C from forested land has increased (less harvesting slash left on the ground).

Because of differences in their growth rates, conversion of slow growing, old-growth forests to fast growing young forests is usually assumed to increase C uptake by vegetation (C sink shown as regrowth in Table 5). But this may be deceptive because the critical factor is the fate of C stored within the forest by continuous accumulation over a long time period, not the annual rate of C uptake (Harmon et al., 1990). Harmon et al. (1990) calculated by using a simple model which compared the growth rates, the amount of slash left in the forest and the final use of wood removed, that the conversion of 5 x 10[6] hectares of old-growth forests to younger plantations in western Oregon and Washington during the last 100 years has added 1.5 to 1.8 Pg of carbon to the atmosphere.

VI. Management Options and C Pools and Fluxes

A number of land use practices have contributed to changes in pools and fluxes of C in Australia. Of significance was the clearing of forest and savanna landscape for agricultural use and plantations. Clearing of primary forest including rainforest for plantations, agriculture or urban development has been much reduced in recent years, but still persists particularly on private lands. Much of the remaining primary forest under public ownership has been placed into reserves while the remainder is managed for multiple purposes including timber production. Often some form of selective logging is practiced so that the original forests with a large component of mature and overmature trees have been replaced by a mosaic of younger uneven-aged regrowth forests. In other cases old-growth forest has been cleared in large areas during conversion to even-aged stands of regrowth forest.

Although a few plantations of softwoods and hardwoods were established in the late 1800s the full awareness of the need for plantation grown trees was not realized in many Australian states until the 1920s. By that time the choice of land was limited because the more fertile land with moderate slopes had already been taken for agricultural use. There was a common belief that pines could be grown successfully on poor soils. Consequently, soil quality often received scant attention when planting sites were selected. Plantations were established on waste crownland, abandoned farmland and on poor quality native forest sites. Often these were on poor coastal sands. Many of the early plantings were failures and sometimes the plantations were abandoned. It was subsequently shown that most of these failed plantations were severely deficient in nutrients such as phosphorus and/or zinc (Ludbrook, 1942; Young, 1940; 1948; Stoate, 1950). The realization of these limitations promoted the establishment of plantations on more fertile cleared primary forest sites. Increasing public pressure against conversion of indigenous forest to plantations led to increased purchase of farming land. In the case of improved pasture, high nitrogen availability sometimes had adverse effects on tree form. As a consequence of these historic changes in the character of land acquisitions the Australian plantation estate consists of a wide range of soil types which have had a variety of prior land uses.

A. Characteristics and Conversion of Old-Growth or Mature Forest

1. C Pools and Growth Rates

There is wide variation in the C content of mature native forests depending upon species, forest structure, soils and climate. The aboveground biomass (tree, understory and litter) of mature eucalypt forests can vary from 63 Mg C ha^{-1} for *Eucalyptus signata* on sand at Stradbroke Island to 360 Mg C ha^{-1} for *E. regnans* in Gippsland (Turner and Lambert, 1996). Tropical and subtropical rainforests can contain 160 to 190 Mg C ha^{-1} (Turner and Lambert, 1996) while temperate forests in New Zealand can be in the range 350 to 500 Mg C ha^{-1} (including roots) (Ford-Robertson, 1997). Carbon in the forest floor in eucalypt forests can vary from 4 to 22 Mg C ha^{-1} (Turner and Lambert, 1996). Root biomass can be in the range of 21% of total biomass in a mature beech/podocarp community (Beets, 1980) to 42% in *E. saligna* growing on sands (Westman and Rogers, 1977). In some ecosystems up to 150 Mg C ha^{-1} can be retained as standing dead trees or coarse woody debris (Stewart and Burrows, 1994).

Climatic factors, especially rainfall and temperature, soil factors and type of vegetation have a marked influence on the amount of SOM. Organic C in the surface soils (0–75 mm) of eucalypt forests in southern New South Wales fall in the range 2.4 to 5.5% C while rainforests in the northern part fall in the range of 4.6 to 14% C (Turner and Lambert, 1996). The range for surface soils (0–100 mm) in 72 rainforest sites in northeastern tropical Queensland was 1.8 to 8.4% C (Spain, 1990). In Papua New Guinea surface soils (0–400 mm) can vary from 3 to 13% C (Bleeker, 1983). Higher levels, 9 to 19% C, have been found in the montane rainforest in Papua New Guinea (Edwards and Grubb, 1977).

The net productivity of old-growth forests is usually considered to be low, sometimes to be at equilibrium (zero net growth) and occasionally, in degrading stands, to be negative. Estimates of total C in stems of indigenous forests of the South Island and Stewart Island, New Zealand, indicate annual changes in the range +2.0 to -4.1% per year with an average of -0.38% per year (Hall and Hollinger, 1997). This amounts to an annual loss of 1.8 Tg C which can thus have a very substantial effect for calculating carbon balance for forests in a country.

2. Losses in Biomass C Due to Conversion

The recovery of commercially valuable products from old-growth forests depends on the nature of the forest and the availability of markets. In the poorest quality forests in Australia there may be little or no recovery, but even in better quality forests only a small proportion of the woody components is usually harvested and the remaining debris (70 to 80% of standing biomass) is left on the site and burned. Harwood and Jackson (1975) estimated that the amount of C in logging slash was 315 Mg ha[-1] in a mixed eucalypt forest with rainforest understorey in Tasmania. Jones (1978) recorded that on average about 250 Mg C ha[-1] was present as fuels in 14 slash disposal fires in western Australia. These values are probably towards the higher end of the range of slash left on the site following clearcut harvest. With the increased extraction of pulpwood, the values usually lie between 100 to 200 Mg C ha[-1]. In the mixed species eucalypt forest at Maramingo in East Gippsland, Hopmans et al. (1993) measured, after recovery of 100 Mg ha[-1] of wood, about 244 Mg ha[-1] of debris. For three different forest types in the Eden woodchip area in southeastern New South Wales, Turner and Lambert (1986) estimated 203 to 229 Mg C ha[-1] in aboveground biomass out of which about 50 Mg C ha[-1] were removed in the harvest leaving 150 to 175 Mg C in the slash. More than 50% of woody biomass was remaining on site. In Westland, New Zealand, a podocarp-hardwood forest was estimated to carry 127 Mg C ha[-1] as aboveground biomass of which 95 Mg C ha[-1] would be harvested in a woodchipping operation (Levett et al., 1985). In a mature beech/podocarp community the aboveground components comprised only 47% of the total ecosystem biomass, but 57% of this was harvested as utilizable logs (Beets, 1980).

Even when selective logging is practiced there can be considerable damage to a forest ecosystem if it is not well managed. For example, after logging *Araucaria cunninghamii* stems greater than 40 cm from a forest in Papua New Guinea it was found that 60% of the remaining *Araucaria* stems and 94% of stems of noncommercial species had been destroyed (Enright, 1978). Thus the amount of slash can be disproportionately large compared to the amount harvested.

Harvest debris can act as a hazard for regrowth forests and plantations. It may be a source of fuel, increasing fire hazard or become a source of infection for disease and insects under certain conditions. Burning of debris will increase CO_2 in the atmosphere. In their review on the amount of C released to the atmosphere in Australian vegetation fires, Cheney et al. (1980) estimated that gross C release may range between 0.08 and 0.25 x 10^{15} g annually. Out of this amount the C released on regeneration of forests may account for 2.3 x 10^{12} g while plantation establishment accounts for 1.4 x 10^{12} g annually. On a per hectare basis, regeneration fires may consume 113 Mg ha[-1] whereas plantation establishment fires release 68 Mg ha[-1]. Both these values are much higher than those due to prescribed burning and wildfires.

When old-growth forests are converted to regrowth forests any burning operation is applied in a broadcast fashion. Similar practices were used for plantation establishment on native forest sites until the 1960s when it was realized that clearing debris into windrows would facilitate future mechanized operations. This procedure often resulted in the redistribution of soil C into the windrows. When windrows were burnt, the degree of combustion was often more complete than for the broadcast burns. Differences in site preparation not only influence loss of carbon from the site but can also affect subsequent rates of plantation growth and carbon fixation. Sometimes effects on plantation growth are small (Hall and Carter, 1980), but large differences can occur, e.g., growth of *Pinus radiata* on scalped areas (between windrows and heaps) was reduced by up to 67% compared with the growth on windrowed or heaped areas and 28% compared to the control (Dyck et al., 1989).

3. Changes in Soil Organic Matter after Conversion

a. Conversion to Regrowth

Harvesting of mature eucalypt forests in southeastern New South Wales, Victoria and Tasmania, is mostly by clearfelling followed by a broadcast slashburn. This can generate a range of soil microsites, from unburnt to ashbeds, which have experienced significant differences in soil temperature and moisture content and therefore in SOM decomposition rates (Khanna et al., 1994). A decrease in SOM was observed on cut and unburnt sites after clearfelling but transfer of organic C from roots to soil created problems in assessing changes on burnt and ashbed microsites (Khanna et al., 1994). Soil C remains highly dynamic in the surface layers for a number of years after clearfelling.

b. Conversion to Plantation

Few studies have been made on the changes in soil properties that take place after plantation establishment. Hamilton (1965) found that in 5 out of 6 cases, organic C had declined after 20 to 30 years growth of *P. radiata* near Canberra, Australian Capital Territory (ACT). Losses in organic C can also be inferred from the 25% reduction in total nitrogen observed when eucalypt forest on deep coastal sand was converted to *P. elliottii* plantations (Waring, 1968). In all 15 cases in the Batlow, N.S.W. district, organic C in the surface (0–7 cm) soils under *Pseudotsuga menziesii* plantations was less than adjacent regenerated eucalypt forest (Turner and Kelly, 1977). On phosphorus deficient second rotation sites with *P. radiata* it has been shown that application of P-fertilizer to highly stocked stands (11960 stems ha⁻¹) can reduce organic matter in the surface 0–2.5 cm from 2.59 to 2.16%, but by adding both N and P can result in an increase to 3.39% (Snowdon, unpublished data). This trend, but of a smaller magnitude, continued to a depth of 15 cm. Nearby, growing *P. radiata* with a clover nurse crop substantially increased soil organic matter (Waring and Snowdon, 1985). A substantial loss in soil organic C has been noted 41 years after a north Queensland rainforest was converted to a hoop pine (*Araucaria cunninghamii*) plantation (Holt and Spain, 1986). Standing litter was higher under the plantation and turnover rates were lower. The decline in organic matter was attributed to the loss of nutrients during conversion.

In some regrowth forests and plantations, prescribed burning of litter and ground vegetation may be carried out to lower the risk of wildfire. Hunt and Simpson (1985) have reported, from a 3-year study assessing the impacts of low intensity prescribed fire on the growth and nutrition of 13-year slash pine stands, that soil organic C in the surface soil (0–2.5 cm) decreases significantly as a result of repeated burning.

c. Conversion to Pasture or Cropland

Clearing land for pasture in Australia commonly involves burning of debris which emits large amounts of CO_2 to the atmosphere. There is controversy as to whether the land clearing leads to long-term (several decades) changes in soil organic matter. Stands of brigalow (*Acacia harpophylla*) have substantially higher soil organic C and N contents than adjacent natural grasslands on similar soil types (Isbell, 1966). Virgin sites on grey clays can have higher organic C than cleared sites (Webb et al., 1977), and on solodised-solonetz and solodic soils, 22 of 31 comparisons showed organic C declined by an average of 20% under pasture or cropping (Graham et al., 1981). In some studies the loss of organic matter after clearing rainforest grown on krasnozems have been inconclusive (Nicholls and Tucker, 1956; Colwell, 1958), but in other cases losses from 20 to 50% of organic C have been found (McGarity, 1959). In one case it has been estimated that, 80 years after clearing, soil organic C produced by rainforest (~6% C) had been reduced to 25% of its original value, but that total soil organic C had increased by 20% under pasture (Skjemstad et al., 1990).

Tillage, and possibly associated erosion, of Vertisols under savanna woodland in central Queensland have resulted in a decrease of organic C from 1.22 to 0.76 % (Webb and Dowling, 1990). Cultivation of red-brown earths (Alfisols) in south Australia has resulted in a decrease of organic C from 2.75 % to 1.56 % after 58 years (Grace et al., 1995). Native pastures in Australia are commonly phosphorus deficient which reduces legume growth. When this is corrected, increases in soil organic C can occur, e.g., near Canberra, Australia, organic C increased from 1.89% to 2.33% after six years (Simpson et al., 1974). Similar increases have been noted in some New Zealand soils (Jackman, 1964), but changes are not always observed (Murata et al., 1995). Clearly, the loss or gain of soil carbon will depend on many circumstances including the nature of the replacement species, frequency of cultivation, inclusion of pasture or fallow in crop rotations, export of nutrients in products and addition of fertilizer.

Slash and burn agriculture is believed to account for a high proportion of forest loss in Papua New Guinea (Nonggor, 1995). Burning and repeated cultivation generally leads to loss of fertility and organic matter (Bleeker, 1983). When the land is abandoned it often reverts to scrubland, but repeated burning can result in the formation of perpetual grasslands which also have lower organic C levels than the forest. Protection of grasslands from fire can result in a reversion to woody vegetation and an attendant increase in organic C. Shifting cultivation accounts for about 20% of annual forest loss in the Philippines (Eusebio, 1995). However, more recently, ecologists have viewed the traditional practice of slash and burn agriculture as a sustainable land use practice in sparsely populated areas.

B. Characteristics of Regrowth and its Conversion

Most old-growth karri forest in western Australia (*E. diversicolor*) have been harvested and converted to regrowth forest in clearcut areas of about 80 ha (O'Connell and Grove, 1991; 1996). The slash remaining after timber extraction is burnt by fires of moderate to high intensity. Even-aged stands develop from natural regeneration or sometimes after artificial seeding or planting. Initially stem density is very high but as the trees develop there is considerable mortality. A dense scrub layer reaching to 5m develops. Legume species persist for 5 to 15 years after fire but some non-legume species persist for at least 40 years. Annual volume increments are estimated to range from 5 to 8 m^3 ha^{-1} while total accumulation of C in aboveground biomass ranges from 4 to 6 Mg ha^{-1}. These growth rates are intermediate between the growth of *E. grandis* plantations (Bradstock, 1981) and naturally regenerated *E. obliqua* (Attiwill, 1979) in eastern Australia. Regular prescribed burns are carried out on a 7 to 9 year cycle. Thinning at 20 years of age can remove 20 to 25 Mg C ha^{-1} of chipwood and small sawlogs. About twice this amount can be removed at a subsequent thinning at about age 50 years.

C. Characteristics of Plantations and their Conversion

Second and later rotation stands are becoming an increasing proportion of the forest state in Australia and New Zealand. The interaction of the original soil resource with subsequent management practices has increased the diversity of the soil resources at the compartment level. In 1995 there were 1.04 million hectares of plantation in Australia and 1.48 million hectares in New Zealand. In Australia eucalypt plantations occupied only about 28,000 ha in 1976, about 110,000 ha in 1992 and are expected to be about 240,000 ha by 2001 (Cromer, 1996).

The growth and yield of pine plantations is well documented (e.g., Lewis et al., 1976). These data can be combined with information from thinning (Carlyle, 1993) and clearfelling studies (Smethurst and Nambiar, 1990) to make estimates of aboveground biomass. When this is combined with estimates of coarse root biomass using equations based on existing allometric relationships (Jackson and Chittenden, 1981; Watson and O'Loughlin 1990) and estimates of fine root turnover (Keith et al.,

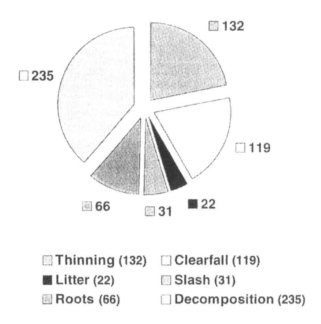

Figure 1. Fate of biomass carbon (Mg ha^{-1}) over a 50-year rotation of *P. radiata* from planting till immediately after clearfelling; this includes wood harvested during thinning and at clearfall, litter, slash and roots retained at the time of clearfall, and organic components which have decomposed during the plantation period.

1997), we were able to develop a simple budget for organic C changes in a Site Quality III stand of *P. radiata*. By the end of a 50-year rotation, about 40% of the total carbon fixed (exclusive of respiration losses) has been harvested by regular thinnings or at final harvest (Figure 1). About 20% is retained as above and belowground litter slash and roots while 40% of the total carbon fixed has already decomposed (i.e., litter, slash, root turnover). While rotations of this length were common in the past, current plantations are considered for rotations of about 30 to 35 years. If there is no change in growth rates, the change to shorter rotations results in a reduction of the total amount of carbon stored per unit area.

Sizes of carbon pools in plantations will depend upon the mix of products for which they are grown, e.g., fuel, pulpwood or sawlogs. Because of their rapid early growth, eucalypts are often grown on short rotations for pulpwood, but the trend in Australia is moving towards the production of high value poles, veneer and sawlogs from eucalypt plantations (Gerrand et al., 1997). Some of the earliest eucalypt plantations in Australia were of *E. grandis* in their native habitat near Coffs Harbour, N.S.W. A number of studies of biomass development have been made in these plantations (Bradstock, 1981; Turner, 1986; Birk and Turner, 1992).

Growth rates of plantations depend upon the species and on the climatic, soil and management conditions under which they are grown. For example, plantations of *E. regnans* on pumice soils in central North Island of New Zealand have a mean annual increment in wood volume about twice that of *E. grandis* at Coffs Harbour, N.S.W. (Frederick et al., 1985). On a high quality site in Tasmania,

E. nitens growth rates were measured at 30 m³ ha⁻¹ yr⁻¹ (Gerrand et al., 1997). More commonly growth rates between 10 and 20 m³ ha⁻¹ yr⁻¹ can be expected. High and sustained rates of productivity are only achieved by efficient site preparation, fertilization, weed management, choice of genetic material, protection from diseases and pests, pruning and thinning (Birk and Turner, 1992; Cromer et al., 1993). Various aspects of past achievements and future opportunities for improvement in pine plantations have been reviewed by Boardman (1974), Boardman and Simpson (1981), Turner (1983), and Nambiar and Booth (1991). Considerable improvement has been made in the productivity of pine plantations and the opportunities for further gains have been identified. There is a high degree of similarity in the soil, nutrient and management problems occurring in pine and eucalypt plantations, but there are several genus- and species-specific issues, e.g., pest control, which may need to be addressed.

Thinning and harvesting operations leave a substantial amount of slash on the site depending upon the type and intensity of the operation. Thinning residues left on the site when a second rotation *P. radiata* stand of 11 years was thinned by reducing the basal area (BA) by 48% had the following biomass in slash residues (Mg ha⁻¹): live needles 5.5, dead needles 3.5, live branches 8.2, dead braches 4.0 and tops 8.3, total 29.5 (Carlyle,1995). Hunter et al. (1989) compared the amount of biomass removed from three different sites of *P. radiata* where three types of thinnings were imposed: waste thinning, production thinning and whole-tree thinning. Amount of residues added to the forest floor ranged from 63 Mg ha⁻¹ to 74 Mg ha⁻¹ in waste thinning and 21 to 36 Mg ha⁻¹ in production thinning whereas whole-tree thinning removed 63 to 113 Mg ha⁻¹ of biomass from the site.

Amounts of residual slash when pine plantations are harvested usually range from 50 Mg and 100 Mg ha⁻¹. For example, Madgwick and Webber (1987) measured logging residue in a *P. radiata* stand to be 15% of the stem+bark removed. The total residue was 97 Mg ha⁻¹ (41 Mg ha⁻¹ of stem+bark, 56 Mg ha⁻¹ of cones+branches +foliage). Flinn et al. (1979) reported that the amount of logging slash on a *P. radiata* site in Victoria was 79.5 Mg ha⁻¹ and the slash burn consumed 84% of the residues. Smethurst and Nambiar (1990) reported on a similar site the logging slash to be 52 Mg ha⁻¹ (74% of that as wood and bark) and 32 Mg ha⁻¹ in the litter on forest floor. Higher values (187 Mg ha⁻¹) of slash residues were reported by Balneave et al. (1991) after logging a 34- year old *P. radiata* stand in New Zealand. The amounts of harvesting residue in Hoop Pine (*Araucaria cunninghamii*) plantations following clearfelling of first rotation plantations was estimated at 206 Mg ha⁻¹ (Constantini et al. 1997).

Slash is an important source of nutrients and organic matter, decomposition plays an essential role in the nutrient cycling processes. Judicious management of slash would promote the long term productivity of a site through its role in the maintenance of soil fertility. However, slash can also act as a hazard for regrowth forests and plantations. The management options for slash include:

1) broadcast burning of slash after redistribution on the site,
2) windrowing without burning,
3) windrowing and then burn, and
4) slash retention on the site by incorporating into the soil (e.g., chopper rolling).

The option to follow will depend upon the amount and type of slash and the requirement of site preparation for subsequent seedling establishment. Option 4 is becoming increasingly common and preferred practice in pine plantations in Australia.

On low fertility soils, retention of slash after thinning or clearfelling plantations can give positive growth responses to subsequent stand (Hopmans et al., 1993; Carlyle, 1995). Such improvements are not universal. Smethurst and Nambiar (1990) found no difference in the growth of replanted *P. radiata* between slash retained and removed 3 years after planting. Alternatively, heavy windrowing and heaping of slash prior to planting can decrease the growth of *P. radiata* on the scalped areas (between windrows and heaps) by up to 67% (compared with the growth on windrowed or heaped areas) an 28% compared to the control (Dyck et al., 1989).

Organic C in podzolised sands is highly dynamic and sensitive to management operations whic influence organic C inputs, decomposition or both. Weeds help to maintain organic C reserves afte clear felling, particularly where logging residues have been burnt. Retention of aboveground loggin residues also helped to maintain soil organic C reserves. Most harvesting and site preparatio: operations result in the loss of the labile carbon pool which represents about 30% of total carbon. Thi pool can be buffered by residue retention and weeds during the period before significant litter inpu from the new crop occurs at which time the pool is replenished. Long-term reductions in soil C occu when there are losses from the recalcitrant pool of soil C. This can occur on scalped sites or i: association with high intensity fires.

D. Characteristics of Grassland and its Conversion

Australia's grasslands have been derived from natural grasslands, savanna, and by clearing o woodlands or forests. Productivity of grasslands varies between climatic zones. In the temperate zone there is usually a marked seasonal pattern of productivity due to the influences of temperature and particularly, drought. Productivity also depends on grass species composition, occurrence of clovers fertilizer history and grazing or harvesting practices. Sown pastures can be 2.5 times as productive as native grazing lands.

Standing biomass can vary considerably. For example, the yield of a *Paspalum notatum* pasture in southern Queensland varied from 0.2 Mg C ha^{-1} to 1.7 Mg C ha^{-1} in winter and summe: respectively (Wilson et al., 1990). In a nearby pasture of *Setaria sphacelata*, biomass was about 3.: Mg C ha^{-1} (Cameron et al., 1989) during summer while pasture roots amounted to 3.0 Mg C ha$^-$ (Eastham and Rose, 1990). Losses of organic C from the pasture sites during conversion to plantation: would be small because fire is normally not used during site preparation for planting. However, n study has been undertaken to follow changes in soil C after plantation establishment.

VII. Life Cycle Analysis of Sequestered C

It is insufficient to consider the effects of changes in land use and forest growing stocks alone in thei: effects on carbon sequestration because the life cycle of forest products may vary once they leave the forest. In effect, the use of wood products postpones natural decomposition so that the carbon sink effect in the forest is extended. Information about product lifetimes is very poor. Harmon et al. (1990: assumed that 45% of harvested wood is converted to long term storage with a 2% annual loss. Dewa and Cannell (1992) assumed that wood products from plantations would be reduced to 95% of thei initial value after one rotation length, and paper and packaging products would have a lifetime of ! years. At equilibrium this model indicates that 16% of carbon would be stored in products. More complex analyses are also made which account for different product mixtures of fuel, paper and solic wood together with decay rates according to latitudinal zones (Winjum and Schroeder, 1997).

VIII. Conclusions

The area under forests and woodlands in Australia has been estimated to be about 157 million ha About 12% of this area is managed for multiple use including wood production, and an even smalle fraction is managed for intensive wood production. Plantations of fast growing eucalypts and pines cover an area of 1.04 million ha. Estimates of mean aboveground biomass density range from 39 Mg ha⁻¹ to 450 Mg ha⁻¹ depending upon the type of forest, plantation and woodland ecosystem. Total aboveground carbon pools for Australian forests have been estimated to be 13,000 Tg, of which more than 60% occur in forests which are not actively managed. The amount of aboveground C in plantations is only a small fraction of the total. This fraction is not expected to change significantly during the next several decades. However, in New Zealand about 20% of the forest area is under plantations which make a significant contribution to the total C stored. Papua New Guinea has the largest area of forest of all the Pacific islands which contain the largest carbon stocks.

Only incomplete estimates on the amount and nature of soil carbon pools are available because all components of C (soil organic matter, coarse and fine root fractions, and inorganic C and charcoal) are not usually measured. In most cases C in the litter layer is not included in soil C estimates. Exclusion of litter and root components in the estimates of soil C can result in significant underestimation. Clearcutting of a primary eucalypt forest causes changes in soil C depending upon the fate of harvested slash, but in most cases a decrease occurs during the first year.

In Australia and New Zealand considerable areas of old-growth native forests have been placed into reserves which will not be used for commercial wood production. While these reserves currently store a considerable quantity of C, their capacity for future storage is uncertain. Fire will play an important role. Exclusion of fire can result in senescence of eucalypts, invasion of rainforest species in wetter areas and eventually the development of a rainforest ecosystem probably with a lower C storage (Gilbert, 1958). If wildfires occur, then considerable C will be lost from the system and C stored in the living component will take many years to recover. Some reserves are also vulnerable to the introduction of exotic pests such as the root rotting fungus, *Phytophthora cinnamomi,* in Australia (Weste, 1994) and possums in New Zealand (Hall and Hollinger, 1997) both of which resulted in degradation of community structure and a decline in C storage capacity. These factors combined with increased vulnerability of some reserves due to fragmentation and small size suggest that there will be a tendency over time for a decline in the quantity of C stored.

During the last few decades the developing market for woodchips and other products has led to the intensification of harvesting in many native forests of the Australasia and Oceania region. In Australia the vast majority of these areas are subsequently managed as regrowth forests. This is not the case in the Pacific islands. For example in the Gogol woodchip project in Papua New Guinea, only 66% of the area clearfelled was scheduled for reforestation or development of plantations while the remainder was to be used for agricultural pursuits (Davidson, 1983).

Once a plantation estate has been established and developed into a system with sustainable yield, there are limited opportunities for changing its carbon storage capacity. While substantial gains in productivity have been made in the past (Nambiar, 1996), e.g., by correction of gross nutrient deficiencies, the opportunities for improvement are declining in well managed plantations. Gains are likely to be offset by other management options such as earlier harvesting. The current tendency is for shorter rotations which will reduce carbon density. The marginal cost for increasing rotation length beyond the optimum is small (Hollinger et al., 1993), but ultimately rotation length will depend on the product mix demand by the market. Increased recovery of smallwood for marketable products would expand the store in offsite products, but the product lifetime would probably be short.

References

ABARE. 1997. *Australian Forest Products Statistics June Quarter 1997.* Australian Bureau of Agricultural and Resource Economics, Canberra. 69 pp.

Attiwill, P.M. 1979. Nutrient cycling in *E. obliqua* (L'Herit.) forest. III. Growth, biomass and net primary production. *Austr. J. Botany* 27:439-458.

Balneave, J.M., M.F. Skinner and A.T. Lowe. 1991. Improving the re-establishment of radiata pine on impoverished soils in Nelson, New Zealand. p. 137-150. In: W.J. Dyck and C.A. Mees (eds.), *Long-Term Field Trials to Assess Environmental Impacts of Harvesting*. Proc. IEA/BE T6/A6 Workshop, Florida, USA Feb. 1990. IEA/BE T6/A6 Report No.5. Forest Research Institute, Rotorua, New Zealand. FRI Bulletin No. 161.

Beets, P.N. 1980. Amount and distribution of dry matter in a mature beech/podocarp community. *New Zealand J. Forestry Sci.* 10:395-418.

Birk, E.M. and J. Turner. 1992. Response of flooded gum (*E. grandis*) to intensive cultural treatments: Biomass and nutrient content of eucalypt plantations and native forests. *For. Ecol. Manage.* 47:1-28.

Bleeker, P. 1983. *Soils of Papua New Guinea*. CSIRO Publishing, Australia.

Boardman, R. 1974. Pine stand improvement in the south-eastern region of South Australia. South Australia Woods and Forests Department, Bulletin No. 21.

Boardman, R. and J.A. Simpson. 1981. Fertilization to optimise productivity. p. 303-317 In: *Productivity in Perpetuity*, Proc. Australian Forest Nutrition Workshop, Canberra, 10-14 August 1981, (Division of Forest Research, CSIRO, Canberra).

Bradstock, R. 1981. Biomass in an age series of *Eucalyptus grandis* plantations. *Austr. Forest Res.* 11:111-127.

Cameron, D.M., S.J. Rance, R.M. Jones, D.A Charles-Edwards and A. Barnes. 1989. Project STAG: an experimental study in agroforestry. *Austr. J. Agric. Res.* 40:699-714.

Carlyle, C. 1993. Organic carbon in forested sandy soils: properties, processes, and the impact of forest management. *New Zealand J. Forestry Sci.* 23:390-402.

Carlyle, C. 1995. Nutrient management in a *Pinus radiata* plantation after thinning: the effect of thinning and residues on nutrient distribution, mineral nitrogen fluxes, and extractable phosphorus. *Can. J. Forest Res.* 25:1278-1291.

Cheney, N.P., R.J. Raison and P.K. Khanna. 1980. Release of carbon to the atmosphere in Australian vegetation fires. p. 153-158. In: G.I. Pearman (ed.), Carbon Dioxide and Climate: Australian Research. Australian Academy Science, Canberra.

Colwell, J.D. 1958. Observations on the pedology and fertility of some krasnozems in northern New South Wales. *J. Soil Sci.* 9:46-57.

Commonwealth of Australia. 1992. Forest and Timber Inquiry. Resource Assessment Commission. Final report Vol. 1, Australian Government Publishing Service. 570 pp.

Commonwealth of Australia. 1997. p. 1-104. In: Australia's First Approximation Report for the Montreal Process. Commonwealth of Australia. Canberra.

Constantini A., J.L. Grimmett and G.M. Dunn. 1997. Towards sustainable managment of forest plantations in south-east Queensland. I. Logging and understorey residue management between rotations in steep country *Araucaria cunninghamii* plantations. *Aust. For.* 60:218-225.

Cromer, R.N. 1996. Silviculture of eucalypt plantations in Australia. p 259-273. In: P.M. Atiwill and M.A. Adams (eds.), *Nutrition of Eucalypts*. CSIRO Publishing Australia.

Cromer, R.N., D.M. Cameron, S.J. Rance, P.A. Ryan and M. Brown. 1993. Response to nutrients in *Eucalyptus grandis*. 1. Biomass. *For. Ecol. Manage.* 62:211-230.

Davidson, J. 1983. Forestry in Papua New Guinea: a case study of the Gogol Woodchip project near Madang. In: Hamilton, L.S. (ed.), *Forest and Watershed Development and Conservation in Asia and the Pacific*. Westview Press, Boulder, CO.

Department of the Environment, Sports and Territories. 1995. Native vegetation clearance, habitat loss and biodiversity decline—an overview of recent native vegetation clearance in Australia and its implications for biodiversity. Biodiversity Series, Paper No. 6, Biodiversity Unit, Department of the Environment Sports and Territories, Canberra.

Dewar, R.C. and M.G.R. Cannell. 1992. Carbon sequestration in trees, products and soils of forest plantations: an analysis using UK examples. *Tree Physiol.* 11:49-71.

Dixon, R.K., S. Brown, R.A. Houghton, A.M. Solomon, M.C. Trexler and J. Wisniewski. 1994. Carbon pools and flux of global forest ecosystems. *Science* 263:185-190.

Dyck, W.J., C.A. Mees, and N.B. Comerford. 1989. Medium-term effects of mechanical site preparation on radiata pine productivity in New Zealand—a retrospective study. p. 79-92 In: W.J. Dyck and C.A. Mees (eds.), Research Strategies for Long-term Site Productivity. *Proc. IEA/BE A3 Workshop*, Seattle, WA, Aug. 1988, IEA/BE Report No. 8. Forest Research Institute, New Zealand Bull. 152.

Eastham, J. and Rose C.W. 1990. Tree/pasture interactions at a range of tree densities in an agroforestry experiment. I. Rooting patterns. *Austr. J. Agric. Res.* 41:683-695.

Edwards, P.J. and P.J. Grubb. 1977. Studies of mineral cycling in a montane rain forest in New Guinea. I. The distribution of organic matter in the vegetation and soil. *J. Ecol.* 65:943-969.

Enright, N.J. 1978. The effects of logging on the regeneration and nutrient budget of *Araucaria cunninghamii* dominated tropical rainforest in Papua New Guinea. *Malaysian Forester* 41:303-318.

Enright, N.J. 1979. Litter production and nutrient partitioning in rainforest near Bulolo, Papua New Guinea. *Malaysian Forester* 42:202-207.

Eusebio, E.C. 1995. Forest ecosystem research and development towards biodiversity conservation in the Phillipines. p. 233-240. In: K.W. Sorensen, G.L. Enriquez, R.C. Umaly and J.T. Kartana. (eds.), *Proc. of the Regional Seminar - Workshop on Tropical Forest Ecosystems Research, Conservation and Repatriation.* SEAMO BIOTROP Bogor, Indonesia.

FAO. 1995. Forest Resources Assessment 1990. Global Synthesis. FAO Forestry Paper 124. Food and Agriculture Organisation, Rome. 44 pp.

Flinn, D.W., P. Hopmans, P.W. Farrell and J.M. James. 1979. Nutrient losses from the burning of *Pinus radiata* logging residue. *Australian Forest Res.* 9:17-23.

Ford-Robertson, J.B. 1997. Carbon balance calculations for forest industries—a review. *New Zealand Forestry* 42:32-36.

Frederick, D.J., H.A.I. Madgwick, M.F. Jurgensen and G.R. Oliver. 1985. Dry matter and nutrient distribution in an age series of *Eucalyptus regnans* plantations in New Zealand. *New Zealand J. Forestry Sci.* 15:158-79.

Grace, P.R., J.M. Oades, H. Keith and T.W. Hancock. 1995. Trends in wheat yields and soil organic carbon in the permanent rotation trial at the Waite Agricultural Research Institute, South Australia. *Austr. J. Exper. Agric.* 35:857-864.

Gerrand, A.M., J.L. Medhurst and W.A. Neilsen. 1997. Research results for thinning and pruning eucalypt plantations for sawlog production in Tasmania. *Forestry Tasmania.* 209 pp.

Gilbert, J.M. 1958. Eucalypt-rainforest relationships and the regeneration of the eucalypts: a report of work carried out under the first Australian Newsprint Mills Forestry Fellowship. Report Austr. Newsprint Mills, Hobart. 123 pp.

Graetz, R.D., M.A. Wilson and S.K. Campbell. 1995. Landcover disturbance over the Australian continent—a contemporary assessment. Biodiversity Series, Paper No. 7, Biodiversity Unit, Department of the Environment Sports and Territories, Canberra.

Graham, T.W.G., A.A. Webb and S.A. Waring. 1981. Soil nitrogen status and pasture productivity after clearing brigalow (*Acacia harpophylla*). *Austr. J. Exper. Agric. Anim. Husb.* 21:109-18.

Hall, M. and P. Carter. 1980. *Pinus radiata* first rotation establishment. *For. Comm. N.S.W. Researc* Report 1979-1980:97-107.

Hall, G.M.J. and D.Y. Hollinger. 1997. Do the indigenous forests affect the net CO_2 emission polic of New Zealand? *New Zealand Forestry* 41:24-31.

Hamilton, C.D. 1965. Changes in the soil under *Pinus radiata*. *Austr. Forestry* 29:275-289.

Harmon, M.E., W.K. Ferrell and J.F. Franklin. 1990. Effects on carbon storage of conversion of ol growth forests to young forests. *Science* 247:699-702.

Harwood, C.E. and W.D. Jackson. 1975. Atmospheric losses of four plant nutrients during a fore fire. *Austr. Forestry* 38:92-99.

Hollinger, D.Y., J.P. Maclaren, P.N. Betts and J. Turland. 1993. Carbon sequestration by Ne' Zealand's plantation forests. *New Zealand J. For. Sci.* 23:194-208.

Holt, J.A. and A.V. Spain. 1986. Some biological and chemical changes in a north Queensland sc following replacement of rainforest with *Araucaria cunninghamii* (Coniferae:Araucariaceae). *Appl. Ecol.* 23:227-237.

Hopmans, P., H.T.L. Stewart and D.W. Flinn. 1993. Impacts of harvesting on nutrients in a eucaly ecosystem in southeastern Australia. *For. Ecol. Manage.* 59:29-51.

Hubble, G.D., R.F. Isbell and K.H. Northcote. 1983. Features of Australian soils p. 17-47 In: *Soil. an Australian Viewpoint*, Division of Soils, CSIRO, Melbourne, Academic Press, London.

Hunt, S.M. and J.A. Simpson. 1985. Effects of low intensity prescribed fire on the growth ar nutrition of a slash pine plantation. *Austr. For. Res.* 15:67-77.

Hunter, I.R., J.D. Graham and J.A.C. Hunter. 1989. Whole-tree thinning of radiata pine — ear effects on growth. In: W.J. Dyck and C.A. Mees (eds.), Research Strategies for Long-term Si Productivity. Proc. IEA/BE A3 Workshop, Seattle, WA, Aug. 1988, IEA/BE Report No.8. Fore: Research Institute, New Zealand Bull.152:213-220.

Isbell, R.F. 1966. Soils of the east Bald Hill area, Collinsville district, north Queensland. Division (Soils, CSIRO, Land Use Series No. 48.

Jackman, R.H. 1964. Accumulation of organic matter in some New Zealand soils under permane pasture. 1. Patterns of change of organic carbon, nitrogen, sulphur and phosphorus. *New Zealar J. Agric. Res.* 7:445-471.

Jackson, D.S. and J. Chittenden. 1981. Estimation of dry matter in *Pinus radiata* root systems. Individual trees. *New Zealand J. For. Sci.* 11:164-182.

Jones, P. 1978. Fuel removal, fuel conditions and seedbed preparation in karri slash disposal burn For. Dept. West Austr. Res. Paper No. 42.

Keith, H., R.J. Raison and K.L. Jacobsen. 1997. Allocation of carbon in a mature eucalypt forest ar some effects of soil phosphorus availability. *Plant Soil* 196:81-99.

Khanna, P.K., K. Jacobsen and R.J. Raison. 1994. Soil carbon and N-mineralization after clearfellin mature Eucalyptus forest in south east Australia. p. 308-309. In: Trans. 15th World Cong. Soil Sc Acapulco, Mexico, Vol. 9—Supplement, International Soil Science Society.

Kuhlbusch, T.A.J. and P.J. Crutzen. 1995. Toward a global estimate of black carbon in residues (vegetation fires representing a sink of atmospheric CO_2 and a source of O_2. *Global Biogeochemic, Cycles* 9:491-501.

Levett, M.P., J.A. Adams, T.W. Walker and E.R.L. Wilson. 1985. Weight and nutrient content (above-ground biomass and litter of a podocarp-hardwood forest in Westland, New Zealand. *Ne Zealand J. For. Sci.* 15:23-35.

Lewis, N.B., A. Keeves and J.W. Leech. 1976. Yield regulation in South Australian *Pinus radia* plantations. South Australia Woods and Forests Department, Bulletin 23.

Ludbrook, W.V. 1942. Fertiliser trials in southern New South Wales pine plantations. *Journal CSI Australia* 15:307-314.

Madgwick, H.A.I. and B. Webber. 1987. Nutrient removal in harvesting mature *Pinus radiata. New Zealand J. For.* 32:15-18.

McGarity, J.W. 1959. The influence of sod-seeded legumes on the nitrogen economy of grassland soils at Lismore, N.S.W. *J. Australian Institute Agric. Sci.* 25:287-293.

Murata, T., M.L. Nguyen and K.M. Goh. 1995. The effects of long-term superphosphate application on soil organic matter content and composition from an intensively managed New Zealand pasture. *European J. Soil Sci.* 46:257-264.

Nambiar, E.K.S. 1996. Sustained productivity in forests is a continuing challenge to soil science. *Soil Sci. Soc. Amer. J.* 60:1629-1642.

Nambiar, E.K.S. and T.H. Booth. 1991. Environmental constraints on the productivity of eucalypts and pine: Opportunities for site management and breeding. p. 215-231. In: P.J. Ryan (ed.), Productivity in Perspective. *Proc. 3rd Australian Forest Soils and Nutrition Conf.*, Melbourne, 7–11 October 1991. Forestry Commission of N.S.W., Sydney.

National Forest Inventory. 1997. National Forest Inventory of Australia. Bureau of Rural Sciences, Canberra.

NGGIC. 1997. National Greenhouse Gas Inventory, Land Use Change and Forestry Sector 1988–1995. Based on workbook 4.2. Commonwealth of Australia.

Nicholls, K.D. and B.M. Tucker. 1956. Pedology and Chemistry of the Basaltic Soils of the Lismore District, N.S.W. CSIRO Soil Publication No. 7.

Nonggor, P. 1995. Conservation, repatriation and sustainable utilization of forest resources in Papua New Guinea: a country report. p. 183-200. In: K.W. Sorensen, G.L. Enriquez, R.C. Umaly and J.T. Kartana (eds.), Proceedings of the Regional Seminar—Workshop on Tropical Forest Ecosystems Research, Conservation and Repatriation. SEAMO BIOTROP Bogor, Indonesia.

O'Connell, A.M. and T.S. Grove. 1991. Processes contributing to the nutritional resilience or vulnerability of jarrah and karri forests in Western Australia. p. 180-197. In: P.J. Ryan (ed.), Productivity in Perspective. *Proc. 3rd Australian Forest Soils and Nutrition Conf.* Melbourne, 7–11 October 1991. Forestry Commission of N.S.W., Sydney.

O'Connell, A.M. and T.S. Grove. 1996. Biomass production, nutrient uptake and nutrient cycling in the jarrah (*Eucalyptus marginata*) and karri (*Eucalyptus diversicolor*) forests of south-western Australia. p. 155-189. In: P.M. Atiwill and M.A. Adams (eds.), Nutrition of Eucalypts. CSIRO Publishing, Australia.

Ryan, M.G., R.M. Hubbard, S. Pongracic, R.J. Raison and R.E. McMurtrie. 1996. Foliage, fine-root, woody-tissue and stand respiration in *Pinus radiata* in relation to nitrogen status. *Tree Physiol.* 16:333-343.

Saulei, S. 1997. Forest Exploitation in Papua New Guinea. *The Contemporary Pacific* 9:25-38.

Simpson, J.R., S.M. Bromfield and O.L. Jones. 1974. Effects of management on soil fertility under pasture. 3. Changes in total soil nitrogen, carbon, phosphorus and exchangeable cations. *Austr. J. Exper. Agric. Anim. Husb.* 14:487-494.

Skjemstad, J.O., R.P. LeFeuvre and R.E. Prebble. 1990. Turnover of soil organic matter under pasture as determined by ^{13}C natural abundance. *Austr. J. Soil Res.* 28:267-276.

Smethurst, P.J. and E.K.S. Nambiar. 1990. Distribution of carbon and nutrients and fluxes of mineral nitrogen after clear-felling a *Pinus radiata* plantation. *Can. J. For. Res.* 20:1490-1497.

Spain, A.V. 1990. Influence of environmental conditions and some soil chemical properties on the carbon and nitrogen contents of some tropical Australian rainforest soils. *Austr. J. Soil Res.* 28:825-839.

Stewart, G.H. and L.E. Burrows. 1994. Coarse woody debris in old-growth temperate beech (Nothofagus) forests of New Zealand. *Can. J. For. Res.* 24:1989-1996.

Stoate, T.N. 1950. Nutrition of the pine. Australia Forestry and Timber Bureau Bulletin 30.

Turner, J. 1983. Maintenance and improvement of forest productivity in Australia. p. 293-304. In: R. Ballard and S.P. Gessel (eds.), *IUFRO Symposium on Forest Site and Continuous Productivity*, Seattle, WA, 22–28 August, 1982. U.S. Department of Agriculture, Forest Service, Pacific Northwest Forest and Range Experimental Station, General Technical Report PNW-163.

Turner, J. 1986. Organic matter accumulation in a series of *Eucalyptus grandis* stands. *Forest Ecol. Manage.* 17:231-242.

Turner, J. and J. Kelly. 1977. Soil chemical properties under naturally regenerated *Eucalyptus* spp and planted Douglas fir. *Aust. For. Res.* 7:163-172.

Turner, J. and M.J. Lambert. 1983. Nutrient cycling within a 27-year-old *Eucalyptus grandis* plantation in New South Wales. *For. Ecol. Manage.* 6:155-168.

Turner, J. and M.J. Lambert. 1986. Effect of forest harvesting nutrient removals on soil nutrient reserves. *Oecologia* 70:140-148.

Turner. J. and M.J. Lambert. 1996. Nutrient cycling and forest management. p. 229-248. In: P.M. Atiwill and M.A. Adams (eds.), *Nutrition of Eucalypts*. CSIRO Publishing, Australia.

Waring, H.D. 1968. The effect of cultural techniques on the growth of *Pinus elliottii* Engelm near Jervis Bay, New South Wales. *Trans. 9th Int. Cong. Soil Sci.* Vol. 3:447-454.

Waring, H.D. and P. Snowdon. 1985. Clover and urea as sources of nitrogen for the establishment of *Pinus radiata*. *Austr. For. Res.* 15:115-24.

Watson, A. and C. O'Loughlin. 1990. Structural root morphology and biomass of three age-classes of *Pinus radiata*. *New Zealand J. For. Sci.* 20:97-110

Webb, A.A. and A.J. Dowling. 1990. Characterization of basaltic clay (Vertisols) from the Oxford land system in central Queensland. *Aust. J. Soil Res.* 28:841-856.

Webb, A.A., B.J. Crack and J.Y. Gill. 1977. Studies on the gilgaied clay soils (Ug 5.2) of the Highworth land system in east-central Queensland. I. Chemical characteristics. *Queeensland J. Agric. Animal Sci.* 34:53-65.

Weste, G. 1994. Impact of *Phytophthora* species on native vegetation of Australia and Papua New Guinea. *Australasian Plant Pathol.* 23:190-209.

Westman, W.E. and R.W. Rogers. 1977. Biomass and structure of a subtropical eucalypt forest, North Stradbroke Island. *Austr. J. Botany* 25:171-191.

Wilson, J.R., K. Hill, D.M. Cameron and H.M. Shelton. 1990. The growth of *Paspalum notum* under the shade of a *Eucalyptus grandis* plantation canopy or in full sun. *Trop. Grasslands* 24:24-28.

Wilson, S.M., J.A.H. Whitham, U.N. Bhathi, D. Horvath and Y.D. Tran. 1995. Survey of Trees on Australian Farms: 1993-1994, ABARE Research Report 95.7, Canberra.

Winjum, J.K. and P.E. Schroeder. 1997. Forest plantations of the world: their extent, ecological attributes, and carbon storage. *Agric. For. Meteorology* 84:153-167.

Young, H.E. 1940. Fused needle disease and its relation to the nutrition of *Pinus*. *Queensland Forest Service Bull.* 13:108.

Young, H.E. 1948. The response of loblolly and slash pine to phosphate manures. *Queensland J. Agric. Sci.* 5:77-105.

Zwartz, D. (ed.). 1997. *New Zealand Official Yearbook 1997*. GP Publications, Wellington. 645 pp.

CHAPTER 4

Organic Carbon Stock in Soils of India

M. Velayutham, D.K. Pal and T. Bhattacharyya

I. Introduction

To sustain the quality and productivity of soils the knowledge of organic carbon (OC) in terms of its amount and quality in soils is essential. This has a relevance in soils of the tropical and subtropical parts of the globe including the Indian sub-continent. The depletion of the ozone layer due to the greenhouse effect has created great concern that led to several studies on qualities, kinds, distribution and behavior of soil organic carbon (SOC) (Eswaran et al., 1993). Global warming and its effect on soils, influencing agriculture specifically with respect to organic matter management in soils, has led to several quantitative estimates for global carbon content in soils (Bohn, 1976, 1982; Buringh, 1984; Kimble et al., 1990; Eswaran et al., 1993; Sombroek et al., 1993; Batjes, 1996).

The first ever comprehensive study of OC status in Indian soils was conducted by Jenny and Raychaudhuri (1960). They studied 500 soil samples collected from different cultivated fields and forests with variable rainfall and temperature patterns. The study confirmed the effects of climate on carbon reserves in both virgin and cultivated soils. However, these authors did not make any estimate of the total carbon reserves in the soils. The first attempt in this area of soil research was made by Gupta and Rao (1994) who reported the OC stock of 24.3 Pg (1 Pg = 10^{15} g) for the soils ranging from surface to an average subsurface depth of 44 to 186 cm with the database of 48 soil series.

The most prudent approach to study SOC, however, would be on an unit area basis for a specified depth interval which requires information on the spatial distribution of soil types, SOC and bulk density (BD). Thus, it would provide a better understanding of the terrestrial reservoir of SOC far beyond the general objectives of C sequestration in soils and the detrimental effects of global warming (Eswaran et al., 1993).

II. Approach to the Quantitative Study of SOC

A. Quality of Data and Their Sources

Difference in sampling methods, exact season of collecting soil samples on different types of landscapes and kinds of vegetation and, above all, the methods of soil analyses in the laboratory determine the quality of SOC data. By and large, the Walkley–Black method for determining SOC has been an acceptable technique to generate SOC data due to easy availability. The BD values are important since it is necessary to convert SOC data by weight to volume. The BD data are not available in routine soil survey reports or in many published literatures. For the present study BD values were generated for those soils where they were not available. In a few reports, however, the BD values were available and were used to compute the total SOC stock.

The relevant data for SOC and BD were drawn from the published and unpublished soil database generated through soil resource mapping of India by the National Bureau of Soil Survey and Land Use Planning, Nagpur in 1:250,000 scale, district soil survey programs in 1:50,000 scale and also from various other sources including research articles (Table 1).

B. Computation of Total SOC Stock

The size of total SOC stock is calculated following the methods described by Batjes (1996). The first step involves calculation of OC by multiplying OC content (g g^{-1}), bulk density (Mg m^{-3}), and thickness of horizon (m) for individual soil profile with different thickness varying from 0–30 cm and 0–150 cm. Earlier studies have indicated that computation of SOC pool to shallow soil depth fails to include a large proportion of the total C content in the soils (Arrouays and Pelissier, 1994). Therefore, a greater soil depth of 0–150 cm was taken into consideration to compute the total SOC in Indian soils. During almost all soil survey programs of India and also while establishing benchmark soils of India (Murthy et al., 1982; Lal et al., 1994) soils were studied and sampled up to a depth of 150 cm. Therefore, these data in terms of OC and BD have been very helpful for the estimation of SOC stock. The total OC content determined by this process is multiplied by the area of the soil unit in the second step. The total SOC content is thus calculated in terms of Pg.

III. SOC Stock in Soil Orders

An attempt has been made to estimate the SOC stock in different soil orders as per Soil Taxonomy (Soil Survey Staff, 1994). This information will have relevance in terms of comparing the SOC stock in soils occurring elsewhere and can also serve as an international reference as far as the wide acceptance of Soil Taxonomy is concerned. Earlier, Eswaran et al. (1993) made similar exercise to compute the soil carbon pool of the world. The SOC stock for India in terms of each soil order is estimated at 0–30 and 0–150 cm depths since such quantitative data reflect the kinds of soil with different amount of OC at different depths (Table 2).

A. SOC Stock in Entisols

This soil order is commonly found in almost all parts of the country encompassing different climatic, geographic and geologic areas. The depth of these soils is often limited to 50 cm in many eroded landscapes. These soils are also commonly observed in young geomorphic surfaces, deserts and coasts. Few Entisols commonly occur in lower topographic situation with an aquic soil moisture regime with the deposition of sulphur. Whatever the geomorphic position of these Entisols, it is observed that except the soils of cold arid region (*Typic Cryorthents*) the soils, in general, are poor in SOC stock (Table 3). It is also observed that the soils viz. *Troporthents, Typic Tropopsamments* and *Typic Tropofluvents* are very poor in OC both in the surface and in subsurface horizons (Table 3). Entisols contribute nearly 7% in the total SOC stock of the Indian soils (Table 2).

B. SOC Stock in Inceptisols

Most of the Indian soils are grouped into the Inceptisols. At subgroup levels various characteristics ranging from moisture regime to sodicity are observed. Large part of the soils in regions of western, central and southern parts of the country are grouped into Vertic intergrades of Inceptisols. By and

Table 1. Sources of data for computing SOC stock in India

1.	Deshpande et al. (1971)	Mollisols from Terai Region
2.	Murthy et al. (1982)	Benchmark Soils of India
3.	Sehgal and Sharma (1982)	Soils of the Northern India
4.	NBSS & LUP (1984)	Soils of Uttar Pradesh
5.	NBSS & LUP (1988)	Benchmark swell-shrink soils of India
6.	Bhattacharyya et al. (1989)	Soils of Western Maharashtra
7.	Bhattacharyya et al. (1992)	Soils of Western Maharashtra
8.	Bhattacharyya et al. (1993)	Soil Formation of Western Maharashtra
9.	Bhattacharyya et al. (1994a)	Soils of Gujarat
10.	Bhattacharyya et al. (1994b)	Soils of Manipur, Meghalaya
11.	Bhattacharyya et al. (1995)	Soils of Brahmaputra valley (Assam)
12.	Bhattacharyya et al. (1996)	Soils of Tripura
13.	Bhattacharyya et al. (1997)	Soil Mineralogy of Western Maharashtra
14.	Bhattacharyya et al. (1998a)	Soils of Arunachal Pradesh
15.	Bhattacharyya and Pal (1998)	Soils of Madhya Pradesh
16.	Lal et al. (1994)	Soils of India
17.	Vadivelu and Bandyopadhyay (1997)	Soils of Lakshdweep
18.	Sehgal et al. (1998)	Agro-Ecological Sub-Region (AESR) of India
19.	Krishnan et al. (1996)	Soils of Kerala
20.	Shivaprasad et al. (1998)	Soils of Karnataka
21.	Natarajan et al. (1997)	Soils of Tamil Nadu
22.	Harindranath et al. (1998)	Soils of Goa

Table 2. Organic carbon stock in different soil orders in India

Soil number	Soil orders	Organic C (Pg)	
		0–30 cm	0–150 cm
1.	Entisols	1.36 (6.5)*	4.17 (6.6)
2.	Inceptisols	4.67 (22.2)	15.07 (24.0)
3.	Vertisols	2.62 (12.5)	8.78 (14.0)
4.	Aridisols	7.67 (36.5)	20.30 (32.3)
5.	Mollisols	0.12 (0.6)	0.50 (0.8)
6.	Alfisols	4.22 (20.0)	13.54 (21.0)
7.	Ultisols	0.14 (0.8)	0.34 (0.5)
8.	Oxisols	0.19 (0.9)	0.49 (0.8)
	TOTAL	20.99	63.19

*Figures in parentheses indicate percent of total organic carbon stock.

Table 3. SOC stock in Entisols

Soil number	Subgroups	Organic C (Pg)	
		0–30 cm	0–150 cm
1.	Typic Cryorthents	0.3200	0.8900
2.	Typic Torripsamments	0.2800	0.8900
3.	Lithic Ustorthents	0.2000	–
4.	Typic Ustorthents	0.0400	0.1300
5.	Typic Ustifluvents	0.1700	0.6400
6.	Typic Ustipsamments	0.0010	0.0130
7.	Typic Tropopsamments	0.0010	–
8.	Typic Tropofluvents	0.0003	–
9.	Typic Tropopsamments	0.0008	0.0300
10.	Typic Udorthents	0.1000	–
11.	Fluventic Sulfaquents	0.0500	0.3100
12.	Typic Sulfaquents	0.2000	1.3100
13.	Troporthents	0.0020	0.0100
	TOTAL	1.3600	4.170

Table 4. SOC stock in Inceptisols

Soil number	Subgroups	Organic C (Pg)	
		0–30 cm	0–150 cm
1.	Typic Ustochrepts	0.610	0.98
2.	Fluventic Ustochrepts	1.910	8.38
3.	Udic Ustochrepts	0.250	1.16
4.	Aeric Haplaquepts	0.400	1.19
5.	Vertic Ustochrepts	0.430	0.47
6.	Vertic Ustropepts	0.130	0.53
7.	Aquic Ustochrepts	0.010	0.03
8.	Fluventic Ustochrepts	0.030	0.17
9.	Natric Ustochrepts	0.030	0.05
10.	Mollic Haplaquepts	0.150	0.44
11.	Dystric Eutrochrepts	0.010	0.05
12.	Vertic Endoaquepts	0.030	0.07
13.	Fluventic Eutrochrepts	0.010	0.04
14.	Typic Haplumbrepts	0.130	0.29
15.	Typic Haplaquepts	0.040	0.14
16.	Typic Dystrochrepts	0.080	0.28
17.	Umbric Dystrochrepts	0.010	0.04
18.	Vertic Haplaquepts	0.080	0.31
19.	Oxic Dystrochrepts	0.310	0.34
20.	Oxic Ustropepts	0.003	0.01
21.	Typic Humitropepts	0.020	0.10
	TOTAL	4.673	15.07

large, the forest soils in the hills of the northeast and south are poor base Inceptisols. These hilly soils are, however, rich in OC especially in the surface layers. Inceptisols in the recent alluvium (*Fluventic Ustochrepts*) are usually characterized by irregular distribution of OC in the soil profile and they are richer in OC than soils developed in the older alluvium (Table 4). Perusal of the data in Table 4 further indicates that in a few Inceptisols with vertic, aquic, natric and oxic subgroups, the total stock of OC at 150 cm depth is surprisingly low in contrast to other subgroups. Inceptisols contribute 22 to 24% share of the total SOC stock (Table 2).

Table 5. SOC stock in Vertisols

Soil		Organic C (Pg)	
number	Subgroups	0–30 cm	0–150 cm
1.	Typic Haplusterts	2.24	7.65
2.	Sodic Haplusterts	0.27	0.09
3.	Udic Haplusterts	0.05	0.74
4.	Sodic Calciusterts	0.03	0.21
5.	Halic Calciusterts	0.03	0.09
	TOTAL	2.62	8.78

C. SOC Stock in Vertisols

Vertisols are extensive in the central and south central parts of India. Among the different subgroups of Vertisols, *Typic Haplusterts* have the highest SOC stock (Table 5) due to larger aerial extent. The next in abundance are the *Sodic Haplusterts* which have more SOC stock than *Udic Haplusterts* and *Sodic/Halic Calciusterts*. This may be attributed to the restriction of leaching due to the development of sodicity in the soil control section. The adverse arid climatic condition in the subgroups other than *Typic Haplusterts* is not favorable for the accumulation of OC. *Udic Haplusterts* cover less area but are under intensive cultivation. This may be the reason for improved OC in these soils. Despite having the largest reactive surface area as well as appreciable spatial distribution, Vertisols constitute only 12 to 14% share of the total SOC stock (Table 2).

D. SOC Stock in Aridisols

Aridisols, in general, are poor in OC due to their high rate of decomposition. A few Aridisols belonging to cold (*Typic Cryorthids*) as well as hot (*Typic Camborthids* / *Natrargids* / *Calciorthids*) arid ecosystem, however, contain substantial amounts of SOC ranging from 32 to 37% (Table 2). This is due to their larger aerial extent in hot arid ecosystems and low rate of OC decomposition in the cold temperature of the higher altitudinal areas of the northern part of the country. On the contrary, the low SOC stock in *Typic Natrargids* is due to less leaching of OC in the subsoils due to sodicity (Table 6).

E. SOC Stock in Mollisols

Mollisols are known for their richness in OC content especially in their surface horizons (Soil Survey Staff, 1975). In India, Mollisols also contain nearly 1% OC in the surface layers. In contrast, C in Mollisols constitutes < 1% share of the total SOC stock (Table 2). This ambiguity is due to the fact that only a small portion of geographical area of the country is covered by these soils, concentrating mainly in the Terai and sub Himalayan regions of the country (Deshpande et al., 1971). Mollisols have recently been reported from Madhya Pradesh (Bhattacharyya and Pal, 1998), Karnataka (Shivaprasad

Table 6. SOC stock in Aridisols

Soil		Organic C (Pg)	
number	Subgroups	0–30 cm	0–150 cm
1.	Typic Cryorthids	5.54000	11.2900
2.	Typic Camborthids	0.60060	1.9430
3.	Typic Natrargids	0.18000	0.7100
4.	Typic Calciorthids	1.35000	6.3500
5.	Ustollic Calciorthids	0.00016	0.0035
6.	Typic Haplargids	0.00030	0.0008
7.	Typic Paleargids	0.00030	0.0009
	TOTAL	7.67136	20.2982

Table 7. SOC stock in Mollisols

Soil		Organic C (Pg)	
number	Subgroups	0–30 cm	0–150 cm
1.	Aquic Hapludolls	0.0800	0.3700
2.	Typic Hapludolls	0.0400	0.1300
3.	Pachic Argiustolls	0.0001	0.0004
4.	Pachic Haplustolls	0.0001	0.0004
5.	Typic Argiudolls	0.0003	0.0005
6.	Typic Argiustolls	0.0003	0.0001
	TOTAL	0.1208	0.5014

et al., 1998), Kerala (Krishnan et al., 1996) and Sikkim (Das et al., 1996). Table 7 shows the SOC stocks in Mollisols.

F. SOC Stock in Alfisols

The maximum SOC stock in different groups of Alfisols is found in *Typic Hapludalfs* (Table 8). Most of these soils occur in sub humid to humid regions of the country. The vegetation on these soils favours the accumulation of OC. Next in abundance in SOC is *Typic Natrustalfs* followed by *Typic Plinthustalfs, Rhodustalfs* and *Lithic Hapludalfs*. The Alfisols contribute 20 to 21% portion of the total SOC stock (Table 2) and rank third after Aridisols and Inceptisols. The relatively higher SOC content of these soils is not only due to larger aerial extent but also due to favorable support of vegetation.

Table 8. SOC stock in Alfisols

Soil number	Subgroups	Organic C (Pg)	
		0–30 cm	0–150 cm
1.	Typic Rhodustalfs	0.3700	0.7200
2.	Typic Natrustalfs	0.8200	3.4100
3.	Aquic Natrustalfs	0.1800	0.2200
4.	Udic Rhodustalfs	0.3400	1.1500
5.	Typic Hapludalfs	1.0600	3.4700
6.	Oxic Rhodustalfs	0.1100	0.3700
7.	Oxic Paleustalfs	0.0400	0.2000
8.	Aeric Ochraqualfs	0.0200	0.0800
9.	Typic Plinthustalfs	0.4600	1.1000
10.	Typic Haplustalfs	0.2800	0.6000
11.	Typic Ochraqualfs	0.0400	0.0600
12.	Typic Paleudalfs	0.1600	0.5100
13.	Ultic Hapludalfs	0.2300	0.6900
14.	Vertic Hapludalfs	0.0200	0.0300
15.	Ultic Paleustalfs	0.0400	0.1000
16.	Typic Kandiustalfs	0.0020	0.0080
17.	Kandic Paleustalfs	0.0140	0.0430
18.	Kanhaplic Rhodustalfs	0.0010	0.0030
19.	Mollic Paleudalfs	0.0020	0.0060
20.	Rhodic Paleustalfs	0.0300	0.0910
21.	Psammentic Paleudalfs	0.0001	0.0006
	TOTAL	4.2191	13.5416

G. SOC Stock in Oxisols

Oxisols reported from India (Murthy et al., 1982) show poor accumulation of OC due to higher decomposition in tropical humid climate and also due to improvisation of mineral matter in these soils (Table 9). These soils contribute < 1% share of the total SOC stock.

Table 9. SOC stock in Oxisols

Soil		Organic C (Pg)	
number	Subgroups	0–30 cm	0–150 cm
1.	Typic Haplorthox	0.19	0.49
	Total	0.19	0.49

IV. Zonation of SOC Stock

The SOC stock in soils of different agro-eco regions (AER) (Sehgal et al., 1992) is shown in Table 10. The data show that the maximum amount of SOC stock is in the surface soils of hot regions covering AERs 4, 5, 6, 7 and 8, followed by cold arid (AERs 1 and 4), hot arid (AERs 2 and 3), and hot subhumid zones (AERs 9 to 15). As a matter of fact, the soils in the hot region contain less OC due to unfavorable climatic condition. When point data of OC in arid to subhumid regions are compared to those of the cold region it is observed that arid and semiarid regions are the real deficient zones in SOC.

A. SOC Stock in Arid Ecosystem

Arid ecosystem is characterized by relatively low rainfall (100–300 mm annual rainfall; Bhattacharjee et al., 1982) which normally does not support dense vegetation and therefore, these soils are expectedly low in OC. This ecosystem is divided into cold and hot on the basis of atmospheric temperature. By and large, the western part of the country is in arid ecosystem which is hot and the soils are poor in OC. Parts of the northern India (Jammu and Kashmir) are covered by cold arid climate where prevailing cool temperature helps store a relatively higher SOC stock. Aridic ecosystem (AERs 1, 2 and 3) constitutes 16% of the total geographic area (TGA) but has a share of 46% of the total SOC stock in 0–30 cm (Table 10).

B. SOC Stock in Semiarid Ecosystem

Major parts of central and south central peninsula extending to western and northwestern parts of the country fall under this ecosystem (AERs 4, 5, 6, 7 and 8) (Table 10). This is characterized by the temperature ranging from 25° to 27°C and annual rainfall from 900 to 1200 mm. This ecosystem is characterized by a type of vegetation which ranges from bushy thorns and grasses to deciduous forest trees. The soils of this ecosystem have low to medium OC content. Geographically semiarid systems cover 36.4% of the TGA and contribute 6.19 Pg SOC stock (Table 10) which represents about 29.5% of the total stock.

C. SOC Stock in Subhumid Ecosystem

The climate in this ecosystem is a transitional phase between humid and semiarid. The average annual temperature of this ecosystem is 22° to 29°C with mean annual rainfall of 900 to 1700 mm. It covers the large part of the Indo-Gangetic Plains and part of the southern Peninsula. Most of these areas are

Table 10. SOC stock in different Agro-Eco Regions (AER) in India

AER no.	Agro-Eco Region (States covered)	Organic C (Pg)		
		Area (m ha)	0–30 cm	0–150 cm
1.	Cold arid ecoregion with shallow skeletal soils (Jammu and Kashmir)	15.2 (4.6)*	5.96	10.47
2.	Hot arid ecoregion with desert and saline soils in the western plains (Punjab, Haryana and Rajasthan)	31.9 (9.7)	2.51	10.43
3.	Hot arid ecoregion with red and black soils (Andhra Pradesh and Karnataka)	4.9 (1.5)	1.21	4.21
4.	Hot semiarid ecoregion with alluvium-derived soils (Uttar Pradesh, Rajasthan, Haryana, Madhya Pradesh, Punjab and Delhi)	32.2 (9.8)	3.88	15.24
5.	Hot semiarid ecoregion with medium and deep black soils (Gujarat, Madhya Pradesh and Rajasthan)	17.6 (5.4)	0.61	1.23
6.	Hot semiarid ecoregion with shallow and medium black soils (Andhra Pradesh, Karnataka and Maharashtra)	31.0 (9.4)	0.69	1.35
7.	Hot semiarid ecoregion with red and black soils (Andhra Pradesh)	16.5 (5.0)	0.55	1.72
8.	Hot semiarid ecoregion with red and loamy soils (Andhra Pradesh, Karnataka and Tamil Nadu)	19.1 (5.8)	0.46	1.41
9.	Hot subhumid (dry) ecoregion with alluvium-derived soils (Bihar, Uttar Pradesh, Haryana, Punjab, Jammu and Kashmir)	12.1 (3.7)	0.15	0.49
10.	Hot subhumid ecoregion with red and yellow soils (Madhya Pradesh and Maharashtra)	22.3 (6.8)	0.62	1.55
11.	Hot subhumid ecoregion with red and yellow soils (Madhya Pradesh, Bihar, Uttar Pradesh, Andhra Pradesh and Maharashtra)	14.1 (4.3)	0.14	0.50

Table 10. continued --

AER no.	Agro-Eco Region (States covered)	Organic C (Pg)		
		Area (m ha)	0–30 cm	0–150 cm
12.	Hot subhumid ecoregion with red and lateritic soils (Andhra Pradesh, Bihar, Madhya Pradesh, Maharashtra, Orissa and West Bengal)	26.8 (8.2)	0.67	1.70
13.	Hot subhumid (moist) ecoregion with alluvium derived soils (Bihar and Uttar Pradesh)	11.1 (3.4)	0.17	0.79
14.	Warm subhumid to humid with inclusion of perhumid ecoregion with brown forest and podzolic soils (Jammu and Kashmir, Himachal Pradesh, Uttar Pradesh and Punjab)	21.2 (6.4)	0.52	1.55
15.	Hot subhumid (moist) to humid (inclusion of perhumid) ecoregion with alluvium-derived soils	12.1 (3.7)	0.42	1.20
16.	Warm perhumid ecoregion with brown and red hill soils (Arunachal Pradesh, Assam, Sikkim and West Bengal)	9.6 (2.9)	0.79	3.76
17.	Warm perhumid ecoregion with red and lateritic soils (Arunachal Pradesh, Assam, Manipur, Meghalaya, Mizoram, Nagaland and Tripura)	10.6 (3.2)	0.45	1.49
18.	Hot, subhumid to semiarid ecoregion with coastal alluvium-derived soils (Andhra Pradesh, Orissa, Pondicherry, Tamil Nadu and West Bengal)	8.5 (2.6)	0.23	0.91
19.	Hot, humid to perhumid ecoregion with red, lateritic and alluvium-derived soils (Maharashtra, Gujarat, Kerala, Karnataka, Goa, Dadra and Nagar Haveli)	11.1 (3.4)	0.84	2.82
20.	Hot, humid to perhumid island ecoregion with red loamy and sandy soils (Andaman and Nicobar and Lakshadweep islands)	0.8 (0.2)	0.12	0.37

*Figures in the parentheses indicate percent of TGA.

rich in vegetation which helps building up the OC status. However, high tropical temperature does not allow appreciable OC stock in soils of this ecosystem (Table 10). Although this ecosystem (AERs 9, 10, 11, 12 and 13) constitutes 24.3% of TGA, it represents only 8.3% share of total SOC stock.

Table 11. Total SOC stock in India

Carbon	Depth range (cm)		
	0–30	0–100	0–150
	Pg		
Soil organic carbon (India)	20.99	47.53	63.19
	(10.14)[b]	(12.08)[b]	–
	(2.98)[c]	(3.16)[c]	–
			–
Soil organic carbon[a] (Tropical regions)	207.00	393.50	
Soil organic carbon[a] (World)	704.00	1505.00	–

[a]Adapted from Batjes, 1996; average values were taken.
[b]% of tropical regions SOC stock.
[c]% of World SOC stock.

D. SOC Stock in Perhumid Ecosystem

This ecosystem experiences high rainfall and low temperature (AERs 16 and 17) (Table 10). The temperature is usually below 20°C with mean annual rainfall between 2000 and 3500 mm. These areas are bestowed with lush green vegetation on the high hills of the northeastern part of the country which allows a huge amount of OC storage in soils. This is evidenced from the SOC stock of 5.9% of the total stock by an area constituting only 6.2% of the TGA (Table 10).

E. SOC Stock in Other Ecosystems (Subhumid to Humid, Subhumid to Semiarid and Humid to Perhumid)

These areas extend from part of the northern portion of the country to the eastern part with characteristic transition of both the ecosystems viz. subhumid to humid, subhumid to semiarid and humid to perhumid. Such areas (AERs 14, 17, 18, 19 and 20) cover 16.1% of TGA and contribute about 10% of the total SOC stock (Table 10).

V. Estimates of SOC Stock in India

The first ever estimate of SOC in India was based on 48 soil series taking into account only a few major soils. The estimated SOC stock was reported to be 24.3 Pg (Gupta and Rao, 1994). The present estimate is based on a relatively large soil data base as indicated in Table 1. All these soils with their geographical distribution in different AERs were used for computing SOC stock in different depths. The total mass of organic carbon stored in the upper 30 cm and 150 cm of the soils in India is 20.99 Pg and 63.19 Pg respectively (Table 11). The SOC stock of Indian soils was compared to the stock for tropical regions and the world (Batjes, 1996). Table 11 shows that SOC stock of Indian soils is 10 to 12% of the tropical regions and about 3% of the total carbon mass of the world. Thus, the share of India in overall SOC stock of the world is not substantial although it covers 11.9% of total geographical area of the world. An attempt has been made to map the SOC stock in different parts of the country (Figure 1). Necessity of such map on carbon stock has recently been felt by Velayutham et al. (1998).

VI. Delineation of Sufficient and Deficient Areas of SOC

Figure 1 shows that the total SOC stock in AERs 1, 2, 3 and 4 are high in the upper 30 cm of the soils. This is in contrast to the basic understanding that high temperature of the arid and semiarid ecosystem does not favor accumulation of OC in soils. This is mainly due to higher geographical area of hot arid ecosystem (Table 12). Moreover, the AER 4 consists of two bioclimatic regions namely dry and moist (Mandal et al., 1996). Semiarid moist zones have soils with higher OC which justifies higher SOC stock. This shortfall in the AER map of India (Sehgal et al., 1992) has, however, been adequately addressed in Agro-Eco Sub Regions map of India (Sehgal et al., 1995).

The SOC stock in hot semiarid to subhumid eco-region of AERs 5, 6, 7, 8, 9, 10, 11 and 12 is moderate to less. This is in spite of higher geographical extent (Table 12) and greater BD values (Table 13) of soils of climate dominated by smectites (Pal and Deshpande, 1987a, 1987b, 1987c). Comparison of point data of SOC in arid and semiarid zones against cold arid to subhumid / perhumid zones indicate that the former areas are the real deficient zones (Figure 2) and therefore the SOC stock is less.

Interestingly, total SOC stock in AERs 13, 14, 15, 16, 17, 18, 19 and 20 is comparatively less than that in arid and semiarid regions (Tables 10 and 12). However, OC content of dominant soils of these areas is well above 1% (Table 13) which suggests that these zones are the sufficient zones in terms of SOC (Figure 2). In view of the OC equilibrium value of 1 to 2% obtained by Saikh et al., (1998), 1% OC has been considered as the tentative boundary limit between sufficient and deficient zones. On the basis of point data it is found that about 70% of the country is deficient in SOC.

VII. Reasons for Low and High SOC

Poor base soils (Dystochrepts/Dystropepts and Ultisols) with almost similar pH and CEC values are prevalent in the states of Tripura, Kerala and Karnataka under typical humid tropical climate (Table 14). The OC content of these soils, however, differs (Bhattacharyya et al., 1996; Krishnan et al., 1996 and Shivaprasad et al., 1998). In terms of SOC stock, Tripura state has an edge over Karnataka, Kerala state being exceptional due to relatively large areas under the Ultisols. The apparent reason for different OC status lies in the fact that Tripura has a cooler winter (mean temperature of 15°C in January) than both Kerala and Karnataka (mean temperature of 25°C in January). It therefore suggests that the cool temperature even for few months (November, December, January and February) can influence the accumulation of SOC.

Similar inference can be drawn from the soils of Maharashtra (Western Ghats) and Madhya Pradesh. The soils of these two states have similar clay content (40 to 45%), clay mineralogy (smectites and smectite/kaolinite interstratified minerals), parent material (basalts) and elevation (~1000 m above msl). This, coupled with the cooler winter of Madhya Pradesh (minimum temperature 7 to 8°C in January), allows the build up of OC in soils as high as to qualify them as Mollisols (Bhattacharyya and Pal, 1998). Such soils are not formed on the Western Ghats due to relatively warm winter temperature (minimum temperature 20 to 22°C in January) (Bhattacharyya et al., 1992) (Table 14).

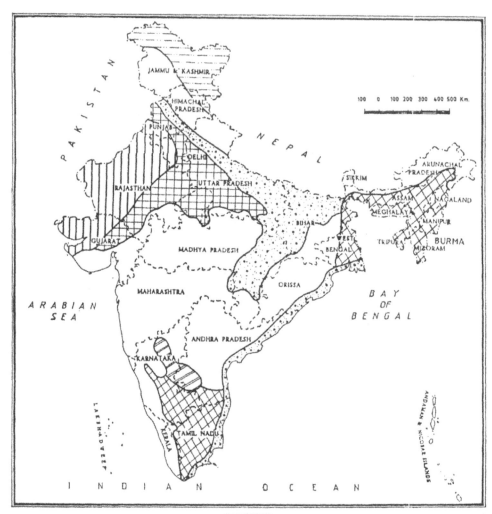

LEGEND

Symbol	SOC stock (Pg)	AERs		
		No.	Area (mha)	% TGA
	< 0.25	9, 11, 13, 18, 20	46.6	14.3
	0-25 - 0.50	8, 15, 17	41.8	12.7
	0.50. - 1.00	5, 6, 7, 10, 12, 14, 16, 19	156.1	47.5
	1.00 - 2.00	3	4.9	1.5
	2.00 - 3.00	2	31.9	9.7
	3.00 - 4.00	4	32.2	9.7
	> 4.00	1	15.2	4.6

Figure 1. SOC stock in Indian soils.

Table 12. Organic carbon stock in Indian soils in different bioclimatic regions

Bioclimatic regimes		Area		Organic C (Pg)			
		Coverage (m ha)	% TGA	0–30 cm	% of total stock	0–150 cm	% of total stock
Arid	Cold	15.2	4.6	5.96	28.4	10.47	16.6
	Hot	36.8	11.2	3.76	17.9	14.64	23.2
Semiarid	Hot	116.4	35.5	6.19	29.5	20.95	33.1
Subhumid	Hot	86.4	26.3	1.75	8.3	5.03	8.0
Subhumid to humid	Warm	21.2	6.4	0.52	2.5	1.55	2.4
	Moist	12.1	3.7	0.42	2.0	1.20	2.0
Perhumid	Warm	20.2	6.1	1.24	5.9	5.25	8.3
Subhumid to semiarid	Hot	8.5	2.6	0.23	1.1	0.91	1.4
Humid to perhumid	Hot	11.9	3.6	0.96	4.6	3.19	5.0

It is important to mention that a cool winter may not always help in higher accumulation of OC in soils. For example, although the northern states of Punjab and Haryana experience cool temperature during winter (minimum temperature 8°C in January), the OC status is low (semiarid hot and subhumid hot in Tables 13 and 14). This is the general scenario in most of the arid and semiarid regions of India. Higher accumulation of OC in soils is related to enriched vegetative cover which is supported by higher rainfall. Such situations are common in the eastern and northeastern parts of the country which represent humid to perhumid agro-ecoregions (Tables 10 and 12). This clearly indicates that the most conducive climatic condition favoring accumulation of OC in soils appears to be humid to perhumid climate punctuated with a cool winter.

It is well known that the major portion of SOC is retained through clay-organic matter interactions indicating the importance of the inorganic part of the soils as substrate to bind the OC. Several studies (Bhattacharyya and Ghosh, 1986, 1994, 1996) have been made on Indian soils to study naturally occurring clay-organic complexes. These experiments show different forms of complexes including free form of OC as aggregates grading to different size and shape. Among the 2:1 phyllosilicate minerals, smectites and vermiculites have the largest specific surface area. These minerals can therefore complex more OC and thus enhance the accumulation of SOC. It is, however, paradoxical that the smectitic soils of India appear mainly in the arid and semiarid parts and are poor in OC content. The effect of higher tropical temperature is thus evident in overshadowing the positive effect of minerals to store OC.

The importance of 2:1 minerals in accumulation of OC is well evidenced in the ferruginous red soils of north eastern, eastern, western and southern parts of the country. The soils of these regions are not dominated by clay minerals of advanced weathering stage (Pal and Deshpande, 1987b; Bhattacharyya and Pal, 1998; Bhattacharyya et al., 1998b; Preeti et al., 1998). Recent studies indicate that even in highly weathered ferruginous soils the presence of smectite and/or vermiculite either in the form of interstratification or in a discrete mineral form is very common (Pal and Deshpande, 1987a, 1987b, 1987c; Pal et al., 1989; Bhattacharyya et al., 1993, 1997, 1998c). Presence of these minerals favors the accumulation of OC in these soils. This suggests that the minerals with higher

LEGEND

Symbol	OC content (%) (0-30 cm)	AERs		
		Nos.	Area (mha)	% TGA
░░░	Sufficient Zones > 1.0	1, 13, 14, 15, 16, 17, 18, 19, 20	100.2	30.5
☐	Deficient Zones < 1.0	2, 3, 4, 5, 6, 7, 8, 9, 10, 11, 12	228.5	69.5

Figure 2. Map showing sufficient and deficient zones of organic carbon in Indian soils.

Table 13. Organic carbon and bulk density values of some representative soils in two selected depths

Ecosystems (Bioclimatic regions)		Soil subgroups	OC (%)		BD (Mg m^{-3})	
			0–30 cm	0–150 cm	0–30 cm	0–150 cm
Arid	Cold	Typic Cryorthid	1.40	0.57	1.50	1.55
	Hot	Typic Calciorthid	0.18	0.05	1.45	1.52
Semiarid	Hot	Typic Natrustalf	0.30	0.18	1.48	1.45
		Vertic Ustochrept	0.66	0.38	1.76	1.74
		Typic Haplustert	0.36	0.31	1.88	1.96
		Chromic Haplustert	0.33	0.33	1.54	1.55
		Sodic Haplustert	0.80	0.30	1.55	1.50
		Oxic Paleustalf	0.48	0.45	1.50	1.53
Subhumid	Hot	Aquic Ustochrept	0.32	0.18	1.43	1.51
		Udic Ustochrept	0.36	0.24	1.52	1.43
		Typic Ustochrept	0.30	0.15	1.21	1.30
		Typic Haplustert	0.41	0.29	1.85	1.95
		Chromic Haplustert	0.64	0.47	1.77	1.78
		Typic Plinthustalf	0.65	0.41	1.80	1.68
		Ultic Paleustalf	0.24	0.12	1.70	1.62
		Typic Hapludult	0.91	0.38	1.49	1.50
Subhumid	Warm	Typic Haplustalf	0.33	0.23	1.54	1.45
	Moist	Vertic Endoaquept	0.62	0.28	1.56	1.68
Perhumid	Warm	Typic Hapludalf	1.33	0.73	1.35	1.35
		Ultic Hapludalf	1.74	1.00	1.35	1.40
		Typic Haplustert	1.00	0.55	1.45	1.55
Subhumid to semiarid	Hot	Vertic Haplaquept	0.70	0.31	1.50	1.50
		Typic Haplorthox	1.63	0.82	1.40	1.45
Humid to perhumid	Hot	Oxic Dystrochrept	2.19	1.37	1.40	1.40
		Typic Dystrochrept	0.84	0.57	1.45	1.45

surface area are also important factors for accumulation of OC in addition to the dominating effect of climate in the humid tropical climate of the country.

Table 14. OC percent and SOC stock of representative soils from different parts of the country

Horizon	Depth (cm)	pH (water)	CEC cmol(+)kg^{-1}	OC (%)	SOC stock (Pg)*
		Typic Dystrochrept - Tripura			
A1	0–10	5.0	5.3	1.6	
B1	10–37	4.7	5.6	1.0	
B2	37–73	4.9	5.8	1.2	0.012
B3	73–120	4.8	7.4	0.8	
B4	120–155	4.8	7.4	0.6	
		Ustic Kandihumult - Kerala			
Ap	0–15	5.1	6.5	1.2	
Bt1	15–39	5.2	6.2	1.0	
Bt2	39–119	5.3	6.6	0.9	0.059
Bt3	119–162	5.2	5.9	0.6	
Bt4	162–205	5.4	5.3	0.5	
		Kanhaplic Haplustult - Karnataka			
Ap	0–14	6.4	6.3	1.2	
AB1	14–34	6.3	5.7	1.1	
AB2	34–50	6.3	5.3	0.6	0.004
Bt1	50–83	5.9	4.9	0.3	
Bt2	83–107	5.2	5.1	0.2	
		Typic Hapludalf - Maharashtra			
A1	0–9	5.7	8.4	1.3	
B1	9–31	5.3	9.8	1.2	
Bt1	31–60	5.3	10.6	1.1	--
Bt2	60–106	5.6	8.6	1.4	
Bt3	106–155	5.6	7.3	0.6	
		Typic Haplustoll - Madhya Pradesh			
A1	0–6	5.9	52.2	3.5	
A2	6–16	5.8	59.8	3.0	
B1	16–37	5.8	59.8	2.0	--
B2	37–74	5.9	67.4	1.2	
B3	74–106	5.6	71.7	0.8	
B4	106–120	5.5	73.9	0.5	
		Typic Haplustalf - Madhya Pradesh			
Ap	0–10	6.2	25.5	1.9	
Bt1	10–30	5.9	29.1	1.8	
Bt2	30–59	5.9	28.8	1.0	--
C1	59–94	6.5	28.8	0.4	
C2	94–131	6.7	36.4	0.4	
		Aquic Natrustalf - Punjab			
A1	0–6	10.3	8.2	0.2	
BA	6–24	9.8	11.8	0.2	
Bt1	24–48	9.6	13.6	0.1	
Bt2	48–73	9.4	14.2	0.14	--
Bt3	73–97	9.4	13.8	0.1	
Bc	97–124	9.4	9.8	0.1	
Ck	124–145	9.3	9.0	0.1	

*SOC stock in 0–150 cm depth.

VIII. Scope of Sequestration of OC in Soils of India

It has already been pointed out that high soil temperature is related to the aridity of the atmosphere (Jenny, 1980). The pioneering work of Jenny and Raychaudhuri (1960) on Indian soils showed that the depletion of SOC at the level of 60 to 70% is due to higher tropical temperature.

It has been hypothesized that increase of OC in soil increases both the surface charge density (SCD) of soil and the ratio of internal / external exchange sites (Poonia and Niederbudde, 1990). With the increase in SOC content, soils show higher preference for Ca over Na due to higher SCD of OC. Conversely, low OC soils will be more prone to sodification than high OC soils (Poonia, 1998). It has been stated that organic exchange sites have a greater preference for Ca than the inorganic exchange sites (Poonia, 1998). While comparing the role of applied and natural organic matter, Poonia et al. (1980) observed that the applied organic matter had a greater preference for divalent cations than the natural organic matter due to inactivation of closely-spaced organic sites over a long-term binding action in natural organic matter. The source of applied organic matter could be reforestation and/or external application of organic manures. It has already been established that with reforestation and other suitable measures for ecological rehabilitation, OC levels of the tropical soils can be increased (Gupta and Rao, 1994).

In India, despite the higher SOC stock in soils of hot arid and semiarid climatic regions, the database indicates overall poor OC content. Besides, the total SOC stock of the country is a meagre 3% when compared with the SOC of the world. This indicates an enormous scope of sequestration of OC in these regions through proper management techniques.

A. Sequestration of OC Vis-á-Vis Soil Type

Understanding the favorable influence of clay mineral type on the formation of clay-humus complex, it is clear that soils containing minerals with higher surface area are the most suitable substrate for sequestrating OC. The significance of nature and content of clay as substrate has also been stressed as the most important factor influencing OC distribution (Arrouays et al., 1995). The accumulation of OC in Indian soils, however, is dependent on the rate of decomposition of OC due to higher temperature in the tropics. Since all such soils are encountered either in arid or in semiarid regions, the continuous depletion of OC must be checked by suitable cultural practices that may help in sequestration of OC. It may be stated that increase in OC content may be possible not only through sequestration but also by other processes viz. decomposition of OM (Gupta and Rao, 1994). Conversely, in soils that do not contain clay minerals having higher surface area the scope of sequestration is reduced. Such soils containing dominant proportions of coarse fractions is common in hot and desert parts of the country.

The current status of SOC in India shows that more than 50% of SOC stock falls under arid (hot) and semiarid (hot) AERs (Tables 11 and 12). These soils are poor in OC (Table 12) and require sequestration. These areas are mostly represented by the Indo-Gangetic Plains, central and western India which are dominated by black soils (shrink-swell soils) and to a lesser extent by the southern peninsular India characterized by ferruginous soils. Climatic adversity of the arid and semiarid tracts create sodicity either throughout the profile or in the subsoils in these parts of the country (Balpande, 1993; Kadu et al., 1993; Balpande et al., 1996). Scarcity of soil moisture followed by precipitation of $CaCO_3$ brings a higher sodium on exchange complex to form sodic soils (Pal et al., 1999). These soils need immediate attention for reclamation for a favorable vegetative growth so important for OC sequestration (Gupta and Rao, 1994). Similar priority needs to be fixed for the central, western and southern parts of the country dominated by black and ferruginous soils. Figure 3 indicates the import-

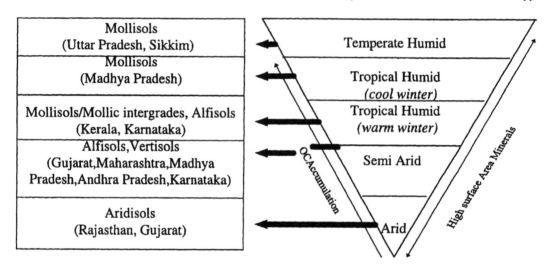

Figure 3. Combined influence of rainfall, temperature and substrate quality in the accumulation of OC in soils.

ance of climate for a favourable OC accumulation in soils of different states of India (Deshpande, et al., 1971; Krishnan et al., 1996; Das et al., 1996; Bhattacharyya and Pal, 1998; Shivaprasad et al., 1998). The influence of rainfall, temperature and substrate quality in the accumulation of SOC thus determines the formation of soils towards Mollisols and/or their intergrades.

B. Sequestration of OC Vis-à-Vis Land Use Planning

The potential for sequestration of OC in the areas dominated by problem soils viz. saline and sodic soils through an appropriate land use planning viz. agriculture, agroforestry and silviculture after taking suitable rehabilitation measures is great (Gupta and Rao, 1994). The prospects of OC sequestration in these soils is high because such soils occupy a major geographical area of the country (Abrol and Bhumbla, 1971).

Application of gypsum followed by cropping in sodic soils of the Indo-Gangetic Plains increased the urease and dehydrogenase activity (measure of biological activity) by about three fold (Rao and Ghai, 1985; Fromm et al., 1993; Batra, 1998). Several studies on Indian soils have established that land levelling, amelioration of sodic soils using gypsum, and other reclamation techniques such as irrigation, drainage of saline soils and cropping pattern using deep-rooted crops and/or growing tolerant multipurpose tree species, can lead to establishment of vegetation cover and thus can increase SOC stock. Reviews on this aspect (Gupta and Rao, 1994) indicate that among the tolerant crop species, grasses (*Leptochloa*) and tree species (*Acacia, Prosopis*) have been found to be producing biomass about 20 Mg ha^{-1} yr^{-1} for tree and 4 Mg ha^{-1} yr^{-1} for grass. A beneficial effect of leguminous crops like *Sesbania* was found in terms of increasing the level of OC in sodic soils.

IX. Management of Organic Inputs and SOC Dynamics

Soil organic carbon is both a source and sink of plant nutrients forming the most important renewable resource in soils (Duxbury et al., 1989). As an ion exchanger it promotes the formation of soil aggregates influencing soil physical properties and also acts as an energy substrate for soil microflora and fauna (Allison, 1973). Due to ever-increasing population, humans have adopted various advanced techniques of external application of fertilizers and pesticides to increase grain yields. The capital-intensive nature and negative environmental impacts of fertilizer and pesticide abuse, however, demand improved management of SOC for sustaining land use systems in the tropics (Fernandes et al., 1997). Besides the impact of SOC on soil productivity, the accelerated decomposition of SOC due to agriculture resulting in loss of carbon to the atmosphere and its contribution to green house effect is considered a serious global problem. It has been estimated that in the early 1980s, land use changes resulted in the transfer of 1 to 2 Pg C yr^{-1} year from terrestrial ecosystem to the atmosphere, of which 15 to 17% carbon is contributed by decomposition of SOC (Houghton and Hackler, 1994).

The conversion of natural forest to an agro-ecosystem results in changes in biological and chemical processes at the plant-soil interface showing an initial decline in SOC content (Bhattacharyya et al., 1993; Saikh et al., 1998). It has been shown that due to the conversion of forest land to agriculture, SOC decreases rapidly in the first year, and at a slower rate thereafter approaching a new equilibrium after 30 to 50 years (Schlesinger, 1986; Houghton et al., 1991; Arrouays and Pelissier, 1994). Recent studies on forest soils (Alfisols of eastern India by Saikh et al., 1998) indicate that OC content of soils sharply declines when put to cultivation. The reduction of OC level is significant even within 5 to 15 years of cultivation. It has also been reported that for all soils, irrespective of their initial carbon levels, there is a tendency to reach some quasi-equilibrium values of 1 to 2%. Since such studies are limited to specific geographical regions, a generalized view about the carbon-carrying capacity of the soils need not be drawn considering the highly variable qualitative nature of the soil substrate.

X. Perspective

As we enter the next millennium, the world is faced with many difficult questions which possibly mankind never confronted in the past. The Bruntland Commission Report, "Our common future," as early as 1987 expressed the significance of global change (Sinha, 1992).

Current arid and semiarid environments prevailing in the central and southern peninsular India are a part of global warming phenomena (Eswaran and van den Berg, 1992). It is in this respect the soils of humid tropics of the Indian subcontinent require immediate attention for better SOC management. Restoration of OC balance and its follow up to enlarge the soil carbon pool by appropriate management techniques should form the strategic perspective for Indian soils. Such efforts not only make the stakers of soils and planner aware about the present situation, but also give a proper guidance for the location of actual areas for possible sequestration to restore OC balance in soils for a better agricultural performance. There appears to be a large scope for the sequestration of OC in black soils (shrink-swell soils) of arid and semiarid climate due to large surface area of the dominant minerals of these soils. However, in the present scenario of different climatic parameters, the rising of temperature and shrinking of annual rainfall in major geographical areas of the country except in the sub-Himalayan region will continue to remain as potential threats for tropical soils of the Indian subcontinent especially in view of global warming phenomenon (Gadgil, 1995). This is not only due to continuous loss of SOC, but also due to natural/human induced degradation which will demand immediate restoration of soil health through C sequestration (Lal, 1994). The process of restoring soil health will ever demand extra resources for the sequestration of OC towards a sustainable agricultural management policy of the nation and will eventually limit agriculture's pace.

Acknowledgement

We are thankful to Dr. C. Mandal, Senior Scientist, NBSS and LUP, Nagpur for her scientific and technical inputs during the preparation of the manuscript. We acknowledge the assistance of Ms. Vaishali for word processing the manuscript. The help rendered by other colleagues are also thankfully acknowledged.

References

Abrol, I.P. and D.R. Bhumbla. 1971. Saline and Alkali soils in India—their occurrence and management. FAO World Soil Resources Report, 41:42-51.

Allison, F.E. 1973. *Soil Organic Matter and Its Role in Crop Production.* Elsevier, Amsterdam. 637 pp.

Arrouays, D. and P. Pelissier. 1994. Modelling carbon storage profiles in temperate forest humic loamy soils of France. *Soil Sci.* 157:185-192.

Arrouays, D., I. Vion, and J.L. Kiein. 1995. Spatial analysis and modelling of topsoil carbon storage in temperate forest loamy soils of France. *Soil Sci.* 159:191-198.

Balpande, S.S. 1993. Characteristics, Genesis and Degradation of Vertisols of the Purna Valley, Maharashtra, Ph.D. Thesis, Punjabrao Krishi Vidyapeeth, Akola (M.S.), India.

Balpande, S.S., S.B. Deshpande, and D.K. Pal. 1996. Factors and processes of soil degradation in vertisols of the Purna valley, Maharashtra, India. *Land Degrad. and Dev.* 7:313-324.

Batjes, N.H. 1996. Total carbon and nitrogen in the soils of the world. *European J. Soil Sci.* 47:151-163.

Batra, L. 1998. Effect of different cropping sequences on dehydrogenase activity of three sodic soils. 1998. *J. Indian Soc. Soil Sci.* 46:370-375.

Bhattacharjee, J. C.,C. Roychaudhury, R. J. Landey, and S. Pandey. 1982. Bioclimatic Analysis of India. NBSSLUP Bulletin, 7, National Bureau of Soil Survey and Land Use Planning, Nagpur, India, 21 pp. + map.

Bhattacharyya, T. and D.K. Pal. 1998. Occurrence of Mollisols - Alfisols - Vertisols associations in Central India—their mineralogy and genesis. Paper presented in National Seminar of Indian Society of Soil Science, 16-19 Nov. 1998, Hisar, India.

Bhattacharyya, T. and S.K. Ghosh.1986. Inorganic amorphous constituents of naturally occurring clay-organic complex. *J. Indian Soc. Soil Sci.* 34:181-184.

Bhattacharyya, T. and S.K. Ghosh. 1994. Nature and characteristics of naturally occurring clay-organic complex of two soils from north-eastern region. *Clay Res.* 13:1-9.

Bhattacharyya, T. and S.K. Ghosh.1996. Naturally occurring clay-organic complex in some soils of India. *Agrochimica* 40:64-72.

Bhattacharyya, T., C. Mondal, and S.N. Deshmukh. 1992. Soils and land use pattern in part of western Maharashtra. *J. Indian Soc. Soil Sci.* 40:513-520.

Bhattacharyya, T., D.K. Pal, and S.B.Deshpande. 1997. On kaolinitic and mixed mineralogy classes of shrink-swell soils. *Aust. J. Soil Res.* 35:1245-1252.

Bhattacharyya, T., D.K. Pal, and S.B. Deshpande. 1993. Genesis and transformation of minerals in the formation of red (Alfisols) and black (Inceptisols and Vertisols) soils on Deccan basalt in the Western Ghats, India. *J. Soil Sci.* 44:159-171.

Bhattacharyya, T., J.L. Sehgal, and D. Sarkar. 1996. Soils of Tripura for Optimising Land Use: Their Kinds, Distribution and Suitability for Major Field Crops and Rubber. NBSS Publication 65 (Soils of India Series 6). National Bureau of Soil Survey and Land Use Planning, Nagpur, India, 154 pp.

Bhattacharyya, T., S. Mukhopadhyay, U. Baruah, and G. S. Chamuah. 1998a. Need of soil study to determine degradation and landscape stability. *Current Sci.* 74:42-47.

Bhattacharyya, T., D.K. Pal, and M. Velayutham. 1998b. Vermiculites as a natural sink for Al^{3+} ions in acid soils of Tripura. Paper presented in National Symposim on Recent Trends in Clay Research for 12th Annual Covention of the Clay Minerals Society of India held at NBSS & LUP, Nagpur, India, 9-10 July 1998.

Bhattacharyya, T., D.K. Pal, and P. Srivastava. 1998c. Role of zeolites in persistence of high altitude ferruginous alfisols of the humid tropical Western Ghats, India. *Geoderma* (in press).

Bhattacharyya, T., S.N. Deshmukh, and C. Roychaudhury. 1989. Soils and land use of Junnar tehsil, Pune district, Maharashtra. *J. Maharashtra Agri. Univ.* 14:1-4.

Bhattacharyya, T., R. Srivastava, J.P. Sharma, and J.L. Sehgal. 1994a. Classification of saline-sodic vertisols in the coastal plain of Gujarat. *J. Indian Soc. Soil Sci.* 42:306-309.

Bhattacharyya, T., T.K. Sen, R. S. Singh, D.C. Nayak, and J.L. Sehgal. 1994b. Morphological characteristics and classification of Ultisols with kandic horizon in north-eastern region. *J.Indian Soc. Soil Sci.* 42:301-306.

Bhattachrayya, T., T.K. Das, P.N. Dubey, U. Baruah, S.K. Gangopadhyay, and Dileep Kumar. 1995. Soil Survey Report of Morigaon district, Assam (1:50,000 scale). Report No. 527, National Bureau of Soil Survey and Land Use Planning, Regional Centre, Jorhat, Assam.

Bohn, H.L. 1976. Estimate of organic carbon in the world. *Soil Sci. Soc. Am. J.* 40:468-470.

Bohn, H.L. 1982. Estimate of organic carbon in the world. *Soil Sci. Soc. Am. J.* 46:1118-1119.

Buringh, P. 1984. Organic carbon in soils of the world. p. 91-109. In: G.M. Woodwell (ed.), *The Role of Terrestrial Vegetation in the Global Carbon Cycle Measurements by Remote Sensing*, SCOPE 23, John Wiley & Sons, New York.

Das, T. H., C. J. Thampi, J. Sehgal, and M. Velayutham. 1996. Soils of Sikkim for Optimising Land Use. NBSS Publication 60b (Soils of India Series). National Bureau of Soil Survey and Land Use Planning, Nagpur, India, 44 pp + 4 sheet soil map (1:100,000 scale).

Deshpande, S.B., J.B. Fehrenbacher, and A.H. Beavers. 1971. Mollisols of Tarai region of Uttar Pradesh, northern India. 2. Genesis and classification. *Geoderma* 6:179-193.

Duxbury, J. M., M. S.Smith, and J.W. Doran.1989. Soil organic matter as a source and sink of plant nutrients. In: D. C. Coleman, J. M. Oades, and G. Uehera (eds.), *Dynamics of Soil Organic Matter in Tropical Ecosystems*. University of Hawaii Press, Honolulu, HI, U.S.

Eswaran, H. and E. van den Berg. 1992. Impact of building of atmospheric CO_2 on length of growing season in the Indian sub-continent. *Pedologie* 42:289-296.

Eswaran, H., E. van den Berg, and P. Reich. 1993. Organic carbon in soils of the world. *Soil Sci. Soc. Am. J.* 57:192-194.

Fernandes, E.C.M., P.P. Motaralli, C. Castilla, and L. Munurumbira. 1997. Management control of soil organic matter dynamics in tropical land-use systems. *Geoderma* 79:49-67.

Fromm, H., K. Winter, J. Filser, R. Hantschel, and F. Besse. 1993. The influence of soil type and cultivation system on the spatial distributions of the soil fauna and microorganisms and their interactions. *Geoderma* 60:109-118.

Gadgil, S. 1995. Climatic change and agriculture—an Indian perspective. *Curr. Sci.* 69:649-659.

Gupta, R.K. and D.L.N. Rao. 1994. Potential of wastelands for sequestering carbon by reforestation. *Curr. Sci.* 66:378-380.

Harindranath, C.S., K.R. Venugopal, N.G. Raghumohan, J. Sehgal, and M. Velayutham. 1998. Soils of Goa for Land Use Planning. NBSS Publication 74b. National Bureau of Soil Survey and Land Use Planning, Nagpur, India, 136 pp. + 2 sheet soil map (1:250,000 scale).

Houghton, R.A., D.L. Skole, and D.S. Lefkowitz. 1991. Changes in the landscape of Latin America between 1850 and 1985. II. Net release of CO_2 to the atmosphere. *For. Ecol. Manage.* 38:173-199.

Houghton, R.A. and J.L. Hackler. 1994. The net flux of carbon from deforestation and degradation in south and Southeast Asia. p. 301-327. In: Dale, V. (ed.), *Effects of Land-Use Change on Atmospheric CO_2 Concentrations : South and Southeast Asia as a Case Study*, Springer-Verlag, New York.

Jenny, H. and S.P. Raychaudhuri. 1960. Effect of Climate and Cultivation on Nitrogen and Organic Matter Reserves in Indian Soils, ICAR, New Delhi, India. 126 pp.

Jenny, H. 1980. *The Soil Resource*. Origin and Behaviour, Springer-Verlag, New York.

Kadu, P.R., D.K. Pal, and S.B. Deshpande. 1993. Effect of low exchangeable sodium on hydraulic conductivity and drainage in shrink-swell soils of the Purna Valley, Maharashtra. *Clay Res.* 12:65-70.

Kimble, J. ,T. Cook, and H. Eswaran. 1990. Organic soils of the tropics. p. 250-258. In: *Proc. Symp. Characterization and Role of Organic Matter in Different Soils*. Soil Sci. 14th, Kyoto, Japan. 12-18 Aug. 1990, Wageningen, The Netherlands,

Krishnan, P., K. R. Venugopal, and J. Sehgal. 1996. Soil Resources of Kerala for Land Use Planning. NBSS Publication 48b (Soils of India Series 10). National Bureau of Soil Survey and Land Use Planning, Nagpur, India, 54 pp. + 2 sheet soil map (1:500,000 scale).

Lal, R. 1994. Sustainable landuse systems and soil resilience. p. 41-67. In: D.J. Greenland and I. Szabolcs (eds.), *Soil Resilience and Sustainable Land Use*, CAB International, Wallingford, U.K.

Lal, S., S.B. Deshpande, and J. Sehgal. 1994. Soil Series of India. NBSS Publication No. 40. National Bureau of Soil Survey and Land Use Planning. 684 pp.

Mandal, C., D.K. Mandal, and C.V. Srinivas. 1996. Defining climate map of India following FAO growing period concepts. *Geogr. Rev. India* 58:243-250.

Murthy, R.S., L.R. Hirekerur, S.B. Deshpande, and B.V. Venkat Rao (eds.). 1982. Benchmark Soils of India. National Bureau of Soil Survey and Land Use Planning, Nagpur, India. 374 pp.

Natarajan, A., P.S.A. Reddy, J. Sehgal, and M. Velayutham. 1997. Soil Resources of Tamil Nadu for Land Use Planning. NBSS Publication 46b (Soils of India Series) National Bureau of Soil Survey and Land Use Planning, Nagpur, India, 88 pp. + 4 sheet soil map (1:50,000 scale).

NBSS & LUP. 1988. Benchmark Swell-Shrink Soils of India—Morphology, Characteristics and Classification, NBSS Publication 19, National Bureau of Soil Survey and Land Use Planning, Nagpur, India.

NBSS & LUP. 1984. Soil Survey and Land Evaluation of ORP, Flood Prone Area, Kandela Village, Kesargang Tehsil, Bahraich District, Uttar Pradesh. Report No. 461 (ICAR).

Pal, D.K. and S.B. Deshpande. 1987a. Parent material, mineralogy and genesis of two benchmark soils of Kashmir Valley. *J. Indian Soc. Soil Sci.* 35:690-698.

Pal, D.K. and S.B. Deshpande. 1987b. Genesis of clay minerals in a red and black complex soils of southern India. *Clay Res.* 6:6-13.

Pal, D.K. and S.B. Deshpande. 1987c. Characteristics and genesis of minerals in some benchmark Vertisols of India. *Pedologie* 37:259-275.

Pal, D.K., G.S. Dasog, S. Vadivelu, R.L. Ahuja, and T. Bhattacharyya. 1999. Secondary calcium carbonate in soils of arid and semi-arid regions of India. In: *Global Climate Change and Pedogenic Carbonates*. (in press). CRC Press, Boca Raton, FL.

Pal, D.K., S.B. Deshpande, K.R. Venugopal, and A.R. Kalbande. 1989. Formation of di- and trioctahedral smectite as an evidence for paleoclimatic changes in southern and central peninsular India. *Geoderma* 45:175-184.

Poonia, S.R. 1998. Sorption / Exchange of some nutrient and non-nutrient cations in soils. 16th Prof. J.N. Mukherjee - ISSS Foundation Lecture, 63rd Annual Convention of ISSS, Hissar, 16-19 November, 1998.

Poonia, S.R. and E.A. Niederbudde. 1990. Exchange equilibria of potassium in soils, V. Effect of natural organic matter on K - Ca exchange. *Geoderma* 47:233-242.

Poonia, S.R., S.C. Mehta, and R. Pal. 1980. Calcium-sodium, magnesium-sodium exchange equilibria in relation to organic matter in the soil. p. 134-141. In: *Proc. Int. Symp. Salt-Affected Soils* (CSSRI), Karnal, India,

Preeti, P., P. Raja, J. Sehgal, and K.S. Gajbhiye. 1998. Study on mineralogy of some selected soils from hot humid to perhumid ecosystem of Kozhikode, Palaghat and Ernakulam areas in Kerala. *J. Indian Soc. Soil Sci.* 46:430-435.

Rao, D.L.N., and S.K. Ghai. 1985. Urease and dehydrogenease activity of alkali and reclaimed soils. *Aust. J. Soil Res.* 23:661-665.

Saikh, H., C. Varadachari, and K. Ghosh. 1998. Changes in carbon, nitrogen and phosphorus levels due to deforestation and cultivation: a case study in Simlipal National Park, India. *Plant. and Soil* 198:137-145.

Schlesinger, W. H. 1986. Changes in soil carbon storage and associated properties with distribution and recovery. p.194-200. In: J.R. Trabalka and D.E. Reichle (eds.), *The Changing Carbon Cycle: a Global Analysis.* Springer-Verlag, New York.

Sehgal, J., D.K. Mandal, and C. Mandal. 1995. AESR Map of India (1:4.3 million). NBSS & LUP, Nagpur, India.

Sehgal, J. D.K. Mondal, C. Mondal, and M. Velayutham. 1998. Agro-Ecological Sub-Regions of India. NBSS Bulletin No. 35, National Bureau of Soil Survey and Land Use Planning, Nagpur, India (unpublished).

Sehgal, J.L. and P.K. Sharma. 1982. Benchmark Soils of Punjab, Soils Bulletin No. 5, Department of Soils, PAU, Ludhiana, India.

Sehgal, J.L., D.K. Mandal, C. Mandal, and S. Vadivelu. 1992. Agro-Ecological Regions of India, Bulletin No. 24, NBSS & LUP, Nagpur, India, 130p.

Shivaprasad, C.R., R.S. Reddy, J. Sehgal, and M. Velayutham. 1998. Soils of Karnataka for optimising land use. NBSS Publication 47 (Soils of India Series). National Bureau of Soil Survey and Land Use Planning, Nagpur, India, 110 pp. ⊦ 4 sheets of soil map (1:500,000 scale).

Sinha, S.K. 1992. Climate change and agriculture. *Indian Farming* 42:13-18.

Soil Survey Staff. 1975. Soil Taxonomy, Agriculture Handbook No. 436. SCS-USDA, U.S. Government Printing Office, Washington, D.C.

Soil Survey Staff. 1994. Keys to Soil Taxonomy, 6th ed., Soil Conservation Service, U.S. Department of Agriculture, Washington, D.C.

Sombroek, W.G., F.O. Nachtergache, and A. Habel. 1993. Amounts, dynamics and sequestrations of carbon in tropical and subtropical soils. *Ambio* 22:417-427.

Vadivelu, S. and A.K. Bandyopadhyay. 1997. Characteristics, genesis and classification of soils of Minicoy Island, Lakshadweep. *J. Indian Soc. Soil Sci.* 45:796-800.

Velayutham, M., N.G. Raghumohan, G. Bourgeon, C.S. Harindranath, C.R. Shivaprasad, and S. Salvador. 1998. Detailed estimates and mapping of soil organic carbon at district level in south India. Paper presented in the 16th World Congress Soil Science - Montpellier, France, 20-26 August, 1998.

Section II.

LAND USE AND CARBON POOL IN SOILS OF AFRICA

CHAPTER 5

Slash-and-Burn Effects on Carbon Stocks in the Humid Tropics

P.L. Woomer, C.A. Palm, J. Alegre, C. Castilla, D.G. Cordeiro, K. Hairiah, J. Kotto-Same, A. Moukam, A. Riese, V. Rodrigues and M. van Noordwijk

I. Introduction

The relative importance of tropical forests in the global carbon cycle has been debated over the past 20 years with several estimates of their contribution to the increase in atmospheric carbon dioxide (Woodwell et al., 1978; Houghton et al., 1987; Detwiler and Hall, 1988; Hall, 1989; Post et al., 1990). Currently there is general agreement, based on land use change data and atmospheric data, that the tropics are a net source of C to the atmosphere, in the range of 1.1 to 2.1 Pg C y^{-1} (Houghton, 1997). The primary cause of this net source is deforestation in the tropical zone, with Asia and Latin America accounting for over 80% of the flux (Houghton, 1995). Sanchez et al. (1994) estimate total tropical deforestation to be 627 million ha, or approximately 40% of the potential humid forest zone, with 120 million ha of these lands subject to shifting cultivation or slash-and-burn agriculture.

Tropical forests are cleared for a variety of reasons that include logging, establishment of plantations and pastures, and slash-and-burn agriculture. The primary cause of deforestation differs by country and even regions within countries (Tomich and van Noordwijk, 1996) but is usually associated with some form of slash-and-burn agriculture, either as the primary driving force or as a consequence of increased access to forests by logging operations and road construction. Farmers practicing slash-and-burn agriculture are clearing forests to produce food and seek improvements in their families' standards of living. In most cases, they are marginalized from society and government support programs and live in relative poverty. Efforts to reduce deforestation and greenhouse gas emissions resulting from deforestation must address these root causes.

In 1991 a global program, Alternatives to Slash-and-Burn Program (ASB), was initiated to address the agronomic, environmental, social and political implications of slash-and-burn (Brady, 1996). The overall goal of the program was to compare the impact of current land use systems in the tropics and to identify alternatives that were sound from an environmental, agronomic and economic perspective. In addition, policies that currently inhibit the adoption of these alternatives were considered. Teams of national and international scientists were established in key locations, referred to as benchmark areas, around the world representing the range in biophysical and socioeconomic environments in which slash-and-burn is practiced.

The environmental impact of slash-and-burn in terms of net CO_2 flux as a result of land use depends on the rates of land use change, the biomass of the vegetation that is cleared and the fate of the carbon within that biomass, the potential for reaccumulation of carbon within subsequent land use systems and the regrowth rates of vegetation. Much of the uncertainty in the values of CO_2 flux from the tropics is a result of inadequate estimates for these parameters (Houghton, 1997). One activity of the ASB project was to characterize the patterns of land clearing and subsequent land use at the different sites and to quantify the changes in carbon stocks associated with land clearing and establishment of different land use systems. Standardized methods were established to measure carbon stocks in the forests, the various land use systems established following slash-and-burn clearing, and promising "best-bet" alternatives at the different sites. These data can be used to calculate the immediate and longer term loss of carbon with slash-and-burn clearing and to take carbon stocks into account in multiple goal evaluation of existing and proposed land use systems. In this chapter we present summary data on carbon stocks in forests and slash-and-burn systems from nine of the ASB sites located in Brazil, Cameroon, Indonesia, and Peru.

II. Estimation of Carbon Stocks in Slash-and-Burn and Alternative Land Uses

Early in ASB activities, a carbon stocks working group was formed among program collaborators and given the responsibility of measuring (1) C stocks in forests undergoing slash-and-burn; (2) C dynamics as these forests are converted to current land uses and (3) the potential to sequester C in alternative "best-bet" land uses. An approach was adopted that nests forests and current land uses within chronosequential transects, where short distances in space substitute for relatively great differ-ences in time (Sanchez, 1987).

A. Institutional Participation and Benchmark Area Selection

Benchmark areas were selected by four national committees in the ICRAF-coordinated global Alternatives to Slash-and-Burn Program (Brady, 1996). The nine benchmark areas reported in this study belong to three floristic zones, the Amazonian, Dipterocarp (S.E. Asia) and Guineo-Congolian (West and Central Africa) Forests (Table 1). The national committees of Brazil, Cameroon, Indonesia and Peru identified a range of benchmark areas that were considered to be typical of current forest conversion by slash-and-burn or deforested lands requiring rehabilitation resulting from past slash-and-burn agriculture (Table 1). Within national research institutions, carbon study research teams were organized and assisted through the development of standardized methods by scientists from ICRAF and the Tropical Soil Biology and Fertility Programme (see Murdiyarso et al., 1994).

B. Transect Position and Land Use Selection

The exact position of the land use chronosequential transects and the selection of land uses within them was based upon the local knowledge and community contacts of national team members. In principle, land use transformation by slash-and-burn occurs in stages beginning with *primary*, managed or mature secondary forests that are felled and burned, *cultivated* and then alternatively placed into a longer-term land use (e.g., pasture or tree plantation) or *abandoned* to natural succession. Team members would discuss these stages at informal meetings with local chiefs or community leaders and in turn be introduced to individual slash-and-burn farmers willing to host the study. The various land uses, their ages and original forest conditions were discussed with farmers during site visits to candidate transects. Similar texture of the topsoil was used as a criteria to assure that all land-use types within a chronosequence were of the same soil type. Care was taken to exclude transects with nonrepresentative soil conditions or abrupt changes in terrain. During the selection process, several

Table 1. Carbon stocks estimated for different land uses within the Alternatives to Slash-and-Burn benchmark areas

Benchmark area/ coordinates	Location/ lead agency	Comments (No. of land uses, chronosequences)
Pedro Peixoto 61.7°W, 10.0°S	Acre, Brazil EMBRAPA[a]	Logged semideciduous Amazonian forest occupied by government-sponsored colonist ranchers and farmers since 1973 (7,1)
Theobroma 62.1°W, 10.1°S	Rondônia, Brazil EMBRAPA	Logged-over semideciduous Amazonian forest occupied by government-sponsored colonist ranchers and farmers since 1979 (7,2)
Ebolowa 11.1°E, 2.5°N	Cameroon IRAD[b]	Unlogged evergreen Guineo-Congolian rain forest occupied by indigenous tribes practicing long-term fallow (6,2)
Mbalmayo 11.7°E, 3.5°N	Cameroon IRAD	Logged moist semideciduous Guineo-Congolian forest occupied by indigenous tribes with large areas of young tree fallow (6,2)
Yaounde 11.4°E, 4.1°N	Cameroon IRAD	Logged-over, drier, peripheral semi-deciduous Guineo-Congolian forest occupied by indigenous tribes and spontaneous migrants practicing bush fallows and continuous cultivation (6,2)
Jambi 102.2°E, 1.5°S	Sumatra, Indonesia AARD[c]	Logged Dipterocarp evergreen forest concession occupied by nonagricultural indigenous tribes and migrants practicing mixed cultivation and rubber agroforests (6,2)
Lampung 108.4°E, 4.5°S	Sumatra, Indonesia AARD	Logged-over Dipterocarp forest occupied by spontaneous migrants cultivating food and market crops under continuous cultivation and agroforests, extensive *Imperata* grasslands (6,2)
Pucallpa 74.5°W, 8.2°S	Ucayali, Peru INIA[d]	Logged-over evergreen Amazonian forest settled by migrant ranchers and farmers with close proximity to city markets (7,1)
Yurimaguas 76.1°W, 5.8°S	Loreto, Peru INIA	Logged Amazonian rain forest settled by farmers with large areas of tree fallow and poor proximity to markets (7,1)

[a]Empresa Brasileira de Pesquisa Agropecuaria, Acre and Rondônia; [b]Institut de la Recherche Agronomique pour Development, Nkolbison; [c]Agency for Agricultural Research and Development; and [d]Instituto Nacional de Investigation Agraria.

slash-and-burn land use types were identified, including original forest (slight human impact), managed forest (selectively logged), recently cleared croplands, bush fallow (less than 5 years following clearing), open-canopy tree fallow (5 to 12 years), secondary forest (18 to 25 years), pasture, *Imperata* grassland, young agroforest or improved fallow (often experimental) and mature agroforest or tree plantation. Whenever possible, "best-bets" were included within the land-use chronosequences, but in many cases it was necessary to collect data from experimental stations, development projects, or other farms. Knowledge of original forest condition, establishment date, and management history was prerequisite for inclusion of "best-bets" within this study.

C. Carbon Pool Measurement

Aboveground carbon was measured for trees, understorey, and surface litter (necromass). Tree diameter measurements were used for estimating tree biomass. Diameter at breast height (DBH) was measured by callipers or diameter tapes and recorded for all trees with diameters greater than 2.5 cm within five quadrates of 4 m x 25 m (Figure 1 A). The positions of the quadrates were assigned by entering well within the individual land use and randomly selecting a direction of the longitudinal axis of the quadrat, and then randomly selecting a new direction in which to place the next quadrat (Figure 1 B). Quadrates were not allowed to "cross-over" one another or to fall outside their intended land use. In Indonesia, part of the data were collected in conjunction with an integrated survey of carbon stocks, biodiversity and greenhouse gas emissions with a sample area of 40×5 m^2. Tree buttressing was corrected by measuring the diameter above the buttress. For trees branching below breast height, the diameter of all branches was measured separately. Only trees with more than one half of their diameter falling within the quadrates were recorded. Tree biomass was estimated with the allometric equation based on tree diameter of Brown et al. (1989) for moist tropical forests: tree biomass (kg tree^{-1}) = 38.4908 - 11.7883*D + 1.1926 * D^2 (adj R^2=0.78). The biomass of fallen dead logs was measured within the quadrates based on volume (length x cross-sectional area) while assuming a density of 0.4 g cm^2. Tree biomass was converted to C by a factor of 0.45.

Understorey biomass, excluding trees with DBH > 2.5 cm, was collected from two 1 m x 1 m subquadrates positioned randomly along the central axis of each 100 m^2 quadrat (Figure 1 C). All vegetation occurring within the borders of the quadrat was cut at ground level and collected. Surface litter, including rotting logs and charcoal, was collected within a 50 cm x 50 cm frame centrally placed within each subquadrat (Figure 1 C). Samples were weighed, subsampled, oven dried at 65° C to constant weight and corrected for moisture content. Live vegetation was assumed to contain 45% C on a dry weight basis and surface litter was ground and analyzed for total organic carbon (Anderson and Ingram, 1993).

A soil and root sample was recovered from an area 20 cm x 20 cm within each subquadrat. The original guidelines recommended excavation to a depth of 40 cm, at 20 cm intervals, but some cooperators chose shallower (15 cm in Indonesia) or greater depths (50 cm in Brazil). Care was taken to recover as much roots as possible during the excavation except in Indonesia where root biomass was not measured. Also during excavation, bulk density measurements were taken at 10 and 30 cm by use of 100 cm^3 rings. All soil and roots from the hole were placed in bags and transported to the laboratory. A subsample was taken for total C analysis. The remaining sample was dispersed in water and passed through a 2 mm sieve; roots were collected from the sieve and washed in water without distinguishing live and dead roots. Roots were oven dried at 65° C to constant weight, weighed, ground and ashed. Ash-corrected dry weight was assumed to contain 0.45% C.

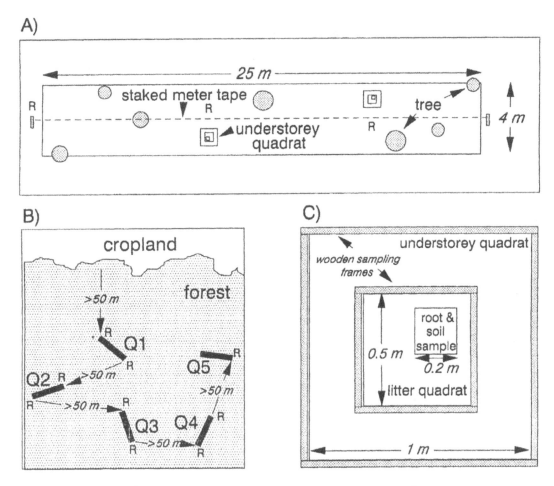

Figure 1. Carbon stock estimation of woody biomass based on five 100 m² quadrates (A) randomly positioned within land uses (B) with understorey, litter and soil measurements collected within sub-quadrates (C).

D. Data Compilation

Additional information on each land use included its chronosequence, benchmark area, geographic coordinates, duration, interval since forest clearing (chronosequential age), soil sampling depth and bulk density. The carbon stock measurements were tree, understorey, surface litter (necromass), root and soil C as t C ha⁻¹. Combined variables were generated including total system C, total above- and below-ground C, relative C stocks with respect to initial forest and proportion of aboveground and soil C. The data were imported into a computer software program, sorted by land use and summary statistics generated. Chronosequences were sorted into types (omitting initial forest C) and carbon sequestration calculated by linear regression. In compiling total system carbon from individual C

Table 2. Chronosequential age, total system C and proportion of aboveground carbon in tropical forests and lands converted by slash-and-burn

Land use	n	Age in sequence (yr)		Total system C (t C ha⁻¹)		Aboveground:total C	
Original forest	10	n/aᵃ		305	(23)	0.72	(0.04)
Managed forest	9	n/a		181	(18)	0.73	(0.05)
Burned and cropped	18	2.1	(0.3)ᵇ	52	(7)	0.23	(0.07)
Bush fallow	17	4.6	(0.2)	85	(9)	0.22	(0.06)
Tree fallow	8	9.4	(0.3)	136	(16)	0.48	(0.08)
Secondary forest	8	19.4	(2.2)	219	(18)	0.61	(0.03)
Pasture	9	10.0	(1.2)	48	(11)	0.20	(0.06)
Imperata grassland	8	13.0	(2.0)	47	(6)	0.05	(0.01)
Young agroforest	10	5.0	(0.7)	65	(10)	0.28	(0.04)
Mature agroforest	19	23.1	(1.6)	130	(11)	0.58	(0.04)

ᵃNot applicable, sequences begin at forest clearing; ᵇdata in parentheses denote standard errors.

pools, no distinction was given to the depth of soil sampling, rather soil values were entered as provided. When soil data were compared between land uses and forest types, however, only data reported for 0 to 15 and 0 to 20 cm were considered requiring that 31 of 116 cases be omitted from analysis.

III. Carbon Stocks in Slash-and-Burn and Land Use Alternatives

Estimates of carbon stocks residing in woody biomass, understorey, surface necromass, roots and soil were collected for 116 sites within 9 benchmark locations. Of these sites, 30 were located in Rondônia and Acre, Brazil, 35 in the forest zone of Cameroon, 30 in Sumatra, Indonesia and 21 in the Peruvian Amazon. Current land uses accounted for 85 observations, with 33% of these placed into a sequence of natural fallow succession, 9% agroforests, 6% pastures and the remainder either forests or cropland (Table 2). "Best-bet" candidates accounted for 31 observations with 68% classified as agroforests, 16% improved fallows, 13% improved pastures and 3% improved managed forests. Because some "best-bet" alternatives in one benchmark site closely resemble current practices at others, no distinction was drawn between them when total system carbon estimates are compared between land uses across benchmark sites.

Carbon stock estimates were grouped by land use and the duration since forest clearance, total system carbon and the proportion of carbon residing aboveground calculated (Table 2). No age was calculated for original or managed (disturbed) forests because the sequence was assumed to begin with forest clearing. When compared by pairwise Tukey t-tests, five classes of carbon stocks emerged. Carbon stocks were significantly greater in original than in secondary ($p = 0.003$) and managed forests ($p < 0.001$). Managed forests, mature agroforests and tree fallows did not significantly differ, nor did tree fallows and bush fallows, or bush fallows from young agroforests, croplands, pastures and *Imperata* grasslands. Carbon dynamics may be inferred from the proportion of carbon residing aboveground (Table 2). Forests and mature agroforests contain greater than 50% of C stocks aboveground, while croplands, grasslands, recovering fallows and establishing agroforests contain less than half.

The forest system C of the original forests (Table 3) contained approximately 200 t aboveground C ha⁻¹ (data not presented) in Cameroon and the Amazon. This estimate appears large compared to

Table 3. Total system carbon in original and managed forest systems

| Forest zone | Total system C | | Comments on management |
	Original	Managed	
	——— t C ha⁻¹ ———		
Amazonian	256 (22)[a]	197 (9)	Selectively logged by settlers
Dipterocarp	433 (n/a)[b]	143 (17)	Logging concessions and agroforests
Guineo-Congolian	308 (26)	179 (14)	Mature cacao agroforests[c]
Overall	305 (23)	166 (11)	

[a]Standard errors in parentheses; [b]based upon single observation; [c]trees are selectively felled and cacao understorey established.

140 t C ha⁻¹ (310 t DM ha⁻¹) reported by FAO (1997) for the same areas estimated by forest inventory methods. However, when the estimates obtained by forest inventory are corrected for trees less than 10 cm DBH (an additional 5%, data not presented), understorey (an additional 3%), and litter layer (an additional 10%, see Figure 2), the value is adjusted to 165 t C ha⁻¹. The original quadratic allometric equation of Brown et al. (1989) for estimating tree biomass that was employed in this study also gives relatively higher values than the more recently derived power function of FAO (1997). Use of the new equation with data from our study gives tree biomass estimates 70 to 95% that of the former equation, bringing the aboveground estimates within the range of those reported above by forest inventory methods. The single estimate for aboveground C in primary Dipterocarp forests of Indonesia in our study, however, is extremely high compared to the 126 to 182 t C ha⁻¹ reported by FAO (1997) for Malaysia. Given the relatively good agreement between aboveground C estimates using our fairly rapid methodology with the estimates from forest inventories, we recommend this method for measuring the C stocks of the vegetation in slash-and-burn areas. It must be noted that the specific allometric equation employed to convert tree dimensions to biomass affects results depending upon the size distribution of trees.

The impact of forest management, other than slash-and-burn, is presented in Table 3. The Amazonian forests were selectively logged by settlers, often with land title (Brazil). The forests in Sumatra were either in the process of logging by large concession (Jambi) or were extensively logged in the past (Lampung). Forest management was heterogeneous in Cameroon with active forest concessions in Mbalmayo and extending toward Ebolowa. Another feature of forest management in Cameroon is the cacao (*Theobroma cacao* L.) forests, where trees are selectively felled and cacao planted as an understorey. This land use is considered as mature agroforests in other tables and figures. Overall, forest management reduced original forest carbon stocks by 46% with greatest losses observed from mechanized logging operations.

It is from these disturbed or logged forests that most slash-and-burn clearing is occuring, the logging practice itself reducing the carbon stocks by more than half. Cropping or pastures further reduce C stocks to 18% and 15%, respectively, that of the orginal forest, or 29% of the disturbed forest.

Soil organic carbon stocks in the 0–15 and 0–20 cm soil layers in different land use categories and forest zones are presented in Table 4. The average of 43 t C ha⁻¹ in the top 15–20 cm of soil in the forest ecosystems (Table 4) is lower than the range of 46 to 69 t C ha⁻¹ reported by Detwiler (1986),

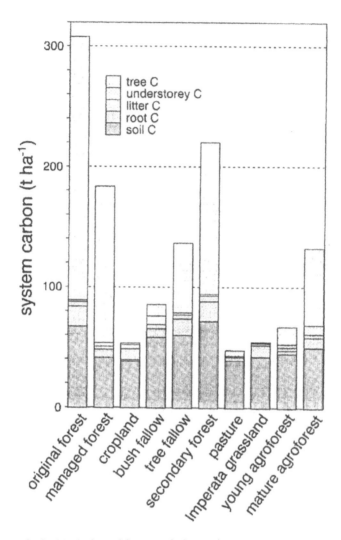

Figure 2. Carbon stocks in 10 slash-and-burn and alternative land uses.

assuming that 45% of the carbon in a 1 m profile is found in the top 20 cm (Moraes et al., 1995). The value is within the range found by Moraes et al. (1995) for undisturbed forests in the Amazon Basin of Brazil. The extremely low values of 31 t C ha^{-1} for the Amazon forests of our study cannot be explained, especially since the forest values are lower than the other land use practices.

Overall, soil C stocks were greatest in forests and agroforests and less in crops, bush fallows and grasslands. Significant differences in soil C with land use was not observed in Cameroon (Kotto-Same et al., 1997), probably reflecting the less intensive use of land in this benchmark area. The topsoil lost 13 to 39% of its carbon during the cropping phase in Cameroon and Indonesia, the larger losses in Indonesia perhaps reflecting the more intensive land use in the colonization areas compared to the traditional slash-and-burn systems in Cameroon. Detwiler (1986) reported averages losses of 40% of the soil carbon in the top 40 cm with cropping. Grasslands, including pastures, lost 21% of the carbon in the topsoil, similar to the 20% loss in pastures reported by Detwiler (1986).

Table 4. Soil organic carbon in different slash-and-burn land uses and forest zones (0–15 or –20 cm[a])

Land use	Forest zone			Mean	Tukey t-test (P)
	Amazonian	Dipterocarp	Guineo-Congolian		
		——— t soil C ha⁻¹ ———			
Forests[b]	30.9	48.1	42.8	42.7	0.03
Crops and bush[c]	38.8	29.5	37.2	35.3	n.s.
Agroforests[d]	22.5	46.7	43.2	39.8	0.02
Grassland[e]	29.6	36.8	–	33.2	n.s.
Mean	31.2	40.5	41.1	38.4	0.01
Tukey t-test (P)[f]	0.01	0.04	n.s.	0.07	

[a]Soil sampling depths are consistent within forest zone; [b]forests include original and secondary forests and tree fallows; [c]crops and bush include all burned and cropped lands and bush fallows; [d]agroforests include all young and mature agroforests; [e]grasslands include all pastures and *Imperata* grasslands; [f]probabilities assigned through Tukey Highest Significant Difference t-test.

The allocation of carbon between woody biomass, understorey, litter, roots and soil is presented in Figure 2. Soil C represents 13% of the forest system carbon and increases to 68% in the cropping systems. A large proportion of system carbon occurs within woody biomass in forests, tree fallows and agroforests and is nearly absent in croplands, pastures and *Imperata* grasslands. Croplands contain higher amounts of litter than do other land uses, which may be largely attributed to fallen and partly combusted woody residues. *Imperata* grasslands contain a large proportion of root biomass. Tree roots are likely to be seriously underestimated by our methods, particularly deeper, structural roots. The shoot:root ratios we obtained were much higher than those reported by Sanford and Cuevas (1996) in fewer but more intensively studied sites.

IV. Carbon Dynamics within Tropical Forest Land Uses

The most rapid loss of system carbon results from felling and burning original and managed forests, with an average loss of 120 t C ha⁻¹ until the end of cropping or 58.3 t C ha⁻¹ yr⁻¹ for a period of 2.1 years (Table 2, Figure 3). Difficulties were encountered in quantifying carbon loss from slash-and-burn during a single year beginning with forest disturbance. On one hand, farmers do not always clear forests during a given year, on the other, burning continues over several years as remnant trees are felled and burned along with remaining woody litter. Pasture establishment in the Brazilian Amazon is often based upon a two stage burning strategy where forests are cut and burned, seeded while land returns to fallow for 2 to 4 years, then cut and burned again resulting in near-complete stands of fire resistant grasses. In general, 80% of the system carbon is lost during the clearing and cropping phase.

Land use following the cropping phase can result in increased carbon losses or carbon sequestration. Following crop abandonement, carbon sequestration in the vegetation and soils of natural fallow succession was 7.9 t C ha⁻¹ yr⁻¹ (Figure 4) and was least in Amazonian bush fallow and greatest in the tree fallows and secondary forests in Cameroon (Table 5). Carbon sequestration rates of various fallow types could not be calculated for all forest zones, owing to the scarcity of older

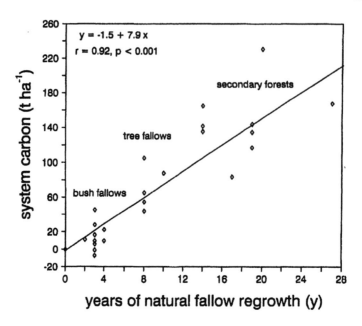

Figure 3. Carbon loss due to forest conversion and recovery in natural fallows and agroforestry systems: all observations plotted.

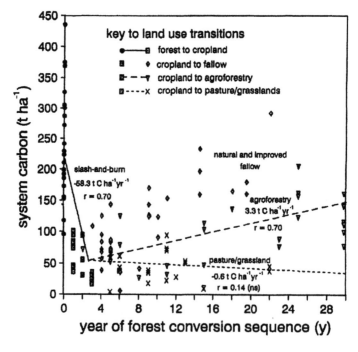

Figure 4. Total system carbon measured along slash-and-burn land use chronosequences in Brazil, Cameroon, Indonesia, and Peru.

Table 5. Carbon sequestration in natural fallows and secondary forests

Forest zone	Natural vegetation		
	Bush fallow	Tree fallow	Secondary forest
		$t\ C\ ha^{-1}\ y^{-1}$	
Amazonian	3.9 (1.0)[a]	–	6.2 (1.3)
Dipterocarp	–	–	6.2 (n/a)[b]
Guineo-Congolian	5.1 (2.6)	8.5 (1.3)	9.3 (0.9)
Overall	4.6 (1.6)	8.5 (1.3)	8.3 (0.8)

[a]Standard errors in parentheses; [b]based upon single observation.

fallows in many locations. These carbon sequestration rates are less in burn fallows and fall within the range of 2 to 9 t C ha^{-1} y^{-1} in vegetation and litter reported by Szott et al. (1994). Houghton (1997) reported lower values of 2 to 5 t C ha^{-1} y^{-1}. Our study found the recovery of carbon in the soil to be 0.2 t C ha^{-1} y^{-1} at a maximim. The relatively high C recovery rates measured in the benchmark sites may help substantiate the claim that regrowth rates in the tropics may be higher than previously estimated (Houghton, 1995) or could be simply be a function of using allometric equations for estimating biomass that were developed for mature forests rather than young secondary forests. Planted tree fallows do not seem to increase carbon sequestration rates above that of the natural fallow (Szott et al., 1994) but might do so in cases where seed banks of trees have been depleted, such as the case of pastures (Uhl et al., 1988). Agroforests sequestered carbon at a lower rate than do natural fallows (3.3 t C ha^{-1} yr^{-1}, r = 0.70). Small amounts of carbon loss continued during the cropland to pasture or *Imperata* grassland sequence (Figure 3).

V. "Bad Bets" and "Best Bets"

Traditional slash-and-burn agriculture as practiced by sparse indigenous populations in large forests and nonmarket settings does not result in large-scale or long-term environmental damage but rather may be viewed as another source of patch dynamics (Kotto-Same et al., 1997). But this form of slash-and-burn was not encountered during our investigations, its closest resemblance being the fallow regeneration practiced by indigenous tribes with poor access to markets in Mbalmayo and Ebolowa, Cameroon. Yet even these farmers have created a rapidly retreating forest margin and reduced fallow intervals. Forest destruction is massive when settlers with poor knowledge of forest resource utilization (Fujisaka et al., 1996) migrate to humid forests and prepare lands for permanent utilization. From the environmental standpoint, attempts to mine system nutrients for annual food crop production or pasture establishment without regard to the need for longer-term input management are poor land use alternatives. The invasion of *Imperata* species into transmigration areas of Indonesia and degraded pastures of Brazil may be regarded as symptomatic expression of poor land management. Another example of poor management of humid forest resources is exhaustive logging where standards for acceptable extraction are progressively lowered until only ill-formed or very small trees remain, as was noted in Jambi and Pucallpa. But the documentation of carbon loss from deforestation "horror stories" was not the intent of our studies, rather we sought to identify opportunities to reduce the deleterious environmental impacts within the tropical forests undergoing rapid transition.

A. Current Forestry and Agroforestry

Owing to the large proportion of system C residing in trees of the humid forests (Figure 2), the obvious opportunities to conserve or sequester carbon involve the protection or re-establishment of trees. In many cases, these involve current management practices, as with the jungle rubber (*Hevea brasiliensis* Muell.-Arg.) agroforestry in Indonesia which contains up to 147 C ha[-1] after 30 years or forest cacao management in Cameroon with 179 t C ha[-1], these systems have remained over many decades. The economic viability of those pursuits dictate the future of carbon storage within those land uses. In the same manner, tropical forest plantations have potential for significant and renewable carbon storage. Although an *Acacia mangium* plantation in Sumatra was projected to contain 200 t C ha[-1], over half that of the original forest 20 years later, these production systems cannot meet the short-term needs of growing and migrating populations of smallholders under current policy environments (Tomich et al., 1997). The economies of many lesser developed nations are dependent upon forest resource utilization, and expectations by more developed nations that tropical forests be set aside for future ecotourists are not realistic, nor consistent with the historical utilization of forests in developed countries.

The potential of more simple agroforests, as compared to complex agroforests such as jungle rubber, to sequester carbon is less well known because of the relatively young age of many of these agroforestry systems. The young (less than 10 y) simple agroforests measured in this study had an average of 65 t C ha[-1], accumlating about 1.5 t C ha[-1] y[-1]. Dixon (1995) reports carbon stocks of 12 to 228 t C ha[-1] for a variety of agroforestry, including silvopastoral, systems in the tropics, the large range depending on the complexity and age of the systems.

B. Deflection from Deforestation

Sanchez (1990) described the socio-economic effects likely to result from widespread tropical deforestation where landless poor with inadequate land management skills and inputs migrate to tropical forests, and speculated that each hectare placed into permanent, profitable agriculture has potential to offset deforestation of 5 to 10 additional hectares by shifting cultivators. He also noted the important role of national policies in deforestation, particularly through government-sponsored settlement programs. The concept of deflection from deforestation through agricultural intensification remains unproven and a provocative issue because of the likelihood that landless poor would "stream" toward new land and agricultural opportunities (Tomich and van Noordwijk, 1996). Harwood (1996) describes the need for capital and technical inputs required within agricultural intensification, suggesting that these remain a barrier to agricultural change unless production problems within current practices arise. Agroforestry systems offer potential for carbon sequestration (Figures 2 and 3) but do these systems deflect from deforestation? In Cameroon, farmers adopted cacao agroforestry but continue to clear forest margins to cultivate annual crops for food and market (Kotto-Same, 1997). Farmers in Theobroma who participated in the carbon dynamics studies were converting degraded pastures to fruit orchards but continuing to clear additional forest for pasture. It may be naive to consider that a new land management will completely substitute for another, rather it will be adapted within the slash-and-burn farming system, or that new agricultural opportunities are sufficient to eliminate the perceived need to clear forests. Another approach to deflection is the requirement that settlers in Theobroma and Pedro Peixoto (Brazil) clear only 50% of their land and manage the remaining forest. This principle is seldom effective because migrant farmers often lack forest management skills, land derives value through clearance and the 50% clearance requirement is seldom enforced (Fujisaka et al., 1996). An important criterion in comparing the environmental benefits of various "best bets" is their potential for deflection; however, criteria for determining deflection must be better defined.

VI. Extrapolation of System C Measurements

The carbon estimates of different land uses in sites of active deforestation have value within themselves in comparing the environmental benefits of different land use alternatives, but may also be used in geographic applications and to validate output of simulation models. Kotto-Same et al. (1997) describe an analysis using the Cameroon data set in which system carbon estimates were substituted into mapping units which had changed from tropical forest to other land uses between 1973 and 1988. This approach indicated that 202 Mt C were lost from 20,000 km^{-2} deforested during 15 years. Similar data extrapolation is being undertaken with data sets from the other benchmark areas.

The carbon estimates may also validate environmental simulations generated by the CENTURY Model (see Parton et al., 1987, 1994). Sitompul et al. (1996) simulated the effects of contrasting land management following forest disturbance in an Ultisol at Lampung including natural fallow regrowth, conversion to sugarcane (*Saccharum* cv. L.) and current rice cultivation practices. Natural fallow rapidly reestablished soil organic C at 50 t C ha^{-1}, but all crop management strategies resulted in continued SOC loss over 25 years, with most loss occurring from the lighter fractions (SOM1C and SOM2C). Sugarcane management resulted in 11% less loss in SOC than did current farmer practices.

Output from CENTURY version 4.0 (Metherell et al., 1993) is presented in Figure 5 where two alternative land managements are compared to an expected scenario of slash-and-burn followed by bush fallow and continuous maize (*Zea mays* L.) cultivation. When basic conservation measures, such as protection of economically important trees, felling along contours to reduce soil erosion and organic inputs to soils are practiced (Kotto-Same et al., 1997), total system C losses are reduced by 24% over 53 years. In contrast, establishment of rubber agroforestry increases the 53 year average total system C by 99 t C ha^{-1} over current practices. The agroforestry scenario may be overestimated as many rubber plantations must be replanted in less than 53 years.

VII. Carbon Dynamics and National Agendas

One of the benefits of this research is to raise the profile of carbon dynamics studies within national research structures. From its inception, the Alternatives to Slash-and-Burn Programme was based upon partnership between international and national research institutes (Brady, 1996) with critical decisions on study locations, land use selection and "best-bet" alternatives as the responsibilities of the lead national agency at each benchmark area (Table 1). In Brazil and Cameroon, lead agencies had little past experience in carbon dynamics measurement and were assisted by national universities and international partners, particularly the Tropical Soil Biology and Fertility Programme and the International Centre for Research in Agroforestry. In Peru, national partners included teams with strong background in forestry, but required assistance in selecting representative land uses within the slash-and-burn chronosequence. Indonesia initiated its carbon studies by sponsoring a workshop attended by representatives from several national institutions and universities where experiences were shared and research plans formalized (Murdiyarso et al., 1994).

Mid-way through these investigations, all national teams were conducting autonomous studies at sites and within land uses of their choice. With this autonomy, some sacrifices in cross-site comparability were experienced as some teams chose to modify soil sampling methods. This sacrifice seems small by comparison to the strengthening of global change agendas within the national structures of countries most affected by tropical deforestation.

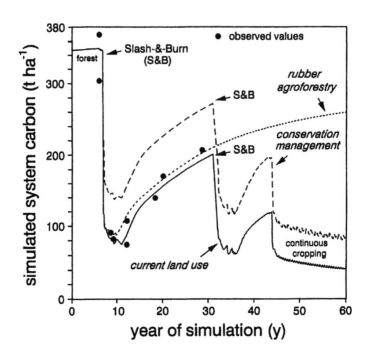

Figure 5. A CENTURY Model simulation of current and alternative land management practices based upon data from Ebolowa, Cameroon.

VIII. Assessment and Complementarity of Rapid Carbon Estimation Procedures

One criticism open to the relatively rapid carbon estimation methods developed for this study is the relative importance of tree biomass (Figure 2) compared to the area and number of trees measured (Figure 1). The allometric equations used for estimating tree biomass were developed from mature forests while many of the systems studied were young secondary forests and agroforests. There is perhaps a need to develop equations for young secondary forests and specific equations for many of the common agroforestry tree species. FAO (1997) recommend power equations for dry, moist and wet forests that may be better suited for calculating the biomass of smaller diameter trees.

No predetermined method of main quadrat alignment (randomization) was found acceptable as this would often lead the team outside of its intended land use or into poorly representative terrain. This led to our selection of field randomization of quadrat direction by "blind-spin-and-toss". In practicing this form of randomization, one must be careful not to attempt targeting larger trees as these often account for a large proportion of woody forest biomass, particularly after previous tosses have established quadrates that contain few trees. In some cases, teams were subject to additional pressures by onlookers who criticized them for failing to measure the largest trees in their forests. In one case, a local chieftain in Ebolowa, Cameroon, insisted that the largest trees in his forest be measured by the research team, and this was done but without entering these data into further analysis. Another source of field confusion was disagreement between local experts and/or farmers in distinguishing primary, managed and mature forests. In general, land managers with clear land tenure (Brazil and Indonesia) appeared to have more concrete opinions on land history, occasionally consulting farm records. In areas with traditional land tenure (Cameroon) or where land operations were previously interrupted by civil unrest (Peru), land histories tended to be more controversial and greater importance was placed upon the insight of the research team.

The carbon dynamics studies reported in this chapter are but one component of the international research agenda of the Alternatives to Slash-and-Burn Programme. Other activities include monitoring greenhouse gases (GHG), and the assessment of above- and below-ground biodiversity. In many cases, carbon dynamics (CD) studies preceded these other investigations and many of the chronosequences and land uses identified by the carbon dynamics team were later characterized in terms of GHG and biodiversity. Some difficulties were encountered in complete interchange of sites between research themes. Aboveground biodiversity (AGB) was assessed by combining plant species with indices of plant functional attributes (Kenyatta, 1997). The minimum sample area deemed necessary for this approach was larger than that for C estimation and many transitional land uses could not be measured. Below-ground biodiversity (BGB) teams focused their attention on the population sizes and diversity with five functional groups of soil organisms; macrofauna, nematodes, rhizobia, mycorrhizae and decomposer communities. These procedures were too time and material intensive to measure all of the sites characterized by the carbon team. Nonetheless, plans are underway to combine the data obtained by the four research groups (CD, GHG, AGB and BGB) into the fuller context of environmental impacts of current slash-and-burn and alternative land uses, and to compare these impacts to the potential of various land uses to alleviate poverty and to deflect from future deforestation.

Acknowledgements

The authors appreciatively acknowledge the financial assistance of the Global Environmental Facility to the Alternatives to Slash-and-Burn Programme permitting our field investigations and to The Rockefeller Foundation for providing financial assistance to the lead author.

References

Anderson, J.M. and J.S.I. Ingram. 1993. *Tropical Soil Biology and Fertility: A Handbook of Methods.* CAB International, Wallingford, UK. 221 pp.

Brady, N. C. 1996. Alternatives to slash-and-burn: a global perspective. *Agricult. Ecosyst. Environ.* 58:3-11.

Brown, S., A.J.R. Gillespie and A.E. Lugo. 1989. Biomass estimation methods for tropical forests with applications to forest inventory data. *Forest Science* 35:881-902.

Detwiler, R.P. 1986. Land use change and the global carbon cycle: the role of tropical soils. *Biogeochemistry* 2:67-93.

Detwiler, R.P. and C.A.S. Hall. 1988. Tropical forests and the global carbon cycle. *Science* 239:42-47.

Dixon, R.K. 1995. Agroforestry systems: sources or sinks of greenhouse gases? *Agroforestry Systems* 31:99-116.

Food and Agriculture Organization of the United Nations (FAO). 1997. Estimating Biomass and Biomass Change of Tropical Forests: A Primer. FAO Forestry Paper 134. FAO, Rome. 55 pp.

Fujisaka, S., W. Bell, N. Thomas, L. Hurtado and E. Crawford. 1996. Slash-and-burn agriculture conversion to pasture and deforestation in two Brazilian Amazon colonies. *Agric. Ecosyst. Environ.* 59:115-130.

Hall, D.O. 1989. Carbon flows in the biosphere: present and future. *J. Geographical Society* 146:175-181.

Harwood, R.R. 1996. Development pathways toward sustainable systems following slash-and-burn. *Agric. Ecosyst. Envir.* 58:75-86.

Houghton, R.A. 1997. Terrestrial carbon storage: Global lessons from Amazonian research. *Ciencia e Cultura* 49:58-72.

Houghton, R.A. 1995. Land-use change and the carbon cycle. *Global Change Biol.* 1:275-287.

Houghton, R.A., R.D. Boone, J.R. Fruci, J.E. Hobbie, J.M. Melillo, C.A. Palm, B.J. Peterson, G.R. Shaver, G.M. Woodwell, B. Moore, D.L.Skole and N. Myers. 1987. The flux of carbon from terrestrial ecosystems to the atmosphere in 1980 due to changes in land use: geographic distribution of the global flux. *Tellus* 39B:122-139.

Kenyatta, C. (ed.). 1997. Alternatives to Slash-and-Burn: Report of the 6th Annual Review Meeting. 17-27 August 1997, Bogor, Indonesia. ICRAF, Nairobi. 124 pp.

Kotto-Same, J., P.L. Woomer, A. Moukam and L. Zapfac. 1997. Carbon dynamics in slash-and-burn agriculture and land use alternatives of the humid forest zone in Cameroon. *Agric. Ecosyst. Environ.* 65:245-256.

Metherell, A.K., L.A. Harding, C.V. Cole and W.J. Parton. 1993. CENTURY: Soil Organic Matter Model Environment. Colorado State University, Fort Collins, CO, U.S.

Moraes, J.L., C.C. Cerri, J.M. Melillo, D. Kicklighter, C. Neill, D.L. Skole and P.A. Steudler. 1995. Soil carbon stocks of the Brazilian Amazon Basin. *Soil Sci. Soc. Am. J.* 59:244-247.

Murdiyarso, D., K. Hairiah and M. van Noordwijk (eds.). 1994. *Modelling and Measuring Soil Organic Matter Dynamics and Greenhouse Gas Emissions after Forest Conversion.* ASB-Indonesia Report No. 1. Bogor, Indonesia. 118 pp.

Parton, W.J., P.L. Woomer and A. Martin. 1994. Modelling soil organic matter dynamics and plant productivity in tropical ecosystems. p 171-189. In: P.L. Woomer and M.J. Swift (eds.), *The Biological Management of Tropical Soil Fertility.* John Wiley & Sons, Chichester, U.K.

Parton, W.J., D.S. Schimel, C.V. Cole and D.S. Ojima. 1987. Analysis of factors controlling soil organic matter levels on Great Plains grasslands. *Soil Sci. Soc. Am. J.* 51:1173-1179.

Post, W.M., T. Peng, W.R. Emmanuel, A.W. King, V.H. Dale and D.L. De Angelis. 1990. The global carbon cycle. *American Scientist* 78:310-26.

Sanchez, P.A. 1990. Deforestation reduction initiative: an imperative for world sustainability in the Twenty-first Century. p. 375-382. In: A.F. Bouwman (ed.), *Soils and the Greenhouse Effect.* John Wiley & Sons, Chichester, U.K.

Sanchez, P.A. 1987. Soil productivity and sustainability in agroforestry systems. p. 205-210. In: H.A. Steppler and P.K.R. Nair (eds.), Agroforestry: a decade of development. International Council for Research in Agroforestry, Nairobi.

Sanchez, P.A., P. Woomer and C.A. Palm. 1994. Agroforestry approaches for rehabilitating degraded lands after tropical deforestation. *JIRCAS International Symposium Series* (Japan) 1:108-119.

Sanford, R.L. and E. Cuevas. 1996. Root growth and rhizosphere interactions in tropical forests. p. 268-300. In: SS. Mulkey, R.L. Chazdon and A.P. Smith (eds.), *Tropical Forest Plant Eco-physiology.* Chapman and Hall, New York.

Sitompul, S.M., K. Hairiah, M. van Noordwijk and P.L. Woomer. 1996. Organic matter dynamics after conversion of forests to food crops or sugarcane: predictions of the CENTURY model. *AGRIVITA* 19:198-206.

Szott, L.T., C.A. Palm and C.B. Davey. 1994. Biomass and litter accumulation under managed and natural tropical fallows. *For. Ecol. Manage.* 67:177-190.

Tomich, T.P. and M. van Noordwijk. 1996. What drives deforestation in Sumatra? p. 120-149. In: B. Rerkasem (ed.), Montane Mainland Southeast Asia in Transition. Chang Mai University, Chang Mai, Thailand.

Tomich, T.P., J. Kuusipalo, K. Mena and N. Byron. 1997. Imperata economics and policy. *Agroforestry Syst.* 36:233-61.

Uhl, C., R. Buschbacher and E.A.S. Serrao. 1988. Abandoned pastures in eastern Amazonia. I. Patterns of plant succession. *J. Ecol.* 76: 663-681.

Woodwell, G.M., R.H. Whittaker, W.A. Reiners, G.E. Likens, C.C. Delwiche and D.B. Botkin. 1978. The biota and the world carbon budget. *Science* 199:141-146.

CHAPTER 6

Crop Residue and Fertilizer Management to Improve Soil Organic Carbon Content, Soil Quality and Productivity in the Desert Margins of West Africa

A. Bationo, S.P. Wani, C.L. Bielders, P.L.G. Vlek and A.U. Mokwunye

I. Introduction

The term crop "residue," with the connotation of something left over that nobody wants, gives a false impression of its value as crop residues perform various functions in the mixed agricultural systems of the desert margins of West Africa. Residues are used as animal feed, as fuel and as construction material. When left in the fields after harvest, crop residues play important roles in nutrient cycling, erosion control and the maintenance of soil physical and chemical properties (Powell and Unger, 1997).

Several studies have concluded that low soil fertility is often the major constraint for production of both food grains and natural vegetation, with phosphorus as the most limiting nutrient in the desert margins of West Africa (Breman and de Wit, 1983; Van Keulen and Breman, 1990; Bationo et al., 1991). Although application of mineral fertilizers increases yields in arable farming, mineral fertilizer alone cannot sustain crop yields in the long run. When mineral fertilizers are combined with organic amendments such as crop residues, productive and sustainable production systems can be obtained (Pieri, 1989; Pichot et al., 1981; Sedogo, 1981).

The Desert Margins zones of West Africa are characterized by low and erratic rainfall, high soil and air temperatures, and soils with poor native fertility. Despite these harsh conditions, rainfed agriculture represents the main livelihood for more than 40 million people in the world's poorest countries (Baidu-Forson and Bationo, 1997). To make matters worse, per capita food production has generally declined in the last two decades.

Demographic and economic factors have contributed to changes in the ecological balance that support extensive crop production. Soil erosion throughout the zone has caused a dramatic loss of the topsoil. The land degradation is aggravated by the activities of people and livestock. Deforestation, overgrazing and burning of crop residue (CR) and fallow vegetation reduce the soil cover and expose topsoil to wind and water erosion.

Presently the increasing population pressure has reduced the availability of land and resulted in reduced ratio of length of fallows to cropping years to the point that shifting cultivation is losing its effectiveness and as a result, soil fertility is decreasing in many areas (Bekunda et al., 1997). In those

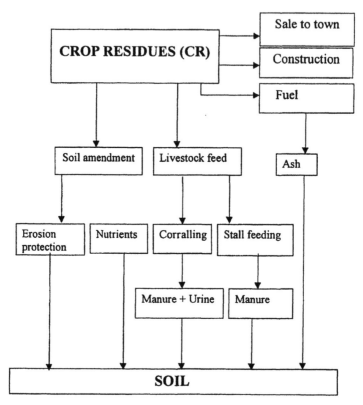

Figure 1. The competing uses of crop residues in the Desert Margins of West Africa.

areas where the length of fallowing has been reduced to less than 15 to 20 years, the traditional low-input farming systems have become unsustainable, low in productivity, and destructive to the environment. Plant nutrient balances are negative for many cropping systems with offtake greater than input, indicating that farmers are mining their soil (Smaling et al., 1997).

Figure 1 gives a schematic representation of the different uses of CR in the Desert Margins of West Africa. Traditionally, many farmers burn whatever is left of their CR once their needs for fuel, animal feed or housing and fencing material have been fulfilled (Bationo et al., 1995).

Although CR can play a vital role in the struggle against land degradation in the Sahel, quantities found on-farm are mostly inadequate for effective mulching. For the Sahelian zone, field experiments showed that the optimum level of CR to be applied to the soil as mulch must be as high as 2000 kg ha^{-1} (Rebafka et al., 1994). However, McIntire and Fussel (1986) reported that on farmers' fields average millet (*Pennisetum glaucum* (L.) R. Br. grain yields are 236 kg ha^{-1} and mean residue yields barely reached 1300 kg ha^{-1}. In a study to determine the availability of CR at farm level in Niger, Baidu-Forson (1995) reported that in the Diantandou district (rainfall 450 mm) an average 1198 kg ha^{-1} of stover was produced at the end of the cropping season, but at the onset of the rains the following year only 250 kg ha^{-1} will still be available on-farm for mulching (Table 1). On average, only 21 to 39% of the mean stover production at harvest time is still available as mulch at the onset of the subsequent rainy season. The results of Powell and Bayer (1985) indicate that at least 50% of the disappearance of millet stover on-farm can be attributed to livestock grazing. This includes virtually 100% of the millet leaves.

Table 1. Summary statistics on millet stover on farms, in three districts of Niger

Year/month of sampling	District	Quantities of stover on farms (kg ha⁻¹)				Number of farms sampled[a]
		Minimum	Maximum	Mean	S.D.	
1991/April						
	Hamdallaye	130	550	283	104	40
	Diantandou	110	790	325	149	52
	Kirtachi	430	1190	755	209	43
1991/October						
	Hamdallaye	570	2730	1090	385	51
	Diantandou	340	3790	1198	554	61
	Kirtachi	700	3560	1618	636	54
1992/March						
	Hamdallaye	90	600	231	96	46
	Diantandou	90	690	251	85	58
	Kirtachi	250	1520	631	249	54

[a]The differences in number of farms sampled reflect changes that resulted from the cultivation of new farms or the abandon of old farms by sample households. (From Baidu-Forson, 1995.)

In the Desert Margin areas, cattle grazing is likely to be responsible for most of the disappearance of CR (Bationo et al., 1995). The increase in crop biomass at farm level is crucial in order to satisfy all the demand for CR. Bationo and Mokwunye (1991) showed that the use of even modest quantities of fertilizers increased stover yields under on-farm conditions. Despite the competing uses (Figure 1) the increased production led to significantly more mulch in subsequent rainy season. The data in Figure 2 clearly indicate that application of small quantities of phosphorus fertilizers and adoption of good cropping systems such as rotation of cowpea with pearl millet will result in a drastic increase in total crop biomass production in the Desert Margins of West Africa.

In this chapter, after a brief presentation of the crop production environment, we will discuss the effect of CR application as mulch on soil conservation, crop productivity, and on soil chemical and biological properties.

II. The Crop Production Environment

A. Climate

Figure 3 gives the geographical distribution of the Desert Margins in West Africa. The rainfall varies from 200 to 600 mm, the ratio of annual rainfall to evapotranspiration from 0.05 to 0.5 PET, while the length of the growing period varies from 75 to 125 days. Rainfall shows a steep north-south gradient associated with the interseasonal movement of the Intertropical Convergence Zone, north and south of the Equator. Rainfall increases at a rate of approximately 1 mm km⁻¹ from north to south. Although in absolute terms rainfall is low only in the northern half of the Desert Margins, the high interannual variability associated with poor distribution of rainfall in space and time during the growing season constitute major limitations for agricultural production (Sivakumar et al., 1993).

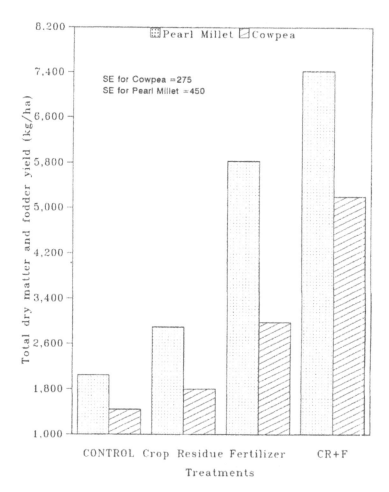

Figure 2. Long-term crop residue management at Sadoré, Niger, in 1996.

Air temperature often exceeds 40°C. Potential evapotranspiration in Niamey amounts to 2294 mm per year and exceeds rainfall in all months except August (Sivakumar et al., 1993). Rainfall is frequently preceded by strong winds with speeds seldom exceeding 15 m s^{-1} (Michels et al., 1995a). Rainfall intensity can exceed 150 mm h^{-1}.

B. Soils

Entisols and Alfisols occupy most of the landscape for rainfed cropping in the Desert Margins of West Africa. Entisols are mainly composed of quartz sand, with a low water capacity and nutrient content. Alfisols have a clay accumulation horizon with low capacity to store nutrients.

The data in Table 2 show physical and chemical properties of selected soils from the West African semiarid tropics. The soils have low organic carbon and total nitrogen content because of low biomass production and a high rate of decomposition. One striking feature of these soils is their inherent low

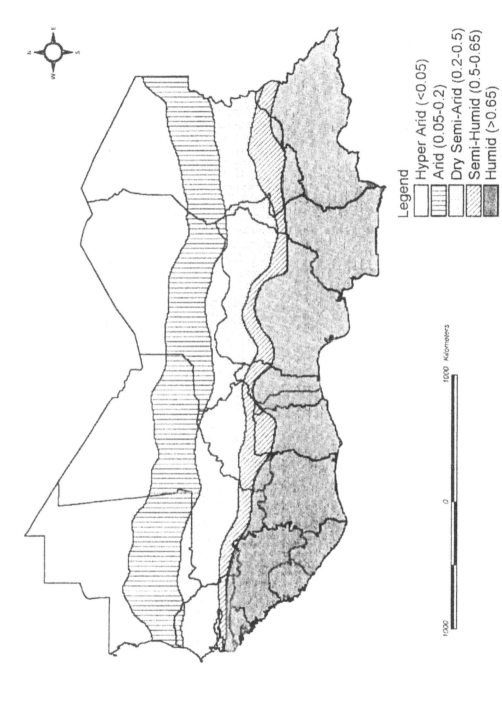

Figure 3. Agro-climate zones of West African Semiarid Tropics (WASAT).

Table 2. Means and ranges of selected physical and chemical properties of West African semiarid soils

Parameter	Range	Mean
pH-H$_2$O (2:1 water:soil)	3.95–7.6	6.17
pH-KCl (2:1 water:soil)	3.41–7.0	5.05
Clay (%)	0.7–13	3.9
Sand (%)	71–99	88
Organic matter (%)	0.14–5.07	1.4
Total nitrogen (mg kg^{-1})	31–226	446
Exchangeable bases (cmol kg^{-1})		
Ca	0.15–16.45	2.16
Mg	0.02–2.16	0.59
K	0.03–1.13	0.20
Na	0.01–0.09	0.04
Exchangeable Al (cmol kg^{-1})	0.02–5.6	0.24
Effective Cation Exchange Capacity (cmol kg^{-1})	0.54–19.2	3.43
Base saturation (%)	36–99	88
Al saturation (%)	0–46	3
Total phosphorus (mg ha^{-1})	25–941	136
Available phosphorus (mg ha^{-1})	1–83	8
Maximum P sorbed (mg ha^{-1})	27–406	109

(From Bationo et al., 1997.)

fertility, which is expressed in low levels of organic carbon (generally less than 0.3%), total and available phosphorus and nitrogen and effective cation exchange capacity (ECEC). The accumulation of organic matter is highly related to total rainfall. The low ECEC is attributed to low organic matter, the low clay content and the kaolinitic mineralogy of the soils. Bationo and Mokwunye (1991), found that the ECEC is more related to the organic matter content than to the clay content, indicating that a decrease in organic matter will decrease the ECEC and then the nutrient-holding capacity of those soils. De Ridder and van Keulen (1990) reported that a difference of 1 g kg^{-1} in organic carbon results in a difference of 4.3 mmol kg^{-1} in ECEC. Apart from low P stocks, the low-activity nature of these soils results in a relatively low capacity to fix added P (Bationo et al. 1995).

The dune soils in the Sahel have very high hydraulic conductivity (100 to 150 mm h^{-1}) and therefore a rapid internal drainage. In the southern half of the Desert Margins zone or in wetter years, this can lead to significant losses of plant nutrients by leaching. Despite their intrinsically high permeability, water losses by runoff can be significant. Such losses of water by runoff can significantly reduce plant available water at critical times as well as lead to severe erosion, often in the form of gullies. These high runoff coefficients result from the sensitivity of the soils to surface crust formation and high intensity rainfall events characteristic of the Sahel. Infiltration rates generally drop below 10 to 15 mm h^{-1} in the presence of surface crusts (Casenave and Valentin, 1989). The soils have bulk densities ranging from 1.4 to 1.7 g cm^{-3} corresponding to a porosity of 36 to 43%, and the plant available moisture can be as low as 7%.

III. Effect of CR Application as Mulch for Soil Conservation

On the sandy soils of the desert margins zone, wind erosion, soil surface crusting, and water erosion, as well as soil compaction constitute the main physical soil degradation processes. In recent years, the effect of CR application on these processes and the associated changes in soil physical conditions have received considerable attention.

As a result of the rapid population growth during the last decades, one has observed a rapid increase in cultivated area, the expansion of agriculture into more marginal areas, and an increased pressure on land through overgrazing and deforestation. Consequently, the area of soil left bare and therefore directly exposed to wind has increased considerably. The effect of these changes in land use on wind and water erosion are aggravated by the sandy nature of the Sahelian soils, which are frequently poorly aggregated, offering little resistance to the erosive forces of the wind. Although wind erosion affects crop growth directly through sandblasting of leaves and the burial of seedlings under sand deposits (Michels et al., 1993; Michels et al., 1995a), it is the process of soil loss that has received most attention.

Extensive research has been carried out on the role of crop residue, mainly millet stover, for alleviating wind erosion-induced soil degradation. On the basis of surface topography measurements, Geiger et al. (1992) have reported relative elevation differences of 150 mm (10 mm of soil = ~ 160 t ha^{-1}) between bare and mulched millet plots after 5 years. The mulch consisted of an annual application of 4 t ha^{-1} of millet stover. The observed elevation difference was attributed to soil loss from the bare soil as well as considerable accumulation of wind blown sediment on the CR plots. Buerkert et al. (1996a) measured absolute soil losses of 190 t ha^{-1} in one year on bare plots, as opposed to soil deposition of 270 t ha^{-1} on plots with a 2 t ha^{-1} millet stover mulch. Michels et al. (1995b) have reported a nearly 50% reduction in wind-transported soil over mulched plots as compared to bare plots. Although the rates of soil loss on bare plots may in some cases have resulted from the combined action of wind and water erosion, the reported rates of soil deposition can only result from the accumulation of wind blown sediment because of the field topographical conditions under which the experiments were carried out. Hence, these observations clearly indicate that the application of 2 t ha^{-1} of millet stover is effective not only for combating wind erosion, but can efficiently trap wind blown sediment eroded in adjacent fields.

Based on agronomic data, the application of 2 t ha^{-1} of CR has been recommended as the optimal rate for the Sahelian zone (Rebafka et al., 1994). For wind erosion control, Michels et al. (1995a) reported that the application of 0.5 t ha^{-1} was ineffective but the application of 2 t ha^{-1} of millet stover residues reduced soil fluxes at 0.1 m height by an average of 47% during the rainy season (Figure 4). Sterk and Stein (1997) observed soil losses of 46 t ha^{-1} in just four dust storms for an application rate of 0.8 t ha^{-1} of millet stover. In general, the efficiency of millet stover for trapping sediment seems to decrease with increasing wind speed (Sterk, 1997; Michels et al., 1995b). Sterk (1997) observed that at wind speeds exceeding 11 m s^{-1}, the application of 1 t ha^{-1} of millet stover may actually enhance wind erosion by increasing near-surface air turbulence. The importance of this phenomenon for higher rates of mulch application has not been determined. However, based on the existing data, the minimum required application rate of millet stover for wind erosion control should probably exceed 1.5 t ha^{-1}.

In an on-farm trial at Banizoumbou, Niger, Bielders (unpublished data) tested the relative effectiveness of broadcast millet stover versus strip applied residue, both at a rate of 2 t ha^{-1}. They observed that broadcast residue was equally effective as strip-applied residue in trapping wind blown sediment. Furthermore, the presence of millet stover considerably modified the soil surface conditions. Whereas almost 50% of the bare plots were occupied by low-permeability erosion crusts, both strip and broadcast residue plots were dominated by structural crusts and aeolian deposits. Mulching reduced the presence of erosion crusts to less than 5% (Bielders, unpublished). In this particular experiment, millet yields on bare plots dropped from 328 kg ha^{-1} in the first year to 78 kg ha^{-1} in the second year. Millet grain yields with the application of CR remained stable at about 500 kg ha^{-1} (Figure 5). Although other processes, including nutrient mining, may have contributed to the rapid decline in soil productivity on bare plots, soil loss by wind erosion and the resulting effects on soil physical and chemical properties is likely to have been a major cause for the observed degradation.

Because the application of CR affects both the soil physical properties and nutrient cycling, it has been notably difficult to establish the relative importance of these components on final crop productivity. In an attempt to separate the physical from the chemical benefits of mulching, Buerkert et al. (1996b) compared a 2 t ha^{-1} millet stover mulch with an equivalent amount of plastic material of similar size and shape as millet stover. Although millet yields in the first year of the experiment were not significantly different in control plots and plots with the polyethylene plastic tubes mulch, grain and dry matter production were significantly lower on bare plots in the second year. Because of the inert nature of polyethylene plastic tubes, the difference between bare and polyethylene plastic tubes-mulched plots may be attributed to a large extent to the protection of the soil surface against wind erosion. The polyethylene plastic tubes-mulched plots gained approximately 5 mm of soil over a one-year period as opposed to a net loss of almost 19 mm of soil on the bare plots. This has resulted in a much lower rate of soil acidification on polyethylene plastic tubes-mulched plots, and the maintenance of higher levels of exchangeable bases. The data in Table 3 gives the relative nutrient balance of a bare compared to a protected (plastic mulch) soil at two depths. In unprotected soil, up to 7.1 kg of available P and 180 kg ha^{-1} of organic carbon are lost from the soil profile within one year.

Compared to bare soil plots, Buerkert et al. (1996b) reported significantly lower values of penetration resistance on polyethylene plastic tubes-mulched plots in the top 5 cm of the soil. These values were not significantly different between polyethylene plastic tubes and millet stover mulched plots, which suggests that the favorable effect of mulching on surface resistance to penetration may be strongly influenced by the maintenance of a loose layer of deposited aeolian sand rather than to biological activity.

Besides improvements in the soil surface penetration resistance, Buerkert et al. (1996b) observed that the presence of a 2 t ha^{-1} millet stover mulch reduced maximum daily soil surface temperatures by 8 °C at 1 cm depth and 4 °C at 5 cm depth. These lower temperatures afforded by mulching can prove decisive for crop emergence and establishment, considering that temperatures of 50 °C are commonly observed in the first centimeter of the soil at the start of the growing season. Because at a rate of 2 t ha^{-1} soil surface coverage by millet stover is less than 10%, Buerkert et al. (1996b) hypothesized that the control of surface temperature by mulching resulted mostly from better moisture storage in and near the stems and modifications in the latent heat fluxes rather than from a shading effect per se.

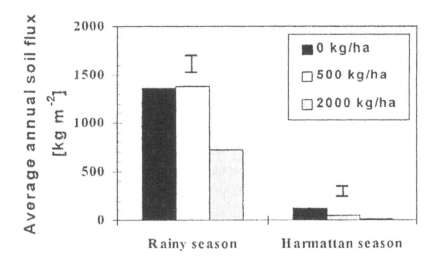

Figure 4. Average annual soil flux (1991–1992) at 0.1 m height as affected by mulching rate during the rainy season and the dry "Harmattan" season; error bar = LSD. (From Michels, 1994.)

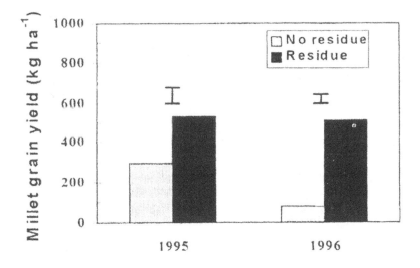

Figure 5. Effect of strip application of millet stover on millet grain yields at Banizoumbou, Niger; error bar = SED.

Table 3. Nutrient balance after two years of erosion and deposition

	Relative losses of a bare compared to a protected (plastic mulch) soil at two depths[a]			
	—————————————kg ha⁻¹ a⁻¹—————————————			
Soil properties	0 to 0.1 m depth	0.1 to 0.2 m depth	Sum	
N (Total)	-14.0	3.0	-11.0	±9.8
P (Bray)	-4.2	-2.8	-7.1	±1.4
P (H$_2$O)	-0.08	-0.06	-0.14	±0.07
K	-3.2	-2.2	-5.4	±2.2
Ca	-18.7	-12.7	-31.4	±9.4
Mg	-1.7	-1.3	-3.0	±2.3
C (Organic)	-115	-65	-180	±8.3
Al (Total)	10.9	11.4	22.3	±7.5
	—————————————Units—————————————			
pH (KCl)	-0.08	-0.05	-0.06	±0.02

[a]Average total dry matter yield of millet in 1992 and 1993 was 1690 kg ha⁻¹ on the bare soil and 1980 kg ha⁻¹ on the protected soil.
(From Buerkert et al., 1996b.)

The application of CR constitutes an effective means for the rehabilitation of degraded land and is frequently practiced by farmers, albeit on a limited scale (Lamers and Feil, 1995). The role of biological activity, and more specifically termites, in this regeneration process stimulated by the presence of organic residue is considerable (Léonard and Rajot, 1998; Mando, 1997), although modification in soil surface conditions as a result of sediment deposition should not be underestimated, as discussed above. Mando et al. (1996) reported that the sole activity of termites doubled cumulative infiltration compared to mulched plots where termite activity had been prevented through the use of an insecticide.

Field rainfall simulation studies on severely crusted soils revealed that the application of 2 t ha⁻¹ of millet stover reduced runoff from 21 to 2% for a 40 mm rainfall event (Bielders, unpublished). Cissé and Vachaud (1987) reported that an increase in soil organic matter from 0.15 to 0.30% had no measurable effect on water infiltration in sandy soils in the Sahelian zone of Senegal. The impact of CR on infiltration into degraded soils therefore first and foremost reflects changes in surface crust conditions. In very sandy soils (>90% sand), the effect of CR additions on the intrinsic hydrodynamic properties of the soil are minimal. Bationo et al. (1993) reported that the addition of CR as mulch had no effect on soil water-holding capacity.

Table 4. Physical and chemical properties of Sadoré soils

Depth (cm)	Sand (%)	Clay (%)	pH	Organic carbon (%)	Available P Bray P1 (mg ha^{-1})	Total bases (Cmol kg^{-1})	Effective CEC (Cmol kg^{-1})
0–18	93.8	2.9	5.2	0.17	3.2	0.9	1
18–30	91	5.2	5	0.12	1.4	0.8	0.8
30–51	87.4	7	4.9	0.14	1	0.9	1.1
51–71	85.2	10.5	4.3	0.12	0.9	0.8	1.1
71–99	84	10.1	4.1	0.1	0.9	0.6	1
99–155	87.5	8.6	4.3	0.08	0.9	0.6	0.7
155–280	86.4	8.3	4.5	0.07	0.9	0.7	0.9

(Adapted from West et al., 1984.)

IV. Effect of CR Application on Soil Productivity

Long-term experiments are a practical means to address the difficult issues associated with quantitative assessment of sustainability in agriculture. In summarizing the results from long-term soil fertility management in Africa, Pieri (1986) concluded that soil fertility in intensive arable farming in the West African semiarid tropics can only be maintained through efficient cycling of organic materials in combination with chemical fertilizers and with rotation of N_2-fixing leguminous species.

In 1983, we initiated at ICRISAT Sahelian Center a long-term crop residue management trial in order to study the sustainability in crop production in the Sahel. The trial consisted of four treatments consisting of a (1) Control; (2) CR residue applied as mulch at 2 t ha^{-1}; (3) Fertilizer application at 13 kg P ha^{-1} and 30 kg N ha^{-1} and (4) Fertilizer applied at 13 kg P ha^{-1} and 30 kg N ha^{-1} with CR applied as mulch at 4 t ha^{-1}. A randomized complete block design with four replication was used in that trial. The data in Table 4 shows the low level of nutrients in the soil profile at the beginning of the experiment. Two years after the beginning of the experiment, while pearl millet grain yield was only 160 kg ha^{-1} in the control plot, the yields were 770 and 1030 kg ha^{-1} respectively for the application of 2 t ha^{-1} of CR and application of fertilizers. In the same year combination of CR and fertilizers resulted in pearl millet of 1940 kg ha^{-1} (Table 5).

The data in Figure 6 for the same long-term experiment over a period of 14 years clearly supports the importance of combining both CR and mineral fertilizers in the Desert Margins of West Africa. In 1993 as an example, the control plot yielded 1238 kg ha^{-1} of total biomass while 3209 and 3482 kg ha^{-1} were obtained respectively from the CR and fertilizers plots. The combination of fertilizers and CR produced a yield of 6393 kg ha^{-1}. This clearly emphasizes the additive effects of CR and fertilizer.

In 1996, surface soil samples were analyzed for pH organic carbon, Bray P1, exchangeable bases and exchangeable acidity, and total N and P at flowering time of pearl millet from the long-term CR management trial described above. At harvest, a stepwise regression was used to determine the effect of the different soil parameters on total dry matter obtained from each plot. Organic carbon alone was

Table 5. Effect of crop residue and fertilizer on pearl millet grain and stover yields at Sadoré, Niger

Treatment	Grain yield (kg ha⁻¹)				Stover yield (kg ha⁻¹)			
	1983	1984	1985	1986	1983	1984	1985	1986
Control	280	215	160	75	n/a	900	1100	1030
Crop residue	400	370	770	745	n/a	1175	2950	2880
Fertilizer	1040	460	1030	815	n/a	1175	3540	3420
Crop residue + fertilizer	1210	390	1940	1530	n/a	1300	6650	5690
$LSD_{0.05}$	260	210	180	200		520	650	870

n/a = Not available (From Bationo et al., 1993.)

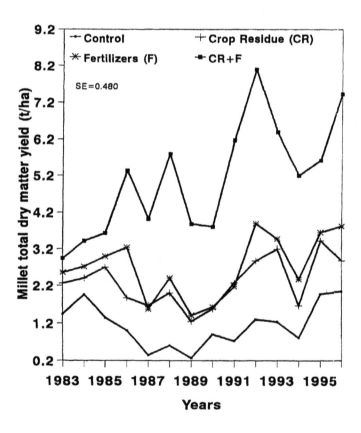

Figure 6. Pearl millet total dry matter yield as affected by different management practices over years.

Table 6. Effect of fertilizer and crop residue mulch in farmer-managed trials on pearl millet grain yields in two villages in Western Niger

Treatments	Banizoumbou Grain yield (kg ha⁻¹)		Karabedji Grain yield (kg ha⁻¹)	
	1996	1997	1996	1997
Control	344	248	550	310
Single superphosphate (SSP)	600	396	703	655
SSP + calcium ammonium nitrate (CAN)	708	521	800	787
SSP + CAN + CR	902	682	1264	1548
Standard error	±36	±28	±42	±37

Three farmers were involved in the trials at each site; single superphosphate (SSP) applied at 13 kg P ha⁻¹; calcium ammonium nitrate (CAN) applied at 30 kg N ha⁻¹; crop residue (CR) (millet stover) applied at 2 t ha⁻¹. (From Bationo, unpublished data.)

responsible for 60% of the yield variation and when available P (Bray P1) was included in the model, the R^2 increased to 76%. The model developed follows:

$$\text{Total dry matter} = -1124 + 149 \text{ Bray P1} + 16771 \times \text{organic carbon. } R^2 = 0.76$$

with total dry matter in kg ha⁻¹, Bray P1 in mg kg⁻¹, and organic carbon in %. This simple model indicates that soil organic carbon is more responsible for pearl millet total dry matter variation than available P. This shows that higher yields can only be maintained when organic amendments are combined with mineral fertilizers. This model also indicates that an increase of soil organic carbon of 0.1 unit will increase total dry matter by 1677 kg ha⁻¹ whereas increase of available P by 1 unit will increase the yield by 149 kg ha⁻¹.

The beneficial effect of CR application should not be attributed only to changes in soil chemical properties but also to physical effects such protection of millet seedlings from sandstorms (Michels et al., 1995c), decreased soil temperature and resistance to penetration (Buerkert et al., 1996b), and increased water infiltration through reduced crust function and increased termite activity (Mando et al. 1996). As indicated in the data in Table 6, the application of CR in farmer-managed trials resulted in substantial yield increase at farm level. Michels et al. (1998) reported that the addition of 2 t ha⁻¹ of CR on several degraded soil increased root length density by a factor of 3 in the top 10 cm, and increased rooting depth from 130 to > 180 cm.

The effect of CR is more substantial in the Desert Margins than in the other agro-ecological zones of West Africa. The data in Tables 7 and 8 for Tara in the Sudanian zone and other sites including the Guinean zone indicate that CR application does not result in the drastic yields increase observed in the Desert Margin areas of West Africa.

For the Sudanian zone, Sedogo (1981) reported negative effects of CR on crop yield and attributed this to N immobilization in the absence of mineral fertilizer application. The contradictory results of

Table 7. Effect of mineral fertilizers and crop residue on pearl millet grain and total dry matter yields at Tara, Niger

Treatment	1990		1991		1992	
	Grain yield	Total dry matter	Grain yield	Total dry matter	Grain yield	Total dry matter
				—kg ha⁻¹		
Control	235	1694	347	2154	102	1193
Crop residue	751	2633	490	2695	140	1673
Fertilizer (NP)	1287	4608	1006	4794	625	3269
Crop residue + fertilizer (NP)	1369	4603	922	4823	616	3427
Crop residue + fertilizer (NPK)	1426	4909	1009	5304	778	3937
Standard error	95	358	101	270	56	238
CV (%)	19	19	27	13	25	18

Crop residue applied as 4 t ha⁻¹ millet stover; N applied as urea at the rate of 30 kg N ha⁻¹; P applied as single superphosphate at the rate of 13 kg P ha⁻¹; K applied as Kcl at the rate of 24 kg K ha⁻¹. (From Bationo et al., 1995.)

Table 8. Effect of different treatments on maize grain yield for different experimental sites of Togo

Treatments	Maize grain yield (kg ha⁻¹)		
	Davié	Amoutchou	Kaboli
Control	1842	2129	1414
Crop residue (CR)	1830	2610	1646
NPK	3513	3971	4014
CR + NPK	3039	3547	3348
Standard error	209	335	402
C.V. (%)	16	21	26

CR = Total residue from the plot applied each year; N = 90 kg N ha⁻¹; P = 26.4 kg P ha⁻¹; K = 60 kg K ha⁻¹. (From IFDC, unpublished data.)

yield reductions and increases due to CR indicate that more research needs to be conducted to investigate the mechanisms of different types of CR effects on crop growth in different zones and to monitor more closely the dynamics of organic matter in the different soil types. For example, the stem borers population (*Coniesta ignefusalis*) may build up and adversely affect crop yields, which is the main reason for burning crop residue by farmers.

Several researchers also reported phytotoxicity problems due to CR application or continuous cereal cultivation (Burgos-Leon, 1979; Pichot et al. 1981; Pieri, 1986). However, it remains unclear to what extent organic acids, lactones, phenols, alkaloids, and turpenoid compounds isolated from the crop residues and produced during decomposition are responsible for yield depressions after CR application (Elliot et al., 1979).

V. Effect of CR Application on Soil Chemical Properties

Table 9 presents the chemical analysis of millet grain and stover. Depending on the rainfall and microbial and termite activity, CR decomposition can be very rapid in the West African semiarid tropics. The decomposition of millet CR in the Sahelian zone at Sadoré and the Sudanian zone at Tara is reported in Table 10. Because of the high termite activity and microbial activities observed at Tara, 99% of CR applied as mulch was decomposed in a single rainy season compared to between 40 and 50% at Sadoré.

In 1996, soil samples were taken from the four replicates after 14 years in the long-term crop residue management trial described above. Sampling depths were 0–10 cm, 10–20 cm, 20–40 cm, 40–60 cm, and 60–100 cm. An adjacent soil under fallow was also sampled in order to study the changes in soil chemical properties as affected by the different management practices. The results are discussed in the following four sections.

A. Phosphorus Parameters

The distribution of available P with depth as a function of treatment is given in Figure 7. Addition of P fertilizers led to a very significant increase in the amount of available P in the topsoil layer (0–10 cm) where available P is about 17 mg P kg^{-1} while available P in the fallow and control plot was about 4 mg P kg^{-1}. As shown by Figure 6, there is almost no difference in total dry matter yield between the fertilizer and the crop residue treatment, and as a result of the importance in biomass production without any addition of P fertilizer, the CR treatment total phosphorus is much lower than in the control plot in the top layer (0–10 cm). The total P content decreased from 60 mg P kg^{-1} in the control plots to 47 mg P kg^{-1} in the CR plots (Figure 8). It is important to note that Kretzschmar et al. (1991) observed that CR increases P availability, probably through complexation of Fe and Al, and reduced P fixation. Bationo et al. (1995) observed a decrease in P sorption with addition of CR. Hafner et al. (1993a) showed that the effect of CR on P uptake by millet results from an increase in root length density in the top 20 cm, and increase in P adsorption per unit root area, which may be due to increased level of available P.

B. Organic Carbon and Total Nitrogen

As compared to fallow, the cultivation of the soils in the Desert Margin areas resulted in a decrease in organic carbon even when inorganic fertilizers are added. The addition of 4 t ha^{-1} of CR resulted

Table 9. Nitrogen, phosphorus, potasium, calcium and magnesium contents of pearl millet when no P is applied and when P is applied as single superphosphate at 13 kg ha^{-1} at Gobery, Niger, rainy season, 1987

	P applied (kg ha^{-1})	
	0	13
Grain yield kg ha^{-1}	471	12.64
Straw yield, kg ha^{-1}	2.248	4,938
N content in grain, %	1.72	1.73
N content in straw, %	0.71	0.80
N uptake by grain, kg N ha^{-1}	8.12	22.30
N uptake by straw, kg N ha^{-1}	15.35	37.22
P content in grain, %	0.18	1.73
P content in straw, %	0.03	0.80
P uptake by grain, kg P ha^{-1}	0.82	22.30
P uptake by straw, kg P ha^{-1}	0.62	37.22
K content in grain, %	0.48	0.44
K content in straw, %	1.77	1.40
K uptake by grain, kg K ha^{-1}	2.10	5.56
K uptake by straw, kg K ha^{-1}	41.44	73.48
Ca content in grain, %	0.14	0.03
Ca content in straw, %	0.09	0.39
Ca uptake by grain, kg Ca ha^{-1}	0.50	0.45
Ca uptake by straw, kg Ca ha^{-1}	4.33	22.58
Mg content in grain, %	0.15	0.13
Mg content in straw, %	0.25	0.21
Mg uptake by grain, kg Mg ha^{-1}	0.65	1.70
Mg uptake by straw, kg Mg ha^{-1}	5.54	10.42

(From Bationo and Mokwunye, 1991.)

Table 10. Decomposition of surface applied millet stover during the rainy season with different soil amendments at two locations, Tara and Sadoré, Niger, 1993

Treatment[a]	Crop residue applied at the beginning of season	Crop residue undecomposed at the end of season
	———————————t ha⁻¹———————————	
Tara (800 mm rainfall)		
Control	4.0	0.05
Crop residue	4.0	0.04
Fertilizer (N, P, K)	4.0	0.06
Crop residue + fertilizer	4.0	0.07
Sadoré (1) (400 mm rainfall)		
Control	4.0	2.3
Crop residue	4.0	2.3
Fertilizer (N, P, K)	4.0	2.5
Crop residue + fertilizer + micronutrients	4.0	2.5
Sadoré (2) (400 mm rainfall)		
No P	2.0	1.0
6.5 P ha⁻¹	2.0	1.1
No P	4.0	1.9
6.5 kg P ha⁻¹	4.0	1.9

[a]The same treatments were applied to plots for the previous three years at Tara, for the previous four years at Sadoré (1), and for the last two years at Sadoré (2); for Sadoré (2), all treatments received CR at the rate of 2 t ha⁻¹ in the previous year.

(From Bationo et al., 1995.)

in the maintenance of soil organic carbon levels. The organic carbon was maintained almost at the same level as in the fallow in the first 20 cm of the soil profile (Figure 9).

The principal source of nitrogen is accumulated in organic matter, hence the distribution of total N as function of depth and treatment was similar to that of organic carbon (Figure 10). In the top layer total nitrogen content was 297 mg kg⁻¹ with the application of 4 t of CR ha⁻¹ but decreased to 153 and 105 mg kg⁻¹ with the application of 2 t ha⁻¹ of CR and in the control plots, respectively.

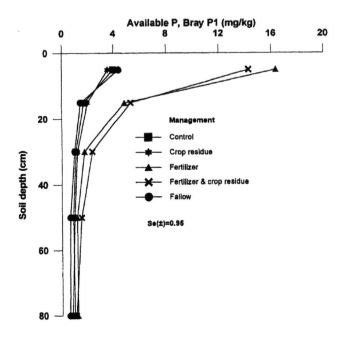

Figure 7. Effect of different management on available P throughout the soil profile at Sadoré, Niger during the 1996 rainy season.

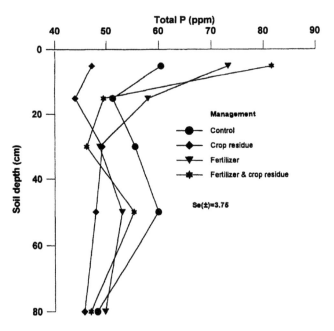

Figure 8. Effect of different management practices on total P throughout the soil profile at Sadoré, Niger during the 1996 rainy season.

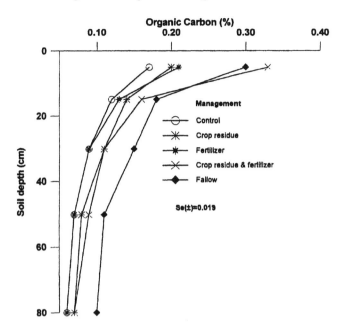

Figure 9. Effect of different management practices on soil organic matter content at Sadoré, Niger during the 1997 rainy season.

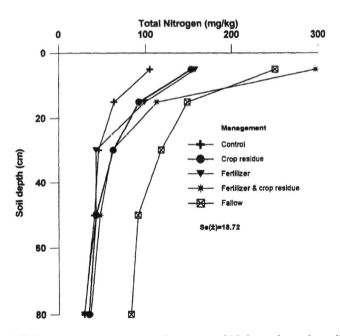

Figure 10. Effect of different management practices on total N throughout the soil profile at Sadoré, Niger during the 1996 rainy season.

C. Acidity Parameters

Continuous cultivation with and without inorganic amendments led to increased acidification (Figure 11). Whereas the topsoil layer pH was 4.82 with 4 t ha^{-1} of CR application in the fallow system, it decreased to 4.12 in the fertilizer treatment. The application of organic amendments led to the chelation of aluminum and a significant reduction in aluminum content in the exchange complex (Figure 12). Aluminum saturation was nil with the application of 4 t ha^{-1} of CR, but it increased to 25% with or without the use of mineral fertilizers in the top layer in the absence of CR.

D. Exchangeables Bases and Effective Cation Exchange Capacity (ECEC)

After 14 years of application of CR, a significant level of exchangeable bases (Figure 13) and ECEC (data not shown) is obtained in the surface soil. In millet, up to 90% of the Ca, Mg and K remains in the straw (Balasubramanian and Nnadi, 1980; Bationo et al., 1993). It is thus still unclear to what extent increases in soil exchangeable bases come from minerals that are carried in dust particles and trapped by mulch residues (Geiger et al, 1992), or from nutrient release from decomposing crop residues. Stahr et al. (1993) reported that up to 20 kg K ha^{-1} can be deposited in the form of mica and feldspar transported by Harmattan wind.

VI. Effect of CR on Soil Biological Properties

Soil is a habitat for a vast, complex and interactive community of soil organisms whose activities largely determine the chemical and physical properties of the soil. Soil microflora plays an important role in the maintenance of soil fertility because of their ability to carry out biochemical transformations and also due to their importance as a source and sink for mineral nutrients (Jenkinson and Ladd, 1981). Organic carbon, the source of energy for heterotrophic microorganisms, is the driving factor for all biological processes in the soil. The decomposition of plant and animal residues in soil constitutes a basic biological process that is brought about by successional population of microorganisms. In the Desert Margins of West Africa, where the soils are fragile and low in organic matter, organic carbon and soil biological processes play a key role in sustaining crop production. They control the release of plant nutrients and thus crop productivity. The following discussions are based on data collected in 1993 in the long-term CR management trials at the ICRISAT Sahelian Center described in the previous section.

A. Soil Respiration and Microbial Biomass C

Soil respiration is a direct measure of the undifferentiated biological activity in the soil. In general, the level of microorganisms in the soil is positively correlated with the level of organic matter (Waksman and Starkey, 1924). Immediately after incorporation into the soil, plant materials are subjected to the transformation and decomposition processes of the heterotrophic microflora and the population of bacteria, fungi, and actinomycetes is increased with application of plant residues and

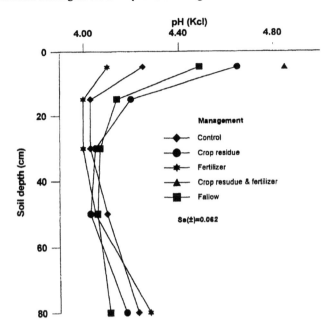

Figure 11. Effect of different management practices on soil pH throughout soil depth at Sadoré, Niger during the 1996 rainy season.

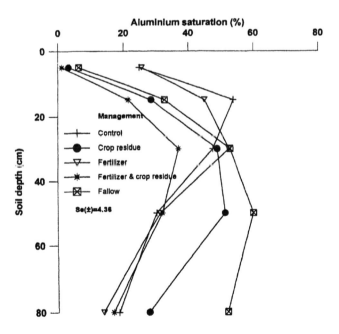

Figure 12. Effect of different management practices on aluminum saturation throughout the soil profile at Sadoré, Niger during the 1996 rainy season.

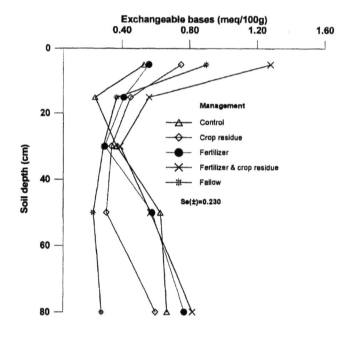

Figure 13. Effect of different management practices on exchangeable bases throughout the soil profile at Sadoré, Niger during the 1996 rainy season.

farm yard manure (FYM) (Gaur et al., 1971; Sidhu and Beri, 1986). In the 1993 season, analysis of surface soil samples showed a significantly higher soil respiration of 26.6 mg C kg^{-1} per ten days with the application of 4 t ha^{-1} of CR as compared to 15 mg C kg^{-1} per 10 days for the control plot at harvest time. Hafner et al. (1993b) showed that the application of CR in the same experiment increased the total number of bacteria and the N_2-fixing bacteria in the bulk soil and rhizosphere soil of pearl millet. Hafner et al. (1993a) also reported an increase of root-length density with CR application which led to an increase in total P uptake. Tien et al. (1979) showed for other crops that diazotrophic bacteria may produce phytohormones such as auxins or giberellins which stimulate lateral root development or root hair formation of most plants.

Maintenance of microbial community through residue management is a means for retaining organic matter and improving nutrient availability in rainfed farming systems. Soil microbial biomass responds more quickly than soil organic matter as a whole to changes in inputs of organic material (Powlson and Jenkinson, 1981). Microbial biomass contains labile fractions of organic C and N which are mineralized rapidly after the death of microbial cells. In the crop residue management trial at Sadoré, biomass C in the soil reached a maximum of 200 mg C kg^{-1} soil at 42 days after sowing (DAS) and decreased considerably at 108 DAS (48 mg C kg^{-1} soil). At 42 DAS, microbial biomass in the soil under pearl millet was significantly higher in treatments that received millet residues with and without fertilizer and in the treatment that received mineral fertilizer alone (190–230 mg C kg^{-1} soil) than was the biomass C in untreated control soil (160 mg C kg^{-1} soil).

Table 11. Effect of crop residue and mineral fertilizer application on mean mineral (NH_4+NO_3) N content and amount of net N mineralised in soil under pearl millet, Niamey, Niger, rainy season, 1993

	Control	Millet residue (CR)	Fertilizer (F)	CR + F	S.E. $\pm$
Mineral N content (mg N kg^{-1} soil)					
Pearl millet	3.50	4.50	3.75	4.90	0.18
Net N mineralized (mg N kg^{-1} soil 10 d^{-1})					
Pearl millet	0.83	1.02	0.96	1.92	0.21

Means of two samplings done at 42 and 102 DAS; soil samples were collected up to 20 cm depth. (Adapted from Wani et al., unpublished data.)

B. Mineral N Content, Biomass N and Net N Mineralization in Soil

Biological turnover of N through mineralization-immobilization leads to an interchange of inorganic forms of N with the organic N. Mineral N in soil is always in a dynamic state as it is used by plants and microorganisms for their growth, and is prone to losses through runoff, deep drainage, denitrification, and volatilization. The decay of organic residues in soil is accompanied by conversion of C and N into microbial tissue and part of the C is liberated as CO_2. As the C:N ratio is lowered below 20, and microbial tissues decompose with synthesis of new biomass, a portion of the immobilized N is released through net mineralization. The quality of organic residues, as reflected in the C:N ratio, is of primary importance in regulating the magnitude of the two opposing processes of mineralization and immobilization. Conversion of organic N to available mineral forms $(NH_4^+ + NO_3^-)$ through biochemical transformations is mediated by microorganisms and is influenced by those factors that affect microbial activity (temperature, moisture, pH, organic matter content and rate of residue application, its lignin content, etc.). A reasonable estimate is that, under conditions favourable for microbial activity, net mineralization of residues will commence after 4 to 8 weeks of decomposition (Stevenson, 1986).

Significantly more mineral N was found in cases of 4 t and 2 t ha^{-1} residue applied than that of the mineral N content of unamended control and fertilizer alone treatments (Table 11). Mean net N mineralization was increased 1.2 times due to application of 2 t ha^{-1} millet residues and 2.3 times with 4 t ha^{-1} residue. Crop residues can thus be used directly for improving soil productivity. Direct incorporation of crop residues in soil along with the appropriate management practices, or use as a mulching material on the soil surface, has also proven to be beneficial in improving soil physical properties and indirectly increasing crop yields in some cases. However, when low-N materials, such as cereal straw, were ploughed into soil, it was found that additional fertilizer N was needed to narrow the C:N ratio in order to avoid adverse effects on crop yields by immobilization of soil N (Bear, 1948; Ganry et al., 1978). In a lysimeter experiment using ^{15}N on a sandy soil in Senegal, the latter authors found that straw incorporation depressed the grain yield by 32%. This effect was mainly due to immobilization of fertilizer N and was alleviated by application of additional N.

Biomass N is an important labile fraction of total soil N which is a readily mineralized upon death of biomass. Microbial biomass N was higher ($p<0.01$) at 108 DAS (7.85 mg kg^{-1} soil) than at 42 DAS (5.2 mg kg^{-1} soil). At 42 DAS, biomass N in soil was significantly higher ($p < 0.05$) by a factor of 1.6

Table 12. Effect of management practices on pearl millet and cowpea root colonization by vesicular-arbuscular mycorrhizae (VAM), 58 DAS, Niamey, rainy season, 1993

Treatment	% root colonization by VAM		
	Cowpea	Pearl millet	Mean
Control	44	39	41.5
Crop residue (4 t ha^{-1})	52	57	54.5
Fertilizer (30:13:25 kg N: P: K)	49	42	45.5
Crop residue + fertilizer	63	42	52.5
S.E. ±	6.0		5.3
Mean	52	45	
S.E.	2.0		

(Adapted from Wani et al., unpublished data.)

to 3.0 for the millet residue alone, fertilizer alone, and residue plus fertilizer treatments as compared to the control treatment. Similar trends for biomass N in these treatments were also observed at 108 DAS. A maximum mean biomass N (9.7 mg kg^{-1} soil) was observed in the case of residue plus fertilizer treatment, followed by fertilizer alone (8.1 mg kg^{-1} soil), residue alone (6.4 mg kg^{-1} soil), and the control (4.3 mg kg^{-1} soil). The C:N ratio of the microbial biomass at 108 DAS was 5.7 and only 3.7 at 42 DAS. At 108 DAS, biomass C:N ratio was higher (9.9) in case of control treatment as compared to the C:N ratios of other treatments (5.2 to 6.25) which received residues or fertilizer alone and in combination. Narrower C:N ratio of microbial biomass indicates the sufficiency of mineral N for microbial biomass growth and also quick release upon mineralization of its tissue. Beneficial effects of residue alone and residue with fertilizer application on mineral N dynamics in soil were very much evident in these trials. These results of mineral N content, net N mineralization in soil, and biomass N content reveal the complexities of N cycling in sandy soils amended with residues with and without fertilizers and its effect on crop yields.

C. Vesicular Arbuscular Mycorrhizal (VAM) Development

In soil, biological activity is enhanced as soon as organic matter is applied and favorable moisture and temperature regimes are there. In addition to the spurt in biological activity which is directly associated with the nutrients cycling, activities of other organisms such as *rhizobia* nodulating legumes and mycorrhizae infecting the crop roots are also influenced. Pearl millet residue application

(4 t ha^{-1}) to sandy soils in Niger increased the root colonization by VAM in pearl millet (57%) as compared to the root colonization in control treatment at 58 DAS (39%). Application of fertilizer alone showed a marginal increase in root colonization by VAM over the control treatment. Application of fertilizer N and P along with residues reduced VAM root colonization to 42% and the residue alone to 57% (Table 12). Another fertility management trial at some sites with pearl millet showed that application of 13 kg P ha^{-1} alone reduced VAM colonization to 41% as compared to the control treatment (50%). Application of pearl millet residues (4 t ha^{-1}) significantly increased VAM colonization of millet roots (66%). Application of 10 t ha^{-1} manure along with crop residues, phosphorus, lime and micronutrient application resulted in a reduction to 37% in VAM colonization. Manure alone did not affect VAM colonization (54%). These results indicate ways to enhance the beneficial effects of residues by manipulating the dose of fertilizer application depending on the requirement of a particular crop. Application of P at doses higher than the crop requirement will suppress root colonization by VAM (Hayman, 1970) as the internal concentration of phosphorus in roots rather than its external concentration in soil controls root colonization by VAM fungi (Menge et al., 1978).

VII. Conclusion

In the Desert Margins of West Africa, the use of CR as mulch plays important roles in nutrient cycling, erosion control, and improvement of soil chemical, physical and biological properties.

Overcoming soil organic matter decline is a prerequisite for a sustainable food production in the region. Continuous cultivation leads to a drastic reduction of soil organic carbon followed by soil acidification and a decrease in exchangeable bases. The addition of 4 t ha^{-1} as mulch maintains soil organic carbon at the same level of an adjacent fallow and drastically increased the efficiency of mineral fertilizers.

It is important to note that the return of CR produced on site corresponds to recycling of a portion of the plant nutrients but will not redress the problem of nutrient depletion. Although CR can play a vital role in the struggle against land degradation in the Desert Margins quantities found on-farm are mostly inadequate for effective mulching. There is need to increase crop biomass at farm level with the use of mineral fertilizers in order to satisfy the multiple uses of CR.

The beneficial effect of CR obtained in the Desert Margins was not observed in the other agro-ecological zones. There is need for more basic research in order to understand this phenomenon by monitoring closely the dynamics of soil organic matter following CR application in those agro-ecological zones.

Future research using labeling techniques should look at the nutrient turnover from organic amendments for a better understanding of nutrient cycling processes and determining the timing of application of organic amendments with the nutrient demands of crops. The issue of socio-economic constraints should be addressed in future research using the participatory approach by involving farmers, researchers and other stakeholders.

References

Baidu-Forson, J. 1995. Determinants of the availability of adequate millet stover for mulching in the Sahel. *J. of Sustainable Agric.* 5:101-116.

Baidu-Forson, J. and A. Bationo. 1997. Saving soil and water in Africa's arid zone. p. 120-124. In: Forum for Applied Research and Public Policy. Winter.

Balasubramanian, V. and C.A. Nnadi. 1980. Crop residue management and soil productivity in savanna areas of Nigeria. p. 106-120. In: Organic Recycling in Africa. FAO Soils Bulletin 43. Food and Agriculture Organization of the United Nations, Rome, Italy.

Bationo, A., T.O. Williams, and A.U. Mokwunye. 1997. Soil fertility management for sustainable agricultural production in semi-arid West Africa. p. 349-367. In: Buzeneh et al. (ed.), Technology Options for Sustainable Agriculture in Sub-Saharan Africa. Publication of the Semi-Arid Food Grain Research and Development Agency (SAFGRAD) of the Scientific, Technical and Research Commission of OAU. Ouagadougou, Burkina Faso.

Bationo, A., A. Buerkert, M.P. Sedogo, B.C. Christianson, and A.U. Mokwunye. 1995. A critical review of crop residue use as soil amendment in the West African semi-arid tropics. p. 305-322. In: J.M. Powell, S. Fernandez-Rivera, T.O. Williams, and C. Renard (eds.), Livestock and Sustainable Nutrient Cycling in Mixed Farming Systems of Sub-Saharan Africa. Volume 2: Technical Papers. Proceedings of an International Conference, 22-26 November 1993. International Livestock Centre for Africa (ILCA), Addis Ababa, Ethiopia.

Bationo, A., B.C. Christianson, and M.C. Klaij. 1993. The effect of crop residues and fertilizer use on pearl millet yields in Niger. *Fert. Res.* 34:251-258.

Bationo, A. and A.U. Mokwunye. 1991. The role of manures and crop residues in alleviating soil fertility constraints to crop production: with special reference to the Sahelian and Sudanian zones of West Africa. *Fert. Res.* 29:117-125.

Bationo A., A.K. Johnson, M. Kone, A.U. Mokwunye, and J. Henao. 1991. Properties and agronomic value of indigenous phosphate rock from West African semi-arid tropics. Paper presented at the Regional Workshop on Phosphorus Cycling in Semi-Arid Africa organized by the Scientific Committee on Problems of the Environments (SCOPE), Nairobi, Kenya, 18-22 March, 1990.

Bear, F.E. 1948. Making compost in the soil. *J. Soil Water Conserv.* 3:131-138.

Bekunda, M., A. Bationo, and H. Ssali. 1997. Soil fertility management in Africa: a review of selected research trials. p. 63-79. In: R. Buresh et al. (eds.), Replenishing Soil Fertility in Africa. SSSA Special Publication No. 51. SSSA, Madison, WI.

Bielders, C.L., K. Michels, and J.L. Rajot. 1998. Evaluation of soil losses by wind erosion under different soil and residue management practices in Niger, West Africa. In: J. Tatarko (ed.), Wind Erosion. An international symposium. *Proc. Intern. Symp.*, Manhattan, Kansas, 3-5 June 1997.

Breman, H. and C.T. de Wit. 1983. Rangeland productivity and exploitation in the Sahel. *Science* 221:1341-1347.

Buerkert, A., J.P.A. Lamers, H. Marschner, and A. Bationo. 1996a. Inputs of mineral nutrients and crop residue mulch reduce wind erosion effects on millet in the Sahel. p. 145-160. In: B. Buerkert, B.E. Allison, and M. von Oppen (eds.), Wind Erosion in West Africa. The Problem and its Control. *Proc. Intern. Symp.*, Univ. Hohenheim, Stuttgart, Germany, 5-7 Dec. 1994.

Buerkert, A., K. Michels, J.P.A. Lamers, H. Marshner, and A. Bationo. 1996b. Anti-erosive, soil physical, and nutritional effects of crop residues. p. 123-138. In: B. Buerkert, B.E. Allison, and M. von Oppen (eds.), *Wind Erosion in Niger*. Kluwer Academic Publishers in cooperation with the University of Hohenheim. Dordrecht, The Netherlands.

Burgos-Leon, W. 1979. Allelopathie induite par la culture du sorgho, origine et détoxication microbienne du sol. Thèse doctorat des Sciences, Université de Nancy I.

Casenave, A. and C. Valentin. 1989. *Les États de Surface de la Zone Sahélienne. Influence sur L'infiltration*. Paris. ORSTOM, Coll. Didactiques. 230 pp.

Cissé, L. and G. Vachaud. 1987. Effet d'un amendement organique sur l'infiltration, les coefficients de transferts hydriques, et l'évaporation d'un sol sableux dégradé du Nord Sénégal. *Hydrol. Ontinent* 2:14-28.

de Ridder, N. and H. van Keulen. 1990. Some aspects of the role of organic matter in sustainable intensified arable farming systems in the West African semi-arid tropics. *Fert. Res.* 26:299-310.

Elliot, L.F., T.M. McCalla, and J.R.A. Waiss. 1979. Phytoxicity associated with residues management. In: W.R. Oschwald (ed.), Crop Residues Management Systems. ASA Special Publication 31. American Society of Agronomy, Madison, WI.

Ganry, F.G., G. Gairaud, and Y. Dommergues. 1978. Effect of straw incorporation on the yield and nitrogen balance in the sandy soil-pearl millet cropping system of Senegal. *Plant and Soil* 50:647-662.

Gaur, A.C., K.V. Sadashivam, O.P. Vimal, and R.S. Mathur. 1971. A study on the decomposition of the organic matter in an alluvial soil, CO_2 evaluation, microbiological and chemical transformations. *Plant and Soil* 35:17-28.

Geiger, S.C., A. Manu, and A. Bationo, 1992. Changes in sandy Sahelian soil following crop residue and fertilizer additions. *Soil Sci. Soc. Am. J.* 56:172-177.

Hafner, H., E. George, A. Bationo, and H. Marschner. 1993a. Effect of crop residues on root growth and nutrient acquisition of pearl millet in an acid sandy soil in Niger. *Plant and Soil* 150:117-127.

Hafner, H., J. Bley, A. Bationo, P. Martin, and H. Marschner. 1993b. Long-term nitrogen balance for pearl millet (*Pennisetum glaucum* L.) in an acid sandy soil of Niger. *Zeitschrift für Pflanzenernährung und Bodenkunde* 156:164-176.

Hayman, D.S. 1970. Influence of soils and fertility on activity and survival of vesicular arbuscular mycorrhizal fungi. *Phytopathology* 72:1119-1125.

Jenkinson, D.S. and J.N. Ladd. 1981. Microbial biomass in soil measurement and turnover. p. 415-471. In: E.A. Paul and J.N. Ladd (eds.), *Soil Biochemistry*. Marcel Dekker, New York.

Kretzschmar, R.M., H. Hafner, A. Bationo, and H. Marschner. 1991. Long- and short-term effects of crop residues on aluminum toxicity, phosphorus availability and growth of pearl millet in an acid sandy soil. *Plant and Soil* 136:215-223.

Lamers, J.P.A. and P.R. Feil. 1995. Farmer's knowledge and management of spatial soil and crop growth variability in Niger, West Africa. *Neth. J. Agric. Sci.* 43:375-389.

Léonard, J. and J.L. Rajot. 1998. Restoration of infiltration properties of crushed soils by mulching. p. 191-195. In: G. Renard, A. Neel, K. Becker, and M. von Oppen (eds.), Soil Fertility Management in West African Land Use Systems. Proc. of the Regional Workshop, University of Hohenheim, 4-8 March 1997, Niamey, Niger. Margnal Verlag, Weikersheim, Germany.

Mando, A. 1997. Effects of termites and mulch on the physical rehabilitation of structurally crusted soils in the Sahel. In: *Land Degradation and Development.*

Mando, A., L. Stroosnijder, and L. Brussard. 1996. Effects of termites on infiltration into crusted soil. *Geoderma* 74:107-113.

McIntire, J. and L.K. Fussell. 1986. On-farm experiments with millet in Niger. III. Yields and economic analyses. ICRISAT Sahelian Center, Niamey, Niger. (Limited distribution).

Menge, J.A., D. Steirle, D.J. Bagyaraj, F.L.V. Johnson, and R.T. Leonard. 1978. Phosphorus concentrations in plants responsible for inhibition of mycorrhizal infection. *New Phytol.* 80:575-578.

Michels, K., C. Bielders, B. Muhlig-versen, and F. Mahler. 1998. Rehabilitation of degraded land in the Sahel: an example from Niger. p. 1287-1293. In: H.-P. Blume, H. Egen, E. Fleischauer, A. Hebel, C. Reis, and K.G. Steiner (eds.), *Towards Sustainable Land Use. Furthering Cooperation between People and Institutions. Advances in Geoecology, Vol. 31 (II).* Catena Verlag, Reiskirchen, Germany.

Michels, K., M.V.K. Sivakumar, and B.E. Allison. 1993. Wind erosion in the Southern Sahelian Zone and induced constraints to pearl millet production. *Agricultural and Forest Meteorol.* 67:65-77

Michels, K., D.V. Armbrust, B.E. Allison, and M.V.K. Sivakumar. 1995a. Wind and wind-blown sand damage to pearl millet. *Agronomy J.* 87:620-626.

Michels, K., M.V.K. Sivakumar, and B.E. Allison. 1995b. Wind erosion control using crop residue: I. Effects on soil flux and soil properties. *Field Crops Res.* 40:101-110.

Michels, K., M.V.K. Sivakumar, and B.E. Allison. 1995c. Wind erosion control using crop residue: II. Effects on millet establishment and yields. *Field Crops Res.* 40:111-118.

Michels, K. 1994. Wind erosion in the Southern Sahelian zone: extent, control, and effect on millet production. Ph.D. Dissertation. Verlag U. Ganer, Stuttgart, Germany.

Pichot, J., M.P. Sedogo, J.L. Poulain, and J. Arrirets. 1981. Evolution de la fertilité d'un sol ferrugineux sous l'influence de fumures minérales et organiques. *Agronomie Tropicale* 36:122-133.

Pieri, C. 1986. Fertilisation des cultures vivrières et fertilité des sols en agriculture paysanne subsaharienne. *Agronomie Tropicale* 41:1-20.

Pieri, C. 1989. Fertilité des terres de savanes. In: *Bilan de Trente Ans de Recherches et de Développement au Sud du Sahara.* Centre de Coopération Internationale en Recherche Agronomique pour le Développement (CIRAD) and Ministère de la Coopération, Paris, France. 443 pp.

Powell, J.M. and Bayer, W. 1985. Crop residue grazing by Bunaji cattle in Central Nigeria. *Tropical Agriculture (Trinidad)* 62:302-866.

Powell, J.M. and P.W. Unger. 1997. Alternatives to crop residues as soil amendments. p. 215-239. In: C. Renard (ed.), *Crop Residues in Sustainable Mixed Crop/Livestock Farming Systems.* International Crop Research Institute for the Semi-Arid Tropics. Patancheru 502324, Andhra Pradesh, India and International Livestock Research Institute. Nairobi, Kenya.

Powlson, D.S. and D.S. Jenkinson. 1981. A comparison of organic matter, biomass, adenosine triphosphate and mineralisable nitrogen contents of plowed and direct drilled soils. *J. Agric. Sci.* 97:713-721.

Rebafka, F.P., A. Hebel, A. Bationo, K. Stahr, and H. Marschner. 1994. Short and long-term effects of crop residues and of phosphorus fertilization on pearl millet yield on an acid sandy soil in Niger, West Africa. *Field Crops Res.* 36:113-124.

Sedogo, M.P. 1981. Contribution à l'étude de la valorisation des résidus culturaux en sol ferrugineux et sous climat tropical semi-aride. Matière organique du sol, nutrition azotée des cultures. Thèse Docteur Ingénieur, INPL, Nancy.

Sidhu, B.S. and V. Beri. 1986. Recycling of crop residues in agriculture. p. 49-54. In: M. Mishra and K.K. Kapoor (eds.), Soil Biology, *Proc. of National Symposium on Current Trends in Soil Biology*, Haryana Agricultural University, Hisar.

Sivakumar, M.V.K., A. Maidukia, and R.D. Stern. 1993. *Agroclimatology of West Africa: Niger.* 2nd ed. Information Bulletin No. 5. ICRISAT, Patancheru, A.P. 502324, India.

Smaling, E.M.A., S.M. Nandwa, and B.H. Jansen. 1997. Soil fertility in Africa is at stake. p. 47-61. In: Buresh et al. (eds.), *Replenishing Soil Fertility in Africa.* SSSA Special Publication 51. SSSA, Madison WI.

Stahr K., K.K. Bleich, A. Hebel, and L. Herrmann. 1993. The influence of organic matter and dust deposition on site characteristics and their microvariability. TP A1. Universitat Hohenheim. p. 25-28. In: University of Hohenheim (ed.), *Standortgemässe Landwirtschaft in Westafrika* -SFB 308. Arbeits und Ergebnisbericht, Zwischenbericht 1991 - 1993.

Sterk, G. 1997. Wind erosion in the Sahelian zone of Niger: Processes, Models and Control Techniques. Ph.D. dissertation. Tropical resource management papers #15, Wageningen Agricultural University.

Sterk, G. and A. Stein, 1997. Mapping wind blown mass transport by modeling variability in space and time. *Soil Sci. Soc. Am. J.* 61:232-239.

Stevenson, F.J. 1986. *Cycles of Soil Carbon, Nitrogen, Phosphorus, Sulfur, Micronutrients.* A Wiley-Interscience Publication, John Wiley & Sons, New York. 380 pp.

Tien, T.M., M.H. Gaskins, and D.H. Hubbell. 1979. Plant growth substances produced by Azospirillum brasiliense and their effect on the growth of pearl millet (*Pennisetum americanum* L.) *Appl. Environ. Microbiol.* 37:1016-1024.

Van Keulen, H. and H. Breman. 1990. Agricultural development in the West African region: a cure against land hunger? *Agric., Ecosystems and Environ.* 32:177-197.

Waksman, S.A. and R.L. Starkey. 1924. Microbiological analysis of soil as an index of soil fertility. VII. Carbondioxide evolution. *Soil Sci.* 17:141-161.

West, L., L.P. Wilding, J.K. Landeck, and F.G. Galhoun. 1984. Soil survey of the ICRISAT Sahelian Center, Niger, West Africa. Texas A&M University: Soil and Crop Sciences Department /TropSoils.

CHAPTER 7

Restorative Effects of *Mucuna Utilis* on Soil Organic C Pool of a Severely Degraded Alfisol in Western Nigeria

R. Lal

I. Introduction

Restoring degraded soils and ecosystems is a high priority especially in the tropics and subtropics where soil degradation is a severe problem due to harsh climate, fragile soils, and prevalence of subsistence or resource-based agriculture (Oldeman, 1994; Lal, 1989; 1997). There are both economic and ecologic/environmental reasons for restoring degraded soils. Economic reasons are related to high demographic pressure, limited availability of per capita land area for agricultural and other uses, and need for finding additional lands that can be converted to agriculture. Ecologic reasons are restoring degraded lands improving the quality of surface and ground waters, improving soil C pool leading to increase in soil quality and contributing to mitigating the greenhouse effect, and enhancement of the aesthetic value of the land.

Soil quality refers to its biomass (economic) productivity and environment moderating capacity (Lal, 1997; Doran and Parkin, 1994; 1996; Carter et al., 1997). More specifically, it is soil's ability to perform specific functions, e.g., agricultural and forestry production, engineering uses, urban and industrial uses, water purification, bioremediation, C sequestration, etc. In general, soil quality declines with conversion from natural to agricultural ecosystem. The rate or intensity of decline is accentuated by soil degradation processes, e.g., reduction in soil C pool, decline in activity and species diversity of soil fauna, decline in percent and stability of aggregation, crusting and compaction leading to accelerated runoff and soil erosion, leaching of cations causing decline in soil pH and acidification, and nutrient imbalance.

Soil degradation is greater in mechanized than manual methods of deforestation (Lal, 1994), in plow-till than no-till (Lal, 1976), in monoculture than legume/pasture based rotations, and with chemical fertilizers alone than with integrated nutrient management involving judicious use of inorganic fertilizers and organic manures (Lal, 1987).

Once degraded, restoration depends on soil resilience and management. Soil resilience, soil's ability to restore its quality while enhancing biomass productivity and improving environmental moderating capacity, depends on soil inherent characteristics and management. Conversion to restorative land use is an important factor. Activity of soil fauna (e.g., earthworms) plays an important role in soil restoration. Earthworms recycle nutrients, incorporate crop residue and biomass in soil, and improve soil structure by increasing aggregation and porosity through burrowing and mixing activity (Lal and Akinremi, 1983; Lal, 1987; 1991). Rate of crop residue mulch has a strong impact of earthworm activity and soil aggregation (DeVleeschauwer and Lal, 1981).

The objective of this chapter is to assess the importance of soil restorative cover crops on soil quality. The latter is as assessed by quantifying changes in soil physical and chemical properties, especially with reference to changes in soil organic carbon content as influenced by the activity of earthworms.

II. Materials and Methods

Field experiments were conducted at the International Institute of Tropical Agriculture (IITA), Ibadan, Nigeria, from 1978 through 1981. The IITA is located in the southwestern region of Nigeria. The bimodal character of rainfall distribution in western Nigeria results in two distinct growing seasons, one from April to July and the other from August to November. Total annual rainfall ranges from 1100 to 1500 mm.

Soils of the experimental site are classified as Alfisols (Oxic Paleustalfs) according to the USDA system and Ferric Luvisols in the FAO system (Moormann et al., 1975). These soils, derived from fine-grained biotite gneiss and schist parent materials, are medium to light textured near the surface with sandy clay to clay B_{2t} horizons and have layers of angular and sub-angular quartz gravel immediately below the surface. The organic carbon content is mostly concentrated in the top 30 cm layer, and the gravelly horizon extends from about 20 to 80 cm below the surface.

Field runoff plots, established in 1972, were used to study the effects of slope and management on runoff and soil erosion from 1972 through 1977 (Lal, 1976; 1981). Physiographic characteristics and management treatments of all 24 plots are described in Table 1. Plots 1 through 5 were established on average slope of about 1%, 6–10 on 5%, 10–17 on 10%, and 18–24 on 15% slope gradient. Cropping systems from 1972 to 1981 are outlined in Table 1. No fertilizer was applied from 1978 to 1981.

From 1972 through 1975, these runoff plots lost topsoil ranging from 0.4 to 36 Mg ha^{-1} for 1% slope, 1.6 to 480 Mg ha^{-1} for 5% slope, 67 to 495 Mg ha^{-1} for 10% slope, and 10 to 524 Mg ha^{-1} for 15% slope (Lal, 1981). In 1978, during the seventh year after their establishment, all runoff plots were sown to Mucuna (*Mucuna utilis*) for soil restoration. Composite soil samples from 0 to 10 cm depths were taken in duplicate from each plot in 1981 to assess effects on soil quality. These samples were air dried, ground and sieved through a 2 mm sieve. Soil samples were analyzed for organic carbon (by dichromate oxidation), total nitrogen (by Kjeldahl digestion), Bray-1 P, and for Ca^{+2}, Mg^{+2}, and K^+ extractable by N ammonium acetate. Particle size distribution from soil sieved through a 2 mm sieve was measured by the hydrometer method. The gravel content (> 2 mm) was recorded separately. Field tests were made for soil bulk density by excavating a known volume of soil, penetrometer resistance by a blunt-tip pocket penetrometer, and infiltration capacity by a double-ring infiltrometer. High and low-energy soil moisture characteristics were determined by using pressure plate extractors and a tension table (ASA, 1986).

Soils of this region are characterized by intense activity of earthworms (*Hyperiodrilus africanus*), especially under native vegetation and soils under shade of a cover crop or mulch. Soils in these runoff plots also exhibited earthworm activity. Therefore, earthworm casts were also collected and analyzed for chemical and physical properties. Soil chemical analyses of worm casts were compared with analyses of surface soil from each plot. Results of analyses of soil sampled in 1981 were compared with results of earthworm casts also collected in 1981, and with that of soil sampled in 1976.

Table 1. Physiographic characteristics and agronomic management from 1972–1976 of runoff plots

Plot #	Slope gradient (%)	Slope length (m)	Tillage methods	Crops 1972-1974	Crops 1975
1	1.80	25	Plow (mulch)	Maize-maize	Soybean-soybean
2	1.30	25	Plow	Maize-maize	Maize+cassava
3	1.10	25	Plow	Uncropped	Uncropped
4	1.00	25	No-till	Maize-cowpea	Soybean-soybean
5	0.88	25	Plow	Cowpea-maize	Cassava
6	4.72	25	No-till	Maize-cowpea	Soybean-soybean
7	5.12	25	Plow (mulch)	Maize-maize	Soybean-soybean
8	5.48	25	Plow	Maize-maize	Maize+cassava
9	5.68	25	Plow	Cowpea-maize	Cassava
10	5.68	25	Plow	Uncropped	Uncropped
11	8.76	25	Plow	Uncropped	Uncropped
12	9.28	25	Plow (mulch)	Maize-maize	Soybean-soybean
13	9.72	25	No-till	Maize-cowpea	Soybean-soybean
14	9.88	25	Plow	Cowpea-maize	Cassava
15	10.12	25	Plow	Maize-maize	Maize+cassava
16	9.33	37.5	Plow	Maize-cowpea	Pigeon peas
17	10.00	12.5	Plow	Maize-cowpea	Pigeon peas
18	19.20	12.5	Plow	Maize-cowpea	Pigeon peas
19	13.44	37.5	Plow	Maize-cowpea	Pigeon peas
20	14.92	25	No-till	Maize-cowpea	Soybean-soybean
21	14.40	25	Plow (mulch)	Maize-maize	Soybean-soybean
22	14.16	25	Plow	Cowpea-maize	Cassava
23	14.36	25	Plow	Maize-maize	Maize+cassava
24	14.64	25	Plow	Uncropped	Uncropped

All plots were sown to maize-cowpea rotation with no-till in 1976 and plow-till in 1977.
All plots were sown to restorative Mucuna (*Mucuna utilis*) cover crop from 1978 to 1981.

Table 2. Antecedent soil chemical properties for different cultural practices

Plot #	pH 1972	pH 1974	SOC content (g kg⁻¹) 1972	SOC content (g kg⁻¹) 1974	Total soil N (g kg⁻¹) 1972	Total soil N (g kg⁻¹) 1974	Bray-P (mg kg⁻¹) 1972	Bray-P (mg kg⁻¹) 1974
1	6.7	6.0	29.0	28.5	2.7	2.4	85.8	6.0
2	6.3	6.0	25.0	16.0	2.3	1.7	96.9	5.6
3	6.5	5.0	27.0	13.0	2.5	1.5	78.4	5.0
4	6.4	5.5	23.0	21.0	2.1	2.2	81.9	5.5
5	6.5	5.8	27.0	17.5	2.1	1.4	100.8	5.8
Mean	6.5±0.1	5.6±0.4	26.2±2.2	19.2±5.9	2.2±0.3	1.8±0.4	88.8±9.7	5.6±0.4
6	6.9	5.7	27.0	16.5	1.9	1.3	44.5	5.7
7	6.9	6.0	23.0	19.0	1.8	1.3	44.5	5.3
8	6.7	5.3	22.0	12.0	1.7	1.1	35.0	5.3
9	6.7	5.8	23.0	16.5	1.9	1.3	31.9	5.8
10	6.4	4.8	19.0	9.5	1.4	1.1	31.9	4.8
Mean	6.7±0.2	5.5±0.5	22.8±2.9	14.7±3.9	1.7±0.2	1.2±0.1	37.6±6.5	5.4±0.4
11	7.0	4.7	21.0	11.5	1.8	1.1	31.9	4.7
12	6.9	6.0	24.0	21.0	2.3	2.0	41.3	6.0
13	6.7	5.8	20.0	22.0	1.6	2.0	47.6	5.8
14	6.8	5.9	20.0	19.0	2.4	1.7	37.8	5.9
15	6.8	5.7	25.0	19.0	2.0	1.3	44.5	5.8
Mean	6.8±0.1	5.6±0.5	22.0±2.3	18.5±4.1	2.0±0.3	1.6±0.4	40.6±6.1	5.6±0.5
20	7.1	5.6	23.0	21.0	2.1	1.9	47.6	5.5
21	7.3	5.8	21.0	25.0	2.1	1.4	44.5	5.3
22	7.1	5.6	23.0	21.0	2.1	1.9	47.6	5.5
23	7.1	5.6	22.0	21.0	1.9	1.9	54.9	5.5
24	7.3	4.2	25.0	11.0	2.4	1.1	53.9	4.2
Mean	7.2±0.1	5.4±0.7	22.8±1.5	19.8±5.2	2.1±0.2	1.6±0.4	49.7±4.5	5.2±0.6

Antecedent soil chemical properties for plots 16–19 and 25–29 are not included in this table. (From Lal, 1976.)

III. Results

Soil chemical properties at the onset of the experiment in January 1972 and three years after in December 1974 are shown in Table 2. Soil pH ranged from 6.3 to 7.3 at the onset of the experiment in 1972. The average soil pH was 6.5 for plots of 1% slope, 6.7 for plots of 5% slope, 6.8 for plots of 10% slope and 7.2 for plots of 15% slope. Soil pH declined rapidly with cultivation and ranged from 4.2 to 6.0 in 1974. The average soil pH was 5.6 for plots of 1% slope, 5.5 for plots of 5% slope, 5.6 for plots of 10% slope, and 5.4 for plots of 15% slope. The most rapid decline in soil pH was observed for the bare uncropped treatments, where the range of decline over the 3 year period was 1.5 pH units for 1% slope plot, 1.6 for 5% slope plot, 2.3 for 10% slope plot, and 3.1 for 15% slope plot. Closely related to temporal changes in soil pH were those of the soil organic carbon (SOC) content.

The SOC content ranged from 20 to 29 g kg^{-1} in 1972 and from 10 to 29 g kg^{-1} in 1974 (Table 2). The mean SOC content declined from 23.5 g kg^{-1} in 1972 to 18.1 g kg^{-1} in 1974, with an average decline of 23%. In comparison with cropped treatments, there was a rapid decline in SOC content of all bare uncropped treatments. For example SOC content of the bare uncropped plot decreased from 27.0 g kg^{-1} to 13.0 g kg^{-1} (52% decline) in plots of 1% slope, 19.0 g kg^{-1} to 9.5 g kg^{-1} (50% decline) in plots of 5% slope, 21.0 g ha^{-1} to 11.5 g kg^{-1} (45% decline) in plots of 10% slope, and 25.0 g ha^{-1} to 11.0 g kg^{-1} (56% decline) in plots of 25% slope.

Temporal changes in total soil nitrogen (TSN) content were similar to that of the SOC content (Table 2). The TSN content ranged from 1.4 to 2.7 g kg^{-1} in 1972 (mean of 2.0 g kg^{-1}) at the onset of the experiment. The mean TSN content was 2.2 g kg^{-1} for plots of 1% slope, 2.0 g kg^{-1} for 5% and 10% slopes, and 2.1% for 15% slopes. The TSN content declined with 3 years of cultivation and ranged from 1.1 to 2.4 g kg^{-1} in 1974. The average TSN content in 1974 was 1.8 g kg^{-1} (18.2% decline) in plots of 1% slope, 1.2 g kg^{-1} (29.5% decline) in plots of 5% slope, 1.6 g kg^{-1} (20% decline) in plots of 10% slope, and 1.6 g kg^{-1} in plots of 15% slope. The overall mean TSN content in 1974 was 1.6 g kg^{-1}. The most drastic decline in TSN occurred in the bare uncropped treatment, which was 2.5 to 1.5 g kg^{-1} (40% decline) in 1% slope plots, 1.4 to 1.1 g kg^{-1} (21% decline) in 5% slope plots, 1.8 to 1.1 g kg^{-1} (39% decline) in 10% slope plots, and 2.4 to 1.1 g kg^{-1} (54% decline) in 15% slope plots.

The mean C:N ratio did not change with cultivation, and was 11.8 in 1972 and 11.7 in 1974. The average C:N ratio in 1972 and 1974, respectively, was 11.9 and 10.7 for plots of 1% slope, 13.4 and 12.3 for plots of 5% slope, 11.0 and 11.6 for plots of 10% slope, and 10.9 and 12.4 for plots of 15% slope. The C:N ratio for bare uncultivated plots for 1972 and 1974 respectively was 10.8 and 8.7 for 1% slope, 13.6 and 8.6 for 5% slope, 11.7 and 10.5 for 10% slope, 10.4 and 10.0 for 15% slope. For bare uncultivated plots, where SOC content declined drastically, the C:N ratio narrowed over the 3 year period.

The data in Table 2 also show a very drastic decline in Bray-P content over the 3 year period. The Bray-P content ranged from 32 to 100 mg kg^{-1} in 1972 and 4 to 6 mg kg^{-1} in 1974.

IV. Soil Chemical Quality from 1974 to 1978

Following a very rapid decline during the first 3 years, soil chemical properties (SOC, TSN and pH) continued to decline in succeeding years (Aina et al., 1977; Lal, 1981). Accelerated soil erosion caused a rapid depletion of SOC and TSN contents and an increase in gravel content of the soil surface. Degradation was the most severe in bare uncropped plots.

It was this severe soil degradation that necessitated conversion from arable land treatments to growing a leguminous cover crop, and to evaluating the impact of 5 years of growing a cover crop in SOC content and C sequestration.

V. Changes in SOC Content and Soil Chemical Quality by 5 Years of Growing a Cover Crop

The data in Table 3 show the impact of historic land use and growing cover crops on soil chemical quality. The mean soil pH increased under cover crop but only slightly. Average soil pH was 5.5 in 1974 and 5.7 in 1981, while pH of some treatments increased to 6.4 and 6.5. The most drastic increase in soil pH occurred in bare uncropped plots ranging from 5.0 to 5.8 for 1% plots, 4.8 to 5.1 for 5% plots, 4.7 to 5.4 for 10% plots, and 5.4 to 5.9 for 15% plots. Increase in soil pH is indicative of recycling of bases from subsoil to the surface through deep-rooted *Mucuna utilis* (Nye and Greenland, 1960). There were also differences in soil pH due to historic cropping systems imposed from 1972 to 1978 especially for plots on 5, 10 and 15% slopes.

There were no improvements in SOC contents even after 5 years of growing *Mucuna utilis* and returning all residue to the soil. The mean SOC content in 1981 was extremely low—6.6 g kg^{-1} compared with 23.5 g kg^{-1} in 1972 and 18.1 g kg^{-1} in 1974. In relation to plots on different slope gradients, mean SOC content in 0–10 cm depth in 1981 was 9.2 g kg^{-1} for 1% slope, 8.0 for 5% slope, 6.8 for 10% slope and 2.5 for 15% slope. It is likely that SOC content continued to decrease with cultivation from 1975 through 1978 (Aina et al., 1977; Lal, 1981), and the decline was exacerbated by accelerated soil erosion in bare uncropped treatments. There were no differences in SOC content among cropping system treatments imposed from 1972 through 1978 for plots of 1, 5, and 10% slope. However, differences in SOC content due to historic cropping systems treatments from 1972 to 1978 were significant at 5% level of probability for plots on 15% slope gradient (Table 3).

There were some changes in TSN content due to growing a leguminous cover crop. The average TSN content was 1.7 g kg^{-1} in 1981 compared with 2.0 g kg^{-1} in 1972 and 1.8 g kg^{-1} in 1974. There were also differences in TSN content among historic cropping systems treatments imposed from 1972 to 1978 for plots on 1% and 10% slope gradients. Regardless of the slope, the least TSN content was recorded for plots with bare uncropped treatment from 1972 to 1978 (i.e., plots 3, 10, 11 and 24).

The C:N ratio in soil analyzed in 1981 was low, probably because of the extremely low SOC content. The average C:N ratio was 3.9 in 1981, compared with 11.8 in 1972 and 11.7 in 1974. The C:N ratio was especially low in bare uncropped treatments and was 5.8 in plot 3 on 1% slope, 5.0 in plot 10 on 5% slope, 4.3 in plot 11 on 10% slope, and 1.5 on plot 24 on 15% slope. The low C:N ratio is an indication of high TSN content due to the biological nitrogen fixation by leguminous *Mucuna* cover and low SOC content. The low C:N ratio in this case is not indicative of improvements in soil chemical or biological quality.

In contrast to SOC content, growing *Mucuna* cover caused notable improvements in Bray-P content (Table 3). The average P content was 48 mg kg^{-1} for plots of 1% slope, 18 mg ha^{-1} for plots of 5% slope, 14 mg kg^{-1} for plots of 10% slope and 16 mg kg^{-1} for plots of 15% slope. Increase in P content due to growing *Mucuna* was probably due to recycling from the sub-soil horizon (Nye and Greenland, 1960) since no P was applied from 1978 to 1981. Relative increase in P content was more for plots of 1% slope which had lower cumulative soil erosion than plots in the other three slope groups.

VI. Role of Earthworm Activity in Soil Quality Enhancement

Earthworm activity plays an important role in soil restoration (Lal, 1987). Establishment of the *Mucuna* cover enhanced activity of earthworms, as was evident by the occurrence of numerous casts of *Hyperiodrilus africanus* under the vegetative cover. Worm casts or ejecta comprise a mixture of soil with organic matter cemented together with intestinal fluids. Therefore, casts are often different in composition than the soil from which they are derived.

Table 3. Historic land use and management effects on C, N and Bray-P level for 0–10 cm depth of soil sampled in November 1981

Plot #	pH	SOC content	Total soil N	Bray-P
		g kg^{-1}		mg kg^{-1}
1	5.2	7.1	1.44	60.3
2	5.8	8.4	1.55	41.8
3	5.8	9.5	1.65	44.9
4	5.6	9.7	2.49	46.2
5	6.0	11.0	2.58	47.6
Mean	5.7	9.2	1.94	48.1
LSD	n.s.	n.s.	0.08**	n.s.
6	5.4	8.3	1.56	22.1
7	5.6	8.0	1.56	21.9
8	5.9	11.6	1.64	16.5
9	5.8	7.5	1.94	14.0
10	5.1	4.8	1.27	16.8
Mean	5.5	8.0	1.59	18.2
LSD	0.4*	n.s.	n.s.	n.s.
11	5.4	1.10	1.10	25.1
12	5.4	1.41	1.41	16.2
13	6.0	1.58	1.58	8.1
14	5.7	1.71	1.71	10.5
15	5.9	1.81	1.81	17.5
16	6.2	1.15	1.15	15.0
17	6.4	1.23	1.23	3.5
Mean	5.8	1.43	1.43	13.7
LSD	0.2**	0.01**	0.01**	n.s.
18	6.1	1.56	1.56	7.6
19	6.2	1.89	1.89	11.9
20	5.7	2.15	2.15	13.9
21	5.9	2.14	2.14	18.3
22	6.5	1.73	1.73	21.7
23	6.0	1.35	1.35	4.8
23	5.2	1.24	1.24	33.5
Mean	5.9	1.72	1.72	16.0
LSD (0.05)	0.3**	n.s.	n.s.	n.s.

n.s., *, ** = non-significant, significant at 10% and 5% level of probability, respectively.

Table 4. Enrichment ratio of soil constituents in earthworm casts compared with soil samples of 0–10 cm depth of samples analyzed in November 1981

Plot #	Enrichment ratio					
	Sand	Clay	Silt	SOC	TSN	Bray-P
1	0.50	1.5	3.0	3.9	3.4	1.1
2	0.69	1.2	2.0	3.1	6.7	2.4
3	0.65	1.2	2.6	2.9	3.0	1.3
4	0.69	1.2	2.7	2.0	1.8	1.0
5	0.65	1.3	2.4	1.4	1.8	1.8
Mean	0.64±0.08	1.3±0.1	2.5±0.4	2.7±1.0	3.3±2.0	1.5±0.6
6	0.84	0.9	2.0	1.7	1.8	1.0
7	0.70	1.2	2.2	2.6	1.9	1.1
8	0.79	1.0	2.0	2.6	1.5	1.9
9	0.77	1.2	2.2	1.7	1.6	1.8
10	0.77	1.0	2.4	4.6	1.8	0.7
Mean	0.77±0.05	1.1±0.1	2.2±0.2	2.6±1.2	1.7±0.2	1.3±0.5
11	0.70	1.3	4.6	4.0	3.2	0.9
12	0.76	1.2	2.9	2.8	2.2	1.1
13	0.84	1.0	1.8	2.2	2.2	3.5
14	0.70	1.1	2.6	1.6	1.9	1.5
15	0.79	1.0	2.1	2.2	1.6	0.9
16	0.79	1.3	2.0	5.6	2.2	1.5
17	0.73	1.3	2.4	2.6	2.6	3.9
Mean	0.76±0.05	1.2±0.1	2.6±0.9	3.0±1.4	2.3±0.5	1.9±1.3
18	0.77	1.4	2.0	2.4	1.7	2.5
19	0.79	1.2	2.5	1.8	2.0	1.8
20	0.64	1.4	2.1	1.7	1.7	1.9
21	0.89	1.0	1.6	1.2	1.2	0.9
22	0.82	1.1	1.7	1.9	1.8	1.7
23	0.95	0.8	1.8	1.8	1.8	3.0
24	0.55	0.9	5.8	3.0	2.0	1.1
Mean	0.77±0.13	1.1±0.2	2.5±1.5	2.0±0.6	1.7±0.3	1.8±0.7

The data in Table 4 show enrichment of some soil constituents and depletion of others as the soil is passed through the guts of earthworms. In terms of the textural composition, earthworm casts have enrichment of clay and silt contents and depletion of sand content. The average enrichment ratio of worm casts (mean of all 24 plots) was 1.18 for clay (compared with 1 for soil) and 2.45 for silt. The enrichment ratio of clay for plots in different slope groups was 1.3 for 1%, 1.1 for 5%, 1.2 for 10%

and 1.1 for 15%. The average enrichment ratio for silt was 2.5 for 1%, 2.2 for 5%, 2.6 for 10% and 2.5 for 15% slope gradient. In contrast to enrichment of clay and silt, there was a depletion of sand content in the worm casts. The average enrichment ratio for sand content was 0.74 with variation among soils on different slope groups with mean values of 0.64 for 1%, 0.77 for 5%, 0.76 for 10% and 0.77 for 15% slope gradient.

The data in Table 4 also show enrichment of SOC, TSN and Bray-P in earthworm casts. The enrichment ratio for SOC ranged from 1.4 to 3.9 with an average of 2.6. The enrichment ratio of SOC for different slope groups was 2.7 for 1%, 2.6 for 5%, 3.0 for 10% and 2.0 for 15% slope gradient. The enrichment ratio of TSN was similar to that of the TSN, and ranged from 1.2 to 6.7 with an average of 2.25. The enrichment ratio of TSN for different slope groups was 3.3 for 1%, 1.7 for 5%, 2.3 for 10% and 1.7 for 15% slope gradient. The enrichment ratio for Bray-P was less than that of SOC or TSN and ranged from 0.7 to 3.9 with an average of 1.6. The average enrichment ratio for Bray-P for soils of different slope groups was 1.5 for 1%, 1.3 for 5%, 1.9 for 10% and 1.8 for 15% slope gradient. Therefore, one of the mechanisms of soil restoration is through enhancement in activity of soil fauna. The activity and species diversity of soil fauna accentuates bioturbation, mixes organic matter in the soil, strengthens cycling of C and plant nutrients, and improves soil structure.

VII. Conclusions

The data presented support the following conclusions:
1. Five years of growing a leguminous cover crop decreased losses due to runoff and soil erosion, improved earthworm activity, increased TSN and Bray-P, decreased soil crusting and improved soil structure.
2. There was no improvement in SOC content of this severely degraded soils even after 5 years of growing a cover crop and returning biomass to the soil.
3. Notable improvements in earthworm activity are indicators of improvement in soil biological quality. It took 2 to 3 years of growing a cover crop before the recolonization by earthworms occurred. The process of restoration and C sequestration in soil is accentuated by the earthworm activity.
4. Rather than the leguminous cover crop alone, establishment of grasses (e.g., *Panicum* spp, *Pennisetum* spp, *Setaria* spp, *Brachiaria* spp) and shrubs (*Gliricidia, Leucaena,* etc.) would have improved SOC content and strengthened recycling mechanism.
5. Addition of plant nutrients, through chemical fertilizers and organic amendments, may be necessary to set in motion the restoration of these highly depleted soils. The SOC content cannot be improved without availability of additional quantity of N, P, S in soils of low inherent fertility.

Acknowledgments

These experiments were conducted under the Farming Systems Program of IITA. Soil chemical analyses were done by the Analytical Service Laboratory of IITA, Ibadan, Nigeria.

References

Aina, P.O., R. Lal and G.S. Taylor. 1977. Soil and crop management in relation to soil erosion in the rainforest of western Nigeria. p. 75-84. In: G. Foster (ed.), *Soil Erosion: Prediction and Control.* SCSA Special Publication 21, Ankeny, IA.

ASA 1986. *Methods of Soil Analysis, Part I and II.* ASA Monograph 9. Madison, WI.

Bezdicek, D.F., R.I. Papendick and R. Lal. 1996. Importance of Soil Quality to Health and Sustainable Soil Management. p. 1-7. Soil Sci. Soc. Amer. Special Publication 49. Madison, WI.

Carter, M.R., E.G. Gregorich, D.W. Anderson, J.W. Doran, H.H. Janzen and F.J. Pierce. 1997. Concepts of soil quality and their significance. p. 1-19. In: E.G. Gregorich and M.R. Carter (eds.), *Soil Quality for Crop Production and Ecosystem Health,* Elsevier, Amsterdam.

DeVleeschauwer, D. and R. Lal. 1981. Properties of worm casts in some tropical soils. *Soil Sci.* 132: 175-181.

Doran, J.W. and T.B. Parkin. 1994. *Defining and Assessing Soil Quality.* p. 3-21. Soil Sci. Soc. Amer. Special Publication 35, Madison, WI.

Doran, J.W. and T.B. Parkin. 1996. *Quantitative Indicators of Soil Quality: a Minimum Data Set.* p. 25-37. Soil Sci. Soc. Amer. Special Publication 49, Madison, WI.

Lal, R. 1976. *Soil Erosion Problems on Alfisols in Western Nigeria and Their Control.* IITA Monograph 1, Ibadan, Nigeria. 208 pp.

Lal, R. 1981. Soil erosion problems on Alfisols in western Nigeria. VI. Effects of erosion on experimental pots. *Geoderma* 25: 215-230.

Lal, R. 1987. *Tropical Ecology and Physical Edaphology.* John Wiley & Sons, Chichester, U.K.

Lal, R. 1989. Soil erosion and land degradation: the global risk. *Adv. Soil Sci.* 11:129-172.

Lal, R. 1991. Soil conservation and biodiversity. p. 89-104. In: D.L. Hawkworth (ed.), *The Biodiversity of Micro-organisms and Invertebrates: Its Role in Sustainable Agriculture.* CAB International, Wallingford, U.K.

Lal, R. 1994. Sustainable land use systems and soil resilience. p. 41-67. In: D.J. Greenland and I. Szabolcs (eds.), *Soil Resilience and Sustainable Land Use.* CAB International, Wallingford, U.K.

Lal, R. 1997. Degradation and resilience of soils. *Phil. Trans. R. Soc. Lond. B.* 352:997-1010.

Lal, R. and O.O. Akinremi. 1983. Physical properties of earthworm casts and surface soil as influenced by management. *Soil Sci.* 135:116-122.

Moormann, F.R., R. Lal, and A.S. Juo. 1975. The soils of IITA. IITA Tech. Bull. #3, 48 pp.

Nye, P.H. and D.J. Greenland. 1960. *The Soil under Shifting Cultivation.* CAB, Harpenden, U.K. 156 pp.

Oldeman, L.R. 1994. The global extent of soil degradation. p. 99-117. In: D.J. Greenland and I. Szabolcs (eds.), *Soil Resilience and Sustainable Land Use.* CAB International, Wallingford, U.K.

CHAPTER 8

Land Use and Cropping System Effects on Restoring Soil Carbon Pool of Degraded Alfisols in Western Nigeria

R. Lal

I. Introduction

The soil organic carbon (SOC) pool constitutes one of the five principal global C pools, others being oceanic, geologic, atmospheric and biotic (Figure 1). Along with the soil inorganic carbon (SIC) pool, the soil C pool is estimated at about 2500 Pg comprising 1550 Pg of SOC (Eswaran et al., 1995) and 950 Pg of SIC (Batjes, 1996). A large part of the increase in atmospheric concentration of CO_2, since the onset of the industrial revolution in 1850, from 280 ppm to 365 ppm is attributed to change in the soil and biotic C pool. The historic loss of C emitted from soil to the atmosphere is estimated at 55 Pg (IPCC, 1995), and may be as much as 100 Pg. An additional 100 to 150 Pg may have been emitted from the vegetation. Soils of the tropics, constituting a major part of the soil C pool, have contributed considerably to the anthropogenic increase in atmospheric CO_2 pool. The rate of decomposition of SOC pool is higher in the tropics than in the temperate climate because of high mean annual temperature (Jenkinson and Ayanaba, 1977; Jenny, 1949; Jenny and Raychaudhuri, 1960). The rate of depletion of SOC in soils of the tropics is exacerbated by the onset of soil degradative processes including decline in soil structure leading to crusting/compaction and accelerated runoff and erosion, reduction in soil biotic activity, leaching of bases and depletion of soil fertility etc. (Lal, 1995; 1996; 1997).

Results of several experiments have shown rapid decline in SOC content when natural ecosystems in the tropics are converted to arable and pastoral land uses (Lal and Kang, 1982). Experiments conducted at IITA have shown declines in SOC content of the surface horizon from 1.7 to 2.0% under the cover of native vegetation to 0.8 to 1.0% within 10 years of cultivation (Lal, 1997). The magnitude and rate of decline of SOC pool are greater in mechanized than in manual clearing (Lal and Cummings, 1979; Hulugalle et al., 1984; Lal, 1997), plow-till compared with no-till (Lal, 1976; 1996), for monoculture compared to mixed cropping and agroforestry systems (Lal, 1989; Kang et al., 1995).

There is a tremendous interest in restoration of degraded soils of the tropics, for both economic and ecologic reasons. Economic rationale for restoration of degraded soils is related to productivity and income. Important among ecologic reasons are C sequestration in soil to mitigate the greenhouse effect and minimizing risks of pollution, contamination and eutrophication of natural water. The objective is to resequester the C in soil lost through land misuse and soil mismanagement, and by

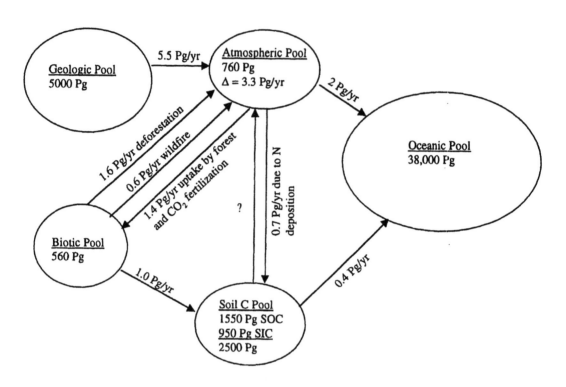

Figure 1. Principal global C pools and fluxes between them (1 Pg = petagram = 10^{15} g = billion tons).

so doing also restore degraded soils. Soil degradation is a widespread problem throughout the tropics and subtropics, and especially so in sub-Saharan Africa. There is a potential to sequester C through restoration of soils and degraded ecosystems in the tropics (Lal, 1997; Lal et al., 1999).

The objective of this chapter is to assess the role of land use, cropping systems and soil management practices on restoration of degraded soils in western Nigeria. The site-specific data on SOC dynamics and soil quality changes will be evaluated for identifying strategies to restore soils for C sequestration and productivity enhancement.

II. Material and Methods

Field experiments, involving twin watersheds began in 1975 at the research farm of the International Institute of Tropical Agriculture (IITA), Ibadan, Nigeria. Details of these experiments are presented elsewhere (Couper et al., 1979; Lal, 1984; 1985). Twin watersheds were managed with no-till and plow-till methods of seedbed preparation for growing 2 crops of maize (*Zea mays*) per year for 6 consecutive years from 1975 to 1980 (Figure 2). During this period, soil under no-till method of seedbed preparation with mechanized sowing and harvesting was severely compacted. Soil bulk density of the surface 0–10 cm layer increased from 1.4 to 1.66 Mg m⁻¹. In comparison, the plow-till watershed underwent severe soil degradation due to accelerated soil erosion, compac-

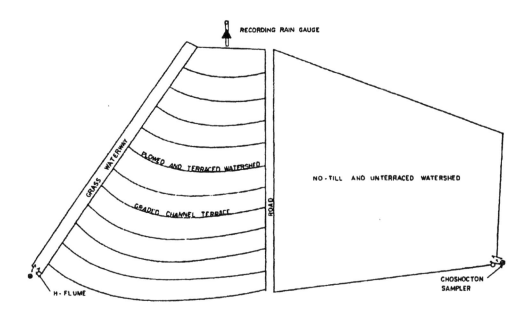

Figure 2. A schematic layout of two watersheds managed with conventional plowing and terraced system (left) and the no-till unterraced method (right); note the location of the recording rain gauge and H-flumes. (Adapted from Lal, 1984.)

tion, and depletion of soil fertility. There were also drastic reductions in maize grain yield during the 5th and 6th year of cultivation. The annual (total of both seasons) maize grain yield for the no-till watershed was 2.8 Mg ha⁻¹ in 1975, 4.5 Mg ha ¹ in 1976, 4.8 Mg ha⁻¹ in 1977, 5.0 Mg ha⁻¹ in 1978, 3.8 Mg ha⁻¹ in 1979 and 3.0 Mg ha⁻¹ in 1980. In comparison, the maize grain yield for the plow-till watershed was 2.7 Mg ha⁻¹ in 1975, 4.0 Mg ha⁻¹ in 1976, 3.9 Mg ha⁻¹ in 1977, 4.0 Mg ha⁻¹ in 1978, 2.9 Mg ha⁻¹ in 1979 and 1.0 Mg ha⁻¹ in 1980.

Because of severe soil degradation, both watersheds were taken out of maize production and put under restorative land use for 3 years from 1981 to 1983. There were four restorative land use treatments imposed on sub-plots within each watershed. These treatments were as follows:

1. Pigeon pea (*Cajanus cajan*) - maize
2. *Leucaena leucocephala* - maize
3. Natural regrowth
4. *Mucuna utilis* - maize - cassava (*Manihot esculenta*)

Leucaena was drilled in rows 4 m apart with maize during the first season in 1980. Pigeon pea was interplanted with maize in the first season using a tall, perennial, acid tolerant variety. Mucuna was interplanted with maize in the first season. All maize plots received 0.5 Mg ha⁻¹ of dolomitic lime, 90 kg ha⁻¹ of N (30 kg at planting as 15-15-15, and 60 kg at 4 weeks after as calcium ammonium nitrate). Maize also received additional fertilizer at planting as 13 kg ha⁻¹ of P (30 kg of P_2O_5) as 15-15-15, 25 kg ha⁻¹ of Mg as $MgSO_4$ and 2 kg ha⁻¹ of Zn as $ZnSO_4$. All six treatments were allocated to two strips as replication for each former tillage treatment. Bulk soil samples (0–10 cm depth) were taken in 1980 and in 1983, air dried, and ground to pass through a 2 mm sieve. Soil samples were analyzed for SOC content by the wet combustion method, total soil nitrogen (TSN) by Kjeldahl and Bray-1 available phosphorus (ASA, 1986). Measurement of soil bulk density was made on undisturbed cores (7.5 cm diameter and 7.5 cm deep), and by the excavation method using a 20 cm x 20 cm quadrant. Soil bulk density was corrected for the gravel content assuming particle density of gravels at 2.65 Mg m^{-3}. Soil pools of SOC, TSN and Bray-P were computed using the corrected bulk density. Soil and plant yield data were analyzed for the analysis of variance using a completely randomized design.

III. Results and Discussion

Results on SOC dynamics are presented below separately for cropping phase and the restorative phase.

A. Cropping Phase (1975-1980)

1. SOC and Nutrient Pools

The data on SOC, TSN and Bray-P pools in the top 0–10 cm layer shown in Table 1 indicate that the SOC pool declined with the cultivation duration. The rate of decline of SOC pool, however, was more drastic for plow-till than no-till watersheds. The magnitude of decline of the SOC pool was 7 Mg ha^{-1} in plow-till compared with 2.8 Mg ha^{-1} in the no-till treatment, with a rate of decline of 1.8 Mg ha^{-1} yr^{-1} and 0.7 Mg ha^{-1} yr^{-1}, respectively. The decline in SOC pool was primarily in the labile fraction. There was also a decline in the TSN pool, and the rate of decline was similar in both tillage systems. The magnitude of decline of the TSN pool was 2.4 Mg ha^{-1} for the no-till compared with 2.7 Mg ha^{-1} for plow-till with corresponding rates of decline at 0.4 Mg ha^{-1} yr^{-1} and 0.45 Mg ha^{-1} yr^{-1}, respectively. The decline in N pool was primarily in the organic fraction, and may have been due to plant uptake, volatilization, leaching, and losses in runoff and soil erosion. The abiotic losses of TSN pool were apparently more in plow-till than no-till watershed.

In contrast to SOC and TSN pools, the Bray-P pool increased in both systems due to application of fertilizers (Table 1). The Bray-P pool increased by 33.4 kg ha^{-1} in no-till and 64.8 kg ha^{-1} in plow-till watersheds, with a rate of increase of 5.6 kg ha^{-1} yr^{-1} in no-till and 10.8 kg ha^{-1} yr^{-1} in plow-till. The data on P-dynamics in these soils show that rate of P application at 13 kg ha^{-1} yr^{-1} was high for the expected yield.

The data on SOC, TSN and Bray-P pools for 10–20 cm depths are shown in Table 2. Comparison of the data in Tables 1 and 2 for the same duration of cultivation show that both SOC and TSN pools in 10–20 cm depth increased for the plow-till compared with the no-till watershed. This may be due to incorporation of crop residue and biomass at the plow depth (15–20 cm) and soil inversion and mixing.

2. Inter-Relationship Among Pools

Inter-relationship among SOC, TSN and Bray-P pools are depicted in Table 3. The C:N ratio increased with cultivation duration from 4 to 8 in no-till and 4 to 6 in the plow-till watershed. In contrast, C:P and N:P ratios decreased with cultivation duration. The C:P ratio decreased from 1750 to 430 (with a rate of decrease of 221 yr^{-1}) for no-till, and from 2040 to 220 (with a rate of

Table 1. Temporal changes in soil organic carbon (SOC), total soil nitrogen (TSN) and Bray-P pools in top 0–10 cm layer

Cultivation duration (yrs)	SOC Pool (Mg ha⁻¹)		TSN Pool (Mg ha⁻¹)		Bray-P (kg ha⁻¹)	
	No-till	Plow-till	No-till	Plow-till	No-till	Plow-till
Cultivation (1975)	20.0	23.3	4.86	5.48	11.4	11.4
2	22.0	23.8	2.45	2.73	34.1	14.3
4	21.6	18.4	3.94	3.08	43.5	14.6
5	19.5	180	2.52	2.59	33.0	57.2
6	19.2	16.7	2.50	2.79	44.8	76.2

Field experiments were discontinued at the end of 1979.
C and nutrient pools are computed using bulk density corrected for gravel content.

Table 2. Tillage effects on soil organic carbon (SOC), total soil nitrogen (TSN) and Bray-P pool for 10–20 cm layer

Cultivation duration (yrs)	SOC Pool (Mg ha⁻¹)		TSN Pool (Mg ha⁻¹)		Bray-P (kg ha⁻¹)	
	No-till	Plow-till	No-till	Plow-till	No-till	Plow-till
5	17.4	20.1	2.16	2.79	13.3	29.3
6	16.6	16.6	1.92	3.03	16.6	53.8

(Recalculated from Lal, 1985.)

Table 3. Temporal changes in C:N, C:P and N:P rates of 0-10 cm layer with cultivation duration

Cultivation duration (yrs)	C:N		C:P		N:P	
	No-till	Plow-till	No-till	Plow-till	No-till	Plow-till
Pre-cultivation (1974)	4.1	4.3	1754	2043	426	481
2	9.0	8.7	645	1664	72	191
4	5.5	6.0	497	1260	91	211
5	7.7	6.9	591	314	76	45
6	7.7	6.0	429	219	56	37

decrease of 303 yr⁻¹) for plow-till watershed. The N:P ratio decreased from 426 to 56 (with a rate of decrease of 62 yr⁻¹) for no-till, and from 481 to 37 (with a rate of decrease of 74 yr⁻¹) for plow-till.

The increase in C:N ratio was apparently due to more rapid decline of the labile C fraction than of the TSN pool. In contrast, the decrease in C:P and N:P ratio was due to both decrease in C and N pools and increase in the Bray-P pools due to applications of phosphatic fertilizers. Therefore, the rate of increase of C:N ratio may be considered as an indicator of soil degradation.

162

Table 4. Effects of 5 years of tillage methods and continuous cultivation of corn on soil organic C content and other soil chemical properties of 0–10 cm and 10–20 cm depth

Soil properties	0–10 cm Plow-till	No-till	10–20 cm Plow-till	No plow	LSD Tillage	Depth	T x D
pH	4.71	5.27	5.00	5.41	***	***	**
SOC (%)	1.35	1.48	1.25	1.05	n.s.	***	***
TSN (%)	0.195	0.191	0.168	0.130	n.s.	***	**
Bray-P (mg kg^{-1})	42.8	25.0	17.7	7.5	***	***	**

n.s.,**, *** = non-significant, significant at 5% and 1% level of probability, respectively.

Table 5. The SOC pool 3 years after implementing the restorative treatments in western Nigeria

Treatment	SOC Pool (Mg ha^{-1}) No-till	Plow-till	Δ SOC (kg ha^{-1} yr^{-1}) No-till	Plow-till
Continuous maize (no-till)	19.9	16.0	+233	-233
Pigeon pea-maize	21.3	24.3	+700	+2533
Leucaena-maize	20.1	18.8	+300	+700
Pre-cultivation	20.0	23.3	—	—
Following 6 years of cultivation	19.2	16.7	—	—

SOC pool is computed using the bulk density corrected for gravel content. (Chapter 7)

3. Tillage Effects on Soil Chemical Quality

Soil chemical quality of 0–20 cm depth, as determined by complete analyses of soil data at the end of the fifth year of cultivation, is shown in Table 4. Decline in soil pH was more for plow-till than no-till treatment, and for surface 0–10 cm depth and 10–20 cm depth. Soil pH in plow-till treatments was about 0.5 unit less than in no-till for both depths. Tillage methods did not have a significant impact on SOC and TSN pools, both of which were signficantly more in the 0–10 cm than 10–20 cm depth. There existed a significant interaction between tillage method and depth for SOC and TSN pools. The SOC content of 0–10 cm depth was more in no-till than plow-till, and vice versa for 10–20 cm depth. The Bray-P level was significantly influenced by tillage methods, depth of sampling and interaction between tillage and depth. Bray-P level was more in plow-till than no-till treatment and in surface 0–10 cm than 10–20 cm depth (Table 4).

B. Restoration Phase

The data on SOC pool at the end of a 3 year restoration period are shown in Table 5. In comparison with continuous no-till maize, the SOC pool increased in all systems that involved establishment of woody perennials. Further, there were differences in the magnitude of increase in the SOC pool in watersheds managed by no-till and plow-till methods during the cultivation phase. The average increase in SOC pool (mean of all three restorative treatments minus the pool in control maize) was 1.2 Mg ha^{-1} for no-till compared with 5.0 Mg ha^{-1} for plow-till watershed.

Thus, the average rate of increase was 0.4 Mg ha^{-1} yr^{-1} for no-till compared with 1.7 Mg ha^{-1} yr^{-1} for plow-till watershed. The higher increase in SOC pool in the plow-till watershed was due to the fact that the soil pool was highly depleted and had a high potential to sequester C. Eventually, soil managed by both tillage systems would attain a similar level of SOC pool at the equilibrium level.

The rate of increase in SOC pool differed among historic tillage and restorative treatments (Table 5). The rate of increase in no-till watershed was 233 kg ha^{-1} yr^{-1} for no-till maize, 300 kg ha^{-1} yr^{-1} for *Leucaena*, 700 kg ha^{-1} yr^{-1} for perennial pigeon pea, and 1000 kg ha^{-1} yr^{-1} for natural regrowth. In contrast, the rate of increase of SOC pool in plow-till watershed was -237 kg ha^{-1} yr^{-1} for continuous maize, 700 kg ha^{-1} yr^{-1} for *Leucaena*, 1100 kg ha^{-1} yr^{-1} for natural regrowth, and 2533 kg ha^{-1} yr^{-1} for perennial pigeon pea.

There are several possible reasons for differential rate of increase in SOC pool in no-till and plow-till watersheds. Increase in SOC pool in continuous maize in no-till watershed was due to application of lime, MgSO$_4$ and Zn. Consequently, crop growth, biomass yield, and amount of crop residue returned to the soil increased. In contrast, decline in SOC pool in plow-till maize was due to continuous degradation of soil by accelerated erosion, frequent drought stress, low stand and poor crop growth, and less amount of biomass returned to the soil.

Natural regrowth was very quickly established in no-till watershed because tree stumps and roots were still alive and reestablished a thick stand even during the first season of establishment. In contrast, there were no perennials in plow-till watershed, where the natural regrowth primarily comprised of grasses and other seasonal weeds (*Talinum triangulare*). Consequently, the biomass production under natural regrowth was substantially more in no-till than plow-till watersheds.

There were also differences in *Leucaena* versus pigeon pea growth in both watersheds. The *Leucaena* was slow to establish with a minimal growth during the first two years. In contrast, perennial pigeon peas established quickly with a large biomass production. In no-till watershed, however, the biomass production under natural regrowth was more than under pigeon pea. In plow-till watershed, the biomass production in pigeon pea was an order of a magnitude more than under natural regrowth.

IV. Conclusions

The data presented support the following conclusions:

1. Continuous cropping of maize, with two crops per year, caused depletion of SOC, TSN pools, and acidification due to leaching of bases. There was also a decline in soil structure leading to crusting, compaction and decline in water infiltrability.
2. Severe soil degradation in plow-till compared with no-till watershed was due to runoff and soil erosion accentuated by plowing.
3. Establishment of woody perennials can lead to soil restoration and enhancement of SOC pool. Appropriate mode of soil restoration and species of perennials may depend on land use history especially the tillage methods.
4. The rate of C sequestration through restoration of degraded soils may range from 200 kg ha^{-1} yr^{-1} to 2500 kg ha^{-1} yr^{-1} depending on the management, and the potential of pool depletion.
5. Establishing perennial pigeon pea is an important strategy for restoration of compacted and degraded soils (Juo and Lal, 1997; Hulugalle and Lal, 1986). In addition to being a legume, it has a deep tap root system that improves soil structure and facilitates deep placement of biomass in sub-soil horizons.

6. Natural regrowth in plow-till watershed was less effective than that in no-till watershed because of the lack of woody perennials. Yet, the rate of SOC sequestration of 1100 kg ha^{-1} yr^{-1} was more than that under *Leucaene*. This high rate was due to the predominance of grasses especially the Gunies grass (*Panicum maximum*). Establishment of grasses in soils of the tropics can lead to SOC sequestration more than that of monoculture legumes (Fisher et al., 1994; 1995; Lal, 1999).

References

ASA. 1986. Methods of Soil Analysis: Chemical Methods. ASA Monograph 9. Madison, WI.

Batjes, N.H. 1996. Total C and N in soils of world. *European J. Soil Sci.* 47:151-163.

Couper, D.C., R. Lal and A.S.R. Juo. 1979. Mechanized no-till maize production on an Alfisol in tropical Africa. p. 147-160. In: R. Lal (ed.), Soil Tillage and Crop Production. IITA Proc. Series 2.

Eswaran, H., E. Van den Berg, P. Reich and J. Kimble. 1995. Global soil carbon resources. p. 27-43. In: R. Lal, J. Kimble, E. Levine and B.A. Stewart (eds.), *Soils and Global Change.* CRC Press, Boca Raton, FL.

Fisher, M.J., I.M. Rao, M.A. Ayarza, C.E. Lascano, J.I. Sanz, R.J. Thomas and R.R. Vera. 1994. Carbon storage by introduced deep-rooted grasses in the South American savannas. *Nature* 371:236-238.

Fisher, M.J., I.M. Rao, C.E. Lascano, J.I. Sanz, R.J. Thomas, R.R. Vera and M.A. Ayarza. 1995. Pasture soils as carbon sink. *Nature* 376:472-477.

Hulugalle, N. and R. Lal. 1986. Root growth of maize in a compacted gravelly tropical Alfisol as affected by rotation with a woody perennial. *Field Crops Res.* 13:33-44.

Hulugalle, N., R. Lal and C.H.H. ter Kuile. 1984. Soil physical changes and crop root growth following different methods of land clearing in western Nigeria. *Soil Sci.* 138:172-179.

IPCC. 1995. Climate change 1995: Impacts, adaptations and mitigation of climate change: scientific-technical analyses. Working Group I, Cambridge Univ. Press, Cambridge, U.K.

Jenkinson, D.S. and A. Ayanaba. 1977. Decomposition of carbon-14 labeled plant material under tropical conditions. *Soil Sci. Soc. Am. J.* 41:912-915.

Jenny, H. 1949. Comparative study of decomposition rates of organic matter in temperate and tropical regions. *Soil Sci.* 68:419-432.

Jenny, H. and S.P. Raychaudhuri. 1960. Effect of climate and cultivation on nitrogen and organic matter reserves in Indian soils, ICAR, New Delhi, India.

Juo, A.S.R. and R. Lal. 1977. Effect of fallow and continuous cultivation on the chemical and physical properties of an Alfisol in western Nigeria. *Plant Soil* 47:567-584.

Kang, B.T. and A.S.R. Juo. 1984. Soil management for intensive crop production. p. 383-393. In: R. Lal, P.A. Sanchez and R.W. Cummings, Jr. (eds.), *Land Clearing and Development in the Tropics.* Balkema, Rotterdam, The Netherlands.

Kang, B.T., S. Hauser, B. Vanlauwe, N. Sanginga and A.N. Atta-Krah. 1995. Alley farming research on high base status soil. p. 25-39. In: B.T. Kang, A.O. Osiname and A. Larbi (eds.), Alley Farming Research and Development, IITA, Ibadan, Nigeria..

Lal, R. 1976. No-tillage effects on soil properties under different crops in western Nigeria. *Soil Sci. Soc. Amer. Proc.* 40:762-768.

Lal, R. 1984. Mechanized tillage systems effects on soil erosion from an Alfisol in watersheds cropped to maize. *Soil Till. Res.* 4:349-360.

Lal, R. 1985. Mechanized tillage systems effects on properties of a tropical Alfisol in watershed cropped to maize. *Soil Till. Res.* 6:181-200.

Lal, R. 1989. Agroforestry systems and soil surface management of a tropical Alfisol. III. Soil chemical properties. *Agroforestry Systems* 8:113-132.

Lal, R. 1995. The role of residue management in sustainable agricultural systems. *J. Sust. Agric.* 6:47-60.

Lal, R. 1996. Deforestation and land use effects on soil degradation and rehabilitation in western Nigeria. II. Soil chemical properties. *Land Degradation and Development* 7:19-45.

Lal, R. 1997. Soil degradative effects of slope length and tillage methods. II. Soil chemical properties. *Land Degradation and Development* 8:221-244.

Lal, R. and D.J. Cummings. 1979. Changes in soil and microclimate after clearing a tropical forest. *Field Crops Res.* 2:91-107.

Lal, R., H.M. Hassan and J. Dumanski. 1999. Desertification control to sequester C and mitigate the greenhouse effect. p. 83-151. In: N.J. Rosenberg, R.C. Izaurralde and E.L. Malone (eds.), *Carbon Sequestration in Soils: Science, Monitoring and Beyond*, Battelle Press, Columbus, OH.

Lal, R. and B.T. Kang. 1982. Management of organic matter in soils of the tropics and sub-tropics. Proc. XIIth Int. Soc. Soil Sci., 8-16 Feb., 1982, New Delhi, India, Vol. 1:115-141.

Section III.

LAND USE AND CARBON POOL IN SOILS
OF TROPICAL AMERICA

Impact of Conversion of Brazilian Cerrados to Cropland and Pastureland on Soil Carbon Pool and Dynamics

D.V.S. Resck, C.A. Vasconcellos, L. Vilela and M.C.M. Macedo

I. Introduction

The Cerrados region occupies an area of 204 million hectares (24% of the Brazilian territory), localized mostly in the midwest region, and in small areas fringing other regions, such as the northeast and north. It is localized between the parallels 5°50' to 21°26' latitude north and south, respectively. "Cerrados" is a denotation for a savanna region where the vegetation is characterized by a high physiognomic diversity. It is called "Campo Limpo" if grass is dominant, "Campo Sujo" if shrubs and trees are dispersed in the grass, "Cerrado latu sensus" when shrubs, trees and grass are equally dominant, and 'Cerrado' when trees are dominant with some trees reaching 6 to 7 m high.

With respect to soil orders, Oxisols (Latosols) predominate, occupying 46% of the region. Other soils include Entisols (Quartz Sand), 15.2% and Ultisols (Podzólics), 15.1% (Adámoli et al., 1985). Latosols and Quartz Sand occur on gently rolling and flat uplands and are developed and managed by intensive mechanization. These soils are predominantly acid, with toxic levels of aluminum and/or manganese, low levels of calcium and magnesium and low cation exchange capacity with a predominance of pH dependent charges and high phosphorus adsorption. The annual rainfall ranges from 900 to 2000 mm. The rainfall distribution is bimodal, without a rainy and a dry season (Figure 1). About 80% of the rainfall occurs from October to March, with a short dry spell of two to three weeks called "veranico" occurring mainly in January and February. The annual mean temperature is 22°C in the south and 27°C in the north. The differences between the maximum and minimum temperatures in the region is around 4 to 5°C, diminishing to 1 or 2°C towards the north region (Amazonic region) (Adámoli et al., 1985).

The Cerrados region contributes significantly to national agricultural and forestry production. However, crop yields are low despite the technological advances being made. Due to inadequate management systems, these soils are easily degraded, leading to decreases in crop yields, losses of organic matter, structure degradation, erosion, high runoff, sedimentation and contamination of the regions rivers.

This chapter describes the contribution of the Cerrados region to the national agricultural production and the impact of the conversion of Cerrados to cropland and pastures. The main focus is on soil carbon stocks and dynamics, soil and water quality and crops and beef cattle yields throughout these past decades.

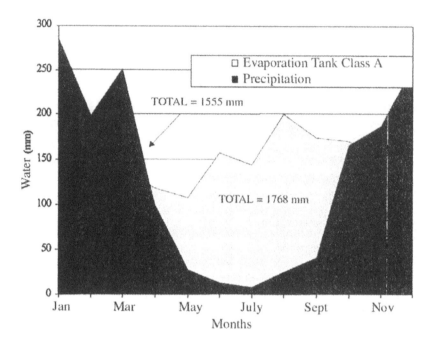

Figure 1. Cerrados region climatological data—average period 1973–1986. (Adapted from EMBRAPA, 1987.)

II. Evolution of Land Use in the Last Three Decades

Approximately 127 million ha of land in the Cerrados region are appropriated for agricultural activities. Since a federal regulation requires that 20% of each farm area be preserved without any deforestation, about 102 million ha are available for cultivation. From that an estimated 12 million ha are occupied with crops, 35 to 40 million ha are cultivated with introduced grass pastures and about 2 million ha have perennial crops (Macedo, 1994; IBGE, 1997), which leaves 55 million ha (54%) covered by natural vegetation but potentially cultivatable.

A. Impact on Grain and Beef Yield

Five states represent 66.3% of the Cerrados region and 15.9% of Brazilian terrritory. These states contribute significantly to national grain and beef cattle production.

For the 21 year period from 1975 to 1996, soybean area increased from 235,736 ha (4% of the total) to 10,115,570 ha (43% of the total), an increase of 4191% (Table 1). Soybean yield also increased by 50%, from 1463 kg ha^{-1} to 2198 kg ha^{-1}. Contribution to national production went from 3% in 1975 to 43% in 1996, a phenomenal increase of 3161%.

The area under corn cultivation increased from 1,527,111 ha (14% of the total land cultivated with corn nationally) in 1975 to 3,296,787 ha (25% of the total land) in 1996, an increase of 116% in area cultivated in corn in the Cerrados region. Corn grain yield was 1542 kg ha^{-1} in 1975 and increased to

Table 1. Evolution of some crops and cattle herds in the Cerrados region, by state, 1975–1996

States	% of region	Crop or cattle herd	Area (ha)	Production (Mg) or cattle head	Yield (kg ha⁻¹)	Area (ha)	Production (Mg) or cattle head	Yield (kg ha⁻¹)
				——— 1975 ———			——— 1996 ———	
Minas	19	Soybean	68,578	80,073	1,168	531,107	992,536	1,869
Gerais		Corn	808,402	1,248,255	1,544	1,378,843	3,538,306	2,566
		Rice	532,793	433,090	813	288,816	498,695	1,727
		Beans	278,859	126,774	455	470,484	340,679	724
		Cattle[a]		11,086,756			12,689,459	
Mato	20	Soybean	325	638	1,960	1,956,148	5,032,921	2,573
Grosso		Corn	59,176	91,705	1,550	542,247	1,513,630	2,791
		Rice	139,609	214,576	1,537	429,086	722,293	1,683
		Beans	16,448	13,647	830	30,619	20,472	669
		Cattle[a]		1,868,830			6,583,993	
Mato	10	Soybean	111,233	156,135	1,404	831,654	2,003,904	2,410
Grosso		Corn	67,260	100,926	1,501	420,005	1,471,871	3,504
do Sul		Rice	443,486	528,092	1,191	87,032	253,096	2,908
		Beans	9,535	7,771	815	18,683	14,544	778
		Cattle[a]		4,365,668			14,188,303	
Goiás	17	Soybean	55,600	73,392	1,320	913,871	2,019,153	2,209
		Corn	589,995	1,182,091	2,004	930,011	3,700,820	3,979
		Rice	806,917	707,086	876	189,897	303,378	1,598
		Beans	207,193	105,018	507	83,742	115,396	1,378
		Cattle[a]		10,704,168			18,580,908	
Distrito	0.3	Soybean	0	0	0	34,733	67,056	1,931
Federal		Corn	2,278	2,536	1,113	25,681	105,284	4,100
		Rice	1,003	1,044	1,041	713	888	1,245
		Beans	1,725	602	349	4,333	7,114	1,642
		Cattle[a]		40,165			123,569	
Total	24	Soybean	235,736	310,238	1,463	4,267,513	10,115,570	2,198
		Corn	1,527,111	2,625,513	1,542	3,296,787	10,329,911	3,388
		Rice	1,923,808	1,883,888	1,092	995,544	1,778,350	1,832
		Beans	513,760	253,812	0,591	607,861	498,205	1,038
		Cattle[a]		28,065,587			52,166,232	
Brazil		Soybean	5,824,492	9,893,008	1,699	10,738,389	23,505,061	2,189
		Corn	10,854,680	16,334,516	1,505	13,391,844	31,993,970	2,389
		Rice	6,821,000	7,781,538	1,466	3,920,590	9,990,310	2,548
		Beans	5,306,270	2,282,466	0,551	4,946,552	2,822,467	571
		Cattle[a]	4,145,916	101,673,753			155,134,073	

[a]Data related to 1993. (Adapted from IBGE, 1997.)

3388 kg ha^{-1} in 1996, i.e., an increase of 120%. Contribution to the total national production of corn was 16% in 1975 (2,625,513 t) and increased to 32% in 1996 (10,329,911 t), i.e., an increase of 293%. The area sown in rice in 1975 was 1,923,808 ha (36% of the total land). This decreased to 995,544 ha (25% of the total land) in 1996, a 52% decrease in the area cultivated in rice. However, rice yield increased 68% (1092 kg ha^{-1} to 1832 kg ha^{-1}) during the same period. Consequently, rice contribution to national grain production dropped from 24% (1,883,888 t) in 1975 to 18% (1,778,350 t) in 1996, with an overall decrease of 36%.

The land area cultivated in beans increased from 513,760 ha (12% of the total land) in 1975 to 607,861 ha (22%) in 1996, with an overall increase of about 18% in 21 years. Crop yield was 551 kg ha^{-1} in 1975 and increased to 1038 kg ha^{-1} (88% of increase), due to cultivation of irrigated beans. The total grain production in 1975 was 253,812 t (11% of the national production) and increased to 498,205 t in 1996 (18% of the national beans production), representing an overall increase of 96%.

Total Cerrados cattle livestock population was 28,065,587 head (28% of the country) in 1975. This increased to 52,166,232 head (34% of the country) in 1993, an increase of 86%.

B. Impact on the Environment

Soils from the Cerrados region contain medium to high contents of organic carbon (Figure 2). Only 17% of these soils have less than 0.87 dg kg^{-1} of organic carbon, and these mostly are Quartz Sands. However, the transition from Cerrados natural vegetation to agroecosystems results in soil organic carbon decay and its consequences, such as soil, water and nutrient losses, soil structure degradation and decrease of water availability.

The use of inappropriate implements to cultivate soils, such as heavy disk harrows, has a negative effect on soil structure. Soil loses its physical protection and organic matter as CO_2 evolved from those agroecosystems, and this diminishes water availability and crop yields.

Oxisols from that region are normally resistant to soil erosion, the intensity of which depends on the type of implement used and on the cover crop. Using only a disk and plowing downhill, Dedecek et al. (1986) found that Dark Red Latosol loses an average 53 t ha^{-1} yr^{-1}, which represents a layer loss equal a 5.3 mm ha^{-1} (Figure 3).

On that same soil, Resck (1981) found nutrient losses in the following order under soybean crop: Ca > Mg > K > P and Al. The runoff enrichment index (amount of nutrients in the runoff divided by that determined in the soil before the experiment) was 14 for phosphorus, 2 for Ca, Mg, and K, and 0.05 for Al. For crops such as corn and rice, nutrient losses would be much higher. Since 10.1 million ha are being cultivated with soybean, 3.3 million ha with corn, and almost one million with rice, these losses can represent a considerable damage to the environment. Management thus must be done to prevent soil organic matter loss and its consequences.

By the year 2025 in the world, probably less than 2 ha will remain per habitant for food and energy production. Particularly in the Amazon and Savannas regions, thousands of hectares of native vegetation are being deforested and converted to pasture for cattle ranching and agricultural settlement.

There is evidence that the temperature in the biosphere is increasing, and this phenomenon can accelerate soil organic matter decomposition, increase CO_2 levels in the air, and cause a further increase in the global warming (Jenkinson et al., 1991).

Changes in the climate pattern can alter the total carbon storage in soil. For example, because of current alterations in CO_2, regions in higher latitudes will became substantially warmer and, probably, dryer. A vicious cycle is possible, with global warming accelerating the decomposition of soil organic matter, CO_2 increasing, and finally increasing global temperature. If world temperatures rises

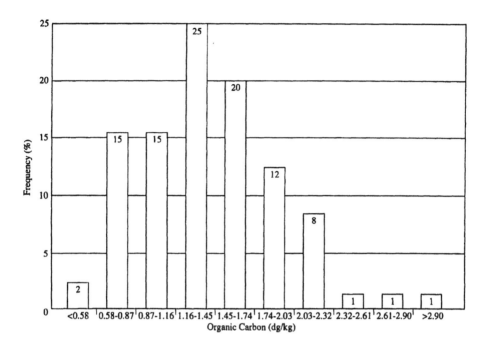

Figure 2. Soil organic carbon content in the Cerrados region. (Adapted from Lopes and Cox, 1977.)

uniformly by 0.03 °C yr^{-1}, the extra release from soil organic matter over the next 60 years will be nearly 61x10^{15}g C. This value is roughly 19% of the CO_2 carbon that will be released by combustion of fossil fuel during the same period (Jenkinson et al.,1991).

III. Dynamics of Soil Organic Carbon

A. Soil Carbon Stock

The amount of organic carbon in soils at a particular time is a function of C decomposition rate, annual C input, soil temperature, soil moisture content, soil type and microbial biomass characteristics. As a result, soil carbon has a high correlation with climate. Post et al. (1982) in a survey based on 2700 soil profiles sampling the soil life zone, showed that global soil organic carbon could be estimated as 1395 x 10^{15} g. For tropical woodland and soils from tropical savannas, with an approximate area of 24 x 10^8 ha, the soil carbon was estimated as 130 x 10^{15} g and the annual input of carbon as 12 x 10^{15} g yr^{-1} (Jenkinson et al., 1991).

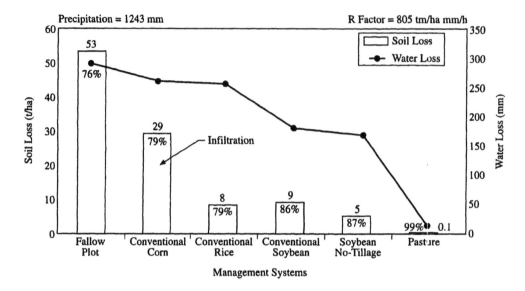

Figure 3. Soil and water losses under different management systems. (Adapted from Dedecek et al., 1986.)

Soil carbon generally increases with increasing rainfall and for any particular level of precipitation decreases with increasing temperature. Roughly 10% of the total soil organic carbon pool can be affected by climate variations.

Grisi (1997) observed that temperate soils have greater C biomass and are affected more by changes in their biomass as air temperature increases than are soils from tropical savannas.

Clayey Latosol under Cerrado natural vegetation, despite being very poor chemically, has high organic carbon content down 120 cm in the profile (Figure 4). The majority of the soil negative charged sites come from the organic matter, represented by H + Al.

After clearing, different land use and tillage systems also may alter soil carbon stock. Undisturbed systems such as long-term pasture, eucalyptus plantation and no-tillage cropping, can accumulate more carbon in a 100-cm profile than natural vegetation of "Cerrados". Corazza et al. (1998) observed lower carbon accumulation in plots where disk plows and heavy disk harrow were used (Table 2).

Since these cultivated systems are at least 12 years old, one can conclude that undisturbed systems can accumulate more carbon than is found in the natural systems of the Cerrado area, (Table 3). Disturbed systems, such as disk plow and heavy disk harrow systems, were not significantly different but accumulated less carbon than undisturbed systems because of the aggregate disruption.

Physical protection of organic matter was destroyed in this case. However, the disk plow has an important role increasing soil pH and bases saturation, besides increasing phosphorus and potassium availability in order to alleviate soil fertility constraints. Disk plow tills soil and enhances soil chemical

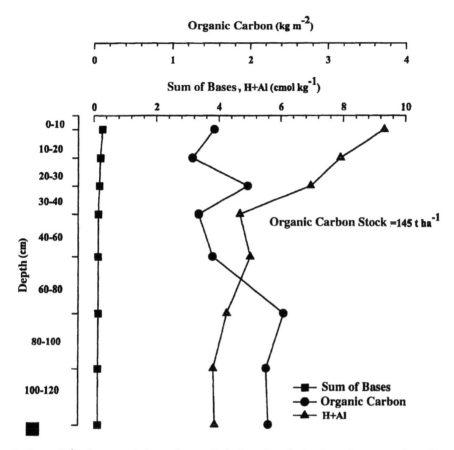

Figure 4. Organic carbon stock in a clayey Oxisol under virgin Cerrado vegetation. (Resck et al., 1997, unpublished data.)

Table 2. Carbon stock in soil layers under different management systems

Soil layer	Management systems					
	No-till	Disk	Harrow	Pasture	Eucalyptus	Cerrados
(cm)	Mg ha[-1]					
0–20	47.35 (31)[a]	37.34 (30)	36.51 (29)	42.18 (28)	44.87 (30)	39.77 (30)
20–40	33.79 (22)	29.25 (23)	28.76 (23)	32.59 (22)	33.50 (23)	30.09 (23)
40–100	73.83 (47)	62.23 (47)	60.00 (48)	75.44 (51)	69.81 (51)	63.72 (47)
Total	154.97 a[b]	128.81 bc	125.28 c	150.22 a	148.18 a	133.59b

[a]Values between parentheses represent C stock percentages in each soil layer.
[b]Same letters are not statistically different—Tukey ($p<0.05$).
(Adapted from Corazza et al., 1998.)

Table 3. Estimates of effects of different management systems on total carbon balance in relation to natural vegetation of Cerrado (clayey Dark-Red Latosol, Oxisol)

System	C stock (Mg ha^{-1})	Time (years)	C stocking rate (Mg ha^{-1} year^{-1})
No-tillage	+21.4	15	+ 1.43
Pasture	+16.6	18	+ 0.92
Eucalyptus	+14.6	12	+ 1.22
Disk plow	- 4.8	15	- 0.32
Heavy disk harrow	- 8.3	12	- 0.69

(Adapted from Corazza et al., 1998.)

reaction, facilitating uniform distribution of lime into the soil profile. This action can reach up to 25 to 30 cm deep under favorable soil moisture and friable consistency, whereas a heavy disk harrow can reach only 12 to 15 cm into the soil profile, cutting the soil surface without mixing. The little difference in soil carbon loss found between disk and an area cultivated by heavy disk harrow for 10 years was caused by a pasture growing in that same place in the last two years, which recovered to a certain extent soil aggregation, thus protecting soil organic carbon from microbial attack (Santos et al., 1996).

Implements such as a rototiller also can destroy organic matter's physical protection and cause a tremendous loss of organic carbon. It is the worst implement for cultivating soils, but largely used for horticultural purposes. In an experiment conducted in a clayey Dark-Red Latosol, 15 species of green manure were grown and then incorporated with a rototiller. In the following year soybean was grown and also incorporated with the same implement. Because of a complete soil structure destruction, large amounts of C-CO$_2$ were lost (21 Mg ha^{-1}), when compared to a degraded pasture field (Figure 5). These green manure species produced an average 6650 kg ha^{-1} $\pm$ 1636 kg.ha^{-1} of dry matter. At the beginning, the organic carbon stock was 38 Mg ha^{-1}.

Cultivating corn or soybean under no-tillage and conventional tillage for 16 years in a clayey Dark-Red Latosol demonstrated that, regardless of cropping type, no-tillage stored 7% more organic carbon in a 40-cm soil layer than disk plow (Figure 6). For all these years, no-tillage could store 101 Mg ha^{-1} into the profile against 94.6 Mg ha^{-1} for conventional tillage, for both crops studied. Both systems stored more carbon than a degraded pasture field colonized by *Brachiaria decumbens* for 17 years consecutively.

B. The Dynamics of Soil Nutrients and Soil Carbon Loss

Soil organic matter is the key to successful and sustainable agriculture in tropical soils. Soil organic matter positively affects structure, aggregation, cation exchange capacity, microbial activity and water holding capacity. Furthermore, soil organic matter is the major soil component that affects availability and mobility of macro and micro-nutrients to plants.

Nicolardot et al. (1994) suggested that roughly 60% of the soil organic matter is very resistant to decomposition but, considering tropical conditions, that value could be less. As mentioned by Jenkinson and Ayanaba (1977), the addition of plant residues (corn and soybean) had a great stimulating influence on the decomposition of native organic matter. The rate of plant residue addition had a great influence upon the rate of residue decomposition. A decrease in biological activity follows exposing fresh residues on the topsoil to a rapid drying process. At the same time, mineralization of the residues is an important source of plant nutrients in farming systems receiving low external inputs.

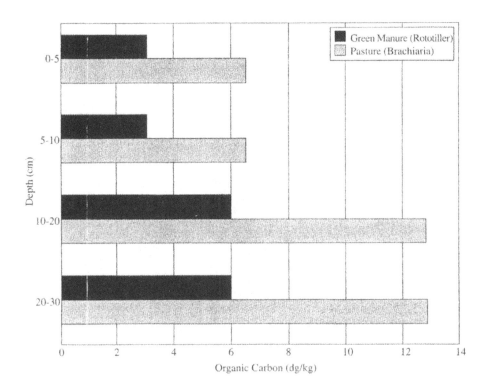

Figure 5. Losses of organic carbon stock in a clayey Oxisol profile by rototiller. (Adapted from Resck et al., 1991.)

Soils with low active decomposition substrates may have soil microbial biomass in a dormant state or resting with low respiration and low turnover rates. The addition of any carbon source into the soil can increase the decomposition rate, and apparently, can accelerate the decomposition of native organic matter. That process was called the "priming effect" by Jenkinson (1966), who considered four mechanisms to explain (1) changes in the rate of native soil organic matter decomposition, (2) non-uniform substrate, (3) changes in soil pH, and (4) changes in microbial activity (numbers of microorganisms, metabolic capacity, and metabolites). Brookes et al. (1987) found evidence that humidified organic matter could provide at least some of the energy required for soil biomass. Joergensen et al. (1990), working with a silty clay loam soil, pH 4.5, showed that the sum of the organic C remaining at the end of incubation was between 99% and 103% of that originally present, without significant losses. Probably the pH could influence the stability of the biomass and the organic matter.

The soil microbial biomass is an important living portion of soil organic matter, more than 500 μm^3, besides living plants, roots, and organisms. It is one of the main sources of organic residues and can be considered the agent responsible for nutrient cycling to plants. Jenkinson and Powlson (1976), in a series of five papers, showed that measurement of the flush of $C-CO_2$ evolved after $CHCl_3$ fumigation could be used as a measurement of soil biomass. Grisi (1997) showed that estimates of biomass of total carbon were 1.22% for savanna soils and 2.08% for forest soils.

Chaussod et al. (1988) developed an experimental procedure to measure the turnover time of C biomass in the soil, by measuring the declining of a C-labelled biomass which developed after addition of C-labelled plant material. As pointed out by Jenkinson and Parry (1989), the internal recycling of biomass metabolites can affect the estimate of turnover time. Wu (1990) described the theoretical

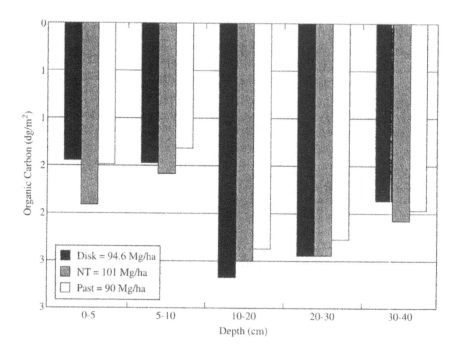

Figure 6. Organic carbon stock under different management systems in a clayey Oxisol. (Adapted from Resck et al., 1997, unpublished data).

basis for a newly developed experimental procedure to determine the biomass turnover time, after showing that internal cycling followed by addition of C-labelled glucose was negligible.

Joergensen et al. (1990) showed that incubation at higher temperatures (35°C) caused the biomass, in a temperate UK grassland soil, to decline rapidly. The biomass declined by 16% of its initial value during incubation at 35°C during 240 days. The corresponding biomass turnover was 139, 62, and 4 days at 15, 25, and 35°C, respectively. The biomass behavior and the suitable methodology for that kind of study at such temperatures in tropical soils is not well known.

A study by Vasconcellos (1998), as summarized in Figures 7 and 8, showed the biomass decreased as the temperature increased.

With glucose additions, the total biomass decreased by 79%, 64%, 79% and 73%, for Janaúba, Capinópolis, Woburn and Pegwell soils, respectively, following incubation at 35oC. Without glucose, the ^{12}C biomass decreased by 43%, 52%, 67%, and 31% for Janaúba, Capinópolis, Woburn, and Pegwell soils. Thus, the biomass which developed following glucose addition (new biomass) seemed to be more sensitive to higher temperatures than the initial biomass.

At 15°C, with glucose addition, Janaúba and Pegwell soils showed the same biomass pattern as the native organic C (see unamended treatment); Capinópolis and Woburn soils decreased the native organic ^{12}C biomass (Table 4). The same behavior was reported by Anderson and Domsch (1985), and Brookes et al. (1987) . The corresponding decline at 35°C was 16 to 45%. It is important to note that

Table 4. Final amounts of ^{12}C biomass after incubation at 15°C and 35°C for 90 and 70 days, respectively

Soil	^{12}C-biomass µg g^{-1} dry soil			
	Unamended soil		Amended soil	
	15°C	35°C	15°C	35°C
Janaúba	112	64	120	34
Capinópolis	305	145	217	93
Woburn	187	62	148	42
Pegwell	654	449	672	221

(Adapted from Vasconcellos, 1998.)

microbial biomass from soils of temperate regions decreased more intensively from changes in temperature than did soils from tropical regions (Grisi, 1997).

Grisi (1997) showed the variability of C biomass of soils under savannas and forest when the incubation occurred at 15°C and 35°C (Figure 9). At 15°C, C biomass increased for up to 10 days of the incubation period, possibly as a consequence of microbial population growth. At 35°C, the biomass increased as well after 10 days, but decreased between 25 and 50 days, partly because of the increasing of extracted C from the unfumigated samples, i.e., low biomass was obtained from the fumigated samples.

Microbial biomass highly depends on crop management practices and soil moisture content, which relate to seasonal changes. Carbon mineralization directly relates to the crop residues-C returned to soil, which also depends on management practices. Apparently, interaction between crop stem and remains of pruning results in a system that oscillates between N immobilization and mineralization.

Henrot and Robertson (1994) measured the C biomass in two humid tropical soils. Both soils showed a similar pattern in total soil organic matter and C biomass decline following the removal of the vegetation. After 3 years, total C and total N were reduced by 20%. C biomass stabilized at 35% of its initial value following tropical forest clearing.

Soil biomass, as a portion of the active organic component of the soil, is considered a sink and source in the decomposition process, as well as potential plant nutrients.

Vasconcellos (1994), comparing the CO_2 evolution rate among soils from tropical savannas and temperate regions, showed the importance of pH and the significant effect of total soil organic carbon on CO_2 evolution. No significance was found for the C/N ratio and soil clay content. From unamended and amended soils amended with ^{14}C labelled glucose (1000 µg g^{-1}soil) and incubated at 15°C and 35°C, total CO_2 evolved was higher in soils from temperate regions (Woburn and Pegwell) than soils from tropical savannas (Janaúba and Capinópolis) (Table 5). In soils with low active decomposition rates, the biomass is dormant or resting with low respiration and low turnover rates. Addition of glucose to soil increased the decomposition rate and apparently accelerated decomposition of native organic matter in tropical savanna soils. This was not observed in the temperate soils studied. It means that microbial activity of tropical soils, with low fertility, can be stimulated by glucose amendment (Table 6).

In relation to the ^{14}C evolved at 15°C, there were not large differences between soils from tropical and temperate regions. From the U.K., Woburn soil has evolved more CO_2 than the others. Total labelled glucose evolved on 63.1%, 69.3%, 73.6% and 60.6% from Janaúba, Capinópolis, Woburn and Pegwell soil, respectively (Table 7).

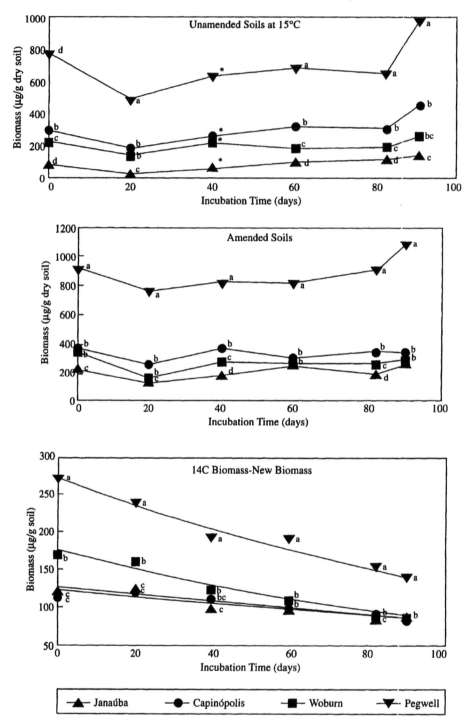

Figure 7. Biomass C variation at 15°C as incubation temperature. Symbols are for observed data and filled lines to fitted points. Biomass values for each individual soil followed by different letters are significantly different (Duncan $p < 0.05$). (Adapted from Vasconcellos, 1998.)

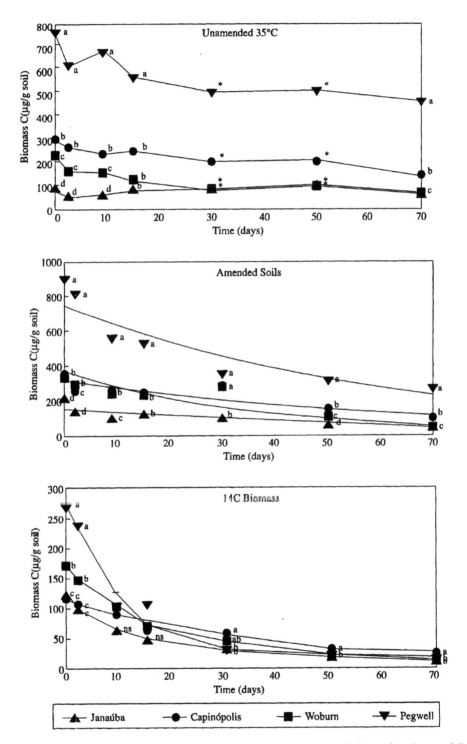

Figure 8. Biomass C variation at 35°C as incubation temperature. Symbols are for observed data and filled lines to fitted points. Biomass values for each individual soil followed by different letters are significantly different, (Duncan $p < 0.05$). (Adapted from Vasconcellos, 1998.)

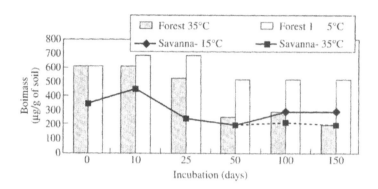

Figure 9. Evolution of C biomass in Brazilian soils over 150 days, at 15°C and 35°C.
(Adapted from Grisi, 1997.)

Table 5. [14]C biomass, C balance at 15°C and 35°C of incubation temperature, in amended and
unamended soils with C-glucose, from tropical and temperate regions

Soils[a]	[14]C biomass		CO$_2$ evolved[b]		Residue	
	15°C	35°C	15°C	35°C	15°C	35°C
			— μg g^{-1} soil —			
Janaúba	92	11	640	683	268	306
Capinópolis	92	19	714	705	194	276
Woburn	100	9	734	795	166	196
Pegwell	167	7	606	691	225	302

[a]Janaúba and Capinópolis are soils from Cerrados of Minas Gerais State, Brazil; Woodurn and
Pegwell are from England; [b]Incubation period 90 days.
(Adapted from Vasconcellos, 1994.)

At 35°C, during the period of 25 to 95 days of incubation, the soils from temperate regions evolved
more [14]C: 67.2%, 70,0%, 78.4% and 68.4% of ^{14}C added at the beginning from Janaúba, Capinópolis,
Woburn and Pegwell soil (Table 6).

As Table 8 shows, the addition of glucose to soil accelerated the turnover of native organic C. That
acceleration was higher at 35°C, mainly in soils from temperate regions.

Table 9 gives results of one experiment carried out in the Cerrados region by Vasconcellos et al.
(1997), where one of the subjects studied was the correlation between evolved CO$_2$ and soil moisture
content. The rate of CO$_2$ evolved from each treatment was affected differently by soil moisture
content, probably due to the quality and quantity of organic matter and biomass imposed by soil

Table 6. CO_2 evolved at 35°C from 25 to 95 days of incubation period

	CO$_2$ evolved		
	Without glucose		With glucose
		μg C g^{-1} soil	
Soil	^{12}C	^{14}C	^{14}C + ^{12}C
Janaúba	371c	160c	567bc
Capinópolis	601c	156b	794b
Woburn	865b	173a	1003b
Pegwell	1578a	200a	1720a
C.V.%	5.1	7.5	12.8
DMS	92	27	27.4

Values followed by the same letter vertically are not significantly different, Duncan ($p<0.05$).
(Adapted from Vasconcellos, 1994.)

Table 7. Total CO_2 evolved at 15°C during a 110 day incubation time

	CO$_2$ evolved		
	Without glucose	With glucose	
		μg C g^{-1} soil	
Soil	^{12}C	^{14}C	^{14}C + ^{12}C
Janaúba	330c	631c	986c
Capinópolis	211c	693b	1088c
Woburn	639b	736a	1319b
Pegwell	1007a	606c	1491a
C.V.%	11	7.5	5.8
DMS	12	26	149

Values followed by the same letter vertically are not significantly different, Duncan ($p<0.05$).
(Adapted from Vasconcellos, 1994.)

Table 8. Glucose effect on native soil carbon

| | Percentage of turnover increase over native C | |
Soil	15°C	35°C
Janaúba	+ 12	+ 27
Capinópolis	+ 47	+ 39
Woburn	- 8	+ 47
Pegwell	+ 22	+ 47

(Adapted from Vasconcellos, 1994.)

Table 9. Correlation between evolved CO_2 and soil water content in a savanna soil Red Latosol (Oxisol), under different management systems

Management		Angles for correlation between CO_2 and soil moisture	Correlation (r value)
Tefrósia sp (green manure)		74.3t	0.703**
No-tillage and crop rotation		39.0t	0.792**
Conventional tillage and crop rotation		23.8t	0.552n.s.
Native savanna		38.0t	0.760**
30% lime requirement[a]	Maize	52.3t	0.755**
	Soybean	69.4t	0.539n.s.
100% lime requirement	Maize	33.4t	0.532n.s.
	Soybean	91.1t	0.863**
100% lime requirement	Maize	55.2t	0.719**
+ gypsum[b]	Soybean	53.8t	0.629*

[a]Lime requirement = CEC (60-actual base saturation)/100; [b]Gypsum = 6 t ha^{-1}.
n.s., *, ** = not significant, and significant at 10% and 5% respectively; t is time in days.
(Adapted from Vasconcellos et al., 1994.)

management. The relationship between C biomass and soil organic carbon derives largely from microbial population size and the amount of degradable C sources in soil. For Clement and Williams (1967), the CO_2 evolution depends on the organic matter C/N ratio.

Vasconcellos et al. (1997) observed a correlation between CO_2 evolution and available nitrogen extracted with 0.5N K_2SO_4. However, this was significant only for soil samples obtained from treatments of *Tefrosia*, maize and soybeans with 100% of the lime requirement (Table 9). *Tefrosia* spp showed high CO_2 evolution, probably because of their biomass activity. Supposedly this was due mostly to the C/N ratio of the senescent leaves and their ability to fix nitrogen. Limestone applied to soybean treatment, with 100% of the lime requirement, increased organic matter decomposition and biomass activity, which could explain the high correlation obtained.

C. Changes in Soil Carbon Pools and the Impact on the Physical, Chemical, Physicochemical and Microbiological Properties of Soil

The destruction of the physically protected pool by implements such as a heavy disk harrow which farmers insist on using because of the short time required to till soil and prepare for seeding, is accentuated, as Figure 10 shows. In a Latosol with more than 70% clay, soybean was grown for 9 years after 2 years with rice. The soil organic carbon increment detected during the first two years of cultivation with rice is not explained solely by the organic carbon increment due to these crop residues. Supposedly most of it came from roots and branches of the natural vegetation (normally, approximately 22 t ha^{-1} of dry matter or 12.8 t ha^{-1} of organic carbon), which in the following years decomposed and were detected by soil analyses.

In the third year, the farmer grew soybean, which requires a higher pH and larger amounts of phosphorus, potassium and micronutrients, besides symbiotic N fixation by bacteria. After 9 years

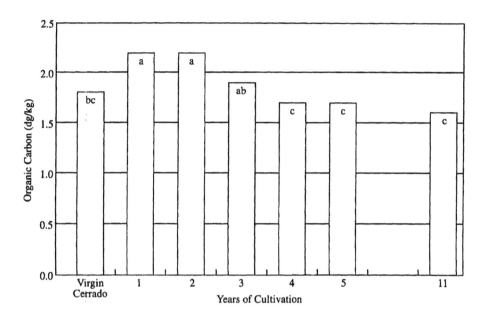

Figure 10. Effect of soybean residues and heavy disk harrow on soil organic carbon—Latosol 70% of clay. (Adapted from Resck et al., 1991.)

with soybean, the percentage of aggregates > 2mm at 0 to 5 cm depth within the implement zone of action dropped from 92% to 62% and the percentage of aggregates < 2 mm (>1, >0.5, >0.25, > 0.10 and < 0.10 mm) increased from 1.6% in average to 8% for all sieve diameters, which shows the destruction of the physically protected organic matter pools and redistribution among lowers sieve diameters. At 20 to 30 cm depth, soil aggregates remained unchanged because the implement cannot reach so deep into the profile.

Evaluating the effect of soil texture in mineralization rates of C and N, Hassink et al. (1991), verified that in sandy soils decomposable organic nitrogen was higher than in loamy and clayey soils, but the same was not observed for C. Two protection mechanisms could explain the difference: the micropore system of clayey soils and the association between amorphous and undefined organic matter with clay particles in sandy soils. Loamy soils seem to have both simultaneously. Physical protection of the soil organic matter can restrict the rate of demposition by microorganisms.

Carbon concentration in soil biomass, as a percentage of total organic carbon, is estimated in the range of 0.8 to 4.0%, as mentioned by Jenkinson and Ladd (1981). Cerri et al. (1985), studying an Oxisol from the Amazon region, showed values between 0.73% for natural vegetation and 0.04% for recently burned forest. Under natural vegetation, microbial biomass was concentrated on the 15 cm of the soil surface. Deforestation and burning were followed by a 2/3 decrease in the microbial biomass, which disappeared completely in the top 10 cm. In soil monitored for three years, after two years of cultivation, the microbial biomass was quantitatively the same as under natural vegetation. The distribution, however, was different in the soil profile. Over time, low soil biomass may cause plant nutritional problems, mainly with nitrogen.

Table 10. Profile statistical analysis—components of the equation

Equation complete	$Y = p_o + p_1 X_L + p_2 X_Q + p_3 X_Q$
Mean	$Y = p_o$
Linear component	$Y = p_1 X_L$
Quadratic component	$Y = p_2 X_Q$
Cubic component	$Y = p_3 X_Q$

(Adapted from Pereira, 1997.)

Table 11. Interpolation coefficients for each depth (average)

Depths (cm)	Coefficients		
	X_L (linear)	X_Q (quadratic)	X_C (cubic)
2.5	-5.8	19.62093	-51.30986
7.5	-3.8	-2.386286	60.80235
15	-0.8	-20.39711	41.29568
25	3.2	-16.41154	-83.03967
35	7.2	19.57402	32.25171

Studies of that nature have not occured in the Cerrados region and are needed. A study was carried out in the "Cerrados" region of Brazil, in the central portion of the country, to estimate the impact of six management systems, including a natural vegetation as a check plot, on the soil macro and microaggregates and organic carbon content. The soil was a typical clayey Dark Red Latosol (53% clay) and the research site was the experimental station of the Cerrados Agricultural Research Center (EMBRAPA-CPAC). Soil samples were collected at five depths and analyzed for water stable aggregates, organic carbon, and other physical and chemical properties. Cation exchange capacity and mean weight diameter also were calculated. The software "Profile" ($p<0.10$) (Cowell, 1978) was used to detect statistical differences among treatments and determination of the coefficients as linear, quadratic and cubic ones (Tables 10 and 11).

The management systems studied were disk plow (disk), Cerrado (Ce) natural vegetation, eucalyptus (Euc), heavy disk harrow (Harrow), pasture (Past), and no-tillage (Notill). All treatments corresponding to the cultivated systems were at least 15 years old (average). The complete equations for these systems were determined and they appear in Table 12, which gives data for macroaggregates only (aggregates > 0.25 mm) according to Tisdal and Oades (1982). Aggregates stability analysis in water was run, and the sum of the percentages of soil aggregates left on sieves > 2mm, > 1 mm, > 0.25 mm were accounted for and expressed as macroaggregates.

Each system affected some of the soil's physical or chemical properties in a different way. Higher aggregate stability was found in the undisturbed systems or less disturbed systems (Ce, Euc, Past) but not in no-tillage system (Figure 11). The disk plow is an efficient implement in the process of homogenization of soil fertility properties.

The mechanism of aggregate formation by continuous pasture seems to be the most efficient. It showed the highest percentage of macroaggregate and intermediate values of organic carbon (Figure 12). No-tillage was the management system which conserved the most organic carbon. Plots prepared

Table 12. Complete equation for 8 to 2 mm sieve—macroaggregates calculations.

Treatments	Equations
Disk	$Y = 91.74^{***} + 0.45 X_L^{****} - 0.04748X_Q^{****} + 0.0055X_C^{*}$
Cerrado	$Y = 96.57^{****} + -0.15 X_L^{****} - 0.000013X_Q^{****} - 0.0077X_C^{*}$
Eucalyptus	$Y = 96.34^{****} - 0.06 X_L^{****} + 0.01913X_Q^{****} - 0.0055X_C^{*}$
H.D. Harrow	$Y = 96.70^{****} + 0.09 X_L^{****} - 0.00375X_Q^{****} + 0.0007X_C^{*}$
Pasture	$Y = 95.36^{****} + 0.25 X_L^{****} - 0.03523X_Q^{****} + 0.0020X_C^{*}$
No-tillage	$Y = 93.50^{****} + 0.08 X_L^{****} - 0.01954X_Q^{****} - 0.0030X_C^{*}$

****, ***, *, significant at 0.01%, 1% and 10%, respectively, by the t test.
(Adapted from Pereira, 1997.)

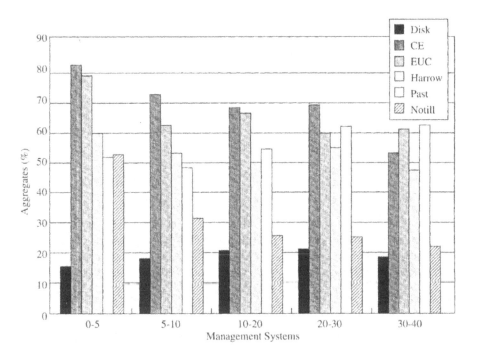

Figure 11. Aggregates > 2 mm in different management systems. (Adapted from Guedes et al., 1996.)

by heavy disk harrow for 10 years, followed by 2 years of pasture, recovered soil properties enough to make them similar to the undisturbed systems. During soil sample preparation, the selection of aggregates from the class between 8 and 2 mm favored the presence of more stable aggregates. These are possibly constituted of more stabilized organic material. The management systems did affect microaggregates in a different way in contrast to findings by Tisdal and Oades (1982).

Another experiment observed the decay of soil organic matter (SOM) with cultivation time and its effect on cation exchange capacity (CEC) in Quartz sands soils (AQ) and in a loamy (LVm) and in one clayey Red Yellow Latossols (LVa). These soils were cultivated during five years in the

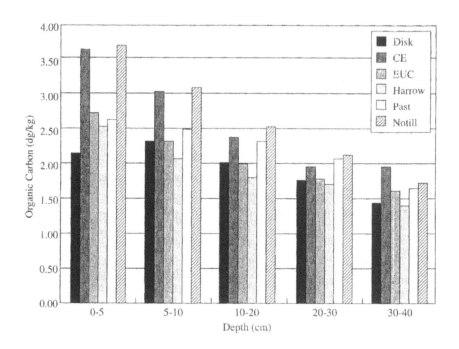

Figure 12. Organic carbon content into aggregates > 2mm of different management systems. (Adapted from Guedes et al., 1996.)

Cerrados region of western Bahia (Silva et al., 1994). An exponential decay model was used, based upon the proposition made by Woodruff (1949) and studied by Dalal and Mayer (1986) with 70 years of observation data in soils cultivated with cereals in south Queensland (Australia).

These observed data were fitted to the following model in order to analyze the soil organic matter behavior with time: $MO_t = MO_e + (MO_i - MO_e) e^{(-kt)}$, where MO_t = organic matter at moment t; MO_e = organic matter at equilibrium; MO_i = initial content of organic matter, k= rate of organic matter loss (% year^{-1}), and t = time.

Simple and multiple regression procedures also were used to estimate the relationship between CEC and SOM and between CEC and SOM + clay content. The rate of loss of SOM was 0.32, 0.30, and 0.24 % yr^{-1}, respectively, in AQ, LVm, and LVma (LVm with clay > 30% and LVa). The half-life ($t_{1/2}$) of SOM of LVma (2.90 yr) was a little higher than those of AQ and LVm (2.16 and 2.31 yr, respectively), and after five years relative losses of SOM from the initial stock were 73, 68, and 45%, respectively, in AQ, LVm and LVma (Figure 13).

The decay in CEC in relation to the initial values was 61%, 53% and 29% for AQ, LVm and LVma, respectively (Figure 14). SOM was responsible for 81% and 75% of the CEC variability in AQ and LVm soils. Clay content was important only for LVma (13%, against 71% of SOM contribution), which implies a different behavior of the clay fraction in these types of soils.

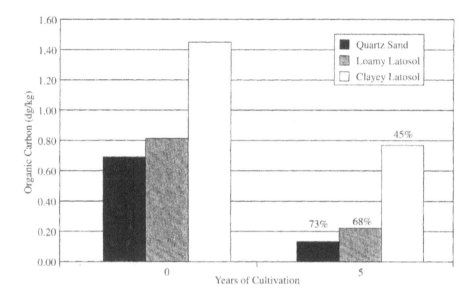

Figure 13. Organic carbon loss after 5 years of soybean cultivation and heavy disk harrow in the Cerrados region - Bahia State. (Adapted from Silva et al., 1994.)

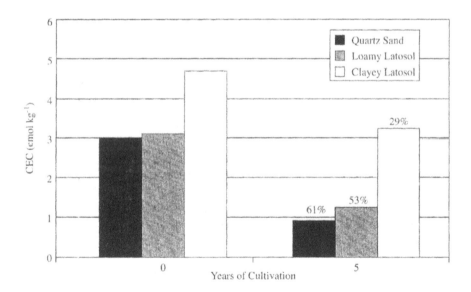

Figure 14. CEC decrease after 5 years of soybean cultivation and heavy disk harrow in the Cerrados region - Bahia State. (Adapted from Silva et al., 1994.)

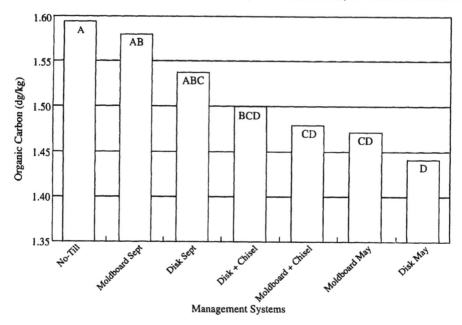

Figure 15. Effect of tillage systems on soil organic carbon content - average over 0 to 40 cm depth. (Adapted from Resck et al., 1997, unpublished data.)

Different tillage systems affect soil organic carbon as Figure 15 shows, not because of the different way each implement affects soil properties, but because of the season when crop residues are incorporated. Incorporating crop residues in September is before planting in the Cerrados region, whereas incorporating in May is after harvest and in this case plowing is repeated in September, right before planting, so the decomposition rate is higher and more CO_2 escapes from the system. In the case of disk plowing, organic carbon decreased drastically and its quality changed greatly since the decomposition rate was lower (-0.005) (Figure 16), a half-life of 12 years in contrast with 5 months for the moldboard plow in May (Figure 17).

Despite the high aggregation found in the pasture system, the CO_2 evolved from that system is higher than in the Cerrado one (Figure 18).

Disk after harvesting, i.e., in May, produces more CO_2 than no-tillage, which does not form aggregates and thus offers no physical protection, although it maintains high organic carbon content (Figures 12 and 15) which, presumably, is localized in the labile pool.

No-tillage has such high organic carbon content, despite the organic carbon location, because it evolves the lowest amount of CO_2 of all the systems studied.

IV. Soil Management for Sustainable Crop and Beef Production

The maintenance and improvement of soil organic matter levels are important concerns in farming systems, more so in savannas with higher mean annual temperature than temperate regions. It is very important to study and adjust adequate soil management to prevent decline in soil organic C, losses in mineralization of N and decreases in crop productivity.

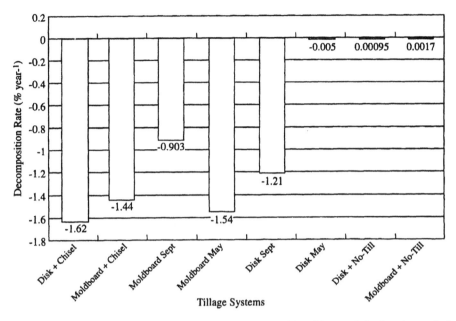

Figure 16. Organic matter decomposition rate in a clayey Dark Red Latosol during the period of 1986 to 1988. (Adapted from Resck et al., 1997, unpublished data.)

Soil management must maintain a harmony among four factors: a) liming to neutralize surface and subsoil acidity and Al^{+3}, b) applying corrective and maintenance fertilizers, c) establishing crop rotation, and d) alternating tillage systems (Resck, 1998). The timing and the intensity of each factor also matters. It is not worthwhile to apply phosphorus to soil, for example, without liming, nor is it worth applying both if cultivating only one type of crop and with only one implement.

Pasture in a crop rotation system is one of the best options for these soils, as Table 13 shows. A farmer in Uberlândia, state of Minas Gerais in 1983, had 1014 ha of cultivated pasture on his farm and no annual crop. His cattle herd size was 1094 animals. In 1985 he started growing annual crops for one, two or three years, cultivating pasture after that. In 1992 he had only 412 ha of pasture after annual cropping and practically the same herd size (1150 animals), corresponding to a stocking rate (animals/ha) of 2.8 compared to 1.1 in 1983.

At the Embrapa Cerrados, in a long-term experiment (22 years), two cultivation systems were studied: the annual crop rotation (10 years with soybean and 2 years with corn, and after that these crops were rotated annually) and annual crop/pasture (with soybean cultivated for 2 years and *Brachiaria humidicola* for 9 years, soybean again for 1 year and, after that having also soybean and corn rotated annually). Soil organic carbon decreased under soybean crop. However, pasture cultivated after 2 years of soybean increased organic carbon content from 1.62 dg kg^{-1} to 2.61 dg kg^{-1} up to 1987. From this point on, both treatments cultivated with soybean-corn rotation using disk plow plus light disk harrow showed a significant decrease in soil organic carbon content (Figure 19).

Undoubtedly pasture can stock organic carbon in soil despite the high respiration rate, which is counterbalanced by the formation and protection of physically protected organic pools in soil due to a highly developed and voluminous deep root system.

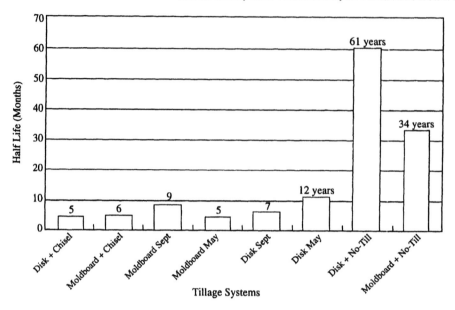

Figure 17. Organic matter half-life under different tillage systems in a clayey Dark Red Latosol. (Adapted from Resck et al., 1997, unpublished data.)

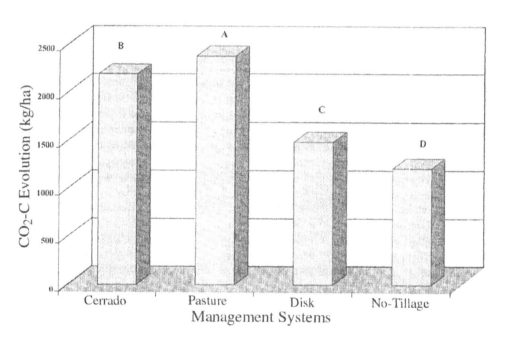

Figure 18. CO_2 evolution of different management systems during 314 days. (Adapted from Resck et al., 1996.)

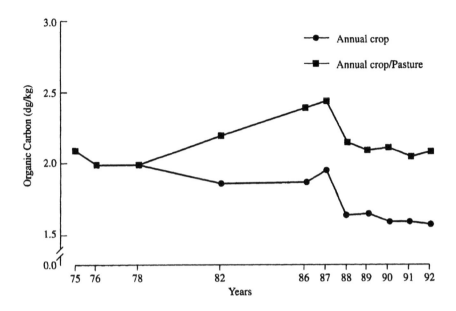

Figure 19. Soil organic carbon dynamics under annual crops and pasture. (Adapted from Sousa et al., 1997.)

V. Conclusions

- Soil organic carbon decomposition rates are high in the Cerrados region, mainly in loamy Latosols (54% of the region) and Quartz Sand (15.2%), depending on the type of tillage implements and crop management.

- In order to increase crop yields without degrading soil organic carbon and its correlated properties, farmers must manage tillage system dynamics or include pasture in the rotation system.

- Pasture seems to have a high capacity to recover degraded soil physical properties and a high capacity to store organic carbon in the soil profile.

- No-tillage systems accumulate organic carbon in the soil profile, but only maintain or improve soil physical properties very slowly, after years of cultivation.

References

Adámoli, J., J. Macedo, L. G. Azevedo, and J. M. Netto. 1985. Caracterização da região dos Cerrados. p. 33-74. In: W.J. Goedert (ed.), Solos dos Cerrados: tecnologias e estratégias de manejo. Planaltina: EMBRAPA-CPAC/São Paulo: Nobel.
Anderson, T.H. and K.H. Domsch. 1985. Determination of ecophysiological maintenance carbon requirements of soil microorganisms in a dormant state. *Biol. and Fertil. of Soils* 1:81-89.

Ayarza, M., L. Vilela, and F. Rauscher. 1993. Rotação de culturas e pastagens em solo de cerrado: estudo de caso. p. 121-122. In: Congresso Brasileiro De Ciência Do Solo, 24., 1993, Goiânia, GO. Anais. Goiânia: SBCS.

Brookes, P.C., A.D. Newcombe, and D.S. Jenkinson. 1987. Adenylate energy charge measurements in soils. *Soil Biol. and Biochem.* 19:21-217.

Cannel, R. and J.R. Finney. 1973. Effects of direct drilling and reduced cultivation on soils conditions for root growth. *Outlook on Agriculture* 7:184-189.

Campos, B.C., D.J. Reinert, R. Nicolodi, J. Ruedell, and C. Petrere. 1995. Estabilidade de um latossolo vermelho-escuro distrófico após sete anos de rotação de culturas e sistemas de manejo de solo. *Revista Brasileira de Ciência do Solo* 19:121-126.

Cerri, C.C., B. Volkoff, and B.P. Eduardo. 1985. Efeito do desmatamento sobre a biomassa microbiana em Latossolo Amarelo da Amazônia. *Revista Brasileira de Ciência do Solo* 9:1-4.

Chaussod, R., S. Houot, G. Guiraud, and J.M. Hetier. 1988. Size and turnover of the microbial biomass in agricultural soils: laboratory and field measurements. In: D.S. Jenkinson and K. A. Smith (eds.), *Nitrogen Efficiency in Agricultural Soils*. Elsevier Applied Science, Amsterdam.

Clement, C.R. and T.E. Williams. 1967. Ley and soil organic matter. *J. Agric. Sci. Camb.* 69:133-138.

Corazza, E.J., J.E. Da Silva, D.V.S. Resck, and A.C. Gomes. 1998. Comportamento de diferentes sistemas de manejo como fonte ou depósito de carbono em relação à vegetação de Cerrado. p. 352-353. In: *Reunião Brasileira De Manejo E Conservação Do Solo E Da Água*, 17, 1998, Fortaleza. Anais. Fortaleza: UFCE , SBCS.

Cowell, J.D. 1978. Computations for studies of soil fertility and fertilizers requirements. Canberra: A.C.T.

Dalal, R.C. and R. J. Mayer. 1986. Long term trends in fertility of soils under continuous cultivation and cereal cropping in southern Queensland. II. Total carbon and its rate of loss from the soil profile. *Australian J. of Soil Res.* 24:281-292.

Dedecek, R.A., D.V.S. Resck, and E. De Freitas. 1986. Perdas de solo, água e nutrientes por erosão em Latossolo Vermelho-Escuro dos Cerrados em diferentes cultivos sob chuva natural. *R. Bras. Ci. Solo* 10:265-272.

EMBRAPA. 1987. *Relatório Técnico Anual do Centro de Pesquisa Agropecuária Dos Cerrados* 1982–1985. Planaltina, DF, EMBRAPA-CPAC. 532 pp.

Grisi, B.M. 1997. Temperature increase and its effect on microbial biomass and activity of tropical and temperate soils. *Revista de Microbiologia* 28:5-10.

Guedes, H.M., D.V.S. Resck, I. da S. Pereira, J.E. da Silva, and L. H. R. Castro. 1996. Characterization of aggregates size distribution of diferent soil management systems and its organic carbon content in a dark-red latosol in the Cerrados region, Brazil. p. 329-333. In: International Symposium on Tropical Savannas: Biodiversity and Sustainable Production of Food and Fibers in the Tropical Savannas, 1. Brasília. Proceedings. Planaltina, EMBRAPA-CPAC.

Hassink, J., G. Lebbink, and J.A. van Veen. 1991. Microbial biomass and activity of reclaimed-polder soil under a conventional or a reduced-input farming system. *Soil Biol. Biochem.* 23:507-513.

Henrot, J. and G.P. Robertson. 1994. Vegetation removal in two soils of the humid tropics: effect on microbial biomass. *Soil Biol. and Biochem.* 26:111-116.

IBGE. 1997. Anuário Estatístico do Brasil - 1997. Rio de Janeiro, IBGE, 57.

Jenkinson, D.S. 1966. Studies on the decomposition of plant material in soil. II. Partial sterilization of soil and the soil biomass. *J. Soil Sci.* 17:280-302.

Jenkinson, D.S. and D.S. Powlson. 1976. The effects of biocidal treatments on metabolism in soil. V. A method of measuring soil biomass. *Soil Biol. and Biochem.* 8:209-213.

Jenkinson, D.S. and A. Ayanaba. 1977. Decomposition of the ^{14}C labelled plant material under tropical conditions. *Soil Sci. Soc. Am. J.* 41:912-915.

Jenkinson, D. S. and J.N. Ladd. 1981. Microbial biomass in soil: measurement and turnover. p. 415-417. In: E.A. Paul and J. N. Ladd (eds.), Soil Biochemistry. Volume 5. Marcel Dekker, New York.

Jenkinson, D.S. and L.C. Parry. 1989. The nitrogen cycle in the Broadbalk wheat experiment: a model for the turnover of nitrogen through the soil microbial biomass. *Soil Biol. Biochem.* 21:535-541.

Jenkinson, D S., D.E. Adams, and A. Wild. 1991. Model estimates of CO_2 emissions from soil in response to global warming. *Nature* 351:304-306.

Joergensen, R.G., P.C. Brookes, and D.S. Jenkinson. 1990. Survival of the soil microbial biomass at elevated temperatures. *Soil Biol. Biochem.* 22:1129-1136.

Lopes, A.S. and F.R. Cox. 1977. A survey of the fertility status of surface soil under 'Cerrado' vegetation in Brazil. *Soil Sci. Soc. Amer. J.* 41:742-747.

Macedo, J. 1994. Prospectives for the rational use of the brazilian cerrados for food production. *Anais da Academia Brasileira de Ciências* 66:159-165.

Nicolardot, B., G. Fauvert, and D. Cheneby. 1994. Carbon and nitrogen cycling through soil microbial biomass at various temperatures. *Soil Biol. Biochem.* 26:253-261.

Pereira, I. S. da. 1997. Efeito de diferentes sistemas de manejo na distribuição de macro e microagregados e nos teores de carbono orgânico em um latossolo vermelho-escuro na região dos Cerrados. Brasília, UnB. Master Thesis.

Post, W.M., W.R. Emanueal, P.J. Zinke, and A.G. Stangenberger. 1982. Soil carbon pools and world life zones. *Nature* 298:156-159.

Resck, D.V.S. 1981. Perdas de solo, água e elementos químicos no ciclo da soja, aplicando-se chuva simulada. Planaltina: EMBRAPA-CPAC. (EMBRAPA-CPAC. Boletim de Pesquisa, 5).

Resck, D.V.S., J. Pereira, and J.E. Silva. 1991. Dinâmica da matéria orgânica na região dos Cerrados. Planaltina-DF: EMBRAPA-CPAC. (EMBRAPA-CPAC. Documentos, 36).

Resck, D.V.S., A.C. Gomes, D.C. Rodrigues, A.J. Santos, and J.E. da Silva. 1996. Influência do uso e manejo do solo na produção de CO_2 em diferentes agroecossistemas na região dos Cerrados. In: Congresso Latino Americano De Manejo E Conservação do Solo E da Água, 11., 1996, Águas de Lindóia. Anais. Água de Lindóia: SLACS-SBCS, (CD-ROM).

Resck, D.V.S. 1998. Agricultural intensification systems and their impact on soil and water quality in the Cerrados of Brazil. p. 288-300. In: R. Lal (ed.), *Soil Quality and Agricultural Sustainability*. Ann Arbor Press, Chelsea, MI.

Santos, M.N. dos, D.V.S. Resck, J.E. da Silva, and L.H.R. Castro. 1996. Influência de diferentes sistemas de manejo no teor de matéria orgânica e no tamanho e distribuição de poros em latossolo vermelho-escuro argiloso na região dos Cerrados, Brasil. p. 372-374. In: Simpósio Sobre O Cerrado: Biodiversidade E Produção Sustentável De Alimentos E Fibras Nos Cerrados, August, 1996. Brasília. Anais. Brasília: CPAC.

Silva, J.E., J. Lemainski, and D.V.S. Resck. 1994. Perdas de MO e suas relações com a capacidade de troca catiônica em solos da região de Cerrados do Oeste Baiano. *Revista Brasileira de Ciência do Solo, Campinas* 18:541-547.

Sousa, D.M.G., L. Vilela, T.A. Rein, and E. Lobato. 1997. Eficiência da adubação fosfatada em dois sistemas de cultivo em um latossolo de cerrado. In: Congresso Brasileiro de Ciência do Solo, 26., 1997, Rio de Janeiro, RJ. Informação, globalização, uso do solo. Anais. Rio de Janeiro: SBCS. (CD-ROM).

Tisdal, J.M. and J.M. Oades. 1982. Organic matter and water stable aggregates in soils. *J. Soil Sci.* 33:141-163.

Vasconcellos, C.A. 1994. Temperature and glucose effects on soil organic carbon: CO_2 evolved and decomposition rate. *Pesquisa Agropecuária Brasileira,* 29:1129-1136.

Vasconcellos, C.A., C.C.M. França, and I.E. Marriel. 1997. Dynamics of soil carbon, nitrogen and microbial biomass in tropical agroecosystems. In: *International Symposium on Plant-Soil Interaction at Low pH*, April, 1996. Belo Horizonte, MG. Proceedings...A. C. Moniz et al. Viçosa: BSSS. 224 pp.

Vasconcellos, C.A. 1998. Temperature effect on carbon biomass in soils from tropical and temperate regions. *Sci. Agric.* 55:94-104.

Woodruff, C.M. 1949. Estimating the nitrogen delivery of soil from the organic matter determinations as reflected by Sanborn Field. *Soil Sci. Soc. Am. Proc.* 14:208-212.

Wu, J. 1990. The turnover of organic C in Reading soil. Reading University. PhD Thesis.

CHAPTER 10

Soil Carbon Accumulation or Loss Following Deforestation for Pasture in the Brazilian Amazon

C. Neill and E.A. Davidson

I. Introduction

Tropical soils play an important role in the carbon (C) cycle of the earth. Human clearing of tropical forests for agriculture has the potential to alter soil C storage. These changes may make soils significant sources of C to the atmosphere (Houghton et al., 1983; Detwiler and Hall, 1988). They are also important indicators of soil fertility and agricultural sustainability (Tiessen et al., 1994). In recent decades, tropical forests have replaced temperate forests and grasslands as the region on Earth experiencing the most rapid land cover conversion. The Brazilian Amazon Basin—the world's largest region of intact tropical forest—is being deforested at the rate of approximately 11,000 to 29,000 km^2 y^{-1} (INPE, 1998; Skole and Tucker, 1993). Cattle pastures now represent the largest single use of deforested lands Basin-wide, and more than half of deforested land is used for cattle pasture at some time (Fearnside, 1987; Serrão, 1992). In some regions, such as central Rondônia or eastern Pará, that figure may be even higher.

Changes to soil C after tropical forest clearing for pasture have important consequences for Basin-wide C storage. The soils that dominate much of the humid Amazon Basin contain significant amounts of C, and high temperature and moisture can rapidly degrade soil C after disturbance. Loss of soil C can influence soil physical and chemical properties, such as water holding and nutrient regenerating capacity (Chauvel et al., 1991; Cassel and Lal, 1992; Neill et al., 1995). Approximately 47 Pg C are contained in the soils of the Brazilian Amazon to depth of 1 m (Moraes et al., 1995). Of that, 21 Pg occur in the top 20 cm where the changes to soil C stocks that follow land use alterations are most rapid.

Despite the enormous scale of pasture creation in the Amazon, there is yet no clear understanding of the direction of the resulting change in soil C stocks. In some locations, C stocks in pastures are lower compared with the original forest (Luizão et al., 1992; Desjardins et al., 1994; Trumbore et al., 1995). In other locations, pasture grass productivity declines in older pastures, but soil C concentrations remain relatively constant (Falesi, 1976; Serrão et al., 1979; Hecht, 1982; Buschbacher et al., 1988). In yet other locations, inputs of C from roots of pasture grasses cause increases in soil C stocks (Cerri et al., 1991; Moraes et al., 1996; Neill et al., 1997).

The factors that influence soil C gains or losses following pasture creation are not well understood. There are still relatively few studies of soil C balance after deforestation for pasture in the extensive area over which pasture agriculture now dominates cleared forest lands. Pasture formation in the Amazon occurs on a variety of soils and in regions that differ in the amount and timing of precipitation. The sequence leading to pasture formation also differs. Some pastures are created by planting grasses directly into forest slash (Moraes et al., 1996). Others are created after one or two years of annual cropping or after a cropping and fallow sequence (Fernandes et al., 1997). Grass species and practices of interplanting with legumes also differ. All of these factors can influence whether a pasture soil will accumulate or lose C. After establishment, pasture management by stocking rate, burning, fertilizing or disking may also affect soil C balance. Few of these practices have been examined by direct experimentation. Distinguishing among these factors may be difficult, as they are often confounded. For example, much of the pasture formation on Oxisols of eastern Pará occurred in the 1960s and 1970s and used predominantly *Panicum maximum* (Serrão et al., 1979), whereas more recent pasture formation in Rondônia, much on Ultisols, is planted to *Brachiaria brizantha* (Neill et al., 1997). Understanding which factors have the most important effects on soil C gains or losses can aid management of existing pastures, guide future development and help to predict the consequences of agricultural development on regional C fluxes.

This chapter compiles information from existing studies that have measured changes to surface soil C stocks following the conversion of forest to pasture in the Brazilian Amazon. It then examines the influences of precipitation, soil texture, pasture age and grass species on soil C balance. This review provides a basis for understanding regional differences in soil C responses and for designing studies that will shed light on mechanisms controlling soil C balance.

II. Methods

Information was compiled from published studies of C in the surface soils from chronosequences that consisted of a reference natural forest and a pasture or pastures of different ages. Twenty-nine pastures in 14 chronosequences from 7 studies are included (Table 1). Chronosequences were centered in Rondônia, the Manaus area and eastern Pará (Figure 1). From each study, the following were determined: pasture age, pasture grass species, sampling depth, soil C stock (kg m^{-2}) in the top 10 to 30 cm depth, soil type, percent soil clay and precipitation. Because pasture ages varied and because studies reported stocks (or C concentration and bulk density from which stocks could be calculated) to different depths, changes to C stocks were calculated in percent per year so that different pasture ages and soil depths could be compared. Soil C stock with C gain or loss was also compared with original soil C stock under forest. These comparisons do not account for the fact that C accumulation or loss rates change over time, with the highest rates occurring during the first several years after clearing and declining thereafter. Errors in C stock associated with changes in bulk density that can occur when sampling is based on a fixed depth (Davidson and Ackerman, 1993; Veldkamp, 1994) were accounted for only when they were reported this way in the original study. This database allowed examination of the relationships between changes in soil C stocks and individual pasture characteristics. For one study (Eden et al., 1990), mean values for multiple pastures classified as young (2 to 4 yr) and old (6 to 25 yr) were used because data were not reported for individual pastures. The change in soil C was plotted against individual factors and used multiple regression (REG procedure of SAS) to determine the best fit to the measured pasture characteristics.

III. Results

Nineteen of 29 pastures accumulated C in surface soils and 10 showed C loss (Figure 2). The majority of pastures gained or lost 5% of surface soil C per year or less (Figure 2). Two pastures lost more than 10% C yr^{-1} and four pastures gained more than 10% C yr^{-1}. Surface soil C stock in the original forest from which pastures were derived was the strongest predictor of pasture soil gain or loss ($p = 0.0005$; $r^2 = 0.37$; $df = 1,27$). Pastures formed on forest soils with high C stocks ($> \sim 6.0$ kg C m^{-2} in the top

Table 1. Chronosequence studies in the Brazilian Amazon that examined soil carbon inventories in surface soils of forests and pastures

Location	Soil	Land use	Grass species	Depth (cm)	C stock (kg m^{-2})	Reference
Manaus AM	Yellow latosol	Forest		0–20	6.60	Cerri et al.,
		1-yr pasture	*Brachiaria humidicola*	0–20	5.40	1991
Manaus AM		Forest		0–20	9.00	Cerri et al.,
		2-yr pasture	*Brachiaria humidicola*	0–20	6.70	1991
		8-yr pasture	*Brachiaria humidicola*	0–20	9.60	
Manaus AM	Yellow latosol	Forest		0–20	7.47	Luizão et al.,
		1-yr pasture	*Brachiaria humidicola*	0–20	6.74	1992
Capitão Poço PA	Kandiudult	Forest		0–20	3.14	Desjardins et
		10-yr pasture	*Pennisetum purpureum*	0–20	2.97	al., 1994
Paragominas PA	Haplustox	Forest	*Panicum*	0–10	2.60	Trumbore et
		23-yr pasture (degraded)	*maximum Brachiaria*	0–10	2.20	al., 1995
		23-yr pasture (managed for 5 yr)	*brizantha*	0–10	2.30	
Maracá RR	Ultisol	Forest				
		Young pasture (2-4 yr)	*Panicum maximum and B. humidicola*	0–10 / 0–10	1.45 / 2.45	Eden et al., 1990
		Old pasture (6-25 yr)	*Panicum maximum and B. humidicola*	0–10	2.51	
Porto Velho RO	Hapludox	Forest		0–30	6.20	Neill et al.,
		7-yr pasture	*Brachiaria brizantha*	0–30	6.10	1997
Cacaulândia RO	Paleudult	Forest		0–30	3.93	Neill et al.,
		8-yr pasture	*Brachiaria humidicola*	0–30	3.21	1997
Nova Vida RO	Kandiudult	Forest		0–30	3.23	Neill et al.,
		3-yr pasture	*Brachiaria brizantha*	0–30	3.46	1997
		5-yr pasture	*Brachiaria brizantha*	0–30	4.38	

Table 1. continued – –

Location	Soil	Land use	Grass species	Depth (cm)	C stock (kg m^{-2})	Reference
		9-yr pasture	*Panicum maximum*	0–30	3.98	
		13-yr pasture	*Panicum maximum*	0–30	3.55	
		20-yr pasture	*Brachiaria brizantha*	0–30	3.88	
		41-yr pasture	*Panicum maximum*	0–30	4.68	
		81-yr pasture	*Brachiaria brizantha*	0–30	5.00	
Nova Vida RO	Paleudult	Forest		0–30	2.74	Neill et al., 1997
		3-yr pasture	*Brachiaria brizantha*	0–30	3.97	
		5-yr pasture	*Brachiaria brizantha*	0–30	3.65	
		20-yr pasture	*Brachiaria brizantha*	0–30	3.92	
Ouro Preto RO	Paleudult	Forest		0–30	2.97	Neill et al., 1997
		8-yr pasture	*Panicum maximum*	0–30	3.85	
		20-yr pasture	*Brachiaria brizantha*	0–30	4.46	
Ouro Preto RO	Paleudalf	Forest		0–30	4.81	Neill et al., 1997
		8-yr pasture	*Panicum maximum*	0–30	4.56	
			Brachiaria brizantha	0–30	5.15	
Vihena RO	Hapludox	Forest		0–30	5.04	Neill et al., 1997
		7-yr pasture	*Brachiaria brizantha*	0–30	4.76	
		14-yr pasture	*Brachiaria brizantha*	0–30	5.43	
Jamarí RO	Hapludox	Forest (mean of two forests)		0–10	2.67	Kauffman et al., 1995;
		4-yr pasture	*Brachiaria brizantha*	0–10	5.21	Kauffman et al., (in press)
		2-yr pasture (from third-growth forest)	*Brachiaria brizantha*	0–10	3.22	

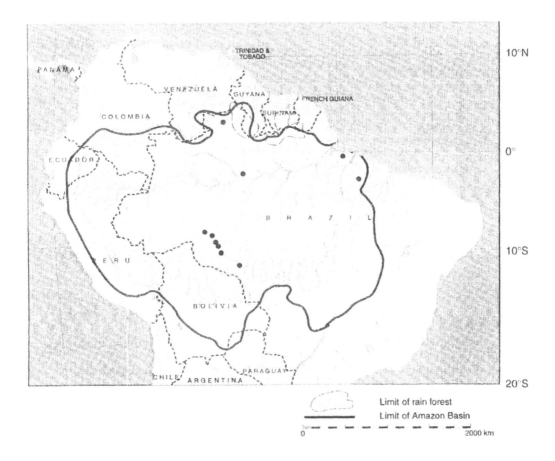

Figure 1. The locations within the Amazon Basin (•) of chronosequence studies that examined soil C stocks after forest clearing for pasture.

20 cm) tended to lose C, while pastures formed on soils with C stocks < ~5 kg C m^{-2} tended to gain C (Figure 3). Young pastures showed the greatest variation in soil C gain or loss (Figure 4), but there was no significant trend toward a higher total change in soil C stock with pasture age ($p = 0.220$, $r^2 = 0.06$, $df = 1,27$). Change in soil C stock was not related to soil clay content ($p = 0.103$; $r^2 = 0.10$; $df = 1,27$) (Figure 5), but soil clay and original forest soil C stock were highly correlated ($p = 0.0001$, $r^2 = 0.73$, $df = 1,22$). Three pastures that showed large C losses occurred on Oxisols with the some of the highest clay contents, but some high clay soils also showed C accumulation (Figure 5). The Ultisols, with clay contents of 13 to 29%, gained C or showed very small C losses (Figure 5). Changes in pasture C stocks were not related to precipitation.

Pasture grass species also showed a relationship with the change in surface soil C stocks ($p = 0.004$; $r^2 = 0.27$, $df = 1,27$). Pastures planted with *B. humidicola* tended to lose C and pastures planted with *P. maximum* and *B. brizantha* tended to gain C (Figure 6). Multiple stepwise regression showed that while original forest soil C accounted for 37% of the variation in the annual percent change in soil C, no other variables added significantly to the model fit (Table 2).

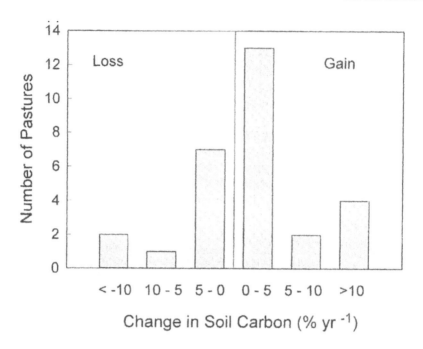

Figure 2. Change in soil C stock for 29 Amazon pastures. Change in soil C stock was the percent change in soil C from the original forest divided by the pasture age in years.

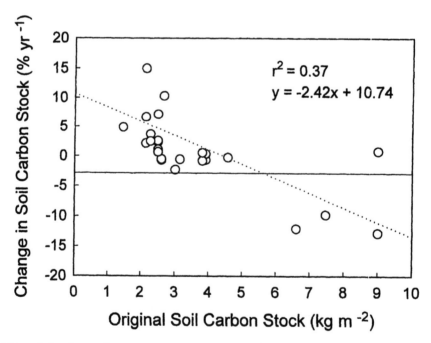

Figure 3. Plot of the change in pasture soil C stock against the soil C stock in the top 20 cm of the original forest. Change in soil C stock was the percent change in soil C from the original forest divided by the pasture age in years. The regression of change in C stock against original C stock was significant ($p = 0.0005$; $r^2 = 0.37$; $df = 1,27$).

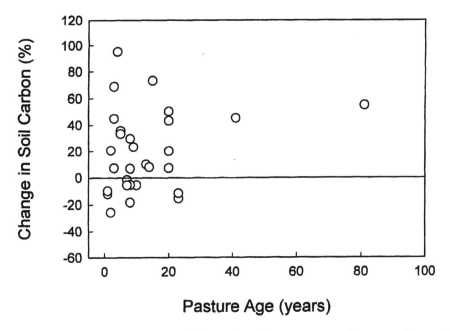

Figure 4. Plot of the change in pasture soil C stock against pasture age in years. Change in soil C was the percent of the original forest C soil stock. There was no tendency for older pastures to gain or lose C.

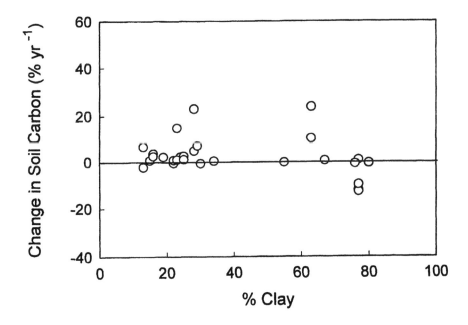

Figure 5. Plot of the change in pasture soil C stock against percent soil clay. Change in soil C stock was the percent change in soil C from the original forest divided by the pasture age in years.

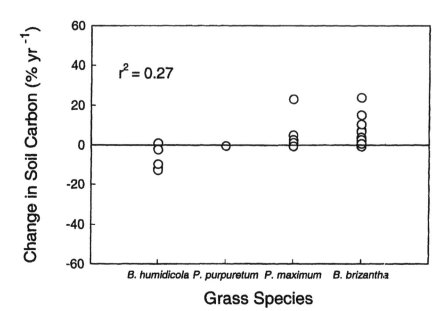

Figure 6. Plot of the change in pasture soil C stock against pasture grass species. Change in soil C stock was the percent change in soil C from the original forest divided by the pasture age in years. A regression of change in C stock against pasture species was significant ($p = 0.004$; $r^2 = 0.27$; $df = 1,27$).

Table 2. Results of multiple stepwise regression of factors influencing the annual percent change in soil C

Variable	Model r^2	F	$p<$
Original C stock	0.37	15.66	0.0005
Grass species	0.38	0.74	0.3990
Precipitation	0.40	0.47	0.4978
Clay percent	0.43	1.27	0.2718

IV. Discussion

A. Controls on Pasture Soil C Balance

The maintenance of soil C depends both on the stability of organic matter derived from the former forest vegetation and the rate of organic matter input from planted pasture grasses. The contribution of forest- and pasture-derived C to total soil C determined from bulk soil $\delta^{13}C$ values invariably shows a pattern of pasture C augmenting a declining pool of original forest-derived C (Desjardins et al., 1994; Trumbore et al., 1995; Neill et al., 1997) Our finding of a trend of C losses from soils with initial high C stocks but C gains from soils with lower initial C stocks suggests that this balance is shifted toward losses in high C, high clay soils.

Rates of decomposition of residual forest-derived C can be generally similar across a range of soil types in Rondônia (Neill et al., 1997), and carbon derived from newly-established pasture vegetation plays an important role in the development of pasture surface soil C stocks. This suggests that factors that increase grass productivity will have the greatest effect on soil C accumulation. This is consistent with our finding that grass species was an important influence on soil C balance. It also indicates that management that increases grass production can be very important in determining whether surface soils gain or lose C after pastures are planted on cleared forest lands. The correlation between grass and soil C suggested that pastures planted with *B. brizantha* and *P. maximum* had a greater likelihood of C accumulation than pastures planted with *B. humidicola*. However, in our small sample the *B. humidicola* pastures were on high clay soils, so the effects of original soil C and grass species were not independent. *B. brizantha* is now the grass of choice throughout most of the Brazilian Amazon, having replaced *B. humidicola* and *P. maximum* in the late 1970s and early 1980s (Smith et al., 1995). *B. brizantha* is now almost exclusively the grass that is planted after pastures are reformed, typically by mechanical scraping, disking and fertilizing with phosphorus (Nepstad et al., 1991; Smith et al., 1995). Trumbore et al. (1995) concluded that differences in C inputs between productive reformed pastures and unproductive degraded pastures caused C gains in reformed pastures and C losses in degraded pastures. Fisher et al. (1994) argued an important role for C inputs to soil from roots of planted grasses in pastures formed from South American savannas.

Other management practices probably contribute to pasture soil C gain or loss. These include burning frequency, stocking rate, grazing rotation pattern, effectiveness of weed control, soil mechanization and fertilization. These could not be evaluated in this study because they were not known, were not reported, or varied with time in the studies examined. It is also possible that grass species covaried with these other factors. For instance, the vigor and greater ground cover afforded by *B. brizantha* may require less burning to control weeds (Smith et al., 1995) or may allow more intensive grazing. In reformed pastures, establishment of *B. brizantha* is accompanied by disking and fertilization (Nepstad et al., 1991).

The reviewed studies covered nearly the entire range of soil textures and soil C concentrations of the dominant Ultisols and Oxisols upon which most of the pasture agricultural development in the Brazilian Amazon has taken place. The range of rainfall (1750 to 2500 mm yr^{-1}) also spans the range in the main regions where pasture agriculture dominates land use within the native moist forest biome. It is not likely that the strength of this analysis could be greatly improved simply by including a wider range of soils and climate regimes.

Little is known about the role of prior land use in determining soil C stocks in tropical pastures, although changes in land use and land management involving pasturing, cultivation, fertilization and abandonment can have long-lasting effects on soil nutrient and organic matter cycles in temperate ecosystems (Davidson and Ackerman, 1993; Jenkinson and Rayner, 1977; Mann, 1986; Burke et al., 1989; Mosier et al., 1991; Reiners et al., 1994; Motzkin et al., 1996). Clearing in the Brazilian Amazon is conducted both by small and by larger landholders (Fearnside, 1987, 1993; Schmink and Wood, 1992). Although pasture is an important component of the use of land by both small farmers and large ranchers, the history of land use on parcels held by small farmers and large ranchers differs. Small farmers typically practice traditional slash and burn agriculture, in which the forest is cut, an annual crop is grown for one or two years, and the land abandoned to fallow for several years (Sanchez 1976; Fernandes et al., 1997). The secondary vegetation that regrows during this fallow is then cut and burned, and another period of crop production follows. When yields decline after one or more cycles of crop and fallow, land is put into permanent pasture. In contrast, large ranchers may plant one cycle of crops after the initial forest clearing, but more typically they place land directly to pasture by planting grass seed into ash after burning (Serrão et al., 1979; Toledo and Serrão, 1981, personal observation). The majority of the evidence suggests that a cultivation phase on deforested lands will decrease soil C stocks. Vitorello et al. (1989), Cerri and Andreux (1990) and Tiessen et al. (1994) all found losses of organic matter after soil cultivation, and Allen (1985) found lower organic matter in cultivated soils than in adjacent forest soils. Losses of soil C in these cultivated soils were most rapid within the first several years after clearing. Larger decreases in soil C are common following more severe disturbances like bulldozing that can accompany land clearing (Alegre and Cassel, 1986). Woomer et al. (Chapter 5) reported that, on average, pastures on small farms in Rondônia that underwent several years of slash-and-burn agriculture had lower surface soil C stocks than the original forest. One pasture in Rondônia that was formed from third-growth forest after three cut-and-burn cycles had higher soil C than in the original forest (J.B. Kauffman, *personal communication*). There is no other information on how the magnitude of potential post-clearing losses from cultivation compare with changes after pasture is planted on lands that have undergone one or more cultivation-fallow cycles.

Disking pastures to control weeds is now common in Pará (Nepstad et al., 1991) and is becoming increasingly frequent in Rondônia (personal observation). Disking alone may decrease surface soil C stocks because it exposes more soil organic matter to decomposition and may mimic the effects of cultivation (Schlesinger, 1986; Davidson and Ackerman, 1993). However, because fertilization with phosphate increases pasture grass production (Serrão et al., 1979; Gonçalves and Oliveira, 1984), increased C inputs to soil from grass roots may counteract any losses from disking. One study suggested that this reformation practice has a positive effect on pasture surface soil C balance (Trumbore et al., 1995).

The fate of existing agricultural lands is an important and little-known component of land cover and biogeochemical change in the Amazon. The legacy of more than two decades of high forest clearing rates is an enormous area of pasture that is now 10 to 20 years old, situated in the developed regions of the Basin. This aging pasture land is poised to play a dominant role in land cover change in the Brazilian Amazon in the very near future. It is likely that many of the biogeochemical changes that will accompany future land use transitions from active pasture to abandoned pasture or to intensified pasture agriculture will be as large and as important as those that follow original forest conversion, but because of complex land use history, addressing these questions with chronosequence studies will be difficult.

B. Chronosequence Studies

Chronosequences are a valuable but imperfect means to address changes to ecosystem C storage after human-induced land use change. Although chronosequences are chosen to eliminate or minimize differences in soil characteristics and site history, they cannot completely eliminate preexisting site differences as a source of variation in measured responses to the land use history of interest (Pickett, 1989). These factors will be most important in locations where soil type and previous land use are more variable or less well known. Another problem in interpretation of results from chronosequence studies of changes to soil C stocks is that changes following land use change are often small compared with total C pools. Within-site variability makes it difficult to determine significant differences in soil C stocks among sites, even when changes of up to 1 kg m^{-2} in surface horizons can be measured (Trumbore et al., 1995; Neill et al., 1997).

Table 3. Design of proposed long-term experiments to examine management effects on the C balance of pasture soils in the Brazilian Amazon

Location	Land use types	Environmental variables	Management variables	Current or emerging land use issues
Eastern Pará	Forest	Soil type	Grass species	Pasture reformation, conversion
	Pasture		Burning frequency	of pastures to row crops, pasture
			Cattle stocking rate	abandoment to secondary forest.
Central Rondônia	Forest	Soil type	Grass species	Pasture reformation, conversion
	Pasture		Burning frequency	of pastures to row crops, slash
			Cattle stocking rate	and burn trajectories.

C. Soil Depth

Only surface soil C stocks were considered in this review. Many soils under *terra firme* moist forests in the Amazon contain significant live root biomass at depths of at least 8 m (Nepstad et al., 1994), indicating inputs of new C derived from current vegetation occur at those depths. In deep soils, C concentrations are low and differences between sites become very difficult to detect, but these low concentrations translate into large C stocks when tallied over the large volumes of these deep soils (Trumbore et al., 1995). Because pasture grasses are typically more shallowly rooted than forest trees, replacement of forest with pasture may both reduce C inputs and accelerate C losses in deep soil (Nepstad et al., 1994). Thus, a pasture soil may gain C at the surface and lose C at depth. The C balance of deep soils of the Amazon has been measured in only one location at Paragominas (Pará). It is likely to be important in determining total soil C balance on the widespread highly-weathered Ultisols and Oxisols where deep-rooted trees occur (Nepstad et al., 1994).

D. How to Improve Understanding of Pasture Soil C Dynamics

Studies of specific management regimes maintained over the long term (many years to decades) are needed to improve our understanding of the consequences of land use and management practices for soil C stocks and related soil C and nutrient cycling processes. Future studies should focus on manipulation experiments that examine common and potentially beneficial pasture management practices. There are several excellent models for these kinds of studies, including the Rothamsted Experimental Station in England (Jenkinson and Rayner, 1977) and the Long Term Ecological Research Program in the United States (Franklin et al., 1990). Activities at two sites are proposed to address the effects of land use change on soil C in the Brazilian Amazon (Table 3). These sites would be located in areas of past and current deforestation where pasture agriculture is widespread, eastern Pará and central Rondônia. Each site would examine core environmental and management variables that have been suggested here and elsewhere to have important effects on soil C stocks and cycling: soil type, pasture grass, stocking rate and grazing management and burning frequency. Each site would also examine other land use management issues that are particularly important in that region, including but not limited to pasture reformation (disking and fertilization), pasture abandonment to secondary forest and pasture conversion to row crop agriculture. Sites could also be established to examine the effects of different, known land use trajectories, such as slash-and-burn agriculture followed by pasture, that are commonly practiced in each region.

Each set of land use manipulations would take place in large plots of several hectares. Where possible, the same manipulations would be conducted on different soil types. Methods used to examine soil responses would be the same at all sites. Each set of plots to test pasture management effects would be accompanied by a reference forest that would be protected from human activity to the maximum extent possible. These controlled experimental plots could become the focus of shorter term experiments on dynamic soil processes, including organic matter turnover, soil fauna, nutrient dynamics and trace gas exchanges with the atmosphere. These studies of soil C dynamics could have great practical benefits because of the crucial role soil C plays in the maintenance of soil structure, water holding capacity, nutrient regeneration, and grass production.

V. Summary

There are still relatively few studies of soil C balance after deforestation for pasture in the extensive area of the Amazon over which pasture agriculture now dominates the use of cleared forest lands. A literature survey of studies where soil C stocks were followed after deforestation for pasture in the Brazilian Amazon showed that 19 of 29 pastures examined accumulated C in surface soils and 10 showed C loss. Surface soil C stock in the original forest was a strong predictor of pasture soil C gain or loss. Pastures formed on forest soils with high C stocks tended to lose C, while pastures formed on soils with lower soil C stocks ($< \sim 5$ kg C m^{-2}) tended to gain C. Pasture grass species also showed a strong relationship with the change in surface soil C stocks. Pastures planted with *Brachiaria humidicola* tended to lose C and pastures planted with *Panicum maximum* and *B. brizantha* tended to gain C. Change in soil C stocks was not related to soil clay content, pasture age or annual precipitation. Long term studies that maintain land in specific controlled management represent the best way to further understanding of the consequences of land use and management for soil C stocks and processes.

Acknowledgment

The support of the NASA Terrestrial Ecology Program and the conference sponsors (USDA-NRC, Ohio State University and EMBRAPA-CPATU) are gratefully acknowledged. We would also like to thank our colleagues at the Centro de Energia Nuclear na Agricultura, MBL and the WHRC whose support and insights have been invaluable for the preparation of this review. These include Carlos Cerri, Brigitte Feigl, Jerry Melillo, Daniel Nepstad, Marisa Piccolo, Paul Steudler and Susan Trumbore.

References

Alegre, J.C. and D.K. Cassel. 1986. Effect of land-clearing methods and postclearing management on aggregate stability and organic carbon content of a soil in the humid tropics. *Soil Sci.* 142:289-295.

Allen, J.C. 1985. Soil response to forest clearing in the United States and the tropics: geological and biological factors. *Biotropica* 17:15-27.

Burke, I.C., C.M. Yonker, W.J. Parton, C.V. Cole, U. Flach and D.C. Schimel. 1989. Texture, climate and cultivation effects on soil organic matter content in U. S. grassland soils. *Soil Sci. Soc. Am. J.* 53:800-805.

Buschbacher, R., C. Uhl and E.A.S. Serrão. 1988. Abandoned pastures in eastern Amazonia. II. Nutrient stocks in the soil and vegetation. *J. Ecol.* 76:682-699.

Cassell, D.K. and R. Lal. 1992. Soil properties of the tropics: common beliefs and management restraints. p. 61-89. In: R. Lal and P.A. Sanchez (eds.), *Myths and Science of Soils of the Tropics*. Soil Science Society of America, Madison, WI.

Cerri, C.C. and F. Andreux. 1990. Changes in organic carbon content of Oxisols cultivated with sugar cane and pasture, based on ^{13}C natural abundance measurement. *Int. Congr. Soil Sci.* 14:98-103. Kyoto, Japan.

Cerri, C. C., B. Volkoff and F. Andreaux. 1991. Nature and behavior of organic matter in soils under natural forest, and after deforestation, burning and cultivation, near Manaus. *For. Ecol. Manage.* 38:247-257.

Chauvel, A., M. Grimaldi and D. Tessier. 1991. Changes in soil pore-space distribution following deforestation and revegetation: an example from the central Amazon Basin, Brazil. *For. Ecol. Manage.* 38:259-271.

Davidson, E.A. and I.L. Ackerman. 1993. Changes in soil carbon inventories following cultivation of previously untilled soils. *Biogeochem.* 20:161-193.

Desjardins, T., F. Andreux, B. Volkoff and C.C. Cerri. 1994. Organic carbon and ^{13}C contents in soils and soil size-fractions, and their changes due to deforestation and pasture installation in eastern Amazonia. *Geoderma* 61:103-118.

Dewtiler, R.P. and C.A.S. Hall. 1988. Tropical forests and the global carbon cycle. *Science* 239:42-47.

Eden, M.J., D.F. M. McGregor and N.A.Q. Viera. 1990. III. Pasture development on cleared forest land in northern Amazonia. *Geogr. J.* 156:283-296.

Falesi, I.C. 1976. Ecosistema de Pastagem Cultivada na Amazônia Brasileira. Centro de Pesquisa Agropecuária do Trópico Umido. Empresa Brasileira de Pesquisa Agropecuária. Boletim Técnico No. 1. 193 pp.

Fearnside, P.M. 1987. The causes of deforestation in the Brazilian Amazon. p. 233-251. In R.E. Dickenson, (ed.), *The Geophysiology of Amazonia: Vegetation and Climate Interactions*. John Wiley & Sons, New York. 526 pp.

Fearnside, P.M. 1993. Deforestation in Brazilian Amazonia: the effect of population and land tenure. *Ambio* 22:537-545.

Fernandes, E. C.M., Y. Biot, C. Castilla, A.C. Canto, J.C. Matos, S. Garcia, R. Perin and E. Wanderli. 1997. The impact of selective logging and forest conversion for subsistence agriculture and pastures on terrestrial nutrient dynamics of the Amazon. *Ciência e Cultura* 49:34-47.

Fisher, M.J., I.M. Rao, M.A. Ayarza, C.E. Lascano, J.I. Sanz, R.J. Thomas and R.R. Vera. 1994. Carbon storage by introduced deep-rooted grasses in the South American savannas. *Nature* 371:236-238.

Franklin, J.F., C.S. Bledsoe and J.T. Callahan. 1990. Contributions of the long-term ecological research program. *BioScience* 40:509-523.

Gonçalves, C.A. and J. R. da C. Oliveira. 1984. Avaliação de Sete Gramineas Forrageiras Tropicais em Porto Velho, RO. Boletim de Pesquisa No. 2, Empresa Brazileira de Pesquisa Agropecuária, Porto Velho, Rondônia.

Hecht, S.B. 1982. Deforestation in the Amazon Basin: magnitude, dynamics and soil resource effects. *Stud. Third World Soc.* 13:61-101.

Houghton, R.A., J.E. Hobbie, J.M. Melillo, B. Moore, III, B.J. Peterson, G.R. Shaver and G.M. Woodwell. 1983. Changes in the carbon content of terrestrial biota and soils between 1860 and 1980: a net release of CO_2 to the atmosphere. *Ecol. Monogr.* 53:235-262.

INPE (Instituto Nacional de Pesquisas Espaciais). 1998. Desflorestamento Amazônia 1995-1997. INPE, São José dos Campos, São Paulo, Brazil.

Jenkinson, D.S. and J.H. Rayner. 1977. The turnover of soil organic matter in some of the Rothamsted classical experiments. *Soil Sci.* 123:298-305.

Kauffman, J.B., D.L. Cummings, D.E. Ward and R. Babbit. 1995. Fire in the Brazilian Amazon: biomass, nutrient pools and losses in slashed primary forests. *Oecologia* 140:397-408.

Kauffman, J. B., D. L. Cummings and D. E. Ward. In press. Fire in the Brazilian Amazon: 2. Biomass, nutrient pools and losses in cattle pastures. *Oecologia.*

Luizão R.C., Bonde T.A. and Rosswall T. 1992. Seasonal variation of soil microbial biomass--the effect of clearfelling a tropical rainforest and establishment of pasture in the central Amazon. *Soil Biol. Biochem.* 24:805-813.

Mann, L.K. 1986. Changes in soil carbon storage after cultivation. *Soil Sci.* 142:279-288.

Moraes, J.F.L. de, B. Volkoff, C.C. Cerri and M. Bernoux. 1996. Soil properties under Amazon forest and changes due to pasture installation in Rondônia, Brazil. *Geoderma* 70:63-81.

Moraes, J.F., C.C. Cerri, J.M. Melillo, D. Kicklighter, C. Neill, D.L. Skole and P.A. Steudler. 1995. Soil carbon stocks of the Brazilian Amazon Basin. *Soil Sci. Soc. Amer. J.* 59:244-247.

Mosier, A., D. Schimel, D. Valentine, K. Bronson and W. Parton. 1991. Methane and nitrous oxide fluxes in native, fertilized and cultivated grasslands. *Nature* 350:330-332.

Motzkin, G. , D. Foster, A. Allen, J. Harrod and R. Boone. 1996. Controlling site to evaluate history: vegetation patterns of a New England sand plain. *Ecol. Monogr.* 66:345-365.

Neill, C., M.C. Piccolo, P.A. Steudler, J.M. Melillo, B.J. Feigl and C.C. Cerri. 1995. Nitrogen dynamics in soils of forests and active pastures in the western Brazilian Amazon Basin. *Soil Biol. Biochem.* 27:1167-1175.

Neill, C., P.A. Steudler, J.M. Melillo, C.C. Cerri, J.F.L. Moraes, M.C. Piccolo and M. Brito. 1997. Carbon and nitrogen stocks following forest clearing for pasture in the Southwestern Brazilian Amazon. *Ecol. Appl.* 7:1216-1225.

Nepstad, D.C., C.R. de Carvalho, E.A. Davidson, P.H. Jipp, P.A. Lefebvre, G.H. Negreiros, E.D. da Silva, T.A. Stone, S.E. Trumbore and S. Vieira. 1994. The role of deep roots in the hydrological and carbon cycles of Amazonian forests and pastures. *Nature* 372:666-669.

Nepstad, D.C., C. Uhl and E.A.S. Serrão. 1991. Recuperation of a degraded Amazonian landscape: forest recovery and agricultural restoration. *Ambio* 20:248-255.

Pickett, S.T.A. 1989. Space-for-time substitution as an alternative to long-term studies. p. 110-135. In G.E. Likens (ed.), *Long-Term Studies in Ecology: Approaches and Alternatives.* Springer-Verlag, New York. 214 pp.

Reiners, W.A., A.F. Bouwman, W.F.J. Parsons and M. Keller. 1994. Tropical rain forest conversion to pasture: changes in vegetation and soil properties. *Ecol. Appl.* 4:363-377.

Sanchez, P.A. 1976. *Properties and Management of Soils in the Tropics.* John Wiley & Sons, New York. 618 pp.

Schlesinger, W.H. 1986. Changes in soil carbon storage and associated properties with disturbance and recovery. p. 194-220. In J.R. Trabalka and D.E. Reichle (eds.), *The Changing Carbon Cycle.* Springer-Verlag, New York. 592 pp.

Schmink, M. and C.H. Wood. 1992. Contested Frontiers in Amazonia. Columbia University Press, New York, USA. 385 pp.

Serrão, E.A.S. 1992. Alternative models for sustainable cattle ranching on already deforested lands in the Amazon. *Ann. Acad. Bras. Cien.* 64 (Suppl. 1): 97-104.

Serrão, E.A.S., I.C. Falesi, J.B. Da Veiga and J.F.T. Neto. 1979. Productivity of cultivated pastures on low fertility soils of the Amazon Basin. p. 195-225. In P. A. Sanchez and L. E. Tergas (eds.), *Pasture Production in Acid Soils of the Tropics*. Centro Internacional de Agricultura Tropical, Cali, Colombia. 488 pp.

Skole, D.S. and C.J. Tucker. 1993. Tropical deforestation and habitat fragmentation in the Amazon: satellite data from 1978-1988. *Science* 260:1905-1910.

Smith, N.H.H., E.A.A. Serrão, P.T. Alvim and I.C. Falesi. 1995. Amazonia: Resiliency and Dynamism of its Land and its People. United Nations University Press, NY. 253 pp.

Tiessen, H., E. Cuevas and P. Chacon. 1994. The role of soil organic matter in sustaining soil fertility. *Nature* 371:783-785.

Toledo, J.M. and E.A.S. Serrão. 1981. Practice and animal production in Amazonia. p. 281-310. In S.B. Hecht (ed.), *Amazonia, Agriculture and Land Use Research*. Centro Internacional de Agricultura Tropical, Cali, Colombia. *428 pp.*

Trumbore, S.E., E.A. Davidson, P.B. De Camargo, D.C. Nepstad and L.A. Martinelli. 1995. Below-ground cycling of carbon in forests and pastures of eastern Amazonia. *Global Biogeochem. Cycles* 9:512-528.

Veldkamp, E., 1994. Organic carbon turnover in three tropical soils under pasture after deforestation. *Soil Sci. Soc. Amer. J.* 58:175-180.

Vitorello, V.A., C.C. Cerri, F. Andreux, C. Feller and R.L. Victoria. 1989. Organic matter and natural carbon-13 distribution in forested and cultivated Oxisols. *Soil Sci. Soc. Amer. J.* 53:773-778.

CHAPTER 11

The Potential and Dynamics of Carbon Sequestration in Traditional and Modified Fallow Systems of the Eastern Amazon Region, Brazil

M. Denich, M. Kanashiro and P.L.G. Vlek

I. Introduction

Between 1980 and 1990, a net area of 34 million hectare of the total of 785 million hectare closed forest in Latin America has been converted into non- or sparsely-wooded land (FAO, 1996). These include areas with permanent agriculture, shifting cultivation with short fallows, pastures, agricultural plantations, or shrub land. The FAO study shows that agricultural activities are the main cause for forest destruction in the tropics. In Brazil, agriculture caused 91% of the deforestation during the 1980s, with a share of 51% by annual and permanent cropping and 40 % by cattle ranching (Amelung and Diehl, 1992). The latter, however, decreased in the 1990s due to the removal of tax advantages for the expansion of pasture land. In tropical America, small-scale farming is estimated to account for about 32% of the deforestation (Houghton et al., 1991).

The conversion of forests into agricultural land is characterized by considerable changes in the biomass stocks. In Brazil, 87% of the biomass losses in forest conversion are caused by agriculture (Amelung and Diehl, 1992) and in a given area, conversion into agricultural land use can lead to reductions of 95% or more of the above-ground biomass and carbon stocks, as compared to pristine rainforests. Such losses cannot be readily replenished by agricultural activities in regions with permanent settlements forming agricultural landscapes.

II. The Northeast of Pará State

During the last decades, the development of agricultural landscapes in Amazonia could be observed in large areas, e.g., along the Transamazon highway or in the state of Rondônia. The Northeast of Pará state (including the so-called "Zona Bragantina"), located East of the urban center of Belém, was settled more than 120 years ago and agriculture has been practiced in some parts of the region since then (Figure 1). The region covers approximately 20,000 km² which amounts to only 2% of the state's area. On the other hand, the region contributes to the state's agricultural production as follows: 19% of the value of the agricultural commodities, 44% of the cassava (*Manihot esculenta*) production, 18% of the maize (*Zea mays*) as well as bean (mainly *Vigna unguiculata*) production and 68% of the passion fruit (*Passiflora edulis*) production. Furthermore, 24% of Pará's small farmers' land is concentrated in the region (Denich, 1996). The importance of the small farmers to the economy of

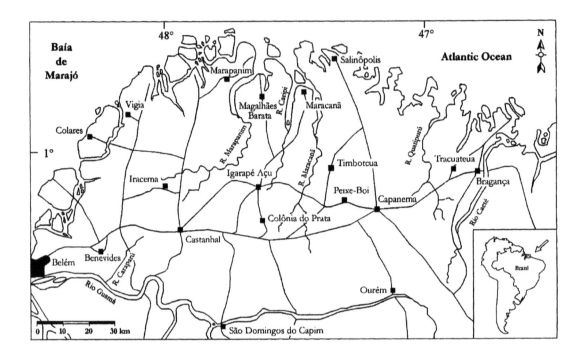

Figure 1. The Northeast of Pará state.

Pará is shown by the fact that they own only 20% of the arable land, but produce 60% of the total plant production value and supply 78% of the rural job opportunities (Denich, 1996).

The small holdings of NE Pará are not only characterized by the size of their property (on average, 25 ha), but also by their socioeconomic situations. Only a few hectares are occupied by crops, with subsistence crops playing an important role in the cropped area. Labor is provided mainly by familiy members. Most of the small holdings are participating in the local and regional markets and the marketed products have a share from 60% up to 90% in the entire production (Hurtienne, 1997, personal communication). Furthermore, they apply fertilizers on a limited scale and have access to some farm machinery. They also have access to extension services and information via the mass media.

In regions where small farmers predominate, the landscape is characterized by their land-use systems. Such an example is the municipality of Igarapé Açu (786 km²), one of the oldest settlement areas of NE Pará: 97% of its farm properties are small holdings (properties with less than 100 ha), which own 74% of the arable land (IBGE, 1991) and, accordingly, its land use and vegetation cover is characterized by the different phases of the rotational cultivation system of the small farmers, such as crop land and fallow land with secondary vegetation of different ages. In Table 1, the land cover classes "initial" (1 to 6-year-old secondary vegetation), "intermediate" (7 to 12-year-old secondary vegetation) and "advanced secondary succession" (13 to 18-year-old secondary vegetation) as well as "annual and semi-permanent crops" mainly represent areas with small-farmer land-use systems. These land cover classes (covering 75% and 77% of the land, respectively) did not change appreciably between 1984 and 1991, indicating relative stability in the small farming of the study region.

Table 1. Vegetation cover and land use (in percent land cover) in the municipality of Igarapé Açu (Pará state); based on satellite imagery of Landsat [TM]

Land cover class	1984	1991
Mature forest	9	5
Advanced secondary succession	21	18
Intermediate secondary succession	32	32
Initial secondary succession	20	24
Annual and semi-permanent crops	2	3
Permanent crops (mainly oil palm)	<1	1
Pasture	5	2
Degraded pasture	8	14
Bare soil	3	1

(Adapted from Watrin, 1994.)

III. The Landscape-Ecological Context of Small Farmer Fallow Systems in Northeast Pará

Biomass accumulation and, consequently, carbon sequestration depend on the agroecological environment as well as the predominant land-use systems and the respective agricultural practices. The agroecological environment in NE Pará is characterized by a mean annual temperature of around 25 °C, and an annual precipitation of 2000 to 2500 mm. The climate is humid with a dry period from September to November. The predominant soils are Ultisols. These soils were described as uniform kaolinitic loamy sands with the following characteristics (0–30cm): $pH(H_2O)$ 4.9–5.1, organic C 0.7 %, total Kjeldahl N 0.04 %, available P (Mehlich extract) 1.7 mg kg^{-1}, effective CEC 13.7 mmol$_c$ kg^{-1}, K 0.3 mmol$_c$ kg^{-1}, Ca 4.9 mmol$_c$ kg^{-1}, Mg 1.2 mmol$_c$ kg^{-1} and Al-saturation 53 % (Denich, 1989). The limiting nutrient for plant production is phosphorus, followed by nitrogen.

The small farmers' land use involves cropping systems with fallow periods and slash-and-burn technologies. The subsistence crops are maize, upland rice (*Oryza sativa*), cowpea (*Vigna unguiculata*), and cassava. The latter has importance as family food and cash crop. In the study region, on average, cassava yields (11,300 kg ha^{-1}) are higher than the mean yields of Pará state (9200 kg ha^{-1}), whereas maize (600 kg ha^{-1}), rice (740 kg ha^{-1}) and cowpea yields (620 kg ha^{-1}) are lower (Pará: 760, 1180 and 660 kg ha^{-1}, respectively; IBGE, 1991), due to the low soil fertility. At present, the most important semi-permanent crop is passion fruit, which is exclusively cultivated as a cash crop. Fertilizer use is common for cowpea and is increasing for the other crops.

After a cropping period of 1½ to 2 years the land is abandoned and the area enters usually into a 3 to 8 years, sometimes slightly longer, fallow period. The relatively short fallow periods are, in part, explained by the demographic pressure, since the population density of the municipality of Igarapé Açu increased from 13 in 1950 to 36 inhabitants per km^2 in 1994 (IBGE, 1957; Diário oficial, 1995). But more important are labor considerations: Land clearing prior to cultivation by manual methods requires a greater amount of work in older than in younger fallow vegetation. As this short fallow practice has become increasingly common over the last several land-use cycles, most of the small farmer land is covered with a low (2 to 5 m high), dense, woody secondary vegetation.

The fallow vegetation after short cropping periods of regularly fallowed fields are different from the stands which develop subsequent to the abandonment of semi-permanent plantations (e.g., passion fruit and black pepper [*Piper nigrum*]), the cultivation of which does not necessarily include fallowing. The abandonment of those plantations is due to the lack of financial means for their re-establishment, or the lack of labor or marketing opportunities. The emerging secondary vegetation is structurally very heterogeneous and consists of a mosaic of tree and shrub islands and grassy patches. The same can be observed if land preparation is carried out with machinery. In both cases the fallow vegetation has lost its regeneration potential which is predominantly vegetative: practically all tree, shrub and woody vine species as well as herbaceous perennials regenerate by resprouting from stumps, roots or rhizomes which have survived the burning and the relatively short cropping period. Prolonged cultivation with repeated weeding (e.g., semi-permanent crops) or cultivation with an implement such as a harrow or plow destroys the root system and reduces the vitality (= biomass accumulation per time unit) of the secondary vegetation.

A similar effect can be observed if the fallow periods in the traditional land use cycle are repeatedly very short (< 3 years). The secondary vegetation loses its vitality and progressively degrades, presumably together with the soil resources. This degradation process is attributed to the fact that the vegetation does not have the chance to recover physiologically from the preceding slash-and-burn impact and that, due to the burn during land preparation, a considerable amount of the nutrient storage of the slashed vegetation is released to the atmosphere. As much as 96% of the standing nitrogen, 76% of the sulfur, 47% of the phosphorus, 48% of the potassium, 35% of the calcium and 40% of the magnesium can be lost in a burn. All the nutrient losses (volatilization, removal by the harvested products, leaching) and gains (atmospheric inputs, fertilization) together result in negative overall nutrient balances for the whole crop/fallow cycle, irrespective of fertilization (Hölscher et al., 1997).

IV. Carbon Storage on Small Farmer Land in Northeast Pará

Land preparation with slash-and-burn technologies, above all, leads to considerable carbon losses to the atmosphere within the few minutes of slash burning. Hölscher (1995) estimates the carbon release at 13,000 kg ha^{-1} during the burn of a 7-year-old fallow vegetation, which corresponds to 98% of the carbon stock of the above-ground biomass. Thus, the above-ground carbon stocks on small farmer land depend on the crops and the following fallow regrowth.

The contribution to carbon storage of annual and semi-permanent crops alone is low, particularly when the time they stay in the field is rather limited (Table 2). A ground cover of 1 to 2 t C ha^{-1} formed by herbs and grasses could be present in addition. The total carbon stocks of the cassava crop are an exception because of the relatively high below-ground biomass of the tubers. Although annual and semi-permanent crops show quite similar above-ground carbon stocks, the latter maintain a standing biomass over a longer period of time. Oil palm (*Elaeis guineensis*), the only true permanent crop in the region, may reach above-ground carbon stocks comparable to that of a fallow vegetation in advanced stages of secondary succession (Table 2). In the region, however, oil palm is not grown by small holdings because of the high initial investments.

The most important carbon sink in small-scale farming regions are the fallow areas. In the study region, the initial and intermediate stages of this fallow vegetation predominate (>50% of the region; Table 1) with above-ground carbon stocks (live and dead biomass) between 3 and about 40 t ha^{-1} (Table 2). Thus, on small farmer land the above-ground carbon stocks lie temporarily between 1 and 25% of those determined for primary forests in the region, which is reported at about 160 t ha^{-1} (Salomão, 1994).

Table 2. Above-ground carbon stocks (t ha^{-1}) of the most important compartments in small-farmer land use as well as an oil palm plantation and a primary forest remnant in NE Pará

Compartment	Carbon stock
Maize (at harvest 4 to 5-month-old)	2.1
Cowpea (at harvest 3 to 4-month-old)	1.6
Cassava (above/below-ground, at harvest 1 to 1½-yr-old)	2.6/5.6
Passion fruit (at harvest time; 1-year-old)	2.6
Black pepper (at harvest time, 2½-year-old)	5.3
1-year-old fallow vegetation (live and dead biomass)	3–5
4-year-old fallow vegetation (live and dead biomass)	8–16
7-year-old fallow vegetation (live and dead biomass)	18–33
10-year-old fallow vegetation (live and dead biomass)	34–41
Oil palm (including fruits; 5 to 8-year-old)	20/57
Primary forest (live and dead biomass)	160

(The crop examples are from fertilized fields; carbon estimates are based on Haag et al., 1973; Kato, 1978; Viegas, 1993; Salomão, 1994; Nunez, 1995; Kato et al., 1998; and own studies.)

The bulk of the carbon stocks in most ecosystems is found in the soil, as was reported elsewhere (e.g., Cerri et al., 1991; Nepstad et al., 1994; Trumbore et al., 1995; Lal et al., 1998a). Similarily, the soil carbon stocks in the agricultural landscape of our study area are 4 to >100 times as high as the above-ground stocks (Tables 2 and 3). The only exception is the primary forest, the above and below-ground stocks of which are roughly the same. These calculations are based on soil carbon stocks which were found down to 6 m depth. Approximately half of the carbon stock is stored at 0–1 m depth and the remaining at 1–6 m (Sommer, 1996).

According to Sommer (1996), charcoal carbon makes up <1 to 3% of the total soil carbon stocks, but this is highly variable. Living roots under crops or fallows and in the primary forest contribute 5 to 15 t ha^{-1} or 3 to 8 % (on average 11 t ha^{-1} or 6%) of the carbon stock. Under semi-permanent cropping, the share of roots is only 1 to 2 t ha^{-1} or <1%. During the fallow period, 40 to 60% of the roots grow between 0 and 1 m. In the passion fruit plantation very few roots could be found in soil layers deeper than 0.5 m and none at depths > 3 m, which is consistent with the lower root mass and presumably lower carbon inputs. It is concluded that land-use systems which include a fallow period can slow down soil carbon losses or maintain soil carbon stocks at comparable levels to those of a primary forest. This is associated with simultaneous root production and litterfall and their turnover: The annual root production under young fallow vegetation (0–1 m) can reach 2.5 t ha^{-1} yr^{-1} (Wiesenmüller, 1997, personal communication) and the annual litter production by a 5-year-old fallow vegetation can be as high as 2.1 t ha^{-1} yr^{-1} and in a 10-year-old fallow 6.3 t ha^{-1} yr^{-1}. On the other hand, semi-permanent land use leads to a rapid decrease of soil carbon in the upper soil layers and, due to very low root biomass, a noticeable decrease of the carbon stocks in the deeper layers can be expected in the long term.

Table 3. Below-ground carbon stocks (t ha^{-1} 6 m^{-1}) of different compartments in small farmer land use as well as an oil palm plantation and a primary forest remnant of NE Pará (means ± 95% confidence intervals; n = number of analyzed soil cores [0–6 m] per site)

Compartment	n	Root-C	Soil-C	Charcoal-C	Total
8 months cropping (maize-cassava)	8	11±5	170±20	2.4±1.6	183±24
15 months cropping (cassava)	4	6±5	174±20	6.3±8.8	186±24
21 months cropping (cassava)	16	12±5	163±7	3.6±1.6	179±10
Passion fruit plantation (2½-year-old)	4	1±2	165±18	1.2±3.1*	168±19
Black pepper plantation (5-year-old)	3	2±2	160±31	3.6±1.9	166±31
1-year-old fallow vegetation	4	5±3	150±16	4.6±5.1	160±22
5-year-old fallow vegetation	20	15±3	197±13	5.5±1.9	217±15
12-year-old fallow vegetation	4	14±6	166±37	3.6±1.8**	183±32
40-year-old secondary forest	4	15±6	185±27	4.3±1.6	204±33
Oil palm plantation (8-year-old)	4	6±3	139±15	0.5±1.7	146±16
Primary forest	6	10±3	186±25	1.7±1.5	197±25

* n = 2; ** n = 3. (Adapted from Sommer, 1996.)

V. Management Options

On a long-term basis, soil carbon stocks on farm land with rotational agriculture including a fallow period do not differ from those of primary forest sites. Two options are available to stimulate carbon sequestration on small farmer land in NE Pará: (i) to raise the soil organic carbon content by supplying plant material to the top soil and (ii) to improve the above-ground carbon storage capacity during the fallow period.

The first option involves mulch technologies and management of crop residues. The positive effects of these technologies are well known and they presumably could contribute to an increase of the soil organic carbon content if the actual content has fallen below its original equilibrium level due to degradation processes. However, little is known about the potential to increase the carbon content beyond the steady-state level of the soil, and how long such a higher level can be maintained. Increasing the soil carbon content requires rates of organic matter additions which exceed decomposition rates. The latter depend on the quality of the added plant material, climate, soil properties and agricultural practices (Beinroth et al., 1996).

The second option is based on fallow management techniques. These techniques are derived from successional processes which lead to floristic changes in plant communities through time. As the fallow vegetation on small farmer land in NE Pará is allowed to grow only a few years, it is effectively maintained in the initial stages of secondary succession. During the short fallow periods, biomass accumulation can be accelerated by introducing fast growing species of the advanced successional stages and improving the growth conditions of the existing vegetation. In succession management such interventions are related to controlled colonization and controlled species performance (Luken, 1990). In the study region about one third of the land (estimation based on Watrin, 1994), such as recently abandoned fields, degraded pastures or still cropped land, could potentially be subject to succession management.

A. Mulch Technology

Mulching in fallow systems is possible only with fire-free land preparation, where it substitutes for slash burning prior to the cropping period, thus avoiding nutrient losses by volatilization and considerably slowing down the carbon loss to the atmosphere. We were able to show in field trials that fire-free land preparation and slash-and-burn gave similar yields as long as fertilizers were applied to both. In the absence of fertilizers the crops in the burned fields benefitted from the nutrients in the ashes, whereas the decomposing mulch tied up nutrients and caused substantial yield depression (Kato et al., 1998). The transformation of the woody fallow vegetation into a manageable mulch was proved to be the most important technical constraint of the fire-free land preparation. In order to further develop those technologies for future adoption by the farmers, practical methods for cutting and chopping the fallow vegetation had to be found. To meet this objective the development of machinery was deemed necessary.

The following demands had been made from the equipment: (1) cutting the woody vegetation near the ground (to avoid stumps which make weeding difficult) without destroying the root system of the vegetation (to assure the regrowth of a vital woody fallow vegetation), (2) chopping the plant material, (3) spreading the chips homogeneously over the field. The construction of the chopper had to be simple and robust. Accordingly, the Institute for Agricultural Engineering of the University of Goettingen (Germany) developed a tractor-propelled bush chopper, which is mounted on the front power lift (Figure 2). Driving forward (1 to 6 km h^{-1}) the vegetation is cut in a width of 2 m by two rotating circular saws (diameter 1 m) and subsequently chopped by two vertical steel helices sitting on the saw-blades. Paddles between the saw-blades and the helices throw out the chopped material towards the back and under the tractor's front wheels. The power demand for chopping a 3 to 4-year-old fallow vegetation is at least 60 kW. The average rate of output is 10 t (8–17 t) chopped fresh plant material per hour, thus, one hectare of 3 to 4-year-old fallow vegetation (3 m high) can be chopped within 4 to 5 hours. During the winter of 1997, a prototype of the chopper was tested in the field and the feasibility of the new technology could already be demonstrated.

B. Fallow Management

Various authors dwell on the potential of soils as carbon sinks and the management options to realize this potential (e.g., Fisher et al., 1994; Lal et al., 1998b). The above-ground biomass is rarely seen as an easily manageable carbon sink. Lugo and Brown (1993) also conclude that the greatest potential for carbon sequestration lies in forest fallows, but emphasize the accumulation of organic carbon in the soil, due to litterfall and root turnover. However, only a slow increase in soil organic carbon with a very weak relation to the age of the secondary forest could be shown.

The above-ground storage of carbon in fallow systems is affected by the short-term biomass accumulation of the crops and by that of the fallow vegetation over longer periods of time. Management options for increasing the carbon storage during the cropping period are limited. In contrast, the fallow period allows a variety of management interventions at different levels of labor and investment. Carbon sequestration through a fallow vegetation is, first of all, a function of time (Table 2) and then of the vitality of the vegetation which in turn depends on the site conditions, the species composition and the land-use history including agricultural practices (plowing, harrowing).

Under the nutrient-poor soil conditions of NE Pará the fallow regrowth was assumed to be limited by low nutrient availability. We, therefore, evaluated whether soil nutrient constraints can be overcome by fertilization (Gehring et al., 1998). The application of 180 kg N ha^{-1}, 106 kg P ha^{-1}, 83 kg K ha^{-1}, 208 kg Ca ha^{-1}, 140 kg Mg ha^{-1}, 100 kg S ha^{-1} and 50 kg ha^{-1} of a commercial mixture of micronutri-

Figure 2. Front view of the bush chopper (prototype).

ents boosted the biomass production of the secondary vegetation within 2½ years sufficiently to be significantly higher than the unfertilized control (Figure 3). Although fertilization is a common intervention in the context of controlled species performance (Luken, 1990), it is usually too expensive as a means of succession management, particularly under the economic conditions of tropical small holdings. In our case each additional ton of carbon sequestered in the fallow biomass would cost about $100 U.S. per hectare within 2½ years, when considering the application of only phosphorus and nitrogen. These were found to be the nutrients primarily limiting the fallow regrowth.

A more feasible means for a medium-term improved carbon sequestration by fallow vegetation may be enrichment plantings. In the study region, such plantings were originally introduced to improve the economic value of fallow areas and to bring additional revenue through products like timber and construction material (Yared and Carpanezzi, 1981; Brienza, 1982; Dubois, 1990). With the short fallow periods currently practiced in NE Pará, the approach of enrichment planting aims at shortcutting the natural succession and at accelerated biomass accumulation of the fallow vegetation. To achieve the latter, fast-growing, nitrogen-fixing leguminous trees are usually used.

Our first experience with enrichment plantings involved the fast growing leguminous tree *Acacia auriculiformis*. Later we included in the experiments *A. angustissima, A. mangium, Clitoria racemosa, Inga edulis* and *Sclerolobium paniculatum* (Kanashiro et al., 1998). *A. auriculiformis* was planted at spacings of 2 m x 2 m (2500 trees per hectare) and 1 m x 1 m (10,000 trees per hectare) in mixed-cropped plots of maize and cassava after the maize was harvested. The plots were left to fallow after the cassava harvest 8 months later and *A. auriculiformis* developed together with the spontaneous fallow vegetation. After 21 months of fallow (29 months after the planting of *Acacia*) the enriched plots had a 2 to 3 fold higher above-ground carbon stock (including live biomass, leaf litter and stand-

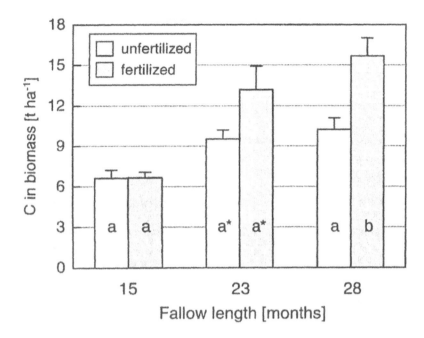

Figure 3. Above-ground carbon stocks (including live biomass, litter and standing dead) of a fertilized and unfertilized fallow vegetation (bars represent means + SE; n = 8; different letters within a pair of bars indicate a significant difference for $p < 0.05$; two-sample t test; * p = 0.07).
(Adapted from Gehring et al., 1998.)

ing dead) than a fallow without enrichment planting of the same age (Figure 4 and 5). The *Acacia*'s contribution to the total carbon stocks was 57% (2 m x 2 m) and 72% (1 m x 1 m), respectively. Moreover, with an enrichment planting spaced at 1 m x 1 m, the above-ground carbon stock was as high as that of 7-year-old fallow plots. Root-carbon stocks were higher under the enriched vegetation than when not enriched, particularly in soil layers between 0.4 and 2 m depth: the total root-carbon stocks were estimated at 4.2 t ha^{-1} over 2 m depth in enriched and 2.5 t ha^{-1} over 2 m depth in a not enriched vegetation, both 21 months old (recalculated from Kanashiro et al., 1998).

Based on the time span the fields lie agriculturally unproductive, the enriched 21-month-old fallows had 3 to 4 times higher mean carbon accumulation rates than the 7-year-old natural fallow (Table 4). Tentatively it was concluded that the highest cost-benefit ratio for biomass and carbon accumulation of an *A. auriculiformis*-enriched fallow is found at spacings of 2 m x 2 m and at a fallow period of two years.

Figure 4. Traditional (right) and enriched fallow with *Acacia auriculiformis* (left) in a field experiment 21 months after abandonment of the cropped plots, NE Pará, Brazil.

VI. Scenario-Based Analyses of Carbon Dynamics in Different Fallow Systems

The above-ground and, to a much lower extent, the below-ground carbon stocks of fallow systems are varying in the course of time. Tentative empirical models of the carbon dynamics were developed (Figure 6A-D) to visualize these dynamics and to make the effects over time of the proposed management options on the carbon stocks comparable.

In the models the following components were included: (i) above-ground biomass (live and dead) of differently aged and enriched fallow vegetation stands (Kato et al., 1998; Kanashiro et al., 1998 and own studies), assuming a linear growth of the vegetation during the fallow period, (ii) above-ground biomass and yields of maize, cowpea and cassava with or without fertilizer application (Kato et al., 1998), assuming a linear growth during the cropping period, (iii) land preparation with and without slash burning and (iv) harvesting. Nonlinear decomposition rates were assumed for the mulched plant material of the land preparation without burning and the crop residues (Denich, 1989). In accordance with Menaut et al. (1993), above-ground biomass was converted into carbon by the factor 0.43. Stumps which remain in the field after cutting the fallow vegetation (up to 5 t C ha^{-1} in both 4 and 10-year-old fallow vegetation; Kato and Kato, 1997, personal communication) and weeds which

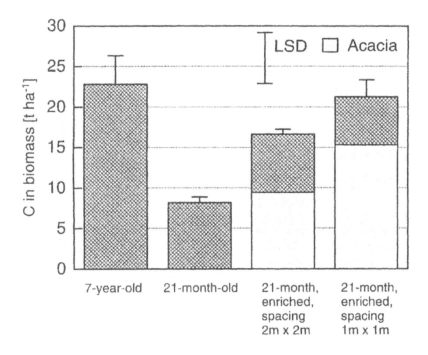

Figure 5. Above-ground carbon stocks (including live biomass, litter and standing dead) of fallow vegetation of different ages with and without *Acacia auriculiformis*-enrichment; bars represent means + SE of total carbon stocks; n = 7; LSD for *p* = 0.05). (Adapted from Kanashiro et al., 1998.)

Table 4. Arithmetic means of the carbon accumulation per year and per planted *Acacia-auriculiformis* individual in fallow vegetation of different ages with and without enrichment planting (means ± 95% confidence intervals; n = 7)

	Arithmetic mean of annual carbon accumulation (t ha⁻¹)	Arithmetic mean of annual carbon accumulation per planted tree (kg tree⁻¹)
7-year-old fallow vegetation	3.2 ± 1.2	-
21-month-old fallow vegetation	4.8 ± 1.0	-
21-month-old, enriched, spacing 2m x 2m	9.5 ± 0.9	3.8 ± 0.3
21-month-old, enriched, spacing 1m x 1m	12.1 ± 3.0	1.2 ± 0.3

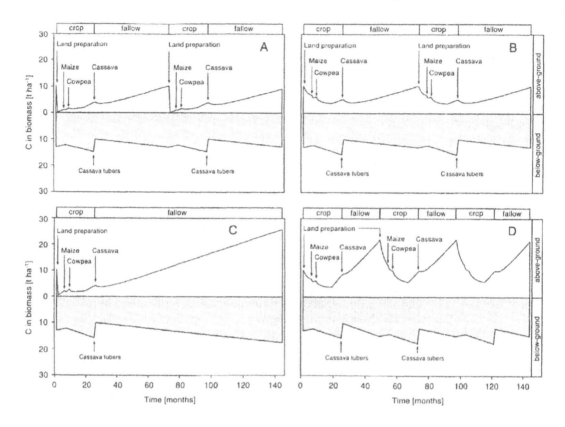

Figure 6. Above-ground and root carbon dynamics in (A) a traditional fallow system (slash-and-burn agriculture, partly fertilized, two land use cycles within 12 years), (B) a mulch-fallow system (not burned, but fertilized, two land use cycles within 12 years), (C) a prolonged fallow system (slash-and-burn agriculture, fertilized, one land use cycle within 12 years) and (D) an enriched fallow-mulch system (not burned, but fertilized, three land use cycles within 12 years); the times of land preparation as well as maize, cowpea and cassava harvest are marked.

grow up during the cropping period (up to 1 t C ha^{-1}) have not been included because reliable data are lacking. From the below-ground carbon stocks only the dynamic compartments have been considered, which are (i) root carbon stocks of the cropping and fallow periods and (ii) cassava tubers' carbon (Kato et al., 1998). Increases and decreases in these stocks were assumed to be linear. Soil organic matter and charcoal carbon stocks (Table 3) were considered relatively stable. Seasonal variations of the root biomass (1 to 2 t ha^{-1}; Wiesenmüller, 1997, personal communication) have not been included. The most drastic changes in the root carbon stocks are the growth and harvest of the cassava tubers.

As might be expected, the carbon stocks of the different fallow systems are characterized by short-term increases and decreases and do not reach a steady-state situation (Figure 6A-D). The 12-years mean of the above-ground carbon stocks of the systems with repeated short fallow periods is about 6 t ha^{-1} (Figure 6A and B), whereas with a prolonged fallow period (10 years) or an enrichment

Table 5. Relative above-ground and root carbon stocks (traditional is set at 1.0) of different crop/fallow cycles and the respective relative protein and carbohydrate production (sum of maize, cowpea and cassava; traditional is set at 1.0) within a 12-year land use period

Land-use system / management option	No. of cycles / duration of one cycle (years)	Relative carbon stock		Protein	Carbohydrate
		Above	Root		
Traditional: burned / crops partly fertilized	2 / 6	1.0	1.0	1.0	1.0
Mulch: not burned / crops fertilized	2 / 6	1.3	1.0	1.7	1.3
10 years fallow: burned / crops fertilized	1 / 12	2.6	1.1	0.9	0.7
Enriched fallow: not burned / crops fertilized	3 / 4	2.5	1.2	2.6	2.0

planting it amounts to about 13 t ha^{-1} (Figure 6C and D). The root carbon stocks are not showing such differences. Their means within a 12-year period lie between 12 and 15 t ha^{-1}. Thus, the long-term relation of the above-ground to the root carbon stocks is 1:2 in situations with repeated cropping and short fallow periods and comes to 1:1 in situations with prolonged or enriched fallows. Considering the total carbon storage of the soil (Table 3) the relation is 1:30 and 1:15, respectively. In rainforest land the relation of above-ground to root carbon is 15:1 and comes to 1:1 if the total soil carbon is considered.

In order to compare the carbon sequestration capacity of the different fallow systems and the contribution of the respective management options the areas under the curves in Figure 6A-B have been calculated as relative measures of the carbon stocks over a given period of time, here 12 years. The area under the curve describing the carbon dynamics in the most common land-use system of NE Pará, below called traditional (Figure 6A), was set at 1.0 and the others were related to this figure (Table 5).

Fire-free land preparation and conversion of the fallow vegetation into mulch leads to above-ground carbon stocks which, in the course of time, are 30% higher than those of traditionally cultivated land (Figure 6B). However, this value is still tentative because we do not know yet whether the substitution of slash burning by microbial decomposition merely results in a slow down of carbon dioxide release. The decomposition process takes place during cropping and is even extended into the subsequent fallow period, so that overall carbon stocks are basically maintained, avoiding the fluctuation of loss/restoration processes typical for slash-and-burn systems.

Prolonged fallow periods (Figure 6C) and fallows enriched with fast growing leguminous trees (Figure 6D) can accumulate 2½-fold the above-ground carbon stock of the traditional fallow system. The root carbon stocks are increased in the prolonged and enriched fallow system by 10 and 20 %, respectively, which account for only <1 to 2 % of the total below-ground carbon stocks. Both management options, however, are coupled with additional costs: The prolonged fallow in form of farm land which is taken out of the agricultural production for a longer period of time and the enriched fallow in the form of planting material, labor and machinery.

Evaluation of management options for enhanced carbon sequestration on farm land has to take agricultural productivity into account. As fire-free land preparation, including mulch techniques, requires the application of fertilizers to get satisfactory yields, we depict only approaches with fertilization (Table 5). To make the comparison easier, the maize, cowpea and cassava production were converted into protein and carbohydrate units and, again, the values of the traditional system were set for reference purposes at 1.0. The productivity of a field within a given period of time is a function of the number of land-use cycles. As a consequence, the fallow system with enrichment plantings gave the most promising results in terms of protein and carbohydrate production (Table 5). It allows three land-use cycles within the period of 12 years due to accelerated biomass accumulation (Figure 5). The mulch-fallow system also results in an increased crop production, which, however, depends solely on fertilization. The production loss due to the prolongation of the fallow period cannot be compensated for in agricultural returns.

VII. Conclusions

Forest or bush fallows as part of the small farmer land use in NE Pará offers prospects with regard to the enhancement of carbon sequestration. As soil carbon stocks in those fallow systems resemble those of rainforest soils and even after decades of rotational land use no dramatic soil carbon decrease could be observed (considering 0–6 m depth), short and medium-term management efforts to increase carbon sequestration should concentrate on the capacity of the above-ground biomass.

Carbon sequestration in an agricultural context are subordinate to the farmer's requirements for food and fiber production as well as his labor and financial resources. If land is available, prolonged fallow periods are easily realized in order to immobilize atmospheric carbon and only require more work in land clearing prior to the next cropping period. Enrichment plantings with fast growing (leguminous) trees are advanced management options which require increased labor and financial inputs, but allow shortened fallow periods. They, therefore, allow higher levels of agricultural production from a given piece of land than the prolonged fallow option. Shortened fallow periods, however, must be combined with mulch techniques and fertilizer applications in order to secure adequate yields and to avoid soil mining and the decrease of soil organic matter. Additionally, fire-free land preparation with mulch techniques avoids nutrient losses during the slash burning and slows down considerably the carbon releases to the atmosphere compared to slash-and-burn techniques. As a consequence, mulching might even contribute to a long-term increase of the soil organic matter content. In agricultural landscapes both prolonged and enriched fallows represent productive agroecosystems with noticeable capacity for short, medium and long-term net carbon sequestration. Those capacities have essentially been fully exploited in terrestrial steady-state ecosystems, such as rainforests.

If stable carbon dioxide levels in the atmosphere are a desireable common good, carbon sequestration capacities in agricultural systems might play a role in future global environmental markets if trade in carbon sequestration services between tropical farmers and regions with an over-production in carbon dioxide are realized. In that case, farmers would be asked to provide land and labor as a contribution to a, basically global, environmental service. However, the farmer himself would benefit from the fallow management with regard to the sustainability of the productivity of his land. According to Smith et al. (1998), small farmers have an environmental awareness and are willing to contribute to resource conservation. However, to ensure that fallow management works as a tool in carbon sequestration, program incentives and subsidies may have to be offered to the farmers through global environmental markets in order to compensate for making land available. In addition, farmers will need access to agricultural inputs (tree seedlings, fertilizer) and appropriate machinery for fire-free land preparation.

If contractual fallow management is considered desireable, land-use planning has to take this management option into consideration. Measures should be taken at the farm and community level to prevent accidental fires in set aside fallows as well as in areas of permanent cultivation. Thus, land-use planning can help harmonize short-term agricultural interests with long-term environmental services and makes land-use decisions as well as carbon sequestration capacities predictable.

The Northeast of Pará state represents a field laboratory for the first steps in realizing carbon sequestration services through fallow management. The predominance of small-scale farming, extensive occurence of fallow vegetation with low vitality, land availability, good infrastructure, support by extension services, market access and the vicinity of the city of Belém with research facilities and planning authorities would make this area eligible for such an experiment.

Acknowledgements

The study was supported by the German Federal Ministry of Education, Science, Research and Technology (BMBF) and the Brazilian National Council for Scientific and Technological Development (CNPq) as part of the bilateral research program "Studies on Human Impact on Forests and Floodplains in the Tropics (SHIFT)". We thank all the participants of the project "Secondary Forests and Fallow Vegetation in the Agricultural Landscape of the Eastern Amazon Region" and the small farmers of the municipality of Igarapé Açu, who contributed in the one way or another to this study.

References

Amelung, T. and M. Diehl. 1992. Deforestation of tropical rain forests - Economic causes and impact on development. Kieler Studien 241, J.C.B.Mohr (Paul Siebeck), Tübingen, Germany.

Beinroth, F.H., M.A. Vázquez, V.A. Snyder, P.F. Reich, and L.R. Perez Alegría. 1996. Factors controlling carbon sequestration in tropical soils - A case study of Puerto Rico. University of Puerto Rico, Mayagüez Campus, Dept. Agronomy and Soils, USDA Natural Resources Conservation Service, World Soil Resources and Caribbean Area Office.

Brienza Jr., S. 1982. *Cordia goeldiana* Huber (freijó) em sistemas "taungya" na região do Tapajós - Estado do Pará. EMBRAPA/CPATU, Belém, Brazil, Circular Técnica 33.

Cerri, C.C., B. Volkoff, and F. Andreaux. 1991. Nature and behaviour of organic matter in soils under natural forest, and after deforestation, burning and cultivation, near Manaus. *For. Ecol. Manage.* 38:247-257.

Denich, M. 1989. Untersuchungen zum Beitrag junger Sekundärvegetation zur Nutzungssystempro- duktivität im östlichen Amazonasgebiet, Brasilien. *Göttinger Beiträge zur Land- und Forstwirtschaft in den Tropen und Subtropen* 46:1-265.

Denich, M. 1996. Ernänrungssicherung in der Kleinbauernlandwirtschaft Ostamazoniens – Probleme und Lösungansätze. *Göttinger Beiträge zur Landund Forstwirtschaft in den Tropen und Subtropen* 115:78-88.

Diário oficial from August 30, 1995.

Dubois, J.C.L. 1990. Secondary forests as a land-use resource in frontier zones of Amazonia. p. 183- 194. In: A B. Anderson (ed.), *Alternatives to Deforestation: Steps Towards Sustainable Use of the Amazon Rain Forest.* Columbia University Press, New York.

FAO. 1996. Forest resources assessment 1990. Survey of tropical forest cover and study of change processes. FAO Forestry Paper 130. 152 pp.

Fisher, M.J., I.M. Rao, M.A. Ayarza, C.E. Lascano, J.I. Sanz, R.J. Thomas, and R.R. Vera. 1994. Carbon storage by introduced deep-rooted grasses in the South American savannas. *Nature* 371:236-238

Gehring, C., M. Denich, M. Kanashiro, and P.L.G. Vlek. 1998. The response of secondary vegetation in Eastern Amazonia to relaxed nutrient availability constraints. *Biogeochemistry*. (In press.)

Haag, H.P., G.D. Oliveira, A.S. Borducchi, and J.R. Sarruge. 1973. Absorcão de nutrientes por duas variedades de maracujá. ANAIS da Escola Superior de Agronomia "Luís de Queiroz" (ESALQ) Piracicaba - SP, Brazil 30, 267-279.

Hölscher, D. 1995. Wasser- und Stoffhaushalt eines Agrarökosystems mit Waldbrache im östlichen Amazonasgebiet. *Göttinger Beiträge zur Land- und Forstwirtschaft in den Tropen und Subtropen* 106:1-133.

Hölscher, D., R.F. Möller, M. Denich, and H. Fölster. 1997. Nutrient input-output budget of shifting agriculture in Eastern Amazonia. *Nutrient Cycling in Agroecosystems* 47:49-57.

Houghton, R.A., D.S. Lefkowitz, and D.L. Skole. 1991. Changes in the landscape of Latin America between 1850 and 1985. I. Progressive loss of forests. *For. Ecol. Manage.* 38:143-172.

IBGE - Instituto Brasileiro de Geografia e Estatística. 1957. Enciclopédia dos Municípios Brasileiros Vol. XIV.

IBGE - Fundação Instituto Brasileiro de Geografia e Estatística. 1991. Censo Agropecuário Número 6, Pará, 1985, Rio de Janeiro.

Kanashiro, M., M. Denich, S. Brienza Jr., and P.L.G. Vlek. 1998 *Can Enrichment Plantings Shorten the Fallow in Eastern Amazonia* ? Agroforestry Syst. (Submitted.)

Kato, A.K. 1978. Teor e distribuicão de N, P, K, Ca e Mg em pimenteiras do reino (*Piper nigrum* L.). Master thesis, Escola Superior de Agronomia "Luís de Queiroz" (ESALQ) Piracicaba - SP, Brazil. 78 pp.

Kato, M.S.A., O.R. Kato, M. Denich, and P.L.G. Vlek. 1998. Fire-free alternatives to slash-and-burn for shifting cultivation in the Eastern Amazon region. 1. The role of fertilizers. *Field Crops Res.*

Lal, R., J.M. Kimble, R.F. Follett, and B.A. Stewart (eds.). 1998a. *Soil Processes and the Carbon Cycle*. Advances in Soil Science, CRC Press, Boca Raton, FL.

Lal, R., J.M. Kimble, R.F. Follett, and B.A. Stewart (eds.). 1998b. *Management of Carbon Sequestration in Soil*. Advances in Soil Science, CRC Press, Boca Raton, FL.

Lugo, A.E and S. Brown. 1993. Management of tropical soils as sinks or sources of atmospheric carbon. *Plant and Soil* 149:27-41.

Luken, J.O. 1990. *Directing Ecological Succession*. Chapman and Hall, London. 251 pp.

Menaut, J.-C., L. Abbadie, and P.M. Vitousek. 1993. Nutrient and organic matter dynamics in tropical ecosystems. p. 213-231. In: P.J. Crutzen and J.G. Goldammer (eds.), *Fire in the Environment: The Ecological, Atmospheric, and Climatic Importance of Vegetation Fires*. John Wiley & Sons, Chichester.

Nepstad, D.C., C.R. Carvalho, E.A. Davidson, P.H. Jipp, P.A. Lefebvre, G.H. Negreiros, E.D. Silva, T.A. Stone, S.E. Trumbore, and S. Vieira. 1994. The role of deep roots in the hydrological and carbon cycles of Amazonian forests and pastures. *Nature* 372:666-669.

Nunez, J.B.H. 1995. Fitomassa e estoque de bioelementos das diversas fases da vegetação secundária, provenientes de diferentes sistemas de uso da terra no nordeste paraense, Brasil. Master thesis, Federal University of Pará, Belém, Brazil.

Salomão, R.P. 1994. Estimativas de biomassa e avaliação do estoque de carbono da vegetação de florestas primárias e secundárias de diversas idades (capoeiras) na Amazônia Oriental, Município de Peixe-Boi, Pará. Master thesis, Federal University of Pará, Belém, Brazil.

Smith, J., S. Mourato, E. Veneklaas, R. Labarta, K. Reategui, and G. Sanchez. 1998. Willingness to pay for environmental services among slash-and-burn farmers in the Peruvian Amazon: Implications for deforestation and global environmental markets. *J. Environ. Econ. Manage.* (Submitted.)

Sommer, R. 1996. Kohlenstoffvorräte in Böden unter intensiv genutzten Sekundärwaldflächen im östlichen Amazonasgebiet. Degree dissertation, University of Goettingen, Germany. 122 pp.

Trumbore, S.E., E.A. Davidson, P.B. Camargo, D.C. Nepstad, and L.A. Martinelli. 1995. Below-ground cycling of carbon in forests and pastures of Eastern Amazonia. *Global Biogeochemical Cycles* 9:515-528.

Viegas, I.J.M. 1993. Crescimento de dendezeiro (*Elaeis guineensis* Jacq.), concentração, conteudo e exportação de nutrientes nas diferentes partes de plantas com 2 a 8 anos de idade, cultivadas em latossolo amarelo distrófico, Tailândia, Pará. PhD thesis, Escola Superior de Agronomia "Luís de Queiroz" (ESALQ) Piracicaba - SP, Brazil.

Watrin, O.S. 1994. Estudo da dinâmica na paisagem da Amazônia Oriental através de técnicas de geoprocessamento. Master thesis, Instituto Nacional de Pesquisas Espaciais (INPE), São José dos Campos - SP, Brazil. 153 pp.

Yared, J. A. G. and A.A. Carpanezzi. 1981. Conversão de capoeira alta da Amazônia em povoamento de produção madeireira: o método do "Recrû" e espécies promissoras. EMBRAPA-CPATU, Belém, Brazil, Boletim de Pesquisa 25.

CHAPTER 12

Greenhouse Gas Emissions from Land-Use Change in Brazil's Amazon Region

P.M. Fearnside

I. Introduction

Deforestation in Brazilian Amazonia releases quantities of greenhouse gases that are significant both in terms of their present impact and in terms of the implied potential for long-term contribution to global warming from continued clearing of Brazil's vast area of remaining forest. The way in which emissions are calculated can have a great effect on the impact attributed to deforestation. Two important indices for expressing the global warming impact of deforestation are net committed emissions and the annual balance of net emissions (or, more simply, the annual balance).

Net committed emissions expresses the ultimate contribution of transforming the forested cover into a new one, using as the basis of comparison the mosaic of land uses that would result from an equilibrium condition created by projection of current trends. This includes emissions from decay or reburning of logs that are left unburned when forest is initially felled (committed emissions), and uptake of carbon from growing secondary forests on sites abandoned after use in agriculture and ranching (committed uptake) (Fearnside, 1997a).

Net committed emissions considers the emissions and uptakes that will occur as the landcover approaches a new equilibrium condition in a given deforested area. Here the area considered is the 13.8 x 10^3 km^2 of Brazil's Amazonian forest that was cut in 1990, the reference year for baseline inventories under the United Nations Framework Convention on Climate Change (UN-FCCC). The "prompt emissions" (emissions entering the atmosphere in the year of clearing) are considered along with the "delayed emissions" (emissions that will enter the atmosphere in future years), as well as the corresponding uptake as replacement vegetation regrows on the deforested sites. Not included are trace gas emissions from the burning and decomposition of secondary forest and pasture biomass in the replacement landcover, although both trace gas and carbon dioxide fluxes are included for emissions originating from remains of the original forest biomass, from loss of intact forest sources and sinks, and from soil carbon pools. Net committed emissions are calculated as the difference between the carbon stocks in the forest and the equilibrium replacement landcover, with trace gas fluxes estimated based on fractions of the biomass that burn or decompose following different pathways.

In contrast to net committed emissions, the annual balance considers releases and uptakes of greenhouse gases in a given year (Fearnside, 1996a). Annual balance considers the entire region (not just the part deforested in a single year) and considers the fluxes of gases entering and leaving the region both through prompt emissions in the newly deforested areas and through the "inherited" emissions and uptakes in the clearings of different ages throughout the landscape. Inherited emissions and uptakes are the fluxes occurring in the year in question that are the result of clearing activity in previous years, for example, from decomposition or reburning of the remaining biomass of the original forest. The annual balance also includes trace gases from secondary forest and pasture burning and decomposition.

The annual balance represents an instantaneous measure of the fluxes of greenhouse gases, of which carbon dioxide is one. Even though the present calculations are made on a yearly basis, they are termed "instantaneous" here to emphasize the fact that they do not include future consequences of deforestation and other actions taking place during the year in question.

The present chapter updates previous estimates of net committed emissions (Fearnside, 1997a) and annual balance (Fearnside, 1996a). The present chapter incorporates additional information on wood density (Fearnside, 1997b), below-ground biomass, Cerrado biomass (Graça, 1997), soil carbon releases (Fearnside and Barbosa, 1998), burning efficiencies, charcoal formation and other factors.

II. Forest Biomass

The average biomass of the primary forests present in the Brazilian Amazon has been estimated based on analysis of published wood volume data from 2954 ha of forest inventory surveys distributed throughout the region (Fearnside, nd, updated from Fearnside, 1994). Average total biomass (including dead and below-ground components) is estimated to be 463 t ha^{-1} for all unlogged mature forests originally present in the Brazilian Legal Amazon. The average aboveground biomass is 354 t ha^{-1}, of which 28 t ha^{-1} is dead; below-ground biomass averages 109 t ha^{-1}. These estimates include wood density calculated separately for each forest type based on the volume of each species present and published basic density data for 274 species (Fearnside, 1997b). The total biomass estimates are disaggregated by state and forest type, allowing use of the data in conjunction with Brazil's LANDSAT-based deforestation estimates, which are reported on a state-by-state basis (Fearnside, 1993, 1997c).

The areas of protected and unprotected vegetation of each type in each state have been estimated (Fearnside and Ferraz, 1995). By multiplying the per-hectare biomass of each forest type by the unprotected area present in each state, one can estimate the biomass cleared if one assumes that clearing within each state is distributed among the different vegetation types in proportion to the unprotected area present. By weighing the biomass by the deforestation rate in each state, the average total pre-logging biomass in areas cleared in 1990 has been estimated to be 433 t ha^{-1}, or 6.5% lower than the average for forests present in the Legal Amazon as a whole (see Fearnside, 1997a). The difference is due to concentration of clearing activity along the southern and eastern edges of the forest, where per-hectare biomass is lower than in the areas of slower deforestation in the central and northern parts of the region.

The values for biomass from "unlogged" forest represent the best estimates for each forest type at the time it was surveyed (in the 1950s in the case of the Food and Agriculture Organization of the United Nations [FAO] forest inventories that comprise 10% of the data and in the early 1970s in the case of the RADAMBRASIL data covering the remaining 90%). FAO data are from Heinsdijk (1957, 1958a,b,c) and Glerum (1960); RADAMBRASIL data are from Brazil, Projeto RADAMBRASIL (1973–1983). There is some reason to believe that the survey teams avoided logged-over locations (Sombroek, 1992). In addition, logging damage was much less widespread at the time of the surveys than it is at present. Logging is progressing rapidly, with the fraction of areas cleared that are logged prior to felling increasing noticeably since the mid-1970s as road access has improved. In addition, logs and wood for charcoal and firewood are sometimes sold *after* the burn.

The biomass reduction due to logging in areas being felled is much higher than the average biomass reduction over the forest as a whole, as the areas being felled generally have the best road access. Much of the biomass reduction from logging will result in gas releases similar to those that would occur through felling: decay of the slash and the substantial number of noncommercial trees that are killed or damaged during the logging process; decay and/or burning of the scrap generated in the milling process, plus a slower decay of wood products made from the harvested timber (see Fearnside, 1995a). With adjustment for logging, areas cleared in 1990 had an average total biomass of 406 t ha^{-1}, of which 249 t ha^{-1} was aboveground live biomass, 59 t ha^{-1} was aboveground dead and 98 t ha^{-1} was below-ground.

III. Greenhouse Gas Emissions

A. Initial Burn

The burning efficiency (percentage of pre-burn aboveground carbon presumed emitted as gases) averaged 38.8% in the 10 available measurements in primary forest burns in Brazilian Amazonia (Table 1). Adjustments for the effect of logging on the diameter distribution of the biomass gives an efficiency of 39.4%.

Charcoal (char) formed in burning is one way that carbon can be transferred to a long-term pool from which it cannot enter the atmosphere. Charcoal in the soil is a very long-term pool, considered to be permanently sequestered in the analysis. The mean of the four available measurements of charcoal formation in primary forest burns in Brazilian Amazonia indicate 2.2% of aboveground carbon being converted to charcoal (Table 1).

Graphitic particulate carbon is another sink for carbon that is burned. A small amount of elemental carbon is formed as graphitic particulates in the smoke; over 80% of the elemental carbon formed remains on the site as charcoal (Kuhlbusch and Crutzen, 1995). Graphitic particulate carbon is calculated by emission factors from the amount of wood combusted. The amount of carbon entering this sink is only 1/13 the amount entering the charcoal sink.

The pre-1970 secondary forest must be considered separately from the primary forest, as these areas are not included in the deforestation rate estimate (13.8 x 10^6 km^2 y^{-1} in 1990). A rough estimate of clearing rate is 713 km^2 y^{-1} (Fearnside, 1996a). Pre-1970 secondary forest is only relevant to the annual balance, not net committed emissions. The amounts of greenhouse gases contributed by clearing of pre-1970 forest are very small.

Greenhouse gas emissions and uptakes are tabulated for a net committed emissions calculation in a "low trace gas scenario" (Table 2) and a "high trace gas scenario" (Table 3). These two scenarios use high and low values appearing in the literature for the emissions factors for each gas in different types of burning (reviewed in Fearnside, 1997a). They do not reflect the doubt concerning forest biomass, deforestation rates, burning efficiency and other important factors.

The initial burn represents 270 x 10^6 t of CO$_2$ gas, or 27% of the gross committed emission of 999 x 10^6 t. Gross emission of a gas refers to all releases of the gas, but not uptakes. The initial burn contribution of CH$_4$ is 0.87–1.07 of 1.18–1.51 x 10^6 t (70–74%), CO is 21–26 of 30–37 x 10^6 t (68–70%) and N$_2$O is 0.05–0.14 of 0.07–0.18 x 10^6 t (71–78%). For NO$_x$ and NMHC, if considered apart from the loss of mature forest sources, represent, respectively, 0.66 of 0.81 x 10^6 t (81%) and 0.58–1.10 of 0.63–1.26 x 10^6 t (87–92%).

B. Subsequent Burns

The burning behavior of ranchers can alter the amount of carbon passing into a long-term pool as charcoal. Ranchers reburn pastures at intervals of 2 to 3 years to combat invasion of inedible woody vegetation. Logs lying on the ground when these reburnings occur are often burned. Some charcoal formed in earlier burns can be expected to be combusted as well. Parameters for transformations of

Table 1. Combustion and charcoal formation studies in Brazil

Location	State	Burn year	Pre-burn aboveground biomass Dry weight (t ha⁻¹)	Carbon (t ha⁻¹)	Burning efficiency % pre-burn C	Net charcoal formation (t C ha⁻¹)	% of pre-burn biomass C	Source
Original forest (first burning)								
Manaus	Amazonas	1984	264.6	130.2	27.6	3.5	2.7	Fearnside et al., 1993
Altamira	Pará	1986	263.0	129.9	41.9	1.6	1.3	Fearnside et al., nd-a
Manaus	Amazonas	1990	368.5	181.7	28.3	3.4	1.8	Fearnside et al., nd-b
Jacunda	Pará	1990	292.4	147.6	51.5			Kauffman et al., 1995
Maraba	Pará	1991	434.6	218.2	51.3			Kauffman et al., 1995
Santa Barbara	Rondônia	1992	290.2	142.1	40.5			Kauffman et al., 1995
Jamari	Rondônia	1992	361.2	178.9	56.1			Kauffman et al., 1995
Manaus	Amazonas	1992	424.4	203.5	25.1			Carvalho et al., 1995
Tomé Açu	Pará	1993	214.2	96.2	21.9			Araújo, 1997
Nova Vida	Rondônia	1994	306.5	142.3	34.6	4.1	2.9	Graça, 1997
Mean			321.9	157.0	39.	3.2	2.2	
Original forest remains (subsequent burnings)								
Apaiu	Roraima	1991	101.2	48.4	30.1	0.6	1.3	Fearnside et al., nd-c
Apaiu	Roraima	1993	96.3	46.1	13.2	0.3	0.7	Barbosa and Fearnside, 1996
Mean			98.7	47.2	21.6	0.5	1.0	
Secondary forest (not including remains of original forest)								
Altamira	Pará	1991	26.1	11.3	25.9	0.1	1.1	Guimarães, 1993
Apiau	Roraima	1991	41.5	17.8	66.5	0.2	1.2	Fearnside et al., nd-c
Apiau	Roraima	1993	6.2	2.8	69.1	0.02	0.8	Barbosa and Fearnside, 1996
Mean			24.6	10.7	53.6	0.1	1.0	
Pasture								
Apaiu	Roraima	1993	8.0	3.4	93.4	0.04	1.1	Barbosa and Fearnside, 1996

Table 2. Net committed greenhouse gas emissions by source for 1990 clearing in the Legal Amazon: low trace gas scenario

Source	Area affected (10^3 km^2)	Emissions (million t of gas)					
		CO_2	CH_4	CO	N_2O	NO_x	NMHC
Forest							
Initial burn	13.8	270	0.87	20.90	0.05	0.66	0.55
Reburns	13.8	57	0.28	8.89	0.01	0.15	0.14
Termites aboveground decay	13.8	17	0.014				
Other aboveground decay	13.8	365					
Below-ground decay	13.8	247					
Cattle (a)	6.1		0.010				
Pasture soil (a)	6.1				0.002		
Loss of intact forest sources and sinks (a)	7.3		0.0003			-0.01	-0.09
Soil carbon (top 8 m)	13.8	43					
Regrowth	13.8	-65					
Forest subtotal		934	1.18	29.79	0.07	0.81	0.63
Cerrado							
Initial burn	5.0	11	0.04	0.85	0.002	0.03	0.02
Reburns	5.0	1	0.01	0.18	0.01	0.003	0.003
Termites aboveground decay	5.0	0.1	0.0001				
Other aboveground decay	5.0	2					
Below-ground decay	5.0	9					
Cattle (a)	5.0		0.008				
Pasture soil (a)	5.0				0.002		
Loss of intact cerrado sources and sinks (a) (b)	5.0		0.0002			-0.0004	-0.004
Soil carbon (top 8 m)	5.0	16					
Regrowth	5.0	-9					
Cerrado subtotal		31	0.05	1.03	0.004	0.03	0.02
Total for Legal Amazon		964	1.23	30.83	0.07	0.83	0.66

(a) Recurring effects (cattle methane, forest soil methane sink, pasture soil N_2O,) summed for 100-year period for consistency with IPCC 100-year horizon calculation.

(b) Intact cerrado source for NO_x and NMHC derived from the forest per hectare emission assuming emission is proportional to the tree leaf dry weight biomass in each ecosystem. Cerrado tree leaf biomass (dry season) = 0.756 t $^{-1}$ha (dos Santos, 1989: 194); forest (at Tucuruí, Pará) = 12.94 t ha^{-1} (Revilla Cardenas et al., 1982).

Table 3. Net committed greenhouse gas emissions by source for 1990 clearing in the Legal Amazon: high trace gas scenario

Source	Area affected (10^3 km²)	Emissions (million t of gas)					
		CO_2	CH_4	CO	N_2O	NO_x	NMHC
Forest							
Initial burn	13.8	270	1.05	26.13	0.14	0.66	1.10
Reburns	13.8	57	0.44	11.32	0.03	0.15	0.25
Termites aboveground decay	13.8	17	0.014				
Other aboveground decay	13.8	365					
Below-ground decay	13.8	247					
Cattle (a)	6.1		0.01				
Pasture soil (a)	6.1				0.002		
Loss of intact forest sources and sinks (a)	7.3		0.0003			-0.01	-0.09
Soil carbon (top 8 m)	13.8	43					
Regrowth	13.8	-65					
Forest subtotal		934	1.51	37.45	0.18	0.81	1.26
Cerrado							
Initial burn	5.0	11	0.04	1.07	0.006	0.027	0.04
Reburns	5.0	2	0.01	0.36	0.001	0.005	0.01
Termites aboveground decay	5.0	0.1	0.0001				
Other aboveground decay	5.0	2					
Below-ground decay	5.0	15					
Cattle (a)	5.0		0.01				
Pasture soil (a)	5.0				0.002		
Loss of intact cerrado sources and sinks (a) (b)	5.0		0.0002			-0.0004	-0.004
Soil carbon (top 8 m)	5.0	16					
Regrowth	5.0	-9					
Cerrado subtotal		37	0.07	1.43	0.009	0.03	0.05
Total for Legal Amazon		971	1.88	38.87	0.18	0.84	1.31

(a) Recurring effects (cattle methane, forest soil methane sink, pasture soil N_2O, hydroelectric methane) summed for 100-year period for consistency with IPCC 100-year horizon calculation.
(b) Intact Cerrado source for NO_x and NMHC derived from the forest per hectare emission assuming emission is proportional to the tree leaf dry weight biomass in each ecosystem. Cerrado tree leaf biomass (dry season) = 0.756 t ha⁻¹ (dos Santos, 1989: 194); Forest (at Tucuruí, Pará) = 12.94 t ha⁻¹ (Revilla Cardenas et al., 1982).

gross carbon stocks are given in Fearnside (1997a), with changes in biomass, aboveground fraction, burning efficiency, charcoal formation and soil carbon release as specified elsewhere in the present chapter. A typical scenario of three reburnings over a 10-year period would raise the percentage of aboveground C converted to charcoal from 2.2% to 2.9%. Parameters for carbon emissions by different pathways as CO_2, CO and CH_4, and for other trace gas emissions are also given in Fearnside (1997a). The calculations are carried out by a program known as "DEFOREST," contained in a series of approximately 150 interlinked spreadsheets.

C. Decay of Unburned Remains

Aboveground decay of unburned remains is calculated using the available studies listed in Fearnside, 1996a). Decay makes a significant contribution to greenhouse gas emissions, and it is apparent that the focus of interest on biomass burning leads many to overlook the contributions of decay. The greenhouse gas emissions from deforestation that have been put forward by official Brazilian government sources (Borges, 1992; Silveira, 1992) are lower than those calculated in the present chapter by a factor of three, mainly because they ignore the inherited emissions, in which decay plays a large role.

Bacterial decomposition and termite activity occur largely over the first decade. Termite emissions of methane from decay of unburned biomass (Martius et al., 1996) are substantially lower than previous estimates (Fearnside, 1991, 1992). This is mainly because estimates of the number of termites in deforested areas indicate that the populations are insufficient to consume the quantity of wood that had previously been assumed. Lower emissions of methane (0.002 g CH_4 per g of dry wood consumed) also contributes to lower emissions from this source, estimated to total only 0.014 x 10^6 t of CH_4 gas from original forest in cleared area in 1990 (Tables 2 and 3).

D. Soils

Conversion of natural forest to the replacement landcover will result in a new equilibrium of soil carbon stocks. Changes under cattle pasture are particularly important because of the dominance of pasture and secondary forest derived from pasture in the replacement landcover. Changes in the surface soil (0–20 cm depth under forest) are important because of higher concentrations of carbon in this layer and because the changes occur more quickly than in deeper layers. Compaction of the surface soil must be corrected for one must consider the layer of soil in the replacement land use that is compacted from the 0–20 cm layer of forest soil (see Fearnside, 1980). The emission calculated here (43 x 10^6 t CO_2) considers the top 8 m of forest soil, but only considers emissions in the first 15 years (Fearnside and Barbosa, 1998). The 1–8 m layer contains a large stock of carbon (Nepstad et al., 1994; Trumbore et al., 1995); unfortunately, data on soil carbon in the 1–8 m layer are only available from one site (Paragominas, Pará). The carbon stock in the deep soil may be drawn down to a new lower equilibrium level over a longer time period because the deep roots of trees in natural forest are a source of carbon inputs to this soil layer, and their replacement by pasture and other less deeply rooted types of vegetation can be expected to shift the balance between carbon inputs and oxidation in the deep soil layer. Transformation of forest to the equilibrium landcover results in emission of 8.5 t C ha^{-1} from the top 8 m of soil, 7.9 t C ha^{-1} of which is from the top 1 m (Fearnside and Barbosa, 1998).

E. Removal of Sources and Sinks in Pre-Clearing Landcover

1. Soil Sink for CH$_4$

The tropical forest soil provides a natural sink for methane, removing 0.0004 tons of carbon per hectare per year (Keller et al., 1986). Clearing the forest eliminates this sink, thereby having an effect equal to a source of the same magnitude.

2. Forest Sources of NO$_x$ and NMHC

The leaves of the forest release 0.0131 t ha^{-1} y^{-1} of NO$_x$ (Kaplan et al., 1988; see Keller et al., 1991) and 0.12 t ha^{-1} y^{-1} of non-methane hydrocarbons (NMHC) (Rasmussen and Khalil, 1988: 1420). No information is available on the releases of these gases from the replacement vegetation. Assuming no releases from farmland, productive and degraded cattle pasture, and releases from secondary forests the same as those from primary forests, the area cleared in 1990 implied loss of fluxes of 0.01 x 10^6 t y^{-1} of NO$_x$ and 0.09 x 10^6 t y^{-1} of NMHC (Tables 2 and 3).

3. CH$_4$ Release by Termites

Termites in the mature forest release methane produced by bacteria that digest cellulose under anaerobic conditions in the insects' abdomens. These emissions will be lost when forest is cleared, but for a long time thereafter these emissions will be more than compensated for by termites that ingest the unburned biomass after clearing. In calculating emissions from termites in the forest, the item of interest is the absolute amount of biomass decaying annually (in t ha^{-1} y^{-1}), rather than the rate (fraction) of decomposition per year. For fine litter the amount can be known directly from data on litter fall rates, since all that falls decomposes and the level of the stock can be assumed to be in equilibrium. For coarse litter such data are unavailable, and the amount decomposing must be calculated from information on the stock and the rate of decomposition. Dead trees in a tropical forest can decay remarkably quickly. The decay constant (k) for decomposition of boles in Panama has been calculated to be 0.461 y^{-1} for trees >10 cm DBH, based on observation after a 10-year interval (Lang and Knight, 1979). Here, however, the lower decay rates measured in slash-and-burn fields are used for all coarse biomass. The amounts of fine and coarse litter are calculated from available studies (Fearnside, 1997a).

4. Possible Carbon Sink in Standing Forest

A possible sink of carbon in "undisturbed" standing forest is not considered in the present calculation. Eddy correlation work (studies of gas movements in air flows inside and immediately above the forest) at one site in Rondônia indicated an uptake of 1.0 ± 0.2 t C ha^{-1} y^{-1} (Grace et al., 1995). This would imply an annual uptake of 366 x 10^6 t C by the 358.5 x 10^6 ha of forest still standing in 1990 in the Brazilian Legal Amazon, and a loss of the annual uptake of 1.4 x 10^6 t C from the 1.38 x 10^6 ha cleared in 1990. Malhi et al. (1996, cited by Higuchi et al., 1997) have estimated an uptake of 5.6 ± 1.6 t C ha^{-1} y^{-1} based on eddy correlation work near Manaus. Higuchi et al. (1997) have estimated an uptake of 1.2 t C ha^{-1} y^{-1} in 3 ha of forest growth measurements over the 1986–1996 period near Manaus. On the other hand, forest growth measurements over intervals of 10 to 16 years in the 1980–1997 period in 32 one-ha plots >300 m from a forest edge at another site near Manaus indicate no net growth whatsoever (W.F. Laurance, personal communication, 1997; see also same data in 36 one-ha control plots >100 m from a forest edge in Laurance et al., 1997).

Research interest in a possible sink in standing forest is intense, and efforts in progress to evaluate data on basal area changes in long-term forest monitoring sites and to extend eddy correlation studies may well indicate the existence of a sink. Given the vast area of standing forest, even a small uptake per hectare would make a significant contribution to the global carbon balance. Large spatial coverage is needed in order to draw conclusions, as uptake at one site may be balanced by emissions at other sites. Time scale is undoubtedly also important: over the long term, "mature" forest cannot continue to grow in biomass, but imbalances over periods of years or decades are still important for understanding global carbon dynamics, including clarification of the "missing sink." An uptake would increase the impact of deforestation by eliminating part of the sink. For example, if the sink were 0.45 t C ha^{-1} y^{-1}, the 1.38×10^6 ha of deforestation in 1990 would eliminate an annual sink of 0.621×10^6 t C, while the annual loss for the 41.6×10^6 ha that had been lost through 1990 would total 18.72×10^6 t C. While the amount of sink loss in a single year's deforestation may appear modest compared to the emissions from forest biomass caused by the clearing, the fact that the sink represents an annual flux rather than a one-time emission means that it would have significant consequences over the long term if the sink can be assumed to have a duration of decades or more.

F. Hydroelectric Dams

One of the impacts of hydroelectric dams in Amazonia is emission of greenhouse gases such as carbon dioxide (CO_2) and methane (CH_4). Existing hydroelectric dams in Brazilian Amazonia emitted about 0.27×10^6 t of methane and 37×10^6 t of carbon dioxide in 1990. The CO_2 flux in 1990 included part of the large peak of release from above-water decay of trees left standing in the Balbina reservoir (closed in 1987) and the Samuel reservoir (closed in 1988). Most CO_2 release occurs in the first decade after closing. The methane emissions represent an essentially permanent addition to gas fluxes from the region, rather than a one-time release. The total area of reservoirs planned in the region is about 20 times the area existing in 1990, implying a potential annual methane release of about 5.2×10^6 t. About 40% of this estimated release is from underwater decay of forest biomass, which is the most uncertain of the components in the calculation. Methane is also released from open water, macrophyte beds, and above-water decay of forest biomass (Fearnside, 1995b, 1997d).

G. Logging

In a typical situation, forests accessible by land or river transportation are logged, reducing their biomass both by the removal of timber and by killing or damaging many unharvested trees. This logged-over forest is later cleared for agriculture or cattle ranching.

The effect of logging is not as straightforward as it might appear. By removing the trunks of large trees, the burning efficiency will increase, as will the average decay rate of the unburned biomass. This is because small-diameter branches burn better and decay more quickly than do large trunks. These changes will partially compensate for the reduction in emissions from lower biomass. In calculations where discounting or time preference weighting gives emphasis to short-term releases, the effect of logging on the impact of deforestation when the logged areas are subsequently cleared will be further reduced, since the large logs removed would have been slow to decay had they been left to be cut in the deforestation process.

IV. Uptake by Replacement Vegetation

A. The Replacement Landcover

A Markov matrix of annual transition probabilities was constructed to estimate landcover composition in 1990 and to project future changes, assuming behavior of farmers and ranchers remains unchanged. Transition probabilities for small farmers are derived from satellite studies of government settlement areas (Moran et al., 1994; Skole et al., 1994). Probabilities for ranchers are derived from typical behavior elicited in interview surveys by Uhl et al. (1988). Six land uses are considered, which, when divided to reflect age structure, results in a matrix of 98 rows and columns.

The estimated 1990 landcover in deforested areas was 5.4% farmland, 44.8% productive pasture, 2.2% degraded pasture, 2.1% 'young' (1970 or later) secondary forest derived from agriculture, and 28.1% 'young' secondary forest derived from pasture, and 17.4% 'old' (pre-1970) secondary forest. This landcover would eventually approach an equilibrium of 4.0% farmland, 43.8% productive pasture, 5.2% degraded pasture, 2.0% secondary forest derived from agriculture, and 44.9% secondary forest derived from pasture. An insignificant amount is regenerated 'forest' (defined as secondary forest over 100 years old). Average total biomass (dry matter, including below-ground and dead components) was 43.5 t ha^{-1} in 1990 in the 410×10^3 km^2 deforested by that year for uses other than hydroelectric dams. At equilibrium, average biomass would be 28.5 t ha^{-1} over all deforested areas (excluding dams) (Fearnside, 1996b). Official sources have recently claimed a massive C uptake in "crops" resulting in zero net emissions from deforestation (*ISTOÉ*, 1997). Such a claim is completely at variance with the results presented here.

Better quantification of carbon sinks such as secondary forests is important for both scientific and diplomatic reasons. From a scientific standpoint, better assessments of carbon flows to these sinks are needed in order to have better estimates of net emissions, and, consequently, better estimates of such quantities as the "missing sink." On the diplomatic side, scientists who work on global warming are frequently criticized for spending almost all of their time and money in measuring carbon emissions rather than sinks, with the implication that it is therefore unsurprising that researchers conclude that carbon emissions are a major problem. Thorough investigation of all possible sinks would preclude use of such arguments by those in search of excuses for refusing to take global warming seriously.

B. Secondary Forest Growth Rates

The growth rate of secondary forests is critical in determining the uptake over the replacement landcover. Most discussions of uptake by secondary forests have assumed that these will grow at the rapid rates that characterize shifting cultivation fallows (e.g., Lugo and Brown, 1981, 1982). In Brazilian Amazonia, however, most deforestation is for cattle pasture, shifting cultivation playing a relative minor role (Fearnside, 1993). Secondary forests on degraded pastures grow much more slowly than on sites where only annual crops have been planted following the initial forest felling.

Brown and Lugo (1990) have reviewed the available data on growth of tropical secondary forests. The available information is virtually all from shifting cultivation fallows. Brown and Lugo (1990: 17) trace a freehand graph from available data for secondary forest stands ranging in age from 1 to 80 years, including biomass for wood (twigs, branches and stems: 13 data points), leaves (10 data points), and roots (12 data points). This has been used to estimate growth rate and the root/shoot ratio for shifting cultivation fallows of different ages. Secondary forests on abandoned pastures grow more slowly (Guimarães, 1993; Uhl et al., 1988). This information on growth rate of secondary vegetation of different origins has been used to calculate uptakes in the landscape in 1990 (Fearnside and Guimarães, 1996).

Table 4. 1990 annual balance of net emissions by source in the originally forested area of the Brazilian Legal Amazon (a): low trace gas scenario

Source	Emissions (million t of gas)						Sink (million t of carbon)	
	CO_2	CH_4	CO	N_2O	NO_x	NMHC	Charcoal carbon	Graphite carbon
Original forest biomass								
Initial burn	269.97	0.87	20.96	0.05	0.66	0.66	3.52	0.20
Reburns	65.95	0.32	10.21	0.01	0.51	0.16	1.05	0.08
Termites aboveground decay	14.60	0.02						
Other aboveground decay	357.08							
Below-ground decay	321.55							
Secondary forest biomass								
Burning (b)	52.06	0.17	4.03	0.010	0.06	0.11	0.25	0.04
Termites above ground decay	0.98	0.001						
Other aboveground decay	21.29							
Below-ground decay	23.60							
Termites in secondary forest		0.003						
Pre-1970 secondary forest biomass								
Initial burning	5.34	0.017	0.419	0.001	0.013	0.012	0.069	0.004
Reburnings	0.85	0.004	0.135	0.0002	0.007	0.002	0.014	0.001
Termites aboveground decay	0.21	0.0002						
Other aboveground biomass	5.21							
Below-ground decay	3.03							
Termites in pre-1970 stands		0.0035						
Pasture burning	(c)	0.07	1.69	0.004	0.12	0.05	0.08	0.02
Hydroelectric dams								
Forest biomass	35.75	0.12						
Water		0.11						
Macrophytes		0.04						

Table 4. continued

Source	Emissions (million t of gas)						Sink (million t of gas)	
	CO_2	CH_4	CO	N_2O	NO_x	NMHC	Charcoal carbon	Graphite carbon
Other sources								
Cattle		0.31						
Pasture soil				0.07				
Loss of intact forest sources and sinks		0.02			-4.24	-0.46		
Loss of natural forest termites		-0.03						
Soil carbon (top 8 m)	56.65							
Total emissions	1233.40	2.04	37.37	0.16	-2.87	0.45	4.98	0.34
Uptake	-28.98							
Net emissions	1204.12	2.04	37.37	0.16	-2.87	0.45	4.98	0.34

(a) Deforestation in originally forested area in 1990 was 1,381,800 ha.
(b) Secondary forest burning includes both initial and subsequent burns for secondary forest from both agriculture and pasture, and for degraded pasture that is cut and recuperated.
(c) CO_2 from maintenance burning of pasture is not counted, as this is re-assimilated annually as the pastures regrow, making the net flux equal to zero. The gross flux in 1990 from this source is estimated at 22 million t of CO_2 gas.

V. Annual Balance of Net Emissions

The sources of emissions and uptakes of greenhouse gases for the annual balance in 1990 are presented in Table 4 for the low trace gas scenario and in Table 5 for the high trace gas scenario. Considering only CO_2, 1218–1233 x 10^6 t of gas were emitted (gross emission) by deforestation (not including logging emissions). Deducting the uptake of 29 x 10^6 t of CO_2 gas yields a net emission of 1189–1204 x 10^6 t of CO_2, or 324–328 x 10^6 t of carbon. Adding effects of trace gases using the IPCC Second Assessment Report SAR global warming potentials for a 100-year time horizon, the impacts increase to 353–359 x 10^6 t of CO_2-equivalent carbon. Consideration of more indirect effects of trace gases would raise these values substantially: the IPCC SAR recognizes some indirect effects for CH_4 but none for CO, which is an important component of emissions from biomass burning. Logging added 224 x 10^6 t of CO_2 gas, plus trace gases that raised the impact to 228–229 x 10^6 t of CO_2 gas equivalent (63 x 10^6 t CO_2 = equivalent C).

In terms of carbon dioxide from the original forest biomass only, 27% of the emission (before deducting uptakes) in the annual balance was from prompt emissions from deforestation in that year, and 73% was from inherited emissions from decay and reburning of unburned biomass left from clearing in previous years. Because of higher inherited emissions in the areas cleared in the years of faster deforestation preceding 1990, the annual balance is higher than the net committed emissions by 27–29% if only CO_2 is considered and by 29–32% if the CO_2 equivalents of other gases are also included. Net committed emissions would be equal to the annual balance that would prevail were deforestation to proceed at a constant rate over a long period.

Table 5. 1990 annual balance of net emissions by source in the originally forested area of the Brazilian Legal Amazon (a): high trace gas scenario

Source	Emissions (million t of gas)						Sink (million t of carbon)	
	CO_2	CH_4	CO	N_2O	NO_x	NMHC	Charcoal carbon	Graphite carbon
Original forest biomass								
Initial burn	269.97	1.05	26.13	0.05	0.66	1.10	3.52	0.24
Reburns	64.95	0.51	12.99	0.1	0.51	0.31	1.05	0.12
Termites aboveground decay	16.02	0.02						
Other aboveground decay	357.09							
Below-ground decay	321.53							
Secondary forest biomass								
Burning (b)	40.24	0.16	3.89	0.008	0.04	0.16	0.23	0.04
Termites above ground decay	0.93	0.0007						
Other aboveground decay	20.25							
Below-ground decay	22.49							
Termites in secondary forest		0.003						
Pre-1970 secondary forest biomass								
Initial burning	5.34	0.021	0.516	0.001	0.013	0.022	0.069	0.005
Reburnings	0.85	0.007	0.170	0.001	0.007	0.004	0.014	0.002
Termites aboveground decay	0.23	0.0002						
Other aboveground biomass	5.31							
Below-ground decay	3.03							
Termites in pre-1970 stands		0.0027						
Pasture burning	(c)	0.08	2.02	0.004	0.11	0.08	0.08	0.02
Hydroelectric dams								
Forest biomass	35.75	0.12						
Water		0.11						
Macrophytes		0.04						

Table 5. continued

Source	Emissions (million t of gas)						Sink (million t of carbon)	
	CO_2	CH_4	CO	N_2O	NO_x	NMHC	Charcoal carbon	Graphite carbon
Other sources								
Cattle		0.29						
Pasture soil				0.07				
Loss of intact forest sources and sinks		0.02			-4.06	-0.44		
Loss of natural forest termites		-0.03						
Soil carbon (top 8 m)	54.43							
Total emissions	1218.37	2.39	45.72	0.25	-2.71	1.23	4.96	0.42
Uptake	-28.98							
Net emissions	1189.39	2.39	45.72	0.25	-2.31	1.23	4.96	0.42

(a) Deforestation in originally forested area in 1990 was 1,381,800 ha.
(b) Secondary forest burning includes both initial and subsequent burns for secondary forest from both agriculture and pasture, and for degraded pasture that is cut and recuperated.
(c) CO_2 from maintenance burning of pasture is not counted, as this is re-assimilated annually as the pastures regrow, making the net flux equal to zero. The gross flux in 1990 from this source is estimated at 21 million t of CO_2 gas.

Table 6. Comparison of methods of calculating the 1990 global warming impact of deforestation in originally forested areas of Brazilian Amazonia in millions of tons of CO_2-equivalent carbon

Gases included	Net committed emissions (Deforestation only)	Annual balance		
		Deforestation only	Logging	Deforestation + logging
Low trace gas scenario				
CO_2 only	255	328	61	390
CO_2, CH_4, N_2O	267	353	62	415
High trace gas scenario				
CO_2 only	255	324	61	386
CO_2, CH_4, N_2O	278	359	63	422

Net committed emissions and the annual balance are compared in Table 6 for the low and high trace gas scenarios, both considering only CO_2 equivalents using the IPCC Second Assessment Report SAR 100-year integration global warming potentials. The emissions from logging are also tabulated. Inclusion of trace gases (using the IPCC SAR 100-year global warming potentials) raises the impact of net committed emissions by 5 to 9%, and of the annual balance by 8 to 11%. Trace gas impacts are likely to increase when the IPCC reaches agreement on additional indirect effects. For example, if the impact of CO calculated using the global warming potential of 2 that was adopted in the 1990 IPCC report (Shine et al., 1990), but dropped in subsequent reports pending agreement, the annual balance would be increased by the equivalent of 75–92 x 10^6 t of CO_2 gas, while inclusion of the additional effect of CO on extending the atmospheric lifetime of CH_4 due to removal of OH radicals (Shine et al., 1990) would further increase this impact.

VI. Conclusions

1. In 1990, the year for baseline inventories under the United Nations Framework Convention on Climate Change, land-use changes in Brazil's 5 x 10^6 km^2 Legal Amazon Region included 13.8 x 10^3 km^2 of deforestation, approximately 5 x 10^3 km^2 of clearing in cerrado, the central Brazilian scrubland that originally occupied about 20% of the Legal Amazon (savanna), 7 x 10^2 km^2 in "old" (pre-1970) and 19 x 10^3 km^2 in "young" (1970+) secondary forests; burning of 40 x 10^3 km^2 of productive pasture (33% of the area present), and regrowth in 121 x 10^3 km^2 of "young" secondary forests. No new hydroelectric flooding occurred in 1990, but decomposition continued in 4.8 x 10^3 km^2 of reservoirs already in place. Logging at 24.6 x 10^6 m^3 was assumed – the 1988 official rate.

2. Unlogged original forests in Brazilian Amazonia are estimated to have an average total biomass of 463 metric tons per hectare (t ha^{-1}), including below-ground and dead components. Adjustment for the spatial distribution of clearing and for logging indicates an average total biomass cleared in 1990 of 406 t ha^{-1} in original forest areas, 309 t ha^{-1} of which is above-ground (exposed to the initial burn). In addition to emissions from the initial burn, the remains from clearing in previous years emitted gases through decay and combustion in reburns. More rapid deforestation in the years preceding 1990 make these inherited emissions greater than they would have been had deforestation rates been constant at their 1990 levels.

3. Estimated net committed emissions (the net amounts of greenhouse gases that will ultimately be emitted as a result of the clearing done in a given year) from deforestation (not including logging emissions or the clearing of cerrado totaled 934 x 10^6 t CO_2, 1.3–1.5 x 10^6 t CH_4, 30–37 x 10^6 t CO, and 0.07–0.18 x 10^6 t N_2O. These emissions are equivalent to 267–278 x 10^6 t of CO_2-equivalent carbon, using IPCC SAR 100-year GWPs. CO_2 emissions include 270 x 10^6 t of gas from the initial burn, 628 x 10^6 t from decay, 57 x 10^6 t from subsequent burns of primary forest biomass, and 43 x 10^6 t C from soil carbon in the top 8 m. The replacement landcover eventually stores 65 x 10^6 C, or 6.5% of the gross emission. The ranges of emissions given above are for low- and high-trace gas scenarios, reflecting the range of emission factors appearing in the literature for different burning and decomposition processes. These scenarios do not reflect the uncertainty of values for deforestation rate, forest biomass, logging intensity and other inputs to the calculation. Some carbon enters sinks though conversion to charcoal (5.0 x 10^6 t C) and graphitic particulate carbon (0.42 x 10^6 t C).

4. The annual balance of net emissions in 1990 (net fluxes in a single year over the entire region) included 1189–1204 x 10^6 t CO_2, 2.1–2.4 x 10^6 t CH_4, 37.4–45.7 x 10^6 t CO, and 0.16–0.25

x 10^6 t N_2O. CO_2 emissions include 270 x 10^6 t of gas from the initial burn, 693–695 x 10^6 t from decay, 65–66 x 10^6 t from subsequent burns of primary forest biomass, and 46–58 x 10^6 t from burning of secondary forest biomass of all ages, 54–57 x 10^6 t CO_2 from net release of soil carbon to 8 m depth (first 15 years only), 224 x 10^6 t from logging and 36 × 10^6 t from hydroelectric reservoirs. Secondary forest regrowth in 1990 removed 29.0 x 10^6 t of CO_2 gas (only 2.4% of the gross emission, excluding hydroelectric and pasture emissions). Pastures release through burning (and assimilate in growth) 21–22 x 10^6 t of CO_2 gas, not counted in the calculations. The effect of deforestation on the annual balance is a net emission equivalent to 353–359 x 10^6 t of CO_2-equivalent carbon, while logging adds 62 x 10^6 t of CO_2-equivalent carbon.

5. The net committed emissions and annual balance of net emissions from land-use change in Brazilian Amazonia in 1990 were both dominated by deforestation. Because deforestation rates declined in the three years immediately preceding 1990, the annual balance from deforestation (i.e., excluding logging) is higher than the net committed emissions.

6. These results indicate that deforestation in Brazilian Amazonia makes a substantial contribution to global warming, and points to the high priority that should be placed on improving the estimates of these emissions and of the uncertainties they contain. Changes in management in the deforested landcover can only compensate for a small fraction of this impact. Therefore, any policy changes that reduce the rate of deforestation would have the greatest potential for reducing the net emission of greenhouse gases from Amazonia.

Acknowledgment

I thank the National Council of Scientific and Technological Development (CNPq AI 350230/97-98) and the National Institute for Research in the Amazon (INPA PPI 5-3150) for financial support. R.I. Barbosa, S.V. Wilson and two anonymous reviewers made useful comments.

References

Araújo, T.M., J.A. Carvalho Jr., N. Higuchi, A.C.P. Brasil Jr. and A.L.A. Mesquita. 1997. Estimativa de taxas de liberação de carbono em experimento de queimada no Estado do Pará. *Anais da Academia Brasileira de Ciências* 69:575-585.

Barbosa, R.I. and P.M. Fearnside. 1996. Pasture burning in Amazonia: dynamics of residual biomass and the storage and release of aboveground carbon. *J. Geophys. Res.* (*Atmospheres*) 101(D20):25,847-25,857.

Borges, L. 1992. Desmatamento emite só 1,4% de carbono, diz Inpe. O Estado de São Paulo, 10 April 1992, p. 13.

Brazil, Projeto RADAMBRASIL. 1973–1983. *Levantamento de Recursos Naturais*, Vols. 1-23. Ministério das Minas e Energia, Departamento Nacional de Produção Mineral (DNPM), Rio de Janeiro, Brazil.

Brown, S. and A.E. Lugo. 1990. Tropical secondary forests. *Journal of Tropical Ecology* 6:1-32.

Carvalho, Jr., J.A., J.M. Santos, J.C. Santos, M.M. Leitão and N. Higuchi. 1995. A tropical rainforest clearing experiment by biomass burning in the Manaus region, *Atmospheric Environment* 29(17):2301-2309.

dos Santos, J.R. 1989. Estimativa da biomassa foliar das savanas brasileiras: Uma abordagem por sensoriamento remoto. p. 190-199 In: IV Simpósio Latinamericano en Percepción Remota, IX Reunión plenaria SELPER, 19 al 24 de noviembre de 1989, Bariloche, Argentina, Tomo 1. SELPER, Instituto Nacional de Pesquisas Espaciais, São José dos Campos, São Paulo, Brazil.

Fearnside, P.M. 1980. The effects of cattle pastures on soil fertility in the Brazilian Amazon: consequences for beef production sustainability. *Tropical Ecol.* 21:125-137.

Fearnside, P.M. 1991. Greenhouse gas contributions from deforestation in Brazilian Amazonia. p. 92-105 In: J.S. Levine (ed.), *Global Biomass Burning: Atmospheric, Climatic, and Biospheric Implications.* MIT Press, Boston, Massachusetts, USA.

Fearnside, P.M. 1992. Greenhouse Gas Emissions from Deforestation in the Brazilian Amazon. Carbon Emissions and Sequestration in Forests: Case Studies from Developing Countries. Volume 2. LBL-32758, UC-402. Climate Change Division, Environmental Protection Agency, Washington, D.C. and Energy and Environment Division, Lawrence Berkeley Laboratory (LBL), University of California (UC), Berkeley, CA, USA.

Fearnside, P.M. 1993. Deforestation in Brazilian Amazonia: the effect of population and land tenure. *Ambio* 22:537-545.

Fearnside, P.M. 1994. Biomassa das florestas Amazônicas brasileiras. p. 95-124 In: Anais do Seminário Emissão x Seqüestro de CO$_2$. Companhia Vale do Rio Doce (CVRD), Rio de Janeiro, Brazil.

Fearnside, P.M. 1995a. Global warming response options in Brazil's forest sector: comparison of project-level costs and benefits. *Biomass and Bioenergy* 8:309-322.

Fearnside, P.M. 1995b. Hydroelectric dams in the Brazilian Amazon as sources of 'greenhouse' gases. *Environ. Conserv.* 22:7-19.

Fearnside, P.M. 1996a. Amazonia and global warming: Annual balance of greenhouse gas emissions from land-use change in Brazil's Amazon region. p. 606-617 In: J. Levine (ed.), Biomass Burning and Global Change. Volume 2: Biomass Burning in South America, Southeast Asia and Temperate and Boreal Ecosystems and the Oil Fires of Kuwait. MIT Press, Cambridge, MA, USA.

Fearnside, P.M. 1996b. Amazonian deforestation and global warming: Carbon stocks in vegetation replacing Brazil's Amazon forest. *For. Ecol. Manage.* 80:21-34.

Fearnside, P.M. 1997a. Greenhouse gases from deforestation in Brazilian Amazonia: net committed emissions. *Climatic Change* 35:321-360.

Fearnside, P.M. 1997b. Wood density for estimating forest biomass in Brazilian Amazonia. *For. Ecol. Manage.* 90:59-89.

Fearnside, P.M. 1997c. Monitoring needs to transform Amazonian forest maintenance into a global warming mitigation option. *Mitigation and Adaptation Strategies for Global Change* 2:285-302.

Fearnside, P.M. 1997d. Greenhouse-gas emissions from Amazonian hydroelectric reservoirs: the example of Brazil's Tucuruí Dam as compared to fossil fuel alternatives. *Environ. Conserv.* 24:64-75.

Fearnside, P.M. nd. Biomass of Brazil's Amazonian Forests (in preparation).

Fearnside, P.M. and R.I. Barbosa. 1998. Soil carbon changes from conversion of forest to pasture in Brazilian Amazonia. *For. Ecol. Manage.* 108:147-166.

Fearnside, P.M. and J. Ferraz. 1995. A conservation gap analysis of Brazil's Amazonian vegetation. *Conserv. Biol.* 9:1134-1147.

Fearnside, P.M. and W.M. Guimarães. 1996. Carbon uptake by secondary forests in Brazilian Amazonia. *For. Ecol. Manage.* 80:35-46.

Fearnside, P.M., N. Leal Filho and F.M. Fernandes. 1993. Rainforest burning and the global carbon budget: biomass, combustion efficiency and charcoal formation in the Brazilian Amazon. *J. Geophys. Res. (Atmospheres)* 98(D9):16,733-16,743.

Fearnside, P.M., P.M.L.A. Graça, N. Leal Filho, F.J.A. Rodrigues and J.M. Robinson. nd-a. Tropical forest burning in Brazilian Amazonia: measurements of biomass loading, burning efficiency and charcoal formation at Altamira, Pará. (in preparation).

Fearnside, P.M., P.M.L.A. Graça and F.J.A. Rodrigues. nd-b. Burning of Amazonian rainforests: burning efficiency and charcoal formation in forest cleared for cattle pasture near Manaus, Brazil. (in preparation).

Fearnside, P.M., R.I. Barbosa and P.M.L.A. Graça. nd-c. Burning of secondary forest in Amazonia: biomass, burning efficiency and charcoal formation during land preparation for agriculture in Apiaú, Roraima, Brazil. (in preparation).

Glerum, B.B. 1960. Report to the Government of Brazil on a Forestry Inventory in the Amazon Valley (Part Five) (Region between Rio Caete and Rio Maracassume), FAO Report No. 1250, Project No. BRA/FO, Food and Agriculture Organization of the United Nations (FAO), Rome, Italy.

Graça, P.M.L.A. 1997. Conteúdo de Carbono na Biomassa Florestal da Amazônia e Alterações após à Queima. Masters thesis in forest sciences, Escola Superior de Agricultura "Luiz de Queiroz", Universidade de São Paulo, Piracicaba, São Paulo, Brazil.

Graça, P.M.L.A., P.M. Fearnside and C.C. Cerri. nd. Burning of Amazonian forest in Ariguemes, Rondônia, Brazil: Biomass, charcoal formation and burning efficiency. *For. Ecol. Manage.* (in press).

Grace, J., J. Lloyd, J. McIntyre, A.C. Miranda, P. Meir, H.S. Miranda, C. Nobre, J. Moncrieff, J. Massheder, Y. Malhi, I. Wright and J. Gash. 1995. Carbon dioxide uptake by an undisturbed tropical rain forest in southwest Amazonia, 1992 to 1993. *Science* 270:778-780.

Guimarães, W.M. 1993. Liberação de carbono e mudanças nos estoques dos nutrientes contidos na biomassa aérea e no solo resultante de queimadas de florestas secundárias em áreas de pastagens abandonadas, em Altamira, Pará. Masters thesis in ecology, Instituto Nacional de Pesquisas da Amazônia/Fundação Universidade do Amazonas (INPA/FUA), Manaus, Brazil.

Heinsdijk, D. 1957. Report to the Government of Brazil on a Forest Inventory in the Amazon Valley (Region between Rio Tapajós and Rio Xingu). FAO Report No. 601, Project No. BRA/FO, Food and Agriculture Organization of the United Nations (FAO), Rome, Italy.

Heinsdijk, D. 1958a. Report to the Government of Brazil on a Forest Inventory in the Amazon Valley (Part Three) (Region between Rio Tapajós and Rio Madeira). FAO Report No. 969, Project No. BRA/FO, Food and Agriculture Organization of the United Nations (FAO), Rome, Italy.

Heinsdijk, D. 1958b. Report to the Government of Brazil on a Forest Inventory in the Amazon Valley (Part Four) (Region between Rio Tocantins and Rios Guamá and Capim). FAO Report No. 992, Project No. BRA/FO, Food and Agriculture Organization of the United Nations (FAO), Rome, Italy.

Heinsdijk, D. 1958c. Report to the Government of Brazil on a Forestry Inventory in the Amazon Valley (Part Two) (Region between Rio Xingu and Rio Tocantins), FAO Report No. 949, Project No. BRA/FO, Food and Agriculture Organization of the United Nations (FAO), Rome, Italy.

Higuchi, N., J. dos Santos, R.J. Ribeiro, J.V. de Freitas, G. Vieira, A. Cöic and L.J. Minette. 1997. Crescimento e incremento de uma floresta amazônica de terra-firme manejada experimentalmente. p. 87-132 In: N. Higuchi, J.B.S. Ferraz, L. Antony, F. Luizão, R. Luizão, Y. Biot, I. Hunter, J. Proctor and S. Ross (eds.), Bionte: Biomassa e Nutrientes Florestais, Relatório Final. Instituto Nacional de Pesquisas da Amazônia (INPA), Manaus, Brazil.

ISTOÉ. 1997. A Versão do Brasil. *ISTOÉ* [São Paulo] 15 de outubro de 1997.

Kaplan, W.A., S.C. Wofsy, M. Keller and J.M. da Costa. 1988. Emission of NO and deposition of O_3 in a tropical forest system. *J. Geophys. Res.* 93:1389-1395.

Kauffman, J.B., D.L. Cummings, D.E. Ward and R. Babbitt. 1995. Fire in the Brazilian Amazon. 1. Biomass, nutrient pools, and losses in slashed primary forests. *Oecologia* 104:397-408.

Keller, M., D.J. Jacob, S.C. Wofsy and R.C. Harriss. 1991. Effects of tropical deforestation on global and regional atmospheric chemistry. *Climatic Change* 19:139-158.

Keller, M., W.A. Kaplan and S.C. Wofsy. 1986. Emissions of N_2O, CH_4 and CO_2 from tropical forest soils. *J. Geophys. Res.* 91:11,791-11,802.

Kuhlbusch, T.A.J. and P.J. Crutzen. 1995. A global estimate of black carbon in residues of vegetation fires representing a sink of atmospheric CO_2 and a source of O_2. *Global Biogeochemical Cycles* 9:491-501.

Lang, G.E. and D.H. Knight. 1979. Decay rates of boles for tropical trees in Panama. *Biotropica* 11:316-317.

Laurance, W.F., S.G. Laurance, L.V. Ferreira, J.M. Rankin-de-Merona, C. Gascon and T.E. Lovejoy. 1997. Biomass collapse in Amazonian forest fragments. *Science* 278:1117-1118.

Lugo, A.E. and S. Brown. 1981. Tropical lands: popular misconceptions. *Mazingira* 5:10-19.

Lugo, A.E. and S. Brown. 1982. Conversion of tropical moist forests: a critique. *Interciencia* 7:89-93.

Malhi, Y., N. Higuchi, A.D. Nobre, J. Grace, R.J. Ribeiro, M. Pereira, A. Marques Filho, A. Culf, J. Massheder, S. Scott and J. Moncrief. 1996. Direct measurements of carbon uptake by Amazonian rain forest. Unpublished report.

Martius, C., P.M. Fearnside, A.G. Bandeira and R. Wassmann. 1996. Deforestation and methane release from termites in Amazonia. *Chemosphere* 33:517-536.

Moran, E.F., E. Brondizio, P. Mausel and Y. Wo. 1994. Integrating Amazonian vegetation, land-use, and satellite data. *BioScience* 44:329-338.

Nepstad, D.C., C.R. Carvalho, E.A. Davidson, P.H. Jipp, P.A. Lefebvre, G.H. Negreiros, E.D. Silva, T.A. Stone, S.E. Trumbore and S. Vieira. 1994. The role of deep roots in the hydrological cycles of Amazonian forests and pastures. *Nature* 372:666-669.

Rasmussen, R.A. and M.A.K. Khalil. 1988. Isoprene over the Amazon Basin. *J. Geophys. Res.* 93:1417-1421.

Revilla Cardenas, J.D., F.L. Kahn and J.L. Guillaumet. 1982. Estimativa da Fitomassa do Reservatório da UHE de Tucuruí. pp. 1-11 In: Brazil, Presidência da República, Ministério das Minas e Energia, Centrais Eletricas do Norte S.A. (ELETRONORTE) and Brazil, Secretaria do Planejamento, Conselho Nacional de Desenvolvimento Científico e Tecnológico, Instituto Nacional de Pesquisas da Amazônia (SEPLAN-CNPq-INPA), Projeto Tucuruí, Relatório Semestral Jan.-Jun. 1982, Vol. 2: Limnologia, Macrófitas, Fitomassa, Degradação de Fitomassa, Doenças Endêmicas, Solos. INPA, Manaus, Brazil.

Shine, K.P., R.G. Derwent, D.J. Wuebbles and J-J. Morcrette. 1990. Radiative forcing of climate. p. 41-68 In: J.T. Houghton, G.J. Jenkins and J.J. Ephraums (eds.), *Climate Change: The IPCC Scientific Assessment.* Cambridge University Press, Cambridge, UK. 365 pp.

Silveira, V. 1992. Amazônia polui com apenas 1,4%. *Gazeta Mercantil* [São Paulo] 29 May 1992, p. 2 and 6.

Skole, D.L., W.H. Chomentowski, W.A. Salas and A.D. Nobre. 1994. Physical and human dimensions of deforestation in Amazonia. *BioScience* 44:314-322.

Sombroek, W.G. 1992. Biomass and carbon storage in the Amazon ecosystems. *Interciencia* 17:269-272.

Trumbore, S.E., E.A. Davidson, P.B. Camargo, D.C. Nepstad and L.A. Martinelli. 1995. Below-ground cycling of carbon in forests and pastures of eastern Amazonia. *Global Biogeochemical Cycles* 9:515-528.

Uhl, C., R. Buschbacher and E.A.S. Serrão. 1988. Abandoned pastures in Eastern Amazonia. I. Patterns of plant succession. *J. Ecol.* 76:663-681.

CHAPTER 13

Land Use Impact on Carbon Dynamics in Soils of the Arid and Semiarid Tropics

B.A. Stewart and C.A. Robinson

I. Introduction

A well known fact is soil organic carbon (SOC) levels decline when land is converted from grassland or forest ecosystems to cropland. This decline is most rapid in the first few years following conversion and then continues at slower rates until a new steady state is reached. After 50 to 100 years of cultivation, SOC levels are often 50 to 60% lower than the initial levels (Cole et al., 1993).

Cultivation increases biological activities in the soil, often because of better aeration. Cultivation exposes fresh topsoil to rapid drying, and after each drying there occurs a burst of biological activity for a few days following rewetting. This is because the drying process releases organic compounds, probably from the breakdown of soil aggregates bound together by humic materials. Considerable organic nitrogen is mineralized as ammonia and later oxidized in large part to nitrates. Other plant nutrients are also made available from the decomposition of soil organic matter (SOM). This is particularly true for phosphorus since much of the phosphorus in soils is present in organic forms. The nutrients released as a result of tillage are readily available to growing plants and increased yields are often obtained. Tillage also increases rainfall infiltration, controls weeds, and often helps control insects and diseases. Therefore, intensive and frequent tillage has commonly been considered essential for good crop production. However, unless the soil organic supply is replenished, the soil degrades. If the degradation continues, the cropping system cannot be sustained. This has a negative effect on the carbon status because a decline in SOM content adds significant amounts of CO_2 to the atmosphere, and fewer crops use less CO_2 from the atmosphere.

The SOM is a major sink for carbon. The SOC is also significantly correlated with soil productivity and soil quality. It acts as a storehouse for nutrients, increases the cation exchange capacity, and reduces the effects of compaction. It builds soil structure and increases the infiltration and retention of water. It serves as a buffer against rapid changes in pH and is an energy source for soil microorganisms; therefore, SOC is of critical importance.

II. Production and Retention of Carbon in Agriculture

Because SOC is of such great importance for soil quality and crop production, coupled with the fact that carbon stored in the soil reduces atmospheric CO_2, there is great interest in developing practices and policies that will lead to increased production and retention of carbon in agriculture. Producers have done a tremendous job of increasing carbon production in crops, but very little has been retained in the soil. Production of food grains and feed crops more than doubled from 1948 to 1994 in the U.S., and the amount of cropland used for production even decreased (Ahearn et al., 1998; Economic

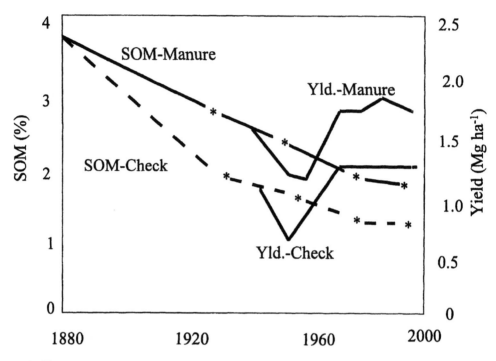

Figure 1. Changes in organic matter and wheat grain yields in a continuous wheat cropping system on manured and control plots in OK. (Drawn from data of Westerman, 1992.)

Research Service, 1997). Therefore, there has been a tremendous increase in the amount of CO_2 removed annually from the atmosphere to grow crops, but there is little evidence that shows SOC has increased. This is certainly the case when intensive cultivation has continued. For example, data from the long-term Magruder plots at Oklahoma State University that have been continuously cropped to wheat (*Triticul aestivum*) since 1892 (Westerman, 1992) are presented in Figure 1. The SOM content decreased over the years in spite of the fact that there had been significant increases in wheat grain yield. Although carbon assimilation was not measured, it is well known that carbon assimilation is closely correlated with grain yields. Even where manure had been added, SOM continued decreasing although it had been fairly constant for the last 10 years indicating that a new equilibrium had been reached. There were also plots that received N,P, and K fertilizers and the yields from those plots were similar to those shown for the manured plots. Although it is somewhat surprising that manure did not increase the SOM, it is important to point out that only enough manure was added to supply adequate nitrogen for wheat production. Beef manure was applied every fourth year at a rate sufficient to add 134 kg ha^{-1} N from 1930–1967 and 268 kg ha^{-1} N from 1967–1991. The important point is that the SOM content continued to decrease even though increasing amounts of crop residue were being returned to the soil. Intensive tillage, however, was used which strongly indicates that if soil organic carbon levels are going to be maintained or enhanced, cropping systems that involve less intensive tillage practices will be necessary.

The effect of tillage on SOM decline was further illustrated by Lamb et al. (1985). They cultivated a native grassland site in western Nebraska for winter wheat production in a crop-fallow rotation under three tillage systems: no-till, stubble mulch, and plow. After 12 years of cultivation, losses of soil N

Table 1. Tillage method effects on soil organic matter (SOM) and total N concentrations with depth in Pullman clay loam, Bushland, TX, 1988

	Tillage method[a]					
	SM		NT-81		NT-79	
Depth	Mean	CI[b]	Mean	CI	Mean	CI
(cm)	SOM (g kg⁻¹)					
0–1	16.5	1.0	21.6	4.9	23.5	3.3
1–2	17.2	1.4	19.8	4.4	18.3	1.6
2–4	16.7	1.5	16.7	2.2	15.7	0.7
4–6	15.8	1.0	14.9	0.9	14.7	0.6
6–8	15.3	0.9	14.7	1.2	14.1	0.3
8–10	14.6	1.0	14.7	1.4	13.9	0.2
10–15	13.4	0.9	14.3	0.6	13.5	0.3
15–20	12.2	0.8	12.9	0.6	12.9	1.2
Prot. LSD[c]	1.2	–	2.7	–	1.5	–
	Total N (g kg⁻¹)					
0–1	1.11	0.09	1.23	0.14	1.40	0.07
1–2	1.08	0.08	1.21	0.16	1.06	0.21
2–4	1.11	0.13	1.07	0.10	1.11	0.07
4–6	1.14	0.15	0.96	0.05	1.02	0.06
6–8	1.11	0.13	0.95	0.15	1.03	0.12
8–10	1.07	0.15	0.99	0.10	0.99	0.07
10–15	1.00	0.12	0.97	0.06	0.97	0.06
15–20	0.94	0.10	0.96	0.05	0.91	0.07
Prot. LSD	0.13	–	0.11	–	0.11	–

[a]Tillage methods: SM = stubble mulch tillage; NT-81 = no-tillage since 1981; and NT-79 = no-tillage since 1979. [b]CI–confidence interval = standard error of the mean × $t_{(0.05)}$. [c]Protected least significant difference at the $p = 0.05$ level.
(Adapted from Unger, 1991.)

from the 0 to 30 cm depth were 3% for the no-till, 8% for the stubble mulch, and 19% for the plow tillage systems. Although not reported, it is assumed that comparable losses of SOC occurred.

Crop residues can be managed in manners that lead to increased SOM levels, thereby sequestering carbon. No-till systems often show increased SOM within the first few years of practice. W.D. Kemper (private communication, 1993) stated that one of the gratifying consequences of no-till management is associated with increases in SOM of the soils which ranged from 100 to over 1000 kg ha⁻¹ yr⁻¹ (approximately 60 to 600 kg ha⁻¹ yr⁻¹ of SOC). The higher rates are usually associated with leguminous winter cover crops whose residues were left on the soil surface. This buildup of SOM is restoring the "nutrient bank" that can help tide plants over periods of deficiency. Also, the sequestration of nitrogen into this accumulating SOM is probably one of the factors decreasing the nitrate concentration of water percolating below no-till fields relative to that below conventionally-tilled fields. Even in semiarid regions where only limited amounts of crop residues are produced, significant increases in SOM occur when no-till cropping systems are adopted. Unger (1991) evaluated the distribution with depth of organic matter in wheat-grain sorghum(*Sorghum bicolor*)-fallow plots (Table 1). The work was done in the Texas High Plains where the annual precipitation averages 465 mm. The no-till fields had higher

SOM levels than the stubble mulch field, although the differences were relatively small and confined mostly to the top 1- to 2-cm soil depths.

The SOC, in addition to being a storehouse for carbon that will reduce the amount of CO_2 in the atmosphere, has many other benefits. Carbon makes up about 60% of the SOM which is of prime importance to mankind. Since the dawn of history, SOM has been considered the key to soil fertility and productivity. Allison (1973) stated that historically the best farmers cultivated their soils frequently even when there were few weeds present. The benefits were ascribed in large part to the dust mulch produced, but Allison said we now know that, aside from weed control, the main reasons that cultivation increased yields was because it released nutrients from the soil, and mostly from the organic portion. Cultivation year after year markedly lowers the SOM content as discussed above, and as this occurs the soil physical properties deteriorate. The soil gradually becomes a continually poorer medium for plant growth. In recent years, farming systems that use significantly less tillage, have been developed that maintain, and sometimes enhance, the SOM level. These systems, in addition to using less tillage, are often based on the use of abundant commercial fertilizer together with the return of crop residues.

Added nutrients for retaining carbon is of critical importance and this is frequently overlooked. An excellent example is the Conservation Reserve Program that began in the late 1980s and resulted in more than 13 million ha of land seeded to grass. Approximately 65% of this land was located in the dryland regions of the Great Plains states. In most cases, these lands were seeded to grass without any added fertilizer and in nearly all cases there were no legumes seeded with the grasses. In order for carbon to be retained in the soil as humus or other forms of SOM, there must be sufficient nitrogen available to "hold" the carbon. As pointed out earlier, nitrogen and carbon have been lost from SOM in somewhat equal proportions. Therefore, if the carbon content of the soil is going to be replenished in the soil to anywhere near its original level, there must be nitrogen additions as well. In dryland regions where crop production has been severely constrained by lack of water, there may be enough nitrate-nitrogen accumulated in the profile to fill the need for awhile; but, for substantial gains there will have to be added nitrogen from either legumes or fertilizer. There may be small additions from rainfall or other atmospheric sources, but this would not be sufficient to retain large amounts of carbon. Soil humus and other forms of soil organic matter generally have a carbon to nitrogen ratio of about 10. Therefore, the build-up of "stable" carbon in the lands enrolled in the Conservation Reserve Program (CRP) is probably less than optimum because of a lack of nitrogen and possibly other nutrients required to form soil humus.

C.A. Robinson (unpublished data) sampled four sites on Pullman silty clay loam, a fine, mixed, thermic Torrertic Paleustoll. One site had been cropped to dryland wheat for more than 50 years with occasional fallow years. Tillage was primarily performed with a tandem disk plow. In recent years, anhydrous ammonia at the rate of 22 kg ha^{-1} N was applied in the fall prior to seeding wheat. The other three sites were grassland – one native grassland field, one previous cropland field returned to grass 37 years prior to sampling as part of the U.S.D.A. Soil Bank program of the 1950s, and one field that had been returned to grass 7 years prior to sampling as part of the CRP initiated by the 1985 U.S.D.A. Food Security Act. The carbon and nitrogen contents of the four fields are presented in Figure 2. The data clearly indicated that substantial amounts of carbon can be retained even under semiarid conditions. The precipitation at the sites averages approximately 500 mm annually. The native grassland site had significantly more SOC and nitrogen at each depth than the CRP site and the wheat site. The Soil Bank site was intermediate. The increases in SOC were mostly in the surface 15 cm, although there were some differnces at lower depths largely associated with differences in bulk density. The results were similar to those of Ibori et al. (1995) on abandoned fields in northeastern Colorado. Others, however, found SOC differences due to tillage were limited to the top 7 cm (Havlin et al., 1990; Potter et al., 1997). Nitrogen accumulations showed similar trends, but there was less

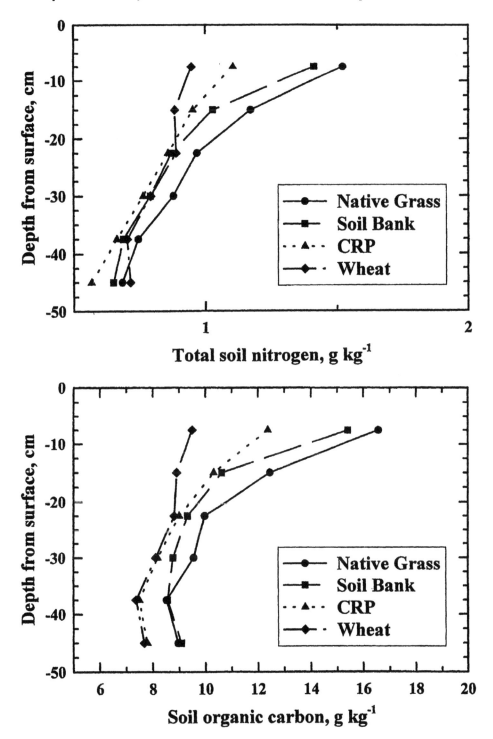

Figure 2. Soil organic carbon (top) and total nitrogen (bottom) by depth as affected by management systems. (Drawn from unpublished data of C.A. Robinson, West Texas A&M University.)

recovery of nitrogen than for carbon, particularly at the lower depths. The CRP site that had been in grass for 7 years recovered 48% of the carbon but only 25% of the nitrogen relative to the Soil Bank site that had been in grass for 37 years. This suggests that the initial rate of carbon accumulation in cropland soils returned to grass is considerably greater than in future years and this rate may be partially constrained by nitrogen as discussed earlier. No legumes were found at any of the sites. Potter et al. (1997) reported 560 kg ha^{-1} yr^{-1} of SOC accumulation during 10 years of no-till continuous cropping wheat systems in the region. Higher rates of carbon accumulation might be expected in revegetated systems, especially in the early years after conversion or transition.

III. Climate Effects

Climate is often the most critical factor determining the sustainability of agricultural systems. Stewart et al. (1991) discussed the effect of climate on maintaining SOC. As temperatures increase and the amounts of precipitation decrease, the maintenance of SOC becomes more difficult. As temperatures increase, SOM decomposition is accelerated, particularly in frequently tilled soils. Not only is the rate of SOM decomposition accelerated under these conditions, the production of biomass is decreased so there are smaller amounts of crop residues and roots available that are necessary to replenish the SOM reserve. Therefore, while it is clear that SOC can be retained in semiarid regions, the potential is less than for soils in more humid regions. Aridisols are the dominant soils in semiarid regions and they account for almost 25% of the area of all soils but contain only about 7% of the SOC (Eswaran et al., 1995).

IV. Conclusions

Increasing carbon sequestration in soils involves increasing the quantity of organic matter returned or added to the soil, or reducing the SOC lost by oxidation or erosion, or a combination of both. Cultivation of soils invariably results in a loss of SOC and this loss is commonly from 25 to 50%. In recent years, there has been a concerted effort by scientists and policy makers to convince producers that crop residues should be managed in such a way that much of the residue remains on the soil surface. This generally involves minimum-tillage or no-tillage systems. The evidence is clear that such practices lead to increased SOC and increased crop production. Therefore, more CO_2 will be used by the crops and more of the carbon used by the crops will be retained in the soil as SOM. Even in semiarid regions, significant amounts of carbon can be sequestered. The key is to use tillage as sparingly as feasible because tillage results in a rapid loss of inherent SOC, and this leads to lower crop yields and smaller residue amounts that are quickly decomposed following tillage. The use of animal manures, where available, can also have positive effects on the SOC balance in semiarid regions.

References

Ahearn, M., J. Yee, E. Ball, and R. Nehring. 1998. Agricultural productivity in the United States. Agricultural Information Bulletin Number 740, Economic Research Service, U.S.D.A., Washington, D.C.

Allison, F.E. 1973. *Soil Organic Matter and its Role in Crop Production*. Elsevier Scientific Publishing Company, Amsterdam, The Netherlands. 637 pp.

Cole, C.V., K. Flach, J. Lee, D. Sauerbeck, and B. Stewart. 1993. Agricultural sources and sinks of carbon. p. 111-119. In: J. Wisniewski and R.N. Sampson (eds.), *Terrestrial Biospheric Carbon Fluxes: Quantification of Sinks and Sources of CO$_2$*. Kluwer Academic Publishers, Dordrecht, The Netherlands.

Economic Research Service. 1997. Agricultural Resources and Environmental Indicators, 1996-97. Agricultural Handbook Number 712, Economic Research Service, U.S.D.A., Washington, D.C.

Eswaran, H., E. Van den Berg, P. Reich, and J. Kimble. 1995. Global soil carbon resources. p. 27-43. In: R. Lal, J. Kimble, E. Levine, and B.A. Stewart (eds.), *Soils and Global Change*. CRC Press, Boca Raton, FL.

Havlin, J.L., D.E. Kissel, L.D. Maddux, M.M. Classen, and J.H. Long. 1990. Crop rotation and tillage effects on soil organic carbon and nitrogen. *Soil Sci. Soc. Am. J.* 54:448-452.

Ibori, T., I.C. Burke, W.K. Lauenroth, and D.P. Coffin. 1995. Effects of cultivation and abandoment on soil organic matter in northeastern Colorado. *Soil Sci. Soc. Am. J.* 59:1112-1119.

Lamb, J.A., G.A. Peterson, and C.R. Fenster. 1985. Wheat fallow tillage systems' effect on a newly cultivated grassland soils' nitrogen budget. *Soil Sci. Soc. Am. J.* 49:352-356.

Potter, K.N., O.R. Jones, H.A. Torbert, and P.W. Unger. 1997. Crop rotation and tillage effects on organic carbon sequestration in the semiarid southern Great Plains. *Soil Sci.* 162:140-147.

Stewart, B.A., R. Lal, and S.A. El-Swaify. 1991. Sustaining the resource base of an expanding world agriculture. In: R. Lal and F.J. Pierce (eds.), *Soil Management for Sustainability*. Special Publication, Soil and Water Conservation Society, Ankeny, IA.

Westerman, R.L. 1992. Efficient use of fertilizers. Agronomy 92-1, Oklahoma State University, Stillwater, OK.

Unger, P.W. 1991. Organic matter, nutrient, and pH distribution in no- and conventional-tillage semiarid soils. *Agron. J.* 83:186-189.

Section IV.

LAND USE AND CARBON POOL IN SOILS OF ASIA AND THE PACIFIC

Impact of Land Use and Management Practices on Organic Carbon Dynamics in Soils of India

A. Swarup, M.C. Manna and G.B. Singh

I. Introduction

The maintenance of soil organic carbon (SOC) in agricultural soils is primarily governed by climate, particularly annual precipitation and temperature, and cropping systems. Although the amount of SOC in soils of India is relatively low, ranging from 0.1 to 1% and typically less than 0.5%, its influence on soil fertility and physical condition is of great significance. Organic carbon level of soils reaches a fixed equilibrium that is determined by a number of interacting factors such as precipitation, temperature, soil type, tillage, cropping systems, fertilizers, the type and quantity of crop residues returned to the soil, and the method of residues management (Ali et al., 1966; Das, 1996; Jenny and Raychaudhuri, 1960; Mathan et al., 1978). Conversion of land from its natural state to agriculture generally leads to losses of SOC. It may take up to 50 years for the organic carbon of soils in the temperate climate to reach a new equilibrium level following a change in management, but this period is much shorter in a semiarid and tropical environment like India. Intensive cropping and tillage systems have led to substantial decreases in the SOC through enhanced microbial decomposition and through wind and water erosion of inadequately protected soils which is often accompanied by a decline in soil productivity. Carbon loss by tillage is caused by greater oxidation of SOC and it ranges from 20 to 50% in soils dominating the semiarid regions of India (Mann, 1986; Mutatkar and Raychaudhuri, 1959). In most parts of tropical cropping systems, little or no agricultural crop residues are returned to the soil which leads to a decline in soil organic carbon (Lal,1986; Post and Mann, 1990). In the *Jhum* agro-ecosystem in northeastern India , high losses occur through volatilization of C and N during the burning process, leading to a reduction in the quantity of these elements in the surface soil layers (Birch and Friend , 1956; Ramakrishnan and Tokey, 1981; Ramakrishnan, 1994). These losses continue throughout the cropping phase. Over several *Jhum* cycles , the extent of soil organic carbon depletion depends upon the length of the cropping period and the ratio of the cropping to the fallow period.

Long- term fertilizer experiments conducted over 25 years in different agro-ecoregions of India involving a number of cropping systems and soil types (Inceptisol, Vertisol, Mollisol and Alfisol) have shown a decline in SOC as a result of continuous application of fertilizer N alone (Swarup, 1998). Balanced use of NPK fertilizer either maintained or slightly enhanced the SOC over the initial values. Application of farmyard manure (FYM) and green manure improved SOC which was associated with increased crop productivity. Considering the nutrient removal by crops and supply through different sources under intensive cropping systems, it is seen that removal is far greater than the supply. It is, therefore, extremely important to maintain SOC at a reasonably stable level, both in quality and quantity, by means of suitable addition of organic materials or crop residues. Jenny and Raychaudhuri

(1960) concluded that fertilization with artificial nitrogen compounds with additional nutrients is expected to enhance greatly the yields and stimulate the root system in the soil, and these will augment the soil humus and bring it to a new higher steady level to the benefit of plant growth and soil management. Soil and crop management practices that are now being perfected in different agro-climatic regions of India represent a new philosophy of land use to sustain maximum crop production per unit time per unit of land with minimum soil degradation. This chapter reviews the work done on various aspects of land use and management practices on organic carbon dynamics in the soils of India.

II. Organic Carbon Stocks of Indian Soils

Soils of India, like most soils of the tropics have long been categorized as low in organic carbon and nitrogen although there are many variations like genetic, morphological, physical, chemical, and biological characteristics, associated with changing physiography, climate and vegetation (Ghosh and Hasan, 1980). The organic carbon reserves of Indian soils, either virgin or cultivated, are higher than those of North America, provided sites having equal annual values of temperature and precipitation are being compared, but they are lower than those of Central America (Jenny and Raychudhuri, 1960). The observed losses in soil organic carbon from managed ecosystems are greater in semiarid environments than in the humid low lands (Table 1). This indicates that a large portion of organic carbon under natural vegetation of arid and semiarid region is less recalcitrant than humid tropical soils. The decline in total organic carbon in the agro-ecosystem was reduced 2 times faster than that of the soil carbon storage in the sub-humid woodland forest and plantations. Jenny and Raychaudhuri (1960) studied the effect of long-term exhaustive practices of soil management and the climate on the organic carbon and nitrogen reserves of Indian soils and concluded that the differentiation of organic C, total N and C/N ratio was a function of temperature, rainfall, and cultivation. A tropical rain forest of India (northeast India and south plateau) is a closed ecological system which represents a highly efficient production form, and a considerable quantity of organic litter falls onto the soil surface and decays, providing food for a highly active soil flora and fauna which ultimately stores more organic carbon than cultivated land.

A perusal of data in Table 1 shows that about 40% of the cultivated soils of the Indo-Gangetic alluvium are calcareous and, more precisely, they contain C as carbonates. Initially, the geographic distribution of these calcareous bodies was conditioned by the flow patterns of the rivers which traverse and erode calcareous strata in the Himalayan Mountains. But in due course of time, as precipitation increased, the portion of calcareous soils declined by 20% at 127 to 152 cm rainfall. This large group of cultivated, alluvial soils are richer in soil organic carbon than non-calcareous soils. This phenomenon may be meaningful if textural and climatic variables are simultaneously taken into account. In the drier section of the Indo-Gangetic Plains the areas of natural vegetation are tiny and far between. They consist of thorny volunteer shrubs on drifting sand dunes and on patches of temporarily abandoned land, and of clumps of wild bunch-grass. The native vegetation comprises brush, small acacias and leguminous broad leaf trees, regularly pruned for cattle feed and intermingled with large specimens of euphoria type bushes. The means of the combined vegetational types show a striking relation to elevation of northwest Himalayan soil. The soils above 1524 m are twice as rich in organic carbon when compared to soils below 1067 m. The surface layer (0–20 cm) is very dark and rich in carbon, and the subsoil is distinctly lower in organic carbon, but there are no visible signs of podsolization.

Under natural vegetation, organic carbon may reach a near-steady state after 500 to 1000 years (Dickson and Crocker, 1953; Jenny, 1950). Cultivation disturbs the natural organic carbon equilibrium. Depletion of soil organic carbon under cultivated field was 23 to 48% of original value. It is documented that the agricultural soils of northwest India exclusive of the Himalayas have lost

Table 1. Influence of elevation, precipitation and temperature on organic carbon status under cultivated and forest lands of India

Location	Elevation (m)	Precipitation (cm)	Temperature (°C)	Carbon (%) Cultivated	Carbon (%) Native
A. Northwest India					
1. Indo-Gangetic Plains		25–51	25	0.3 ± 0.033	0.59 ± 0.211
		53–76	24–26	0.45 ± 0.038	0.91 ± 0.113
		79–102	23–25	0.43 ± 0.037	——
		104–127	23–24	0.55 ± 0.049	1.40 ± 0.157
		130–152	23	0.35 ± 0.059	1.24 ± 0.297
2. Northwest Himalayan					
Dehra Dun-Mussoorie	457–1067	216–224	23–20	1.44 ± 0.145	1.81 ± 0.270
	1067–1524	216–224	20–17	2.01	3.53
	1524–2134	216–224	17–14	3.37 ± 0.365	3.99 ± 0.346
Sima	2195	155	13	2.91 ± 0.386	4.48 ± 0.258
3. Northwest India					
Sriganganagar		25	25	0.33	——
Meerut		74	24	0.50	——
Biharigarh		122	24	0.51	——
Mohan		145	23	2.64	——
B. Northeast India					
Tista-Bramhaputra plain		249–389	24	1.37 ± 0.119	2.32 ± 0.160
Assam Hill and valleys		129–1080	24–17	1.26 ± 0.182	1.56 ± 0.166
Himalayan ranges					
	610–1311	300–315	21–17	3.18 ± 0.221	4.82
	914–1158	218	19–18	2.23 ± 0.219	——
	1524–2316	295–330	16–12	3.58 ± 0.315	6.63 ± 0.695
C. Southeast India				0.32 ± 0.0025	
Modurai-Kodaikanal	610	104	26		1.73
mountain transect	945	117	24		3.23
	1280	130	22	2.31	
	1524	137	20		5.59
	1768	148	18		6.94
	2103	158	16	6.24	
D. West coast of India					
Dry coastal region		56–472	27	0.74 ± 0.134	——
Humid coastal region		180–223	27	1.89 ± 0.272	1.86 ± 0.212
E. Deccan Plateau and ad-jacent mountains					
Mysore-Bangalore area		79–86	25–23	0.52 ± 0.036	1.68 ± 0.230
Nagpure-Bellary area		51–124	27	0.55 ± 0.124	1.09 ± 0.170
Western Ghats-Nilgir hills		130–917	24	1.25	2.59

(Adapted from Jenny and Raychaudhuri, 1960.)

about one half to two thirds of their original organic carbon content. Northeast India consists of Tista-Brahmaputra plains, Assam Hill and valleys, and lesser Himalayan regions. Tista-Brahmaputra plains consist of relatively recent river deposits which have been mapped as new alluvium and occasionally older, dissipated terraces which have weathered into reddish soils. The mean annual precipitation varies from 249 to 389, 129 to 1080 and 219 to 338 cm under Tista-Brahmaputra plains, Assam Hill and valleys and lesser Himalayan ranges, respectively. If the mean annual temperatures of Himalayan range are plotted against elevation, a nearly perfect straight line results with a negative gradient of $0.062^{0}C$, per 30.5 m . The organic carbon values of these soils are two to three times higher than those of the cultivated soils of the Indo-Gangetic alluvium presumably because of higher rainfall (254–356 cm) and finer textures of the soil. For cultivated soils the mean percentage value for carbon of Assam Hill and valleys is 1.26 ± 0.128.

In the regions of southeast India the mean carbon content of cultivated soils of foot hills and plains is about 0.45%, and the values are nearly identical with those from the soils of the Indo-Gangetic plains having corresponding rainfall. The soils covered with native vegetation that are predominant of acacias, euphorbias, thorny shrubs, patches of poor stands of grass, and bare spots are largely unsuitable for agricultural production. The mean organic carbon content is 0.76 ± 0.076%. Organic carbon increased with increase in elevation under native vegetation and in cultivated fields of Madurai-Kodaikanal mountain transect.

The regions of the west coast of India, called Malabar and Kanara coast section, are laterite plateaus carrying bare iron stone crusts of panzer-like hardness and impenetrability. Along the entire dry coastal region the range of annual rainfall is enormously increased from 56 to 472 cm and the mean annual temperature is about 27 °C. The organic carbon average is 0.74 ± 0.134. The mean organic carbon percent in non-paddy soils of humid coastal regions is about 0.92 ± 0.155 which is significantly lower than in paddy soils (C = 1.89 ± 0.272). The organic carbon content in cultivated land is approximately 1/2 that of the native vegetation.

III. Factors Affecting Soil Organic Carbon Dynamics

Soil organic carbon equilibrium is governed by a number of interacting factors such as temperature, moisture, texture, quality and quantity of organic matter, methods of application, soil tillage and cropping systems. Maintenance of soil organic carbon is an important tool for crop productivity and sustainability. Other important benefits of SOC in low-input agro-ecosystems are retention and storage of nutrients (Gaur, 1990; Russel, 1973), increased buffering capacity (Swift and Sanchez, 1984), better soil aggregation (Oades, 1984; Sahoo et al., 1970), improved moisture retention, increased cation exchange capacity (Ali et al., 1966; Biswas et al., 1961; Biswas,1982; Sahoo et al., 1970), and acting as a chelating agent (Stevenson, 1982). The addition of organic carbon improves soil structure, texture and tilth (Biswas et al.,1971; Gupta and Nagarajarao, 1982), activates a very large portion of inherent microorganisms (Goyal et al.,1993), and reduces the toxic effects of pesticides (Gaur and Prasad, 1970; Gaur, 1975).

A. Soil Type

Soil type is one of the important parameters that regulates soil organic carbon status of the soil. The major soil groups of India broadly fall into five groups viz., alluvium derived soils (Inceptisol and Entisol: 74.3 million ha), black soils (Vertisol: 73.2 million ha), red, yellow and laterite soils (Alfisol and Oxisol: 87.6 million ha), and soils of desert regions (Aridisol: 28.7 million ha). The extent of

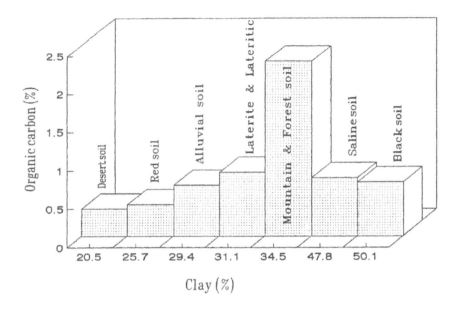

Figure 1. Relationship between clay and organic carbon content of soils.
(Modified from Ali et al., 1966.)

clay aggregation is a direct controlling factor in organic carbon dynamics . Ali et al. (1966) observed that organic carbon content increased with clay content under desert, red, alluvial, laterite and lateritic soil, saline and black soil, except mountain and forest soil which had the highest organic carbon at 34.5% clay (Figure 1). This could possibly be due to continuous deposition of unhumified organic carbon in these soils. Irrespective of climatic factor, increased amounts of sand , coarse loam, or gravelly sandy loam decrease the organic carbon content which may be ascribed to less microbial proliferation and aggregation for carbon restoration.

B. Rainfall and Temperature

Temperature and rainfall exert a significant influence on the decomposition of soil organic carbon and crop residues. A rise in the mean annual temperature reduces the level of SOC of cultivated soil in the humid region (Figure 2). Higher temperature activates the soil microbial population to a greater extent than plant growth. In temperate climates, the soils are several times richer in organic carbon than warmer climate (Table 2). High rainfall and low temperature are conducive to accumulation of organic carbon in soils while high temperature and low rainfall decrease it. Forest soils have higher organic carbon content than the cultivated soils, where the crop residues are not returned to the soil (Jenny and Raychaudhuri, 1960).

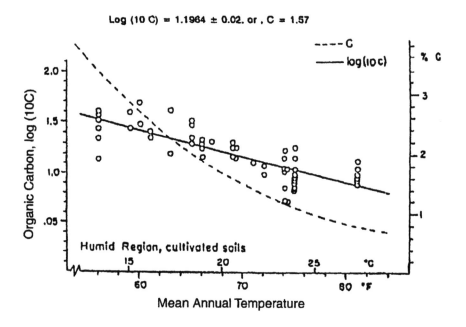

Figure 2. Carbon-temperature function in the humid region (129–223 cm); solid curve = logarithmic function; dashed curved = absolute function. (Adapted from Jenny and Raychaudhuri, 1960.)

C. Soil Organic Carbon Turnover and C:N Ratio

The distribution of soil organic carbon within different pools is an important consideration for understanding the dynamics of soil organic carbon and their diverse role in ecosystems (Jenkinson, 1990; Jenkinson and Rayner, 1977; Parton et al., 1987; Van Veen et al., 1984). So far literature on soil organic carbon changes in the tropics did not throw much light on the carbon functional pools which are highly sensitive indicators of soil fertility and productivity. The distribution of soil organic carbon and its turnover rate into five functional pools may be made for its true representation: (A) structural litter pools mainly consist of straw, wood, stems and related plant parts. The C:N ratio varies around 150:1. These are high in lignin content. (B) metabolic pools comprise particles of leaves, bark, flowers, fruits and animals manure. This fraction releases mineral nitrogen as it is decomposed with loss of CO_2. (C) active pool fraction consists of microbial biomass C and its metabolites. The C:N ratio is around 5 to 15. This fraction provides mineral nutrients and life to the soil. Besides soil microbial carbon, light fraction of organic carbon and water extractable carbon are also active pools of soil organic matter. (D) slow decomposable organic fraction is comparable to mature compost having C:N ratio around 20:1. It makes temporary stable humus in soil which is slowly decomposable, and (E) passive pool of organic carbon which is highly recalcitrant with C:N ratio of 7:1 to 9:1. It is resistant with other types of organic fractions in soil.

Parton et al. (1987) observed that (1) active pools of SOC consist of living microbes and microbial products along with soil organic matter with a short turnover time (1 to 5 yr); (2) a pool of C and N (slow pool) that is physically protected and / or in chemical forms has more resistance to decompo-

Table 2. Temperature effect on soil organic carbon within selected moisture belts (isohytes) in India

Areas	Mean annual temperature (°C)	Organic carbon (%)
Dry region (50–75 cm annual rainfall)	23–24	C = 0.58 ± 0.028
Indo-Gangetic alluvium		
Jaipur Hills, Mysore plateau		
Bhavanagar, Kanya Kumari, Bellary	27–29	C = 0.54 ± 0.075
Semi-humid region (75–100 cm annual rainfall)		
Mysore plateau	23–24	C = 055 ± 0.054
Southeast coast	29	C = 0.30 ± 0.031
Humid region (125–225 cm annual rainfall)		
1. Cultivated soils		
Simla	7	C = 2.91 ± 0.386
Upper Mussoorie	7–8	C = 3.37 ± 0.365
Dehra Dun-Rajpur	8–11	C = 1.44 ± 0.145
Assam	12	C = 1.26 ± 0.182
2. Forest soils		
Simla	7	C = 4.48 ± 0.258
Upper Mussoorie	7-8	C = 3.99 ± 0.346
Dehra Dun-Rajpur	8–11	C = 1.81 ± 0.270
Assam	12	C = 1.56 ± 0.166
Perhumid region (250–500 cm annual rainfall)		
1. Cultivated soils		
Darjeeling	6–8	C = 3.58 ± 0.315
Tista-Brahmaputra plains	12	C = 1.35 ± 0.119
Malabar coast	14	C = 2.06 ± 0.492
2. Forest soils		
Darjeeling	6–8	C = 6.63 ± 0.695
Mercara (Coorg)	8	C = 2.85 ± 0.352
Tista-Brahmaputra plains	12	C = 2.32 ± 0.160
Malabar coast	14	C = 2.46 ± 0.354

(Adapted from Jenny and Raychaudhuri, 1960.)

Table 3. Soil organic carbon functional pools, their turnover and composition

Functional pools	Turnover time (years)	Composition
Metabolic litter	0.1– 1.0	Cellular compounds, shoot and root biomass
Structural litter	1–5.0	Lignin, cellulose, hemicellulose, polyphenol
Active pool	1–5.0	Soil microbial biomass carbon, soluble carbohydrates, exocellular enzymes
Slow pool	20–40	Particulate organic (50 μm–2.0 mm)
Passive pool	200–1500	Humic acid, fulvic acid, organo mineral complex

(Adapted from Parton et al., 1987.)

Table 4. Carbon pools of subhumid, semiarid tropical, and arid ecosystems under different land management practices

	Sal forest[a]	Mixed forest[a]	Pearl millet[b] wheat fallow	Pearl millet-mustard-sunflower[b]	Soybean-wheat[c]-fallow land
Soil organic C (g m^{-2})	2854	2530	1063	1104	1144
Carbon input (g C m^{-2} yr^{-1})	250	250	281	365	265
Turnover (yr^{-1})	11.41	10.12	3.78	3.02	4.47
Soil microbial biomass C (g m^{-2})	87	90	42.30	31.70	50.8
C input/soil microbial biomass C	2.87	2.78	6.64	11.51	15.74

(Adapted from [a]Srivastava and Singh, 1989; [b]Chander et al., 1997; [c]Manna et al., 1996.)

sition, with an intermediate turnover time (20 to 40 yr); (3) a fraction that is chemically recalcitrant (passive SOC) with the longest turnover time (200 to 1500 yr); (4) a structural pool that has 1 to 5 year, and (5) metabolic pools that have 0.1 to 1 year turnover time (Table 3). Research efforts are needed to generate information on SOC turnover in different soils of India.

The soil microorganisms are believed to play a major regulatory role for organic carbon dynamics. The abiotic factors (temperature, moisture, soil type, nature, and quantity of residues) may have a large impact on soil organic carbon dynamics by virtue of their effect on microbial activities. In subhumid and semiarid tropical ecosystem involving cropping systems, the turnover rate of SOC decreased by 3.8 fold compared to forest land (Table 4). The rate of mineralization depends on tillage, residues management, cropping practices, and erosion.

Materials having wide C:N ratio are resistant to decomposition thereby restricting the supply of nitrogen for organisms and eventually resulting in increased immobilization and reduced supply of nitrogen to plants. The C:N ratio of the decomposed materials range from 10 to 15 : 1 , which is better for supplying nutrients to plants (Hajra et al., 1994; Manna et al.,1997). Immediate application of the

Table 5. Effect of farmyard and green manure on soil organic carbon under different land use systems

Land use	Treatment Material	Mg ha^{-1}	Organic C (%)	References
Alluvial maize-wheat	Control	-	0.51	Biswas et al., 1971
(15 years)	FYM	69.7	2.49	
Medium black cotton-	Control	-	0.56	Khiani and More, 1984
sorghum (45 years)	FYM	6.2	1.14	
Black soil Ragi-	Control	-	0.30	Mathan et al., 1978
cowpea-maize (3 years)	FYM	25	0.64	
Red soil rice-rice	Control	-	0.43	Hegde, 1996
(10 years)	50% from inorganic + 50% through green manure (*Sesbania aculeata*)	-	0.90	
Sodic soil rice-wheat	Control	-	0.44	Manna et al., 1996
(3 years)	FYM	16	0.54	
Sodic soil rice-wheat	Fallow-rice-wheat	-	0.23	Swarup, 1998
(7 years)	Green manure (*Sesbania aculeata*) rice-wheat	-	0.37	

cereal straws (C:N ratio 80 to 120 :1) into soils resulted in rapid immobilization of nutrients. Therefore, the C:N ratio of organic materials is more important for degradation than for residues.

D. Rate and Method of Crop Residues, Fertilizers and Manure Application

The importance of organic manuring in Indian agriculture has been known since ancient times. Whether the organic matter status of soil can be built up under tropical conditions of India has been often debated. The rate of organic matter application to soils varies according to the crop, climate, availability, quality of organic matter, and in turn, affects the SOC dynamics.

The crop residues management and their impact on crop yields and soil properties have been recently reviewed (Prasad and Power, 1991). Traditionally, most of the farmers unload manure in small piles or heaps and leave it for 5 to 6 months or so before it is spread in the fields. During this process plant nutrients are lost due to exposure to the sun and rain by volatilization or leaching. To derive the maximum benefit, organic materials should be applied during land preparation and incorporated into the soil with adequate moisture about two to three weeks before sowing the crop. Results on the impact of diverse organic materials on soil organic carbon are provided in Table 5. From different observations it may be concluded that continuous application of FYM and green manure substantially improved the SOC under different soils and cropping systems employed (Biswas et al., 1971; Hegde, 1996; Khiani and More, 1984; Manna et al., 1996; Mathan et al., 1978; Swarup,

1998). Under tropical and subtropical climatic conditions, seasonal applications are necessary to obtain good results. The rate of organic matter application to soil will be determined by the amount of nutrients that can be utilized by the crops (Gaur, 1992). About 25 Mg ha^{-1} of FYM is recommended under intensive irrigated cropping conditions for sugarcane (*Saccherum officinarum* L.), potato (*Solanum tuberosum* L.) and rice (*Oryza sativa* L.), 10 to 15 Mg ha^{-1} for irrigated or rainfed crops where potential rainfall is medium to heavy (about 125 cm yr^{-1}) and 5 to 7 Mg ha^{-1} in dry areas where the mean annual rainfall is about 25 cm (Gaur, 1992). In dry land farming areas application of 25 Mg ha^{-1} of compost can give a significant increase in crop yield. The method of application or placement of residues will also regulate the rate of decomposition.

IV. Land Use Management Effect on Organic Carbon

The increasing cost of fertilizers, the gradual removal of subsidies for fertilizers and the increasing concern of likely pollution of ground waters by fertilizers through leaching from soils are discouraging the use of fertilizers (Kanwar,1994). Intensive interaction exists between infrastructural land use and development on one hand and agriculture, forestry, filtering, buffering, and transformation activities and soil as a gene reserve on the other hand, especially in urban, preurban and forest areas around the world (Blum, 1994). The encouraging results obtained by integrated use of fertilizers, organic and green manures and biofertilizers are providing the leads for future strategies for rational use for continuously enhancing productivity without detriment to the environment.

Potential sources of organic materials available locally to supplement plant nutrients include cereal crop residues, sugarcane trash, animal manures, compost, water hyacinth and etc. (Sarkar et al.1994). Placement of manure once in four years at 0.20 m below the set seedbed rows at the rate of 10 Mg ha^{-1} was more remunerative for improving SOC (Gupta and Venkateswarlu, 1994).

Direct incorporation of crop residues in the field as a means of quick disposal is not always practicable because it delays the transplantation of the following crops, produces phytotoxic compounds for young seedlings, restricts the root respiration by rapid CO_2 evolution, and rapidly immobilizes plant nutrients (Adhikari et al., 1997). The management of crop residues as an integral part of soil fertility management must be recognized as an essential component of sustainable agriculture, but the question is whether, under India's situation, there is really any significant crop residues left after meeting the feed needs of cattle. There is a widespread consensus that there is limited scope for opening up new lands for agricultural use in India.

Virmani (1994) noted that according to some environmentalists, we need to reduce our agricultural area from the current 140 million ha to 110 million ha by the year 2000 to 2010. The area thus released can be used for alternate land use, most likely for forestry because of the low proportion of land area under forest cover (11%) against the required (23%). This would also help in increasing diversity of biological species and improving the quality of environment. Alternate land use systems, viz., agroforestry, agro-horticulture and agro-silviculture are more remunerative for SOC restoration as compared to sole cropping systems. Das and Itnal (1994) reported that organic carbon content was about double in agro-horticultural and agro-forestry systems as compared to sole cropping (Table 6).

The dominant land use system in the northeastern Hill region of India is *Jhum*. In this system, farmers grow mixed crop for a period of one to two years on fallow vegetation. Then land is subsequently allowed for natural vegetation or kept fallow. During this period, soil fertility is restored. This practice helps to optimize resource utilization and ensures adequate return for the farmers. Sometimes local farmers adopt bench terraces as an alternative to *Jhum*. The economic return is often insufficient to meet the cost of these additional external inputs.

Table 6. Organic carbon content in soil after six years of plantation with different land use options

System	Organic C (%)	
	0–15 cm	15–30 cm
Sole cropping	0.42	0.37
Agroforestry	0.71	0.73
Agro-horticulture	0.73	0.74
Agro-silviculture	0.38	0.56

(Adapted from Das and Itnal, 1994.)

V. Impact of Management Practices on SOC Dynamics

A.Tillage and Residues Management

The use of reduced tillage practices and substantial addition of crop residues enhance soil fertility and increase crop yield. The bulk of straw from crop residues is used as main source of fodder for cattle feed. Sometimes it is used as packaging materials, for making paper, for building boards, and as thatching material for huts. India generates annually about 338 million tons of farm residue in which cereals, pulses and oil seeds are 70, 4 and 10%, respectively (Manna and Biswas, 1996).

A common soil management practice, aimed at optimizing the soil physical environment for plant growth, is to incorporate crop residue which results in increased moisture storage, rooting depth, soil organic carbon and nutrient status, and reduced bulk density, soil temperature, soil compaction, runoff and soil erosion (Bhaskar et al., 1995; Das, 1996; Rao et al., 1994).

In cultivated soils, several reduced tillage management systems have been proposed to minimize SOC (Das, 1996; Gupta et al., 1995). In long-term studies covering 45 years, the type of tillage operations and use of organic manure, compost or other farm wastes improved soil physical conditions, available nutrient status(N, P and K), organic carbon content and substantially increased crop productivity (Khiani and More, 1984).

Rao et al. (1994) reported that beneficial effect of FYM on grain yield of rice on a sandy clay loam soil was due to increased soil organic carbon from 0.66 to 1.10% and consequent reduction in bulk density (Table 7). Bairathi et al. (1974) observed that legume crop residues increased the organic C, N, and available P and decreased the bulk density.

Surface residues management and reduced tillage provide a favorable environment for native micro flora and fauna. Among the intermediate products of decomposition, many polysaccharides including microbial gum are especially significant in the formation of soil aggregates by their binding action on the soil particles (Tomar et al., 1992). Incorporation of residues of black gram (*Phaseolus mungo*), wheat (*Triticum aestivum* L.), Indian mustard (*Brassica juncea*) and groundnut (*Arachis hypogaea* L.) increased the organic carbon content in lateritic soil from the initial value of 0.34% to 0.73, 0.59, 0.65 and 0.68 %, respectively (Das, 1996).

The SOC losses can be reduced by several tillage operations such as zero tillage, reduced tillage, stubble mulching and conventional plowing. The effect of conservation tillage on soils and environments are difficult to generalize. In the humid and subhumid tropics, mechanical tillage has more adverse than beneficial effects. On the otherhand, in the arid and semiarid tropics, mechanical tillage is often beneficial. In black soil deep plowing, incorporation of stubble mulch and fertilizer application are essential. In alluvial soil, application of residue mulches or compost often improves

Table 7. Effect of long-term use of manure and fertilizer on soil physical properties and yield of rice (after 12 years)

Manurial treatment	Organic C (%)	Bulk density (g cm^{-3})	% aggregates >0.25 mm	% mean weight	Grain yield (q ha^{-1})
Control	0.70	1.74	76.3	1.5	20.6
N alone 100%	0.70	1.72	74.7	1.5	16.9
NP 100%	0.70	1.59	81.4	1.6	35.6
NPK 100%	0.75	1.63	80.9	1.6	38.9
NPK 100% + FYM	1.10	1.53	86.5	1.8	60.0

(Adapted from Rao et al., 1994; fertilizer applied: 60 kg N, 8.75 kg P, 25 kg K ha^{-1} for each crop for 100% NPK; FYM applied: 15 Mg ha^{-1} yr^{-1}.)

crop yield. Conservation tillage approaches can be widely improved in low input agriculture systems. Such information is not available for subhumid tropical ecosystems. Long-term and large scale agro-ecosystem studies are required to assess the effectiveness of various conservation tillage systems on organic carbon pools.

B. Cropping System

For sustainability of intensive cropping systems it is desirable not to grow a particular crop or a group of crops on the same soil for a long period. As a corollary, productivity of soil can be prolonged if crops are changed over seasons or years. This principle led to the concept of cropping systems and has been known and practiced since vedic times in India and since the prechristian era in the west. Why modern agriculture has ignored it is an enigma. The largest area in India is under rice and is grown in wet as well as dry seasons. Under intensive cultivation, two or three crops are grown per year and the grain yield of 10 to 14 Mg ha^{-1} are achieved per year (Sinha and Swaminathan, 1979; Prasad, 1983). Most of the cropping systems are under cereal-cereal (rice-rice), cereal-cereal-cereal [rice-wheat-maize (*Zea mays* L.)] and cereal-cereal-legume [maize-wheat-green gram (*Phaseolus radiatus*) or fingermillet (*Eleusine coracana* Gaertn)-wheat- gram(*Cicer arietinum* L.) or pearl millet (*Pennisetum typhoides* L.)-wheat-green gram]. The first, second and third crops are being produced during the wet seasons (July–October), winter seasons (November–April), and summer seasons (May–June), respectively. The adoption of these intensive cropping systems depends on irrigation facilities, climatic conditions and introduction of high yielding cultivars and management.

Crop sequence in Indo-Gangetic plains needs to be explored by introducing legume crops in the system in different ways, viz. replacing rice or wheat crop by a legume crop, i.e., rice by pigeonpea (*Cajanus cajan* L.) in summer or wheat by lentil (*Lens culinaris* Medikus) in winter, or introducing a summer green manure crop dhanchia (*Sesbania aculeata* (Wild) Poir) after the harvest of wheat and before planting of rice (Prasad and Goswami,1992) . Singh et al. (1996) showed that over a period of 5 years the net change in SOC was negative under cereal-cereal sequences, whereas in other sequences having legume component the changes were positive. Mixed or intercropping systems are also advantageous in many ways when cereals or millets are mixed or intercropped with legumes. For example, growing pea (*Pisum sativum* L.) or pigeonpea as compared with maize, and green gram or black gram as compared with soybean (*Glycine max* L. Merril.) increased organic carbon (Sharma et al., 1986). In a typical black soil (Vertisol), continuous cropping and manuring increased organic carbon content by 20 to 40 % over a period of three years (Mathan et al., 1978).

C. Cultural Practices

Adoption of suitable cultural practices can substantially increase crop production and also affect SOC content in all ecosystems. The most important cultural practices are cultivation methods, fertilizer management, green manuring and organic manures and liming for acid soils. The effect of cropping systems, tillage and shifting cultivation are addressed earlier in this chapter. SOC loss can sometimes be minimized by using short duration crop plants and wide row spacing. Short duration high yielding crops take less time because of their faster growth. Narrow row spacing usually results in more efficient use of available water and increased activity of organisms, thereby increasing loss of SOC.

In the Indo-Gangetic plains of northern India where irrigation facilities are adequate, rice-wheat is the widely practiced crop rotation. However, a decline in the productivity of this system has been observed. Constraint analysis has identified the declining of SOC level as a probable reason (Hobbs and Morris, 1996). During wet season about 10 to 15 Mg ha^{-1} of well decomposed farmyard manure or compost are applied 2 to 4 weeks before planting of rice. But it should be applied as close the time of crop establishment as possible. This will avoid losses of nutrients due to exposure to sunlight and immediate immobilization of available plant nutrients by flux of native flora and fauna. Under intensive cropping and imbalanced fertilizer use particularly N alone, SOC content declined irrespective of cropping system and soil type (Figures 3 and 4). Balanced use of NPK either maintained or slightly improved organic carbon over the initial values. Beneficial effect of 10 to 15 Mg ha^{-1} of farmyard manure in improving organic carbon over control, N, NP and NPK fertilizer was much more pronounced on Vertic Ustopept (Coimbatore), Chromstert (Jabalpur) and Haplustert (Bhubaneswar). However, organic carbon content declined from the initial level on Eutrochrept (Barrackpore) due to intensive farming which appeared to be associated with the attainment of a stable equilibrium possibly due to enhanced oxidation of SOC (Jenny and Raychaudhuri, 1960). These studies lead us to understand that under conditions of high temperature, organic carbon content can not be raised to a very high level, but we must ensure its regular turnover. High productivity and sustainability can thus be ensured through regular inputs of organic and inorganic fertilizers.

In India different plant species have been used as green manure crops namely, dhanchia, sunnhemp (*Crotolaria juncea* L.), senji (*Meliotus alba* Medik), cowpea (*Vigna sinensis* L.), green gram, cluster beans (*Cyamopsis tetragonoloba* L.), khesari (*Lathyrus sativus*) etc. Invariably, incorporation of green manure crops restored organic carbon status of soil and crop productivity (Singh et al., 1990; Singh et al., 1996; Swarup, 1998). Liming is a practice in acid soils to restore productivity and SOC. Sarkar et al. (1989) reported that application of lime along with NPK raised the pH of acid soils from 5.5 to 6.3 and organic carbon from 0.53 to 0.61% during 28 years maize-wheat cropping (Table 8). Continuous application of N,NP and NPK decreased the soil pH over initial value but increased organic carbon.

VI. SOC and Restoration of Degraded Lands

Land degradation means reduced potential productivity of soils at a fixed level of output. Jha (1995) reported that as per estimates of National Wasteland Development Board of India the extent of degraded lands is around 158.06 million hectares (Table 9). With the thrust on increase in food grain

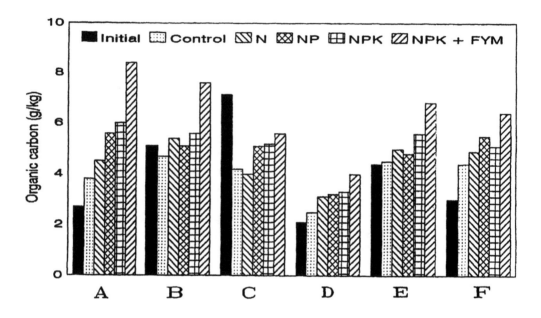

Figure 3. Effect of continuous cropping and fertilizer use on organic C content of Inceptisols. (Adapted from Swarup, 1998.)
A = Bhubaneswar (Haplaquept): rice-rice (1973–1994); B = Hyderabad (Tropaquept): rice-rice (1972–1995); C = Barrackpore (Eutrochrept): rice-wheat-jute (1972–1996); D = Ludhiana (Ustochrept): maize–wheat–cowpea fodder (1971–1996); E = Delhi (Ustochrept): maize-wheat-cowpea fodder (1971–1996); F = Coimbatore (Vertic Ustopept): fingermillet-maize-cowpea fodder (1972–1995).

productivity, there is a limit for increasing fodder and forage production from the present agricultural land. Wastelands can be utilized for fuel or fodder plantation and agroforestry systems to increase the productivity of animal feed and the fertility status of the soils. Inclusion of trees in the agroforestry system enables synchronized release of nutrients from the decaying plant residues with the requirement for nutrient uptake by the crops.

In a 5 year study, Gupta (1995) reported that soil organic carbon increased from the initial value of 0.44% to 0.95, 0.94, 0.88, 0.80 and 0.76% under *Dalbergia sissoo*, *Pongamia sps.*, *Leucaena leucocephala*, *Acacia nilotica* and *Dalbergia latifolia*, respectively. In a silvipastoral system of land management, improved pasture species are grown along with tree species. The selection of tree species could be either for timber or for fuel and fodder. Combining trees with grasses and legumes also helps to conserve soil and improve SOC. Gupta (1995) showed that 7 years of continuous cropping under *Leucaena*, *Acacia nilotica*, *Albizzia lebbek*, and *Albizzia procera* resulted in 13 to 56% increase in soil organic carbon over the open grass (C = 0.60).

In a 3 year field experiment using two reclamation technologies, namely growing karnal grass (*Leptochloa fusca*) as a biological reclamation or applying gypsum at 14 Mg ha^{-1} as a chemical amendment for different cropping system, the average exchangeable sodium percentage was reduced from 95 to 47.5, and the increase in organic carbon was 64% compared with the original soil (Table 10). Forest species differ widely in restoring organic carbon and improvement of alkali soils. Twenty-

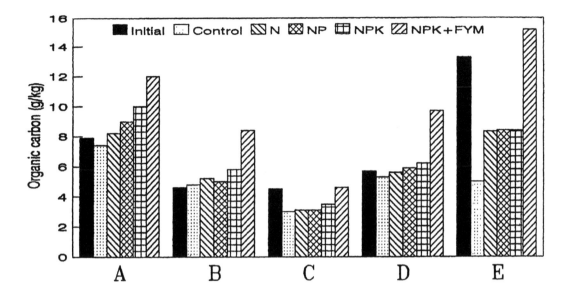

Figure 4. Effect of continuous cropping and fertilizer use on organic C of soils.
(Adapted from Swarup, 1998.)
A = Palampur (Hapludalf): maize-wheat (1973–1995); B = Bangalore (Haplustalf): fingermillet-maize-cowpea fodder (1986–1996); C = Ranchi (Haplustalf): soybean-wheat (1973–1996); D = Jabalpur (Chromustert): soybean–wheat–maize fodder (1971–1996); E = Pantnagar (Hapludoll): rice-wheat-cowpea fodder (1972–1996).

year old plantations of *P. juliflora, A. nilotica, E. tereticornis, A. lebbek* and *Terminalia arjuna* dropped the soil pH from 10.2 to 8.01, 9.03, 9.1, 8.67 and 8.15, respectively. Organic carbon content of the profile increased several fold over the original soil , the highest being under *Prosposis* and lowest under the *Eucalyptus* trees (Table 11).

VII. SOC and Sustainable Agriculture

India has a long tradition of using organic manures to maintain and improve soil fertility. At present sustainable agriculture is the most prevailing topic world wide. The rapidly increasing population, shrinking suitable land resources for crop production and increasing concern for declining soil fertility and environmental degradation create an urgency for enhancing and sustaining productivity of land in India. Judicious management of renewable native soil and water resources and accelerated use of inputs like chemical fertilizers and organic and biological resources can meet this challenge. Land clearing or deforestation and cultivation induce a lower equilibrium level of SOC because removal of the harvested material is much faster than incorporation of crop residues into soil. Raising the crop productivity also results in improving the organic carbon content of the soil through greater contribution of root biomass. Continuous cultivation also brings the carbon and nitrogen ratio of the soils in equilibrium with the environment. Dynamic strategies therefore need to be developed for maintaining and enhancing productivity and sustainability in agriculture.

Table 8. Effect of organic, inorganics and liming on pH and organic carbon on an Alfisol after 28 years of a maize-wheat cropping system

Treatments	pH	Organic C (%)
Control	5.8	0.536
N	3.8	0.558
NP	3.9	0.614
NPK	4.0	0.640
NPK + lime	6.3	0.608
FYM	5.9	0.732
Initial (1956)	5.5	0.529

(Adapted from Sarkar et al., 1989.)

Table 9. Categories of land under different types of wasteland in India

Number	Category	Area (million hectares)
1	Water eroded	73.60
2	Degraded forest	40.00
3	Riverine	2.73
4	Ravines and gullies	3.97
5	Shifting cultivation	4.36
6	Sand dunes	7.00
7	Water logged	6.00
8	Saline/alkaline wasteland	7.50
9	Wind eroded	12.90
	Total	158.06

(Adapted from Jha, 1995.)

VIII. Future Research Needs

1. Research for characterization of different functional pools of carbon should be undertaken in different soil types and cropping systems.
2. Research work should be undertaken for evaluating the effect of surface applied organic materials on carbon sequestration, soil water conservation, and water stability aggregates which greatly influence SOC dynamics.

Table 10. Chemical composition of alkali soil (0–15 cm) under two reclamation technologies after three years of reclamation

Treatment	pH	ESP	Total organic carbon (%)	Dehydrogenase (μg TPF g^{-1} soil)	Microbial biomass carbon (mg kg^{-1} soil)
KIF	9.7	60	0.37	122.9	166.5
KID	9.5	50	0.33	96.9	152.5
K2F	9.5	50	0.45	136.5	149.0
K2D	9.3	40	0.53	118.3	178.8
Mean	9.5	50	0.42	118.7	161.7
KG	9.3	40	0.44	112.1	203.3
SG	9.5	50	0.39	103.0	214.2
RG	9.4	45	0.38	77.6	214.2
DG	9.4	45	0.41	91.5	194.3
Mean	9.4	45	0.41	96.1	206.5

Karnal grass grown for forage for one (KIF) or two (K2F) years without soil amendment; Karnal grass grown for one (KID) or two (K2D) years without soil amendment with the harvested grass left to decompose on the site; Karnal grass plus gypsum at 14 Mg ha^{-1} followed by winter clover (KG); sorghum (SG), rice (RG) or *Sesbania aculeta* (DG) plus gypsum at 14 Mg ha^{-1} followed by winter clover. (Adapted from Batra et al., 1997.)

Table 11. Effect of 20 years tree growth on the properties of an alkali soil

Tree species	Soil depth (cm)	pH	EC (ds m^{-1})	Organic C (%)
Acacia nilotica	0–15	8.4	0.25	0.85
	0–120	9.0	0.53	0.55
Eucalyptus tereticornis	0–15	8.5	0.44	0.66
	0–120	9.2	0.60	0.33
Prosopis juliflora	0–15	7.3	0.51	0.93
	0–120	8.0	0.41	0.58
Terminalia arjuna	0–15	7.9	0.32	0.86
	0–120	8.2	0.45	0.58
Albizzia lebbek	0–15	7.9	0.32	0.62
	0–120	8.7	0.51	0.47

Original soil properties; pH 10.2–10.5; EC 1.75–0.45 ds m^{-1}; OC 0.12–0.24%. (Adapted from Singh, 1994.)

3. There is a need for research on the introduction of legumes or forage crops in the context of judicious land management under different crop rotations for improving soil physical health, and maintaining SOC levels.
4. Strategies need to be developed for subsurface placement of organic residues to improve SOC level in different soils.
5. It is essential to evaluate management practices necessary for the successful introduction and establishment of plantation or alley cropping for higher rate of litter production.

IX. Summary and Conclusions

The cause of low level of soil organic carbon in Indian soils is primarily high temperature prevailing throughout the year. Agricultural activities invariably result in loss of soil organic carbon. Judicious application of bulky organic manures and balanced fertilizer help in restoring the organic carbon status of soil. An improvement in the level of SOC is possible if the rate of productivity increases through added nutrients in the intensive cropping systems. Thus higher fertilizer application rates not only increase yield but also improve root biomass thereby distributing organic carbon deeper in the soil. Increasing efforts are being made to enhance the level of organic carbon in soils by using different quality of organic matters viz., farmyard manure, green manure, crop residues, compost etc. but these have proved insufficient because removal through plant biomass is much faster than feedback to the soil. SOC usually declines when intensive tillage practices are followed which stimulate microbial decomposition of organic litter. However, reduced tillage practices can minimize SOC loss. Cropping systems involving forage or legumes invariably conserve SOC and maintain a greater biological nutrient pool.Tree plantations and perennial grasses of selected exotic species offer possibilities for increasing SOC. Cultivation of fast growing trees with arable crops under agro-horticulture or agro-silviculture systems help in improving soil organic carbon content.

References

Adhikari, T., M.C. Manna, and A.K. Biswas. 1997. Organic matter improves soil health- an overview. *Indian farming*. 47:11-14.

Ali, M.H., R.K. Chatterjee, and T.D. Biswas. 1966. Soil moisture tension relationship of some Indian soils. *J. Indian Soc. Soil. Sci.* 14:51-62.

Bairathi, R.C., M.M. Gupta, and S.P. Seth.1974. Effect of different legume crop residues on soil properties, yield and nutrient uptake by succeeding wheat crop. *J. Indian Soc. Soil Sci.* 22:304 -307.

Batra, L., A. Kumar, M.C. Manna, and R. Chhabra.1997. Microbiological and chemical amelioration of alkaline soil by growing karnal grass and gypsum application. *Exptl. Agric.* 33:389-307.

Bhasker, A., C. Paulraj, B. Rajkannan, S. Avudainayagan, S. Poongothai, R. Natesan, K. Appavu, K.K. Mathan, and R. Perumal. 1995. Twenty five years of soil physics research in Tamilnadu (1967–1992), Tamil Nadu Agriculture University, Coimbatore.

Birch, H.F. and M.T. Friend. 1956. The organic matter and nitrogen status of East African soils. *J. Soil Sci.*7:156-167.

Biswas, T.D., S.K. Gupta, and G.C. Naskar. 1961. Water stability aggregates in some Indian soils. *J. Indian Soc. Soil Sci.* 9: 299- 307.

Biswas, T.D., B.L. Jain, and S.C. Mandal. 1971. Cumulative effect of different levels of manures on the physical properties of soil. *J. Indian Soc. Soil. Sci.* 19:31-34.

Biswas, T.D. 1982. Management of soil physical conditions for soil productivity. *J. Indian Soc. Soil Sci.* 30: 427-440.

Blum, W.E.H. 1994. Sustainable land use and environment. p. 21-30. In: *Management of Land and Water Resources for Sustainable Agriculture and Environment.* Indian Society of Soil Science, New Delhi.

Chander, K., S. Goyal, M.C. Mundra, and K.K. Kapoor. 1997. Organic matter, microbial biomass and enzyme activity of soils under different crop rotation in the tropics. *Biol. Fertil. Soils.* 24:306-310.

Das, D.K. 1996. Coordinator's report (1994–96). AICRP on Tillage requirements of major Indian soils for different cropping systems, Indian Agricultural Research Institute, New Delhi. p. 1-65.

Das, S.K. and C.J. Itnal. 1994. Capability based land use systems: role in diversifying dryland agriculture. In: Soil management for sustainable agriculture in dryland areas. *Bull. Indian Soc. Soil Sci.* 16:92-100.

Dickson, B.A. and R.L. Crocker. 1953. A chronosequence of soils and vegetation near Mt.Shasta, California,I and II. *J. Soil. Sci.* 4:142-154.

Gaur, A.C. and S.K. Prasad. 1970. Effect of organic matter and inorganic fertilizers on Plant parasitic nematodes. *Indian. J. Ent.* 32:186-188.

Gaur, A.C. 1975. Detoxication of lindane by farmyard manure. *Indian J. Agric Sci.*40:329-332.

Gaur, A.C. 1990. *Phosphate Solubilizing Microorganisms as Biofertilizer.* Omega Scientific Publishers. New Delhi. 176 pp.

Gaur, A.C. 1992. Bulk organic manures and crop residues. p. 36-51. In: H.L.S. Tandon (ed.), Organic manures, recyclable wastes and biofertilizers. Fertilizer Development and Consultation Organization, New Delhi.

Ghosh, A.B. and R. Hasan. 1980. Nitrogen fertility status of soils of India. *Fert. News.* p. 19-24.

Goyal, S., M.M. Mishra, S.S. Dhankar, K.K. Kapoor, and R. Batra. 1993. Microbial biomass turnover and activity of enzymes following the applications of FYM to field soil with and without previous long term application. *Biol. Fertil.Soils* .15:60-64.

Gupta, R.P and Y. Nagarajarao. 1982. Soil Structures and its management. p. 60-76. In: *Review of Soil Research in India. Part I.* Indian Society of Soil Science, New Delhi.

Gupta, R.P., P. Agrawal, and S. Kumar. 1995. Highlight of research 1989-91. ICAR, All India Coordinated Research Project on improvement of soil physical conditions to increase agricultural production of problematic areas. S.P.C. *Bull. 11*, Indian Agricultural Research Institute, New Delhi.

Gupta, R.K. 1995. Multipurpose trees. In: R.K. Gupta (ed.), Multipurpose trees for agroforestry and wasteland utilization. pp. 331-335. Center for Research on Environmental Applications, Training and Education, Dehra Dun, India.

Gupta, J.P. and J. Venkateswarlu, 1994. Soil management and rain water conservation and use-V. Aridisols. In: Soil Management for Sustainable Agriculture in Dryland Areas. *Bull. Indian Soc. Soil Sci.* 16:66-77.

Hajra, J.N., N.B. Siha, M.C. Manna, N. Islam, and N.C. Banerjee. 1994. Comparative performance of phosphocomposts and single super phosphate and response of green gram (*Vigna radiata* L. Wilezek). *Tropical Agric* (Trinidad).71:147-149.

Hegde, D.M. 1996. Integrated nutrient supply on crop productivity and soil fertility in rice (*Oryza sativa*)-rice system. *Indian J. Agron.* 41:1-8.

Hobbs, P. and M. Morris. 1996. Meeting South Asia's future food requirements from rice-wheat cropping systems: priority issues facing researches in the post green revolution era. NRG Paper 96-01. Mexico, D.F. CIMMYT.

Jenkinson, D.S. and J.H. Rayner. 1977. The turnover of soil organic carbon in some of the Rothamsted classical experiments. *Soil Sci.* 123:298-305.

Jenkinson, D.S. 1990. The turnover of organic carbon and nitrogen in soil. *Philosophical transactions*, Royal Society of London 239:361-368.

Jenny, H. 1950. Causes of the high nitrogen and organic matter content of certain tropical forest soils. *Soil Sci.* 69:63-69.

Jenny, H. and S.P. Raychaudhuri. 1960. Effect of climate and cultivation on nitrogen and organic matter reserves in Indian soils. Indian Council of Agricultural Research., New Delhi, p 1-125.

Jha, L.K. 1995. Forage and fodder production through agroforestry. p.101-135. In: L.K. Jha (ed.), *Advances in Agroforestry*, School of Agricultural Sciences and Forestry, North Eastern Hill University, Mizorum, India.

Kanwar, J.S 1994. Management of soil and water resources for sustainable agriculture and environment. p. 1-10. In: *Management of Land and Water Resources for Sustainable Agriculture and Environment*, Indian Society of Soil Science, New Delhi.

Khiani, K.N and D.A. More. 1984. Long term effect of tillage operation and farmyard manure application on soil properties and crop yield in a Vertisol. *J. Indian Soc. Soil Sci.* 32:392-393.

Lal, R. 1986. Soil surface management in the tropics for intensive land use and high and sustained production. *Adv. Soil Sci.* 5:1-97.

Mann, L.K. 1986. Changes in soil organic carbon storage after cultivation. *Soil. Sci.* 142:279-288.

Manna, M.C. and A.K. Biswas. 1996. Earthworms as compost agents. *Yojana*. 40:34-35.

Manna, M.C., J.N. Hajra, N.B. Sinha, and T.K. Ganguly. 1997. Enrichment of compost by bioinoculants and mineral amendments. *J. Indian Soc. Soil Sci.* 45:831-833.

Manna, M.C., T.K. Ganguly, and P.N. Takkar. 1996. Influence of farm yard manure and fertilizer nitrogen on Vam and microbial activities in field soil (Typic Haplustert) under wheat. *Agric Sci. Digest.* 16:144-146.

Mathan, K.K., K. Sankaran, N. Kanakabushan, and K.K. Krishanamoorthy. 1978. Effect of continuous rotational cropping on the organic carbon and total nitrogen content in a black soil. *J. Indian Soc. Soil Sci.* 26:283-285.

Mutatkar, T. and S.P. Raychaudhuri. 1959. Carbon and nitrogen status of soils of Arid and Semi arid regions of India. *J. Indian Soc. Soil. Sci.* 7:255-262.

Oades, J.M. 1984. Soil organic matter and structural stability: Mechanisms and implications for management. *Plant and Soil* 76:319-337.

Parton, W.J., D.S. Schimel, C.V. Cole, and D.S. Ojima. 1987. Analysis of factors controlling soil organic matter levels in great plains grassland. *Soil Sci. Soc. Am. J.* 51:1173-1179.

Post, W. and L.K. Mann. 1990. Changes in soil carbon and nitrogen as a result of cultivation. In: A.E. Bouwman, (ed.), *Soil and the Green House Effect*. John Wiley & Sons. Chichester, U.K.

Prasad, R. 1983. Increased crop production through intensive cropping systems. p. 331-322. In : U.C. Holmes and W.M. Tahir (eds.), *More Food from Better Technology*, FAO, Rome.

Prasad, R. and J.F. Power. 1991. Crop residues management *Adv. Soil. Sci.* 15:205-251.

Prasad, R. and N.N. Goswami. 1992. Soil fertility restoraration and management for sustainable agriculture in south Asia. *Adv. Soil Sci.* 17:37-77.

Ramakrishnan, P.S. and O.P. Tokey. 1981. Soil nutrient status of hill agro-ecosystems and recovery pattern after slash and burn agriculture (Jhum) in north-eastern India. *Plant and Soil* 60:41- 44.

Ramakrishnan, P.S.1994 .The Jhum agroecosystem in northeastern India: a case study of the biological management of soils in a shifting agricultural system. p. 189-207. In : P.L.Woomer and M.J. Swift (eds.), *The Biological Management of Tropical Soil Fertility*.

Rao, M.S., R.R. Prabhu, C.N. Rao, and G. Jayasree. 1994. Soil physical constrains and their management for increasing crop production in Andhra Pradesh. Highlight of research (1967-94), Andhra Pradesh Agricultural University, Andhra Pradesh.

Russel, E.W. 1973. *Soil Condition and Plant Growth* (10th ed.). Longman, London, UK/New York, USA.

Sahoo, R., A.K. Bandyopadhya, and B.B. Nanda. 1970. Effect of organic manures with and without drainage on rice yield, nutrient uptake and soil aggregation. *J. Indian Soc. Soil Sci.* 18:51-55.

Sarkar, A.K., B.S. Mathur, S. Lal, and K.P. Singh. 1989. Long-term effect of manure and fertilizers on important cropping systems in sub-humid, red and laterite soils. *Fert. News.* 34:71-80.

Sarkar, A.K., V.C. Srivastava, and B.S. Mathur. 1994. Soil management and rain water conservation and use-1. Alfisols under medium rainfall. In: Soil management for sustainable agriculture in dryland areas. *Bull. Indian Soc. Soil Sci.* 16:33-40.

Sharma, C. P., B.R. Gupta, and P.D. Bajpai. 1986. Residual effect of leguminous crops on some chemical and microbiological properties of soil. *J. Indian Soc. Soil Sci.* 34:206-208.

Singh, Gurbachan. 1994. Afforestration and agroforestry for salt affected soils. p. 260 -281. In: D.L.N. Rao, N.T. Singh, R.K. Gupta, and N.K. Tyagi (eds.), Salinity Management for Sustainability Agriculture, Central Soil Salinity Research Institute, Karnal, India.

Singh, S., R. Prasad, B.V. Singh, S.K. Goel, and S.N. Sharma. 1990. Effect of green manuring, blue-green algae and neem cake coated urea on wetland rice (*Oriza sativa*). *Biol. Fertil. Soils.* 9:235-238.

Singh, Y., D.C. Chaudhary, S.P. Singh, A.K. Bharadawaj, and D. Singh. 1996. Sustainability of rice (*Oryza sativa*) - wheat (*Triticum aestivum*) sequential cropping through introduction of legume crops and green manure crops in the system. *Indian J. Agron.* 41:510-514.

Sinha, S.K. and M.S. Swaminathan. 1979. The absolute maximum food production potential in India-an estimate. *Curr Sci.* 48:425-429.

Srivastava, S.C. and J.S. Singh. 1989. Effect of cultivation on microbial carbon and nitrogen in dry tropical forest soil. *Biol.Fertil. Soils.* 8:343-348.

Stevenson, F.J. 1982. *Humus Chemistry, Genesis, Composition, Reactions.*, Wiley Interscience, New York.

Swarup, A. 1998. Emerging soil fertility management issues for sustainable crop productivity in irrigated systems. *Proc. National Workshop on Long-term Soil Fertility Management through Integrated Plant Nutrient Supply System*, April 2-4, Indian Institute of Soil Science, Bhopal, India.

Swift, M.J and P.A. Sanchez. 1984. Biological management of tropical soil fertility for sustainable productivity . *Nature and Resources* 20:1-10.

Tomar, S.S., G.P. Tembe, and S.K. Sharma . 1992. Effect of amendments on soil physical properties and yield of crops under rainfed upland conditions of paddy-wheat cropping. p. 116- 119. In: Proc. National Seminar on Organic Farming, Jawaharlal Nehru Krishi Viswa Vidyalaya, Indore.

Van Veen, J.A., J.N. Ladd, and M.J. Frissel. 1984. Modelling of carbon and nitrogen turnover through the microbial biomass in soil. *Plant and Soil.* 76:257-174.

Virmani, S.M. 1994. Soil management: use of crop models and GIS as tools for prioritizing sustainable dryland agriculture research needs. In: Soil Management for Sustainable Agriculture in Dryland Areas. *Bull. Indian Soc. Soil Sci.* 16:101-106.

CHAPTER 15

Soil Organic Matter Dynamics and Carbon Sequestration in Australian Tropical Soils

R.C. Dalal and J.O. Carter

I. Introduction

Australian tropical and subtropical region referred to in this chapter covers the area from $10°$ south to $30°$ south and scans the climate from humid and isothermic in the eastern and northern coastal zones to arid inland with hot summers and cold winters. Rainfall varies from 4000 mm in the northeastern coastal zone to less than 250 mm in inland central Australia. More than 60% of the rainfall, often exceeding 90% in northern Australia, is received from November to April. Mean annual temperature varies from $17°C$ in the southern area to $27°C$ in the northern area, while potential evapotranspiration (PET) exceeds mean annual rainfall in all areas except a small area on the northeast coast. Dominant native vegetation of the region are forests including rainforest, open woodlands, savannas and grasslands while the dominant soils are Alfisols and Vertisols, with significant areas of Lithosols, Oxisols and Solodosols.

Temperature and moisture (net effect of rainfall and PET) are the major determinants of plant biomass production and, along with soil type, determine the soil organic matter (SOM) content as well as the rate of change of SOM under different land uses. Since the SOM content depends upon the relative rates at which organic materials are added to the soil and lost from it through decomposition, it can mathematically be expressed as follows (Bartholomew and Kirkham, 1960):

$$d\ SOM/dt = A - k\ SOM \qquad \text{Eq. 1}$$

$$\text{or} \quad SOM_t = SOM_0 \exp(-kt) + A/k\ [1 - \exp(-kt)] \qquad \text{Eq. 2}$$

where SOM_0 and SOM_t are the SOM contents initially (t = o), and at a given time, t, respectively. A is the rate at which organic matter is returned to the soil and k is the rate of loss of SOM. Since SOM is heterogeneous and its various components decompose at different rates, Equation 2 is modified to:

$$SOM_t = SOM_1 \exp(-k_1 t) + A_1/k_1\ [1 - \exp(-k_1 t)] + ... + SOM_n \exp(-k_n t) + A_n/k_n$$

$$[1 - \exp - (k_n t)] \qquad \text{Eq. 3}$$

where SOM_1, SOM_n refer to different components of organic matter, such as microbial biomass (labile), light fraction (active, humified SOM) and heavy fraction (passive, inert SOM) (Dalal and Mayer, 1986c, 1986d, 1987a and 1987b; Parton et al., 1987, 1996; Jenkinson and Rayner, 1977; Jenkinson et al., 1992) and k_1 and k_n refer to the rate of loss of individual components. When A_t = o, at time, t, Equation 3 is reduced to $SOM_t = SOM_1 + ... + SOM_n = SOM_0$ and at time, $t = \infty$, $SOM_t = A_1/k_1 + ... + A_n/k_n = SOM_e$, which may be simplified to give:

$$SOM_e = A/k \qquad \qquad \text{Eq. 4}$$

where SOM_e is organic matter content at steady state after a long period of consistently similar soil management (Bartholomew and Kirkham, 1960) or native vegetation. The simplified form of Equations 2 and 3 is:

$$SOM_t = SOM_e + (SOM_o - SOM_e) \exp(-kt) \qquad \qquad \text{Eq. 5}$$

where k is the overall rate of loss (decomposition) of various organic matter components. The turnover period of SOM is then reciprocal of the k value. Equations 4 and 5 are almost exclusively used here to describe the SOM dynamics in the Australian tropical soils. However, SOM dynamic models consider not only the effect of organic matter addition but also environment, soil matrix, cultural practices and some soil degradation processes such as soil erosion (Jenkinson and Rayner, 1977; Van Veen and Paul, 1981; Parton et al., 1987, 1996).

Soil organic matter is essential in maintaining physical, chemical and biological fertility of tropical soils, especially those containing either low clay content or low activity clays such as kaolinite. Since turnover of SOM is much faster in tropical soils than temperate soils, CO_2 fluxes to the atmosphere could make significant contribution to the greenhouse effect, especially after land clearing and use of inappropriate management practices. On the other hand, tropical soils could sequestrate significant amounts of atmospheric C and thus mitigate greenhouse effect. Since Australia contains large areas of tropical soils, contribution of these soils to CO_2 emission and C sink, and hence global change could be substantial. We present in this chapter C fluxes in these soils and provide estimates of C loss and C sequestration from agricultural and pastoral soils of the Australian subtropical and tropical regions.

II. Soil Organic Matter Dynamics in Natural Ecosystems

The natural ecosystems account for almost 90% of SOM (Post et al., 1982). However, estimates of total SOM content vary widely. The estimates of SOM vary from a low level of 700×10^{15}g C (Bolin, 1986) to a high of 2946×10^{15}g C (Bohn, 1976), most estimates converging towards a narrow range of 1300 to 1600×10^{15}g C (Eswaran et al., 1993). The tropical and subtropical regions account for almost 30% of the total SOM (Table 1).

Based on Holdridge life zones (variables: temperature, rainfall/precipitation, P and potential evapotranspiration, PET), Post et al. (1982, 1996) found that the SOM content was closely related to the annual PET/P ratio. Where PET/P was 1, SOM content was about 100 t C $ha^{-1}m^{-1}$ and decreased as the PET exceeded P and increased as P exceeded PET (Post et al., 1982). Thus, SOM content is generally lower in semiarid tropical zone than in subhumid tropical zone (Table 1).

The ecosystems of northern Australia should be regarded as a continuum of temperature and rainfall overlaying a wide variety of soils. Tree density varies from none in arid areas (predominantly grasslands) to rain forest pockets along the East Coast.

In an analysis of 4000 soil organic carbon samples from Queensland, Carter et al. (1998) found that organic C concentration decreased with increasing air temperature but increased with increasing rainfall (Figures 1a and 1b). They also observed that long term mean Normalised Difference Vegetation Index (NDVI) from the National Oceanographic and Atmospheric Administration (NOAA) satellite, can be used as a surrogate for net primary production as it integrates both precipitation and evaporation over an extended period (Figure 1c).

Table 1. Amount of organic C (C_e), annual addition (A), and rate of loss (k) and period of turnover (1/k) of organic C in soil under natural ecosystems in warm Holdbridge life zones

Life zone	Total amount[1] (C_e) ($\times 10^{15}$ g C)	Annual addition[2] (A) ($\times 10^{15}$ g C yr^{-1})	Turnover rate[3] (k) (Pg C Pg^{-1} C yr^{-1})	Turnover period (1/k) (yr)
Warm desert	2.02	0.6	0.030	33.0
	1292e+13			
Tropical desert bush		0.1	0.050	20.0
Tropical woodland and savanna (semiarid)		11.5	0.089	11.2
Tropical forest very dry (arid)		1.7	0.077	13.0
Tropical forest dry (semiarid)		1.1	0.046	21.7
Tropical forest moist (subhumid)		13.2	0.220	4.5
Tropical forest wet (humid)		15.3	0.196	5.1
Tropical total[4]	3351394	43.5	0.130	7.7
World total		75.8	0.054	18.5

[1]Post et al. (1982); [2]Jenkinson et al. (1991); [3]Calculated from Equation 4, assuming organic C is at steady state in natural ecosystems, that is, dc/dt = 0, and A = k C_e or k = A/C_e; [4]Tropical and subtropical wetlands account for additional 100–150 $\times 10^{15}$ g C, and thus total organic C in soils of world is about 1550 $\times 10^{15}$ g (Eswaran et al., 1993).

A. Organic Matter Contents in Forest Soils

In subhumid and humid tropics, organic C contents of soils under rainforest vary from 2.8% (alluvium) to 6.6% (on basaltic soils) in 0–0.1m soil depth (Bruce, 1965; Spain, 1990; Turton and Sexton, 1996). Usually, the soil under open forest contains less organic C than under rainforest (Turton and Sexton, 1996). However, total amounts of organic C in soil profiles of the rainforest vary widely. Spain et al. (1989) estimated that total amount of organic C in three basaltic soil profiles varied from 103 to 234 t C ha^{-1} in 0–1.2m depth, while Gillman et al. (1985) estimated 148 t C ha^{-1} in 0–0.6m depth of soil formed from granitic parent materials, with more than 50% of organic C concentrated in the top 0–0.3m layer. Turton and Sexton (1996) observed well-defined environmental gradients for canopy openness, depth of leaf litter, soil pH, soil temperatures and understorey solar radiation levels across rainforest-open forest boundaries in northeast Queensland, but organic C concentrations were least affected. However, Spain (1990) observed that organic C concentrations were closely related to the site air temperature and clay content:

$$\text{Organic C (\%)}= \exp (2.7 -0.08 \pm 0.02 \text{ Temp} + 0.014 \pm 0.002 \text{ Clay content}) \qquad \text{Eq. 6}$$

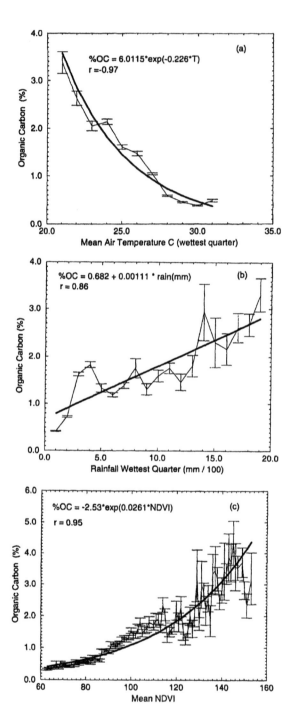

Figure 1. Relationship between soil organic C (0–0.1m depth) and (a, top) air temperature, (b, middle) rainfall, and (c, bottom) Normalised Difference Vegetation Index (NDVI) as a surrogate for temperature, rainfall and potential evapotranspiration. (Adapted from Carter et al., 1998.)

Table 2. Total amounts of organic C in the soil profiles under a forest (*Acacia harpophylla*), a woodland (*Eucalyptus populnea* and *Dicanthium sericeum*) and a grassland (*Dicanthium sericeum*) in a contiguous semiarid tropical region

Soil depth (m)	Cumulative organic C (t C ha^{-1})		
	Forest	Woodland	Grassland
0–0.1	22.1	17.5	13.7
0–0.2	36.6	29.1	23.8
0–0.3	47.4	39.5	32.6
0–0.6	71.1	66.5	59.5
0–0.9	88.9	84.2	83.7
0–1.2	103.8	95.5	103.9

(Adapted from R.C. Dalal, unpublished data.)

Thus, the relationship between temperature (and rainfall) and organic C reflects a predominantly biological and environmental control of the balance between plant biomass productivity (e.g., C input) and C decomposition rates, which are modified by the soil's clay content.

B. Organic Matter Contents in Open Woodland Soils

Woodlands comprise almost half of semiarid tropical and subtropical regions in Australia. The best known woodlands are Mulga-grass communities (*Acacia aneura* and *Ermophila mitchellii*) and Popular box-Queensland blue grass communities (*Eucalyptus populnea* and *Dicanthium sericeum*). A comparison of the amounts of soil organic C between a brigalow forest (*Acacia harpophylla*) and a woodland (*E. populnea* and *D. sericeum*) in a contiguous area showed that although the soil under forest contained more organic C in the top 0–0.3m depth, the total amounts in the 0–1.2m depth were essentially similar under both vegetation (Table 2).

C. Organic Matter Contents in Grassland Soils

An analysis of soil organic C concentration by dominant understorey species showed significant variation between these plant species (Table 3). Lowest organic C contents were found in the soil under *Triodia basedowii* (0.2%) but the highest C contents were found under *Imperata cylindrica* (3.8%) in the 0–0.1m layer; the former grass community is dominant in the semiarid region while the latter predominantly occurs in the humid region, usually following the clearing of wet sclerophyll forest. Thus, the differences in these organic C contents again reflect largely the climate effects in addition to soil properties such as plant available phosphorus concentration (Carter et al., 1998). In a similar temperature and rainfall zone, the amounts of organic C in 0–1.2m depth were similar in soils under grassland, woodland and forest (96 to 104 t C ha^{-1}), although in the 0–0.1m layer, the amounts of organic C were in the order: forest>woodland>grassland (Dalal and Mayer, 1986b) (Table 2). The latter observations are similar to those of Maggs and Hewett (1993) for organic C contents in soils under primary rainforests, secondary rainforest and derived grasslands.

Since depth distribution of organic C differs in soils under different vegetation, it is essential that the soil profile for the bulk of the vegetation rooting zone or down to at least 1m depth should be sampled for the total organic C budgets for both total C sinks and sources and turnover rates while considering the effect of native vegetation.

Table 3. Organic carbon concentration (0–0.1 m) of soils dominated by various understory species

Grassland species	Sites	Organic C (%)			
		Mean	Minimum	Median	Maximum
Aristida aramata	38	0.70	0.22	0.69	1.30
Aristida jerichoensis	25	0.74	0.43	0.70	1.30
Artistida latifolia	28	0.55	0.33	0.50	1.00
Astrebla elymoides	19	0.49	0.31	0.45	0.91
Astrebla lappacea	88	0.55	0.26	0.49	1.30
Astrebla pectinata	45	0.43	0.23	0.42	0.70
Astrebla squarrosa	11	0.44	0.22	0.47	0.57
Atriplex spongosa	12	0.36	0.20	0.34	0.70
Sclerolaena lanicuspis	16	0.34	0.20	0.34	0.57
Bothriochloa decipiens	23	1.78	0.64	1.90	2.90
Bothriochloa ewartiana	15	1.19	0.43	1.18	2.20
Carissa ovata	8	1.25	0.90	1.10	1.90
Cenchrus ciliaris[*]	15	1.35	0.50	1.20	2.70
Chloris gayana[*]	13	2.48	1.20	2.40	4.10
Cymbopogon refractus	8	1.41	0.40	1.25	2.60
Dactylocentium radulans	18	0.46	0.20	0.40	1.20
Dichanthium sericeum	36	0.89	0.35	0.59	2.80
Enneapogon avenaceus	12	0.45	0.27	0.49	0.58
Eragrostis eripoda	12	0.33	0.20	0.32	0.50
Eremophila mitchellii	17	1.14	0.30	0.81	3.00
Eragrostis setifolia	21	0.57	0.27	0.52	1.10
Heteropogon contortus	183	1.38	0.26	1.10	6.08
Heteropogon triticeus	16	0.62	0.04	0.55	2.10
Imperata cylindrica	13	3.83	1.30	2.60	10.30
Iseilema membranaceum	26	0.45	0.30	0.43	0.70
Isilema vaginiflorum	51	0.46	0.23	0.42	0.94
Ophiurous exaltatus	15	1.08	0.70	1.00	1.40
Panicum maximum[*]	30	3.31	1.40	3.30	7.30
Salsola kali	17	0.37	0.20	0.40	0.70
Sorghum nitidum	13	1.29	0.72	1.20	2.40
Sporobolus actinocladus	27	0.45	0.15	0.41	0.97
Themeda triandra	83	1.89	0.13	1.40	6.67
Triodia basedowii	7	0.21	0.10	0.20	0.30
Triodia pungens	22	0.56	0.24	0.53	1.30

*Species commonly used in improved pastures.
(Adapted from Carter et al., 1998.)

D. Organic Matter Decomposition Rates and Turnover Periods

Organic matter turnover in soil at a steady-state can be estimated from Equation 4, especially in natural ecosystems where it can reasonably be assumed that organic matter additions balance organic matter losses (Table 1). Total amounts of organic C in different zones, reported in Table 1, are the estimates of Post et al. (1982, 1996) and annual additions of organic C were calculated using Rothamsted model (Jenkinson et al., 1991). The Rothamsted model provided about 10 to 15% higher estimates of net primary production (that is, organic C input) than earlier estimates (Buringh, 1984) although, Long et al. (1989) measured 2 to 5 times higher net primary production in natural grass ecosystems of the tropics.

The calculated decomposition rates of organic C (or SOM) vary from 0.03 Pg C Pg^{-1} C yr^{-1} in arid desert to 0.2 Pg C Pg^{-1} C yr^{-1} in subhumid and humid tropical forests although in semiarid tropical forests the decomposition rate of organic C is only 25% of that in the subhumid tropical forests. Rate of SOM decomposition in semiarid woodlands and savannas in the tropics (and subtropics) is less than 0.1 Pg C Pg^{-1} C yr^{-1}, a value between dry tropical forest and wet tropical forest.

The period of turnover of SOM is about 2.5 times faster in the tropical and subtropical regions than that for the whole terrestrial ecosystem (Table 1), and in the tropical region alone, it is about 4 times faster than in the temperate region (Jenkinson and Ayanaba, 1977). In the subhumid tropical forest, the rate of turnover of SOM is less than 5 years. It is more than 2 times faster than in semiarid tropical woodlands and grasslands and 4 times faster than in semiarid tropical forest, a reflection of favorable temperature and moisture both for plant biomass production and decomposition.

III. Soil Oganic Matter Dynamics in Derived Agro-ecosystems

Land clearing and development often leads to reduced organic matter levels in soil. This is primarily due to changes in temperature, moisture fluxes and soil aeration, to exposure of new soil surfaces resulting from aggregate disruption, to reduced input of organic materials, and frequently to increased soil erosion. This decline might be attributable to two primary causes: reductions in C inputs and increases in soil temperature. Clearing rain forest considerably alters microclimate and results in solar radiation reaching ground level during day light. Measurement of changes in microclimate shows that soil surface (0–0.02m) temperatures increase by as much as 3°C (Du Pont, 1997). This increase in temperature alone could increase the rates of organic matter decomposition.

There are no detailed studies to fully quantify the effects of land clearing on the amounts and dynamics of organic matter in the soil profile. In the majority of studies of tree clearing in central Queensland there is a significant decline in the organic C concentration of surface soils. For example, Webb et al. (1977) measured a decrease in organic C (0–0.1m depth) of a Vertisol from 2.56% in virgin soils to 2.17% (15% loss) after clearing of brigalow forest in the Callide area. In northern Queensland, Cogle et al. (1995) found that organic C (0–0.1m depth) in an Oxisol declined from 2.38% in natural woodland to 1.26% after clearing (47% loss), although period after clearing is not known. However, a change in organic C concentrations of the soil profile down to 1m depth is rarely measured in most studies.

A. Soil Organic Matter Loss from Cultivated Lands

Cultivation of a soil that previously supported native vegetation or pasture generally leads to reduced level of SOM (Haas et al. 1957; Russell, 1981; Odell et al., 1984; Dalal and Mayer, 1986b; Rasmussen and Collins, 1991). The organic matter level of a cultivated soil eventually attains a steady-state, where

Table 4. Rates of organic matter loss and its turnover period from cultivated soils

Location	Soil	Cropping system	Rate of loss ($\%C\ \%C^{-1}\ yr^{-1}$)	Turnover period (yr)
Illinois, USA[1]	Inceptisol	Continuous corn	0.024	41.7
Adelaide, Australia[2]	Alfisol	Wheat	0.025	40.0
Narayen, Queensland, Australia[2]	Vertisol	Sorghum	0.050	20.0
Southern Queensland, Australia[3]	Vertisol	Wheat	0.041–0.178	5.6–24.4
Kansas, USA[4]	Argiustoll	Wheat-fallow	0.062	16.1
Assam, India[5]	Ultisol	Unshaded tea	0.099	10.1
Congo, Africa[1]	Ultisol/Alfisol	Food crops	0.330	3.0
Nigeria, Africa[6]	Ultisol	Food crops	0.393	2.5
Nigeria, Africa[7]	Alfisol·	Maize	0.637	1.6
Southern Queensland, Australia[3]	Alfisol	Wheat	1.211	0.8
World cultivated lands[8]	All soils	All crops	0.061	16.4

[1]Adapted from Lathwell and Bouldin, 1981; [2]Russell, 1981; [3]Dalal and Mayer, 1986b; [4]Hobbs and Thompson; 1971; [5]Gokhale, 1959; [6]Calculated from Obi, 1989; [7]Calculated from Lal and Kang, 1982; [8]Calculated from Jenkinson et al., 1991.

rate of formation of new SOM from organic residues (plant and crop residues, roots and root exudates, organic wastes, manures and green manures) equals rate of SOM decomposition, provided that soil and crop management practices remain essentially similar over a long period.

1. Overall SOM Loss

The rates of loss and turnover period of SOM in different soils vary considerably (Table 4). Organic matter from Ultisols/Alfisols turns over much faster than from Inceptisols and Vertisols. At least part of the effect is due to decomposition being more rapid under the warmer and moister conditions existing in the Ultisols/Alfisols and partly due to soil matrix (e.g., soil clay content).

The rates of loss of SOM from the cultivated tropical and subtropical soils are generally within the range of those of natural ecosystems in these regions (Tables 1 and 4) except in Alfisols and Ultisols which have much higher rates of SOM loss, especially within the first 8 to 20 years after land clearing.

In a number of these studies reported in Table 4, effect of change in bulk density with the loss in SOM on depth of sampling is not considered, hence equivalent sampling depths may differ from actual sampling depths, so these estimates of rates of SOM loss should be considered approximate.

These considerations were taken into account in SOM studies by Dalal and Mayer (1986a, 1986b) who investigated the long-term trends in fertility of Vertisols and Alfisols under continuous cultivation and cereal cropping in the semiarid subtropical region (southern Queensland, 148–152°E and 26–29°S). All soils lost organic C when cultivated and the loss of organic C increased with increasing period of cultivation. The soils differed, however, in the rates of loss, periods of turnover and amounts of organic C inputs required to maintain SOM at steady state (Table 5).

Dalal and Mayer (1986b) observed that, in the contiguous region (similar rainfall and temperature), initial organic C contents (virgin soils) as well as organic C contents after 20 years of cultivation for cereal cropping were closely correlated with the mean annual precipitation. However, the rate of increase in organic C values for each mm of precipitation was almost twice in soil under native vegetation as

Table 5. Initial values, steady-state values, overall rate of loss, rate of addition and period of turnover of organic C in six soil series (0–0.1 m depth)[a]

		Initial C C_o (t C ha^{-1})	Steady-state C C_e (t C ha^{-1})	Loss rate k (t C ha^{-1} yr^{-1} / t C ha^{-1})	Addition rate[b] A (t C ha^{-1} yr^{-1})	Turnover period (1/k) (yr)
Waco clay	Typic Pellusterts	13.71	8.26	0.065	0.54	15.4
Thallon clay	Typic Chromusterts	7.52	4.41	0.069	0.30	14.5
Langlands-Logie clay	Typic Chromusterts	22.07	7.76	0.080	0.62	12.5
Cecilvale clay	Typic Chromusterts	17.51	10.18	0.180	1.83	5.6
Billa Billa loamy clay	Typic Chromusterts	13.64	8.27	0.259	2.14	3.9
Riverview sandy loam	Rhodic Paleustaffs	15.69	9.38	1.224	11.48	0.8

[a]Dalal and Mayer, 1986b; [b]calculated from Dalal and Mayer (1986a,b,c,d,e) using Equation 4.

compared with that in cultivated soils (4.8 x 10^{-3} C% mm^{-1} vis 2.9 x 10^{-3} C% mm^{-1}). This was primarily due to the less amount of organic residues produced and returned to the soil, and/or accelerated rate of organic matter decomposition in cultivated soils. Increased incidence of erosion may also be a contributory factor (Van Veen and Paul, 1981; Bouwmann, 1990; Cook et al., 1992). In wider regions, however, precipitation/potential evapotranspiration ratio determines the plant biomass production (Post et al., 1982, 1996; Parton et al., 1987).

In the six soil series studied, Dalal and Mayer (1986b) recorded a 20-fold difference in overall rate of organic C loss and the period of turnover of SOM (Table 5). The rate of organic C loss declined exponentially as the clay content in soil increased (Figure 2), irrespective of clay mineralogical composition. Similarly, overall rate of loss of organic C declined as aggregation in soil increased, thus reducing the accessibility of organic C to microorganisms and degradative enzymes and other organic matter decomposers. It is, therefore, primarily the clay that protects SOM from decomposition when soil is cultivated (Dalal and Mayer, 1986b; Skjemstad et al., 1986; Skjemstad and Dalal, 1987; Parton et al., 1996). Using spectroscopic techniques, including ^{13}C nuclear magnetic resonance spectroscopy, Skjemstad et al. (1986) showed that the chemical nature of SOM remained essentially similar in spite of 60% C lost upon cropping. Poly-methylene and methoxyl-C remained the dominant groups of organic C throughout the C loss. Similar results have been obtained by others (Capriel et al., 1992; Skjemstad et al., 1996). Skjemstad and Dalal (1987) observed that besides the association of SOM with clay, SOM stability against decomposition in cultivated soils may be due to increased aromaticity as well as shorter, more highly branched alkyl chains, at least in some Vertisols (Pellusterts).

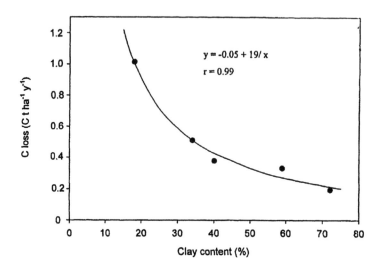

Figure 2. Decrease in rate of loss of organic C (0–1.2m depth) with increasing clay content in five soil series. (Drawn from the data of Dalal and Mayer, 1986b.)

Dalal and Mayer (1986b) also observed that the rate of organic C loss in the six soil series decreased as the ratio of organic C/urease activity, probably a measure of relative accessibility of organic C to proteolytic and other degradative enzymes. This concept is somewhat akin to that of soil micropores small enough to be inaccessible to microorganisms, hence SOM in that environment remain protected from decomposition.

The equations that have been used above to describe the trends in soil organic carbon under continuous cultivation are also helpful in understanding why it is so difficult to restore SOM in some soils, especially coarse-textured soils, if it is allowed to decline to low levels. There is a relationship between content of SOM and the annual rate of addition of SOM to the soil at steady state ($SOM_e = A/k$), that is, the rate of input, A, equals the product of SOM_e and k under constant soil and crop management practices. These values are shown for Vertisols and Alfisols in Table 5 and Figure 2. The rate of C input required in maintaining organic C content at steady state varied as a reciprocal of the clay content (Figure 2). Thus, large C inputs are required to maintain organic C levels in coarse-textured Alfisols as compared with fine-textured Vertisols.

The content of soil SOM that can be maintained ultimately depends on the production of plant biomass and/or addition of farmyard manure and organic wastes. It is apparent, therefore, that SOM declines under continuous cropping for at least two reasons: rates of decomposition are increased due to cultivation (increase in surface temperature) and annual C inputs are lower under a cropping system than they were under a grassland or forest ecosystem.

2. Loss of SOM from Various Soil Fractions

As mentioned above, a more complete understanding of the dynamics of SOM requires knowledge of the various components of organic C and their rates of decomposition (Equation 3) such as soil microbial biomass C, active or light fraction C, particle-size C, and passive C.

Table 6. Rates of loss of organic C from various C fractions of 0–0.1 m layer of Waco soil series (Typic Pellusterts)[a]

Organic C fractions	Initial C C_o (kg ha^{-1})	Steady state C C_c	Proportional rate of loss $(C_o-C_c)/C_o$ (%)	(kg C ha^{-1} yr^{-1}/ kg C ha^{-1})	Turnover period (yr)
Whole soil	1.3 7e+28	8.260 130e+25	4.07e+13	0.065	15.4
Microbial biomass				0.338[b]	2.9
Light fraction, <2 Mg m^{-3}				0.365	2.7
Heavy fraction, >2 Mg m^{-3}				0.033	30.0
Sand-size fraction, 0.002-2 mm				0.096	10.4
Silt-size fraction, 0.002-0.02 mm				0.039	25.6
Clay-size fraction, <0.002 mm				0.057	17.5

[a]After 70 years of continuous cultivation and cereal cropping; values from Billa Billa soil series. (Adapted from Dalal and Mayer, 1986b, 1986c, 1986d, 1987b.)

Returns of reduced amounts of organic residues to cultivated soils would rapidly decrease the amount of microbial biomass C. Since microbial biomass is closely related to a suite of biological and biochemical activities in soil (Dalal and Mayer, 1987b), the reduced amount of microbial biomass upon cultivation of previously fertile soils (Table 6) leads to soil fertility depletion in three ways: reduced source of labile N, P and S; reduced retentive capacity (immobilization) for mineral N in most soils, and possibly P and S in Ultisols, Oxisols and Alfisols; and reduced rate of mineralization and thus lower bioavailability of nutrients (Dalal, 1998). Furthermore, continuous cultivation for cereal cropping, especially monoculture may reduce not only the total amount but also may alter the composition of microbial biomass which could lead to biological degradation (increased incidence of soil-borne pathogens).

The light fraction (<2 Mg m^{-3}), determined using bromoform-ethanol mixture of 2 Mg m^{-3} density (Dalal and Mayer, 1986d), contains most of the partly decomposed plant residues, microfloral and microfaunal debris (Ford et al., 1969) and some microbial biomass (Ladd et al., 1990). It contains about 25% of SOM although it varies from 1 to 85% (Greenland and Ford, 1964; Christensen, 1992; Gregorich and Janzen, 1996), depending on the nature of ecosystem, vegetation, climate and soil characteristics.

The rate of loss of light fraction C in cultivated soils is very rapid and essentially similar to that of microbial biomass C (Table 6) unless this fraction also contains charcoal C originating from anthropogenic burning of vegetation (Skjemstad et al., 1996). In five fine-textured (>30% clay) soil series, the rate of loss of light fraction C on cultivation was 2 to 11 times faster than from heavy fraction (>2Mg m^{-3}) although similar rates of C loss from both fractions were observed in a coarse-textured soil (Dalal and Mayer, 1986d). Rates of SOM loss in these soils have been associated with soil aggregation and clay content,$^{'}$ which provide protection against SOM decomposition by microorganisms and proteolytic and degradative enzymes. The coarse-textured soil, therefore, may provide less protection to SOM irrespective of its fraction size.

Upon cultivation, more than 60% of the light fraction SOM can be lost from soil within 20 to 70 years (Table 6) (Dalal and Mayer, 1986d; Skjemstad and Dalal, 1987). Similar to the rapid rate of loss of microbial biomass from soil on cultivation, the proportion of light fraction to total SOM provides an earlier and a sensitive indication of the consequences of different soil management practices than the total

SOM (Christensen, 1992; Gregorich and Janzen, 1996). Moreover, since crop residues and other organic materials in soil maintain active microbial biomass as well as light fraction, non-return of these materials deprives soil of its labile and active fraction of organic C.

About 75% and 90% of SOM resides in the heavy fraction >2 Mg m^{-3} ('slow' or soil-matrix associated) in virgin and cultivated soils, respectively (Dalal and Mayer, 1986d; Christensen, 1992); the higher proportion of the heavy fraction in the latter is due to the rapid loss of the light fraction from soil on cultivation.

The turnover of SOM from the heavy fraction is much slower (30 years) than that from the light fraction (2.9 years) (Table 6). In fine-textured soils, SOM in the heavy fraction is afforded greater protection than that in the light fraction. The ratio of the rate of loss of light fraction C/heavy fraction C generally increases with the clay content (Dalal and Mayer, 1986d), indicating a clearer separation between the relatively accessible light fraction and relatively protected heavy fraction SOM by the clay fraction. Corroboration of these results was provided by ^{13}C nuclear magnetic resonance spectroscopic studies of SOM fractions from the virgin and cultivated Waco and Langlands-Logie soil series (Skjemstad et al., 1986; Skjemstad and Dalal, 1987) and organic C dynamics in various particle-size fractions (Dalal and Mayer, 1986c). The light fraction SOM is located at the surface of soil aggregates and is, therefore, readily accessible to the soil organisms which are confined to aggregate surfaces and extra-aggregate space. The heavy fraction SOM, however, is contained in intra-aggregate regions, and thus relatively better protected from soil organisms. This concept is supported by Elliott (1986) who found that disintegration of macro-aggregates released readily decomposable SOM.

About 50% of SOM is found in the clay-size fraction (Tiessen and Stewart, 1983; Dalal and Mayer, 1986c; Balesdent et al., 1988; Bonde et al., 1992) and the remainder is distributed between sand-size and silt-size fractions, the proportion of SOM between these two fractions depends upon land use, climate, season, silt and clay content, and cultivation.

The sand-size SOM declines rapidly upon cultivation (Table 6), mostly due to oxidation, but also due to disintegration to enter silt-size or clay-size fractions (Christensen, 1992). The sand-size SOM, which is essentially similar to the light-fraction SOM (Table 6; Dalal and Mayer, 1986d; Christensen, 1992) and microbial biomass constitute the 'labile' SOM in the CENTURY model (Parton et al., 1996). Therefore, the rapid and extensive loss of the 'labile' SOM upon cultivation may result in greater depletion in soil fertility than is apparent from a comparison of total SOM contents of virgin and cultivated soils.

The rate of SOM loss from the silt-size fraction is much slower than from the sand-size and the light fraction (Table 6). Dalal and Mayer (1986c) observed that the proportion of SOM in the silt-size fraction of five Vertisols and one Alfisol remained essentially similar (26 to 27%) upon cultivation (20 to 70 years) of virgin soils, although the sand-size fraction decreased from 26% to 12% overall, and the clay-size fraction increased from 48% to 61%.

The rate of loss of organic C from the clay fraction is generally lower than that from the other fractions and the whole soil (Table 6) (Dalal and Mayer, 1986b, 1986c, 1986d; Christensen, 1992; Parton et al., 1996), resulting in an increasing proportion of total SOM (often >60%) remaining in the clay fraction with cultivation. The clay fraction, therefore, provides protection to SOM or makes it relatively inaccessible through aggregation and micropore formation against microbial and enzymic attack by forming aggregates, for clay-size organic matter separated by soil dispersion mineralises its C and especially N much faster than from the other size fractions (Elliott, 1986; Christensen, 1992).

The overall turnover period of SOM and its various fractions in cultivated soils rarely exceeds 50 years (Tables 1, 4, 5 and 6). However, the passive, inert or stable SOM may have turnover periods ranging from several hundred to a few thousand years (Jenkinson and Rayner, 1977; Parton et al., 1987; Balesdent et al., 1988; Skjemstad et al., 1990; Paul et al., 1997).

Using ^{14}C dating technique, the turnover period of SOM in the resistant fraction has been estimated to vary from 900 years to 9035 years (Jenkinson and Rayner, 1977; Paul et al., 1997). These estimates are based on a number of assumptions inherent in the ^{14}C dating techniques. The two basic assumptions, the constancy in the ^{14}C activity of the atmosphere during the last few thousand years, and minimum fractionation of ^{14}C during physical mixing and chemical and biological transformations leading to the stable and resistant SOM, are rarely met. It is known that the ^{14}C specific activity of the atmosphere has varied quite considerably over the last 7000 years, and from 1890 to 1954, the ^{14}C specific activity of the atmosphere gradually depleted due to the burning of almost ^{14}C-free fossil fuels, so called the Suess effect. From 1954 to 1964 it increased by 60% due to atmospheric nuclear testing, and since then it has gradually been declining. The other assumptions involved in ^{14}C dating and likely errors from charcoal to estimate the turnover period of the stable SOM have been discussed by Jenkinson et al. (1992). Similar assumptions are involved in using ^{13}C natural abundance as a tracer (Balesdent et al., 1988), besides it is restricted to SOM studies where complete changeover from either C_3 to C_4 or versa has occurred.

Using ^{13}C natural abundance technique, Balesdent et al. (1988) measured the turnover period of the stable SOM (about 50% of total SOM) in a temperate Alfisol to be 600 to 1400 years. Skjemstad et al. (1990) measured much shorter turnover periods for top layer but longer C turnover periods at lower depths, that is, 60 years for organic C from 0–0.075m depth, 95 years for 0.075–0.15m depth and 276 years for the 0.6–0.8m depth. In a tropical Oxisol under sugarcane for 50 years, the turnover period of SOM from native vegetation in silt and clay fractions appears to be less than 150 years (137–144 years, calculated from Vitorello et al., 1989) although not all SOM in these fractions form stable SOM (Dalal and Mayer, 1986c).

Notwithstanding the limitations of ^{14}C dating and ^{13}C natural abundance techniques and the difficulty in isolating stable C fraction from SOM, the turnover period as well as the proportion of even the passive, inert or stable SOM would decrease from the temperate to the tropics. Therefore, the postulated periods of 1000–2000 years in the CENTURY model (Parton et al., 1996) and 2871 years in the Rothamsted model (Jenkinson and Rayner, 1977) for the stable SOM need to be modified considerably to model successfully the SOM dynamics in the subtropical and tropical regions. In the CENTURY model the temperature and moisture index and texture modify these base rates. For many Australian soils recalcitrant C such as charcoal also needs to be considered (Skjemstad et al., 1996).

3. Modelling SOM Turnover

Modelling SOM dynamics in an ecosystem integrates physical, chemical and biological processes involved in SOM turnover and therefore, it may assist in developing a better understanding of SOM in biogeochemical and atmospheric environments. They are useful in formulating and testing hypotheses and in establishing the relative importance of parameters.

Most SOM models assume that different components of organic matter differ in their rates of decomposition and turnover periods (Jenkinson and Rayner, 1977; Van Veen and Paul, 1981; Parton et al., 1987, 1996) although most of these are only conceptual components and therefore, currently, not all the conceptual fractions of SOM have been experimentally verified.

The application of the CENTURY model to soil organic matter dynamics in the Billa Billa soil series (Typic Chromusterts) under continuous cultivation and cereal cropping is presented in Figure 3. The active, slow and passive components of SOM were taken as light (<2 Mg m^{-3}) or sand-size fraction, heavy fraction (2–2.4 Mg m^{-3}), and heavy fraction >2.4 Mg m^{-3}, respectively. The magnitudes of decline in total soil organic C and slow component were reasonably close to the experimentally determined values although the dynamics of the active component of SOM (Dalal and Mayer, 1986c, 1986d) were poorly simulated (Dalal and King, 1994).

The SOM models were generally developed and validated in temperate regions. Parton et al. (1989) discussed at least three deficient areas of understanding, which could make applicability of these models

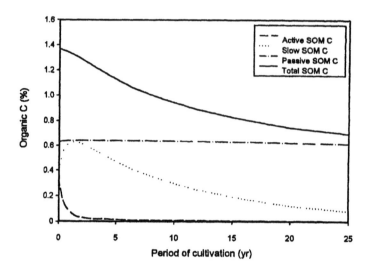

Figure 3. Simulation (CENTURY model) of decline in soil organic C in a Billa Billa soil series (Typic Chromustert) under continuous tillage and cereal cropping. (Adapted from data of Dalal and Mayer, 1986b, 1986c, and 1986d.)

to the SOM of tropics and subtropics more appropriate. These include: (1) functional relationships of soil texture, clay mineralogy, and especially Fe and Al on SOM dynamics; (2) clay mineralogy and parent material on the formation of passive SOM fractions; and (3) low pH and Al toxicity on microbial transformations.

Probert et al. (1995) applied CENTURY model to simulate changes in total N (as a measure of organic C) in a Vertisol used for winter cereal cropping for 20 years. It simulated total N levels satisfactorily for no-till and stubble retained systems but poorly for conventional-tilled and stubble burned systems. Obviously, considerable improvements in initialization and parameterization of models are required to simulate organic matter dynamics in subtropical and tropical environments.

As mentioned earlier, there is an urgent need to experimentally verify the various functional components of SOM used in these models. Unfortunately, methods used to fractionate SOM in different aggregate-size and density fractions are poorly described and thus, comparative evaluation of the importance of aggregate-size and density distribution of SOM on SOM dynamics is made extremely difficult. It is imperative, therefore, to standardize the methods of determining the fractions of SOM based upon soil aggregate pore size and density and then to ascertain whether these fractions of SOM from long-term cultivated soils are biologically meaningful in SOM turnover in subtropical and tropical soils.

4. Magnitude of SOM Loss from Cultivated Soils

Bouwman (1989) showed that SOM decomposition in the tropics was three to four times faster than in temperate regions and that both soil type (presumably soil texture) and soil environment influenced SOM decomposition. Simulated losses of organic-C in the tropical soils, without erosion losses were 31 to 50% after 50 years and 43 to 63% after 100 years of continuous cultivation without assuming any steady-state SOM. The simulated loss of SOM was 40 to 80% when water erosion also considered. Soil erosion caused an extra loss of at least 7% after 100 years. As expected, fine-textured soils were less prone to SOM decomposition than soils low in clay but may be more prone to erosion losses, especially on steep slopes. These simulated SOM decomposition trends are similar to those of Jenkinson and Rayner (1977) and Parton et al. (1987) although SOM decomposition in many tropical and subtropical soils is much faster than the estimates of Bouwman (1989).

Russell (1981) measured decline in organic C of 38% in a Vertisol in a subtropical and semiarid region after 14 years of continuous cultivation for wheat, and Standley et al. (1990) measured 27% loss in soil organic C from a similar soil after 7 years of continuous cultivation for sorghum. Lal and Kang (1982) reported 53% loss in organic C from a tropical Alfisol after 8 years of cultivation. Srivastava and Singh (1989) measured 57% loss in soil organic C after 15 years of cultivation, and Martin et al. (1990) recorded similar losses using ^{13}C natural abundance measurements, 59% after 16 years from a tropical Oxisol. Skjemstad and Dalal (1987) measured 61% of organic C loss from a Vertisol (Langlands-logie soil series) after 35 years of continuous cultivation whereas from a Brazilian Oxisol, organic C loss, using ^{13}C natural abundance, was estimated to be 50% after 12 years of cultivation (Cerri et al., 1985). Chan et al. (1997) measured 41% loss in total organic C (0–0.1m depth) when an originally grassland Vertisol was cultivated for 16 years in northern New South Wales, Australia. Bridge and Bell (1994) observed 59% loss in total organic C (0–0.1m) from an Oxisol after more than 50 years of cultivation in semiarid subtropical Queensland, whereas Cogle et al. (1995) measured a 56% loss in organic C after about 10 years of cropping in subhumid tropical Queensland; thus substantiating the rapid decomposition of organic C in a moist tropical environment (Table 1).

Organic C loss from particle-size <0.05 mm was generally about half (43 to 48%) compared with the larger particle-size fractions, <0.5 mm (89 to 97%) in a tropical Oxisol after 16 years (Martin et al., 1990). Skjemstad and Dalal (1987) measured almost complete loss of organic C from <2.0 Mg m^{-3} fraction, 52% from 2.0–2.4 Mg m^{-3} fraction and none from >2.4 Mg m^{-3} fraction, with an overall loss of 61% from a Vertisol after 35 years of cultivation.

Therefore, the magnitude of not only total SOM loss but that of various fractions vary widely even in soils of the subtropical and tropical regions and thus the consequences of SOM loss on soil fertility depletion may vary in their effects and magnitude considerably.

B. Soil Organic Matter Loss from Grazing Lands

Action of domestic herbivore, macropods and feral animals results in a reduction in grass biomass and soil surface cover. The low digestibility of tropical grasses means that about 50% of the carbon is returned to the system in manure, so even in heavily grazed systems some recycling of above ground biomass occurs. In a recent qualitative survey of the grazing lands of northern Australia it was found that there was significant degradation in the form of changes in soils and species composition (Tothill and Gillies, 1992).

There are few studies with detailed measurements of soil organic C dynamics for grazing on native grasslands. In a study of fence line effects, Ash et al. (1995) recognized 3 degradation states. A progressive decline in organic C content occurs with increasing degradation, with 30% decrease in Organic C content from A class to B class and a further 20% decrease from B class to C class. Overgrazing leads to loss of soil surface cover, which makes soil prone to soil erosion and hence significant soil and organic C loss. Miles (1990) observed significant soil loss in mulga lands (Alfisols), where 50% of soil organic C resides within the top 0.1m layer. A small soil loss can have significant impacts both on plant production and organic matter content of the soil.

On the other hand, from a study of surface soils in a series of long duration (>10 years) grazing trials, Howden et al. (1995) found some sites in which surface organic C concentration increased with grazing by domestic animals and others with little change or a decrease in organic C concentration with grazing intensity. This variation may reflect changes in plant root distribution due to change in plant species, e.g., perennial grasses have 5 times as much root biomass as an annual crop (Dalal et al., 1995). In native grasslands, grazing may to some extent replace the effects of termites and fire. The major difference is that at high stocking rates grazing removes significant amounts of green leaf material whereas fire and termites only affects the dead plant biomass component of the grass sward. Excessive consumption of green material may change species composition and reduce root biomass, which is a major component of grassland net primary production, and contributes to SOM. Thus, Bowman et al. (1997) measured higher organic C in a poor Mitchell grass (*Astrebla lappacea*), which had lower stocking rate than in a good Mitchell grass pasture.

In the tussock grasslands of northern Australia, soil organic C is concentrated in the soil directly under tussock bases. Therefore loss of tussock grass basal area may reduce soil organic C content. Simulation studies with the CENTURY model suggest that the soil organic C in grazing systems is strongly influenced by grazing pressure (Parton et al., 1987). However, a systematic study of grazing effects on soil organic C is required to quantify relationships between pasture utilization and soil organic C.

Fire may have direct and indirect effects of soil organic C storage. These include: removal of C and nitrogen in burnt fuel; feed back effects on vegetation structure; and addition of biologically inactive C (charcoal) to the soils. Charcoal is a common component of Australian soils and is even found in rain forest soils. This charcoal results from partial burning of woody material but might also include much finer material derived from grass leaf. A small study of the black residue remaining after burning of C_4 grass pastures indicated that most of the material was of an inorganic nature with <1% composed of carbonized material (Carter et al., 1998). If this material was all carbon then the input of this material to the soil was estimated at about 45 kg C ha^{-1} from burning of 4800 kg of dry matter. However, in the long term (over centuries) substantial amounts of this material might accumulate in the soil profile (Skjemstad et al., 1996) if not removed and redeposited by soil erosion.

Loss of soil organic C in tropical grasslands exposed to both high intensity rainfall and high levels of evaporation could significantly change water balance if infiltration rates are reduced. This has an adverse effect on plant biomass production, with reduced organic C inputs, which further reduces infiltration rates and soil organic C. In northern New South Wales, Chan et al. (1997) found that water stability of surface soil significantly decreased with decreasing organic C content.

Bruce (1965) observed a 30% decline in organic C after 22 years of a grass pasture system (*Panicum maximum*) on an Oxisol in the humid tropics in north Queensland (Figure 4). Inclusion of a legume (*Centrosema pubescens*) reduced the soil organic C loss in the top 0.15m layer. It would be interesting to investigate whether C losses could have been smaller if total amount of organic C in the whole soil profile were measured since C input by grasses decreases less with depth than under forests (Table 2; Carter et al., 1998).

IV. Management of Soil Organic Matter in Australian Tropics

The contributions of organic matter to soil productivity in the Australian tropics include one or more of the following aspects: (1) SOM is a major natural source of plant nutrients, e.g., N, P and S. (2) It is a major C and energy source for microorganisms. (3) It contributes significantly to cation exchange capacity of soil, especially in low-activity clay soils. (4) SOM is an important sink of nutrients that would otherwise be lost through leaching, especially in coarse-textured soils. (5) It provides buffering capacity to soil, thus reducing the deleterious effects of low pH and high acidity. (6) It provides polyvalent anions to complex strongly with polyvalent metal cations, e.g., Fe, Al and the transition metal cations, thus not only providing micronutrient metal cations in plant available form but also detoxifying heavy metal

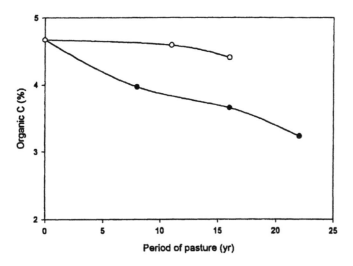

Figure 4. Soil organic concentration (0–0.075m depth) of an Oxisol after 22 years of grass pasture (-l-) (*Panicum maximum*) and after 16 years of grass and legume (-m-) (*Centrosema pubescens*) pasture. (Redrawn from Bruce, 1965.)

cations, especially in acidic Alfisols, Oxisols and Ultisols. (7) SOM also reduces the adsorption of P (and S) by soil colloids. (8) It promotes soil aggregation and thus improves water infiltration and water-use efficiency, and thus reduces soil erosion. (9) It retains pesticides, and provides microbial biodiversity to degrade pesticides, and (10) SOM provides a significant potential sink for atmospheric C, and thus plays an important role in the greenhouse effect. Restoration of organic matter in cropping and grazing lands is therefore required for sustainable agriculture and C sequestration.

A. Organic Matter Restorative Practices for Cropping Lands

Sustaining or enhancing soil productivity depends at least partly on soil and crop management practices that maintain or increase soil organic matter. Management practices such as no till and crop residue retention (Lal, 1989), crop rotation, pasture leys (Hossain et al., 1996), and fertilizers (Cogle et al., 1995) and manure applications may influence organic matter levels in soil.

1. No-Till and Crop Residue Retention

No-till (NT) and crop residue retention reduces the loss of organic matter in soil. For example, Dalal (1989) measured higher organic C contents in a Vertisol under NT and crop residue retention than under conventional till (CT); a positive interaction of tillage practice x crop residue x N fertilizer application was observed after 13 years. These increases in organic C occur in the top 0–0.025m or 0–0.05m layers in Vertisols, after 18 years of no-till practice (Dalal et al., 1991a; Dalal et al., 1995). Thus, NT practice enhances SOM stratification even in a Vertisol although total amounts of SOM may be similar to that in the CT practice.

Table 7. Effect of no-till (NT) for 4–5 years and conventional till (CT) on organic C contents of a tropical Oxisol and a subtropical Oxisol

Soil depth (m)	Organic C (%)			
	Tropical Oxisol[a]		Subtropical Oxisol[b]	
	CT	NT	CT	NT
0–0.05	0.89	1.35[*]	1.35	1.88[*]
0.05–0.10	0.77	1.15[*]	1.44	1.58
0.10–0.20	0.61	0.88	1.37	1.34

[a]From Cogle et al., 1995; [b]Bell et al, 1997, following fertilized kikuyu treatment; [*]Asterisk indicates significant difference in organic C between NT and CT.

Similarly, in Oxisols, NT practice enhances or at least reduces the decline of organic C concentrations in the top layers both in subtropical and tropical regions (Table 7). Thus, the effect of the NT practice on soil organic matter may be greater in the tropical than in the temperate regions (Lal, 1989). Besides providing C inputs, surface residue cover also reduces raindrop impact and enhances water infiltration, which may increase plant biomass production, and hence increased organic C inputs to soil.

2. Crop Rotations

Crop rotations that include legumes and/or grasses generally enhance or reduce loss of soil organic C in many environments. Whitehouse and Littler (1984) observed substantial increases in organic C after 2 to 4 years of lucerne + prairie grass pasture in a Vertisol; organic C increased from 1.18% to 1.37% after 4 years of pasture growth (0–0.15m depth). Dalal et al. (1995) measured the rate of increase in organic C of 0.13 year^{-1} or 650 kg C ha^{-1} year^{-1} in a Vertisol under grass + legume pasture (purple pigeon grass, *Setaria incrassata;* Rhodes grass, *Chloris gayana;* lucerne, *Medicago sativa;* and annual medics, *M. scutellata and M. truncatula*) for 4 years (Figure 5); most of the increases in organic C occurred in the top 0.05m layer. Skjemstad et al. (1994) reported an increase of 550 kg C ha^{-1} year^{-1} in a Vertisol under Rhodes grass in a similar environment. Increase in organic C under pasture was attributed to the much higher input of C through the grass root biomass (10 t root dry matter ha^{-1} year^{-1}) compared with continuous wheat cropping (2 t root dry matter ha^{-1} year^{-1}), aboveground residue inputs were essentially similar (2.5 t residues ha^{-1} year^{-1}) (Dalal et al., 1995). Chan (1997) observed similar effects of pasture on organic C concentrations in a Vertisol in northern New South Wales. He also found that almost 70% of the organic C increases could be attributed to the increase in particulate organic C.

Wheat cropping after the pasture phase resulted in organic C decline but even after 4 years remained above the organic C concentrations in the continuous wheat cropping treatment (Dalal et al., 1995). However, the decreasing organic C trends during the cropping phase were slower than the relatively faster increase in organic C in soil under the grass + legume phase earlier, presumably due to the increased plant C input from higher crop yields compared with continuous wheat cropping.

While grass + legume pasture phase had the positive effect on soil organic C, two-year rotations of lucerne-wheat, medic-wheat, and especially chickpea (*Cicer arietinum*)-wheat after 8 years had a negligible effect on organic C concentrations. The plant C inputs, especially from root biomass, in these treatments were about 50% of those from the grass + legume pasture (Dalal et al., 1995). Therefore, the relatively small amount of C input and rapid rate of turnover of legume C in these short legume rotations does not increase organic matter in Vertisols. Holford (1990) who observed no increase in organic C after 4 years of lucerne pasture on a Vertisol in northern New South Wales confirms this. Similarly,

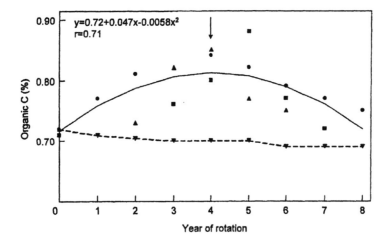

Figure 5. Organic C concentration trends in soil (0–0.1m depth) under 4 years of grass + legume pasture followed by 4 years of wheat. ●, Grass + legume 1986–89; ■, grass + legume 1987–90; ▲, grass + legume 1987–90; ▼, continuous conventional till wheat. (Adapted from Dalal et al., 1995.)

Skjemstad et al. (1994) observed that organic C loss from a fertile Vertisol was faster under black gram (*Vigna mungo*) than sorghum cropped for 11 years—a consequence of lower plant C input from the former (0.2 t ha⁻¹ year⁻¹ aboveground dry matter) than from the latter (0.5-0.75 t ha⁻¹ year⁻¹ dry matter).

In a subtropical Oxisol, Bell et al. (1997) measured similar amounts of organic C in the top 0–0.3m depth after 4 years of two grass pastures (*Pennisetum clandestinum* and *C. gayana*) and in the adjoining soil under continuous cultivation although the bulk density was significantly lowered by the grass pastures. However, Cogle et al. (1995) found that 8 to 10 years of grass pasture established after clearing a tropical Oxisol from woodland reduced organic C loss from 22 t C ha⁻¹ under cultivation to 13 t C ha⁻¹ under grass pasture. Organic C loss was further reduced to only 8 t C ha⁻¹ when a legume (*Stylosanthes hamata*) was also grown in the grass pasture for 4 years. These results corroborate the findings of Bruce (1965) who observed that only a small loss of organic C (7%) occurred from an Oxisol (0-0.15m depth) in the humid tropics after land clearing and establishment of grass + legume pasture as compared with grass only pasture (23%) after 16 years (Figure 4).

In summary, significant increases in organic C or retarding organic C loss from Australian tropical soils can be achieved by crop rotations that include a grass + legume pasture phase.

3. Fertilizers and Manures

A direct consequence of organic matter decline is the concomitant decrease in nutrient supply from soils especially that of nitrogen. In most Vertisols, decreased N supply with organic matter decline could be accounted for in the amounts of N removed in the produce (Dalal and Mayer, 1986e). However, N losses through leaching (Moody, 1994; Dalal and Probert, 1997), denitrification and soil erosion (Rose and Dalal, 1988) do occur. Comparative estimates of N losses through product removal and soil erosion are given in Table 8. In cultivated lands, N removed in the produce is by far the most important process of N loss. This N loss can be remedied by application of N (and other nutrients) from fertilizers, manures and other sources.

Table 8. A comparison of depletion rates of soil N by the processes of N removal in produce and soil erosion.

Source	Removal in produce	Removal in soil erosion
	— kg N ha^{-1} yr^{-1} —	
Terrestrial lands	3.3[a]	1.46[d]
Cultivated lands	30–40[b]	6.35[e]
Alfisols	28.00[c]	6.35
Vertisols	46.00[c]	6.35

[a]Calculated from mean grain-N concentration (22.5 kg t^{-1}) and total food grain production (1952.224 x 10^6 t, FAO, 1992) and total land area (13.075 x 10^9 ha); [b]grain-N removal rate from cultivated lands (1.461 x 10^9 ha yr^{-1}); [c]calculated from Dalal and Mayer, 1986e; [d]assuming N concentration of top soil layer, 0.075% and sediment enrichment ratio of 2; [e]assuming 50% of sediment-N loss from cultivated lands.

Dalal (1992) observed that N fertilizer application for 22 years to a Vertisol reduced the soil N loss as compared with nil fertilizer application. Even N rates exceeding the crop requirements could not reverse soil N decline although the high N rates increased nitrate-N accumulation in the soil profile. For the first 13 years, however, Dalal (1989) observed a positive effect on organic C from the NT practice, residue retained and N application treatment (34.5 t C ha^{-1} vis. 35.8 t C ha^{-1}).

Similarly, Russell (1981) and Skjemstad et al. (1994) found that N applications (100kg N ha^{-1} year^{-1}) to sorghum (*Sorghum bicolor*) grown on a fertile Vertisol for 11 years did not arrest organic C decline. Even on a fertility-depleted Vertisol, Dalal et al. (1995) found no significant increase in organic C from 75 kg N ha^{-1} year^{-1} application for 8 years as compared with nil N fertilizer application. Again, the N application resulted in nitrate-N accumulation in the soil profile.

On Oxisols and Alfisols, fertilizer application may result in increased organic matter in soil although Cogle et al. (1995) did not report any significant differences in organic C concentrations between the fertilized treatments (80 kg N + 23 kg P ha^{-1} year^{-1} applied to sorghum for 4 years) and the unfertilized treatments. Moody (1994) reported 60% decline in organic C in developed Oxisols compared to the relatively undisturbed Oxisols, in spite of the fact that many of the developed Oxisols had been fertilized.

It appears that in the semiarid environment, C inputs from annual crops are limited by rainfall and are not sufficient enough to meet C lost from soil even though crop yields are significantly increased with fertilizer applications.

Manure applications increase organic C in Vertisols (E. Powell, personal communication) and Oxisols (M. Bell, unpublished data) although long-term benefits of manure applications in arable cropping in the Australian tropical soils have not been measured. High-lignin amendments like farmyard manure, which are more recalcitrant to decomposition than plant residues, may result in higher soil organic C concentrations in these soils.

B. Organic Matter Restorative Practices for Grazing Lands

Management practices, which lead to improvement in pasture growth, would likely lead to increased C inputs and hence to increased organic matter in soil. These management practices include use of exotic pasture species, legumes and fertilizer application, modification of stocking rates, and changes in fire frequency. Tree planting, tree regrowth and thickening when appropriately managed can also lead to increase in organic C in the soil.

Pasture systems range from those extensively modified for intensive dairy production to native systems with low level grazing by domestic stock. Since the areas of grazed pasture far exceed the area of crops in northern Australia (200 million ha vs. 5 million ha), even very small changes in soil organic C can result in very large C source/sink and fluxes over the extent of these systems. The current understanding of organic C forms, fluxes and distribution in grazed systems in the Australian tropical soils is poor compared to the extensively studied croplands. Nevertheless a review of the fragmented knowledge base does provide some insights into system function.

1. Introduction of Exotic Pasture Species

The Australian grazing lands carry numerous exotic pasture species, including grasses (Table 3), medicago species, numerous acacias (shrubs and trees), siratro and stylo species. In semiarid subtropics, Chan et al. (1997) found that organic C in degraded Vertisols (Typic Chromustert) could be increased by restoration of pasture with barrel medic (*Medicago truncatula*) and Mitchell grass (*Astrebla lappacea*). Organic C concentrations increased from 1.3% to 1.6% (0–0.05m depth) after 4 years. Bruce (1965) found that inclusion of a legume, *Centrosema pubescens*, to a grass pasture (*Panicum maximum*) increased organic C from 3.66% under grass only to 4.40% under grass + legume in the top 0–0.075m depth (Figure 4) while the corresponding values in the 0.075–0.15m depth, were 2.73% and 3.14% in a humid Oxisol after 16 years; presumably, grass dry matter production was limited by N supply. It was estimated that the legume supplied almost 90 kg N ha^{-1} year^{-1} (Bruce, 1965), which may have increased pasture yield and hence C inputs to the soil.

Increase in organic C concentrations due to the inclusion of legumes in grass pastures is mainly due to N$_2$ fixed by the legume (Fisher et al., 1994), which is subsequently utilized by the associated grass species (Myers and Robbins, 1991). Thus, application of N fertilizer to grass pasture increases both the pasture growth and organic C concentration in the soil (Graham, 1987).

It is a common practice to use N fertilizer in intensively grazed pastures for the dairy industry. Significant additions of nitrogen fertilizer in the dairy pastures increased soil organic C concentrations up to the N application of 150 kg N ha^{-1} year^{-1} (Cowan et al., 1995) although organic C declined when fertilizer N rates were increased above 150 kg N ha^{-1} year^{-1} to as high as 600 kg N ha^{-1} year^{-1} over six years, presumably due to reduced plant biomass production from moisture stress, and hence lower C inputs to the soil.

The pastures typically consisting of *Setaria* and *Pennisetum clandestinum* species are fertilized with both nitrogen and phosphorus (Kerr, 1993). There have been significant losses of soil organic C in surface soils in some of these systems which are now typically 50 to 120 years post tree clearing. Typical organic carbon concentrations are now only 20 to 75% than those of the natural systems.

The native high N status of soils cleared from *A. harpophylla* and other leguminous native vegetation have been used to establish improved grass pastures (Graham et al., 1981), e.g., buffel grass (*Cenchrus cilliaris*), *Chloris gayana*, *Paspalum spp* and *Panicum spp*. For example, a Vertisol cleared from *A. harpophylla* maintained original organic C concentrations under *P. maximum* for at least 11 years; in fact, there was a general trend in increase in organic C with the period of pasture growth (Skjemstad et al., 1994). However, buffel grass or green panic pasture productivity declines after a certain period when N supply becomes limited due to N immobilized in roots and standing biomass (Myers and Robbins, 1991). Nitrogen application or inclusion of a legume restores pasture productivity (Graham, 1987), therefore, it may increase soil organic C concentration.

Woody *Stylosanthes spp.* and *Leuceana spp.* are commonly established to improve the nutrition value of native and improved grass pastures. They are commonly established on soils of good phosphorus status. The majority of legumes established tend to be perennial exotic species suited to grazing. The

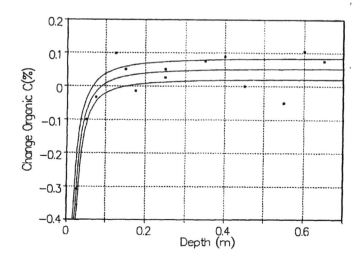

Figure 6. Proportional changes in organic C concentrations with soil depth under pasture about 10 years after establishment of woody legumes (*Stylosanthes* and *Leuceana*) relative to that under native pasture without legumes. (Adapted from Carter et al., 1998.)

currently available species are generally limited to wetter areas. There is potential to increase the areas of these legumes with greatest potential in areas where the vegetation was historically a woody legume such as *A. harpophylla*. An examination of some old woody legume stands (> 10 years) showed a decline in organic carbon concentration in the surface layers but an increase in carbon concentration at depth (Carter et al., 1998) (Figure 6). This indicates that soil under pasture should be sampled to at least to 1m depth for organic C dynamics (sink, source, and fluxes) similar to the cropping lands.

2. Tree Clearing

Tree clearing is perhaps the most recognizable change in the grazing ecosystems. In Queensland alone the current tree clearing rate is about 262,000 ha year[-1] and has probably been this high or higher for the previous 30 years.

The resultant soil organic C concentration depends on the C inputs from the net primary production of trees verses the following pure C_4 grass pastures, litter quality (lignin / nitrogen ratios), grazing removal of biomass, and soil temperature. In woodland, even at high stocking rates the tree canopy is retained to shade the soil while in the grazed situation almost the entire ground cover may be removed with overgrazing, thus causing high temperatures as well as exposing the surface for soil erosion (Beckman, 1991). Soil surface temperature moderation, as a function of system biomass is an important component of the CENTURY model (Parton et al., 1996). These temperature changes may extend deeper into the soil profile (Lewis and Wang, 1998).

The ability to restore soil organic C in grazing lands is unknown although the transition from cropped land to grazed land usually results in significant soil organic C storage (Dalal et al., 1991b; Chan et al., 1997). The C input by a perennial woody legume into the grazing system appears to increase the soil organic C-store by storing it at lower depths (Carter et al., 1998). In higher rainfall regions in the tropics

(>1000 mm per annum), for example Brazil and Hawaii, planting of deep rooted grasses after tree clearing has resulted in SOM increases (Townsend and Vitousek, 1995; Neil et al., 1997). Unfortunately, very little area of the Australian tropics receives rainfall in excess of 1000 mm per annum.

Tree planting is a possible mechanism for increasing soil organic C. However, in degraded rainforest soils in 25 year old pastures in Costa Rica, only 3 of 11 tree species planted in an experiment produced increases in soil organic C (Fisher, 1995).

3. Fire Frequency and Stocking Rate

In analysis of eight short-term enclosed plots from grazing for a period of about 10 years but regularly burnt showed no significant difference in soil organic C (Carter et al., 1998). This suggests that grazing induced changes relative to a more natural regular burning regime may be small if grazing is conservative, although charcoal C is likely to accumulate in soil. Modification of fire regime may directly and indirectly affect soil organic C concentration. If reduction in fire frequency leads to increasingly woody vegetation then soil organic C may increase due to changes in microclimate and inputs of more lignified litter, or decrease if utilization of plant biomass by animals increases.

It is not a common practice to adopt land management strategies specifically aimed at building soil organic C or soil fertility in grazing lands. It is likely that reductions in stocking rate to maintain ground cover and a desirable species composition also positively impact on soil organic C through increased root and litter biomass and cooler soil temperatures. Reducing stocking rates without increasing fire frequency would appear a logical way to increase soil organic C storage although case studies indicating the actual C changes in the entire soil profile are lacking.

V. Potential Carbon Sequestration in the Australian Tropical Soils

Carbon emission and sink calculations have three major components: area of change, age of activity, and rate of change in C per unit area. Each of these components currently has considerable uncertainty (Gifford et al., 1992; NGGIC, 1997b). There have never been good statistics or mapping of timing and location of land use change. The age and area of C changes could potentially be mapped at high resolution from LANDSAT satellite imagery, which dates from the early 1970s. To date, most mapping has been used to determine the current (after 1990) land use or potential use rather than providing spatially explicit time series data.

At least 25 years of history is needed to quantify the area and timing of land use change so the majority of soil organic C changes can be accounted for. For many land use changes, delayed release of soil organic C continues at a rate which declines exponentially with time so a large proportion of emissions occur within a 20 to 30 year time frame. The initial soil organic C amounts and rates of decline are poorly known for most land use changes in most Australian climate regimes. The effect of tree clearing for grazing and post clearing land management is a significant type of land use change where historic areas, ages and soil organic C fluxes are unknown or poorly known. This leads to great uncertainty, 75 to 80% around a potentially large amount of carbon flux (14.4 Tg C for 1995) in the national inventory of CO_2 emissions (NGGIC, 1997a, 1997b).

Soil organic C loss under cultivation is better known with C decomposition curves established for a number of soil types within one zone of similar climate (Dalal et al., 1986b). However, to estimate C losses from all cultivated lands it is necessary to generalize these relationships across climate and clay content and then apply these relationships to areas of cultivation of known age. The lack of suitably detailed soil attribute maps, and cultivation age maps makes this process difficult.

Table 9. Potential for C sequestration in cropping soils and improved grazing lands (143 m ha) in the Australian tropics

Cropping system[a]	Cropped area (m ha)	C sink (Tg C)	Pasture system[e]	Pasture area (m ha)	C sink (Tg C)
From CT to ZT	2.5	11511119	Midgrass	19.5 (6.5)	147 (69)
From CT to grass	1.25		Tropical tallgrass	9.5 (3.0)	86 (41)
+ legume ley (4 years)			Monsoon tallgrass	7.9 (1.2)	48 (11)
Fertilizers	(2.5)[b]		Spinifex	6.3 (2.1)	34 (23)
Sugarcane trash retention[c]	0.50				
Other systems[d]	0.75				
Total	5			43.2 (12.8)	315 (144)

[a]Estimated from Dalal (1989) and Dalal et al. (1995); [b]fertilizer applied to ZT area; [c]estimated from Sutton et al. (1996); [d]including crop residue retention in cotton farming (Conteh et al., 1998); [e]Adapted from Ash et al. (1995), assuming improvement in pastures from B class to A class; the estimates from C class to B class are shown in parentheses.

Decomposition of SOM results in CO_2 and CH_4 emissions from soil. Detwiler and Hall (1988) estimated that CO_2 emissions from tropical soils were 110–250 Tg C year[-1]; Australian tropics accounting for about 5–10%. The emissions of CH_4 related to agricultural activity follow: 60–140 Tg C year[-1] from rice cultivation (a small fraction from Australia), 40–160 Tg C year[-1] from wetlands (about 1% from Australia), 30–70 Tg C year[-1] from landfills (about 2% from Australia), 65–100 Tg C year[-1] from ruminants (about 10% from Australia), and 50–100 Tg C year[-1] from biomass burning (Bouwman, 1990). Land clearing in the Australian tropics removes about 15 Tg C year[-1] (0.3 m ha year[-1] x 100 t dry matter ha[-1] x 0.5 C).

The size of these organic C emissions and sinks might be determined on a statistical basis by detailed sampling of an adequate number of paired sites of known age. To produce maps of emissions, calibrated models which account for plant growth, climate, land management, and effect of soil properties on organic C loss rates would need to be combined with maps detailing the temporal and spatial nature of land use change and maps of soil properties. Based on the land area cultivated, it is estimated that the improved practices have the potential to increase C sinks by about 19 Tg C in the soil (Table 9) over 10 years or C sequestration rate of 0.4 t C ha[-1] yr[-1]

In grazing lands other uncertainties relate to changes in grazing pressure, fire regime and woodland thickening (Burrows, 1995). These processes occur at very large scales and together effect the entire land area of tropical Australia. It may be that most grazing and fire induced changes have already impacted and that grazing land soils have largely re-equilibrated over the last 100 or more years of European settlement. Even very small ongoing changes in soil organic C become significant sources or sinks of atmospheric C if they occur over millions of hectares. For example, Ash et al. (1995) estimated that improved grazing management (from class B to class A, by reducing stocking rate) to increase the perennial grass component could sequester approximately 320 Tg C into the top 10 cm of soil and further 140 Tg C by rehabilitating degraded pasture lands from class C to class B (Table 9). Since P and S deficiencies are widespread in tropical savanna and woodlands (Ahern et al., 1994), P and S fertilization to these lands can increase C sink by 0.5 t C ha[-1] yr[-1], similar to that southern Australia (Gifford et al., 1992). This practice can increase C sequestration by 28 Tg C yr[-1] in over 56 m ha or 280 Tg C over ten years since Carter et al. (1998) found a positive relationship between available P and soil organic C contents in Queensland soils. Scurlock and Hall (1998) estimated that tropical grasslands and savanna of the world could increase annual soil C sink by 500 Tg C.

While the magnitude and direction of these changes is currently uncertain, the small amount of data available suggests the need for spatially and temporally comprehensive studies (Carter et al., 1998). Sampling procedures adapted to the detection of small changes over large areas need to be refined, and there is an urgent need for a permanent soil-monitoring network.

It may be difficult to separate the results of historic land use change from climate change and CO_2 fertilization. A combination of measurement and process modeling will be needed to differentiate the effects of these processes.

VI. Summary and Conclusions

The challenge for sustainable land management is to balance the needs for the production of food and fiber and other land uses with the necessity to maintain sufficient land cover to minimize the adverse impact of soil degradation processes while maintaining or enhancing the organic C sink in soil to mitigate greenhouse gases emissions.

Current agricultural practices for food and fiber production in the Australian tropics have inevitably led to the loss of soil organic matter from soil. Until recently, both food and fiber production systems involved intensive cultivation, which resulted in lack of plant residue soil cover; utilizing the native soil fertility, which led to declining yields and mainly using monoculture systems, which again have resulted in decreasing yields and hence lower C inputs to the soil (Dalal and Mayer, 1986b).

Sustainable cropping practices are, therefore, essential for maintaining crop productivity while maintaining or enhancing organic matter in soil (Cogle et al., 1995; Dalal et al., 1995; Bell et al., 1997; Chan et al., 1997). Except for the introduction of grass + legume pastures or fertilized grass pastures, little is known about other soil and crop management practices to enhance organic matter in soil. In some soils, no-till practices may increase SOM, but in the semiarid tropics and subtropics, rainfall limits the amounts of C inputs from annual crops in excess of SOM losses from decomposition and other soil degradation processes.

Land clearing for grazing is still occurring in the Australian tropics, often resulting in substantial surface SOM losses. Grazing of tropical grasslands and savannas has been practiced on an *ad hoc* basis, often incompatible with the safe carrying capacity of the land, and this is made worse by extreme climate variability. This results in overgrazing (or overstocking), and pasture degradation and disappearance of surface plant residue cover, degraded lands and loss of organic C from soil (Ash et al., 1995).

Introduction of exotic pasture species, especially legumes have enhanced pasture productivity of grassland pastures and often has resulted in increased organic C in soil (Bruce, 1965). Improving pasture productivity through fertilizer inputs, renovation, reducing stocking rates, and eliminating or reducing fire frequency could lead to increase in pasture production and hence increase in organic C in the soil.

Land use change and forestry account for about 12% of total net greenhouse gas emissions in Australia (NGGIC, 1997b). However, there is a great uncertainty about this figure since the data on C sink and source sizes and their turnover rates in the Australian tropical soils, especially in grazing lands are scarce. Studies should be undertaken to quantify the size of the aboveground biomass returned to the soil, the size of the belowground C including roots and charcoal, and the rates of turnover (C fluxes) of both belowground and aboveground C pools. Since C components differ in their turnover rates, understanding of the nature of these components through improved techniques will lead to improved management practices and hence to enhanced C sink to mitigate C emissions in the atmosphere. This approach should also lead to improved simulation of C dynamics in the tropical soils and better prediction of the short and long term consequences of land and vegetation management practices on soil organic matter management in the Australian subtropical and tropical regions.

Acknowledgments

We thank Cecelia McDowall and Michelle Perry for assistance with the literature search.

References

Ahern, C.R., P.G. Shields, N.G. Enderlin, and D.E. Baker. 1994. Soil fertility of central and north-east Queensland grazing lands. Department of Primary Industries, Queensland, Information Series Q194065, Brisbane.

Ash, A.J., S.M. Howden, and J.G. McIvor. 1995. Improved rangeland management and its implications for carbon sequestration. p. 19-20. In: N.E. West (ed.), Rangelands in a Sustainable Biosphere. Proceedings of the Fifth International Rangeland Congress. Salt Lake City, UT.

Ayanaba, A., S.B. Tuckwell, and D.S. Jenkinson. 1976. The effects of clearing and cropping on the organic reserves and biomass of tropical forest soils. *Soil Biol. Biochem.* 8:519-525.

Balesdent, J., G.H. Wagner, and A. Mariotti. 1988. Soil organic matter turnover in long-term field experiments as revealed by carbon-13 natural abundance. *Soil Sci. Soc. Am. J.* 52:118-124.

Bartholomew, W.V. and D. Kirkham. 1960. Mathematical descriptions and interpretations of culture induced soil nitrogen changes. Transactions, *7th Int. Congress of Soil Science*, Communications Vol. 111:471-477.

Beckman, R. 1991. Measuring the cost of clearing. *Ecos.* 67:13-15.

Bell, M.J., B.J. Bridge, G.R. Hatch, and D.N. Orange. 1997. Physical rehabilitation of degraded krasnozems using ley pastures. *Aust. J. Soil Res.* 35:1093-1113.

Bohn, H.L. 1976. Estimate of organic carbon in world soils. *Soil Sci. Soc. Am. J.* 40:468-470.

Bolin, B. 1986. How much CO_2 will remain in the atmosphere? The carbon cycle and projections for the future. p. 93-156. In: B. Bolin, B.R. Döös, J. Jäger, and R.A. Warrick (eds.), *The Greenhouse Effect, Climatic Change and Ecosystems*. SCOPE Vol. 29. John Wiley & Sons, New York.

Bonde, T.A., B.T. Christensen, and C.C. Cerri. 1992. Dynamics of soil organic matter as reflected by natural ^{13}C abundance in particle size fractions of forested and cultivated Oxisols. *Soil Biol. Biochem.* 24:275-277.

Bouwman, A.F. 1989. Modelling soil organic matter decomposition and rainfall erosion in two tropical soils after forest clearing for permanent agriculture. *Land Degradation and Rehabilitation* 1:125-140.

Bouwmann, A.F. 1990. Land use related sources of greenhouse gases - present emissions and possible future trends. p. 154-164. In: *Land Use Policy*. Butterworth & Co.

Bowman, A.M., D.J. Munnich, K.Y. Chan, and J. Brockwell. 1997. Factors associated with density of Mitchell grass pastures in north-western New South Wales. *Rangeland J.* 19:40-56.

Bridge, B.J. and M.J. Bell. 1994. Effect of cropping on the physical fertility of Krasnozems. *Aust. J. Soil Res.* 32:1253-1273.

Bruce, R.C. 1965. Effect of *Centrosema pubescens* Benth. on soil fertility in the humid tropics. *Queensland J. Agric. Anim. Sci.* 22:221-226.

Buringh, P. 1984. Organic carbon in soils of the world. p. 91-109. In: G.M. Woodwell (ed.), *The Role of Terrestrial Vegetation in the Global Carbon Cycle. Measurement by Remote Sensing*. SCOPE. Vol. 23. John Wiley & Sons, New York.

Burrows, W.H. 1995. Greenhouse revisited- an alternative viewpoint on land use change and forestry from a Queensland perspective. *Climate Change Newsletter* 7:6-7.

Capriel, P., P. Harter, and D. Stephenson. 1992. Influence of management on the organic matter of a mineral soil. *Soil Sci.* 153: 122-128.

Carter, J.O., S.M. Howden, K.A. Day, and G.M. McKeon. 1998. Soil carbon, nitrogen and phosphorus and biodiversity in relation to climate change. In: Evaluation of the Impact of Climate Change on Northern Australian Grazing Industries, Final Report for the Rural Industries Research and Development Corporation on project DAQ 139A.

Chan, K.Y. 1997. Consequences of changes in particulate organic carbon in Vertisols under pasture and cropping. *Soil Sci. Soc. Am. J.* 61:1376-1382.

Chan, K.Y., W.D. Bellotti, and W.P. Roberts. 1988. Changes in surface soil properties of Vertisols under dryland cropping in a semiarid environment. *Aust. J. Soil Res.* 26:509-518.

Chan, K.Y., A.M. Bowman, and J.J. Friend. 1997. Restoration of soil fertility of degraded vertisol using a pasture including a native grass (*Astrebla lappacea*). *Tropical Grasslands* 31:145-155.

Cerri, C.C., C. Feller, J. Balesdent, R.L. Victoria, and A. Plenecassagne. 1985. Application du tracage isotopique natural en ^{13}C á l'étude de la dynamique de la matiére organique dans les sols. *C.R. Aca. Sci., Paris.* 300:423-428.

Christensen, B.T. 1992. Physical fractionation of soil and organic matter in primary particle size and density separates. Adv. *Soil Sci.* 20:1-90.

Cogle, A.L., J. Littlemore, and D.H. Heiner. 1995. Soil organic matter changes and crop responses to fertiliser under conservation cropping systems in the semi-arid tropics of North Queensland, Australia. *Aust. J. Exp. Agric.* 35:233-237.

Conteh, A., G.J. Blair, and I.J. Rochester. 1998. Soil organic carbon fractions in Vertisol under irrigated cotton production as affected by burning and incorporating cotton stubble. *Aust. J. Soil Res.* 36:655-667.

Cook, G.D., H.B. So, and R.C. Dalal. 1992. Structural degradation of Vertisols under continuous cultivation. *Soil Tillage Res.* 22:47-64.

Cowan , R.T., K.F. Lowe, W. Ehrlich, P.C. Upton, and T.M. Bowdler. 1995. Nitrogen-fertilised grass in a subtropical dairy system 1. Effect of level of nitrogen fertiliser on pasture yield and soil chemical characteristics. *Aust. J. Exp. Agric.* 35:125-135.

Dalal, R.C. 1989. Long-term effects of no-tillage, crop residue and nitrogen application on properties of a vertisol. *Soil Sci. Soc. Am. J.* 53:1511-1515.

Dalal, R.C. 1992. Long-term trends in total nitrogen of a vertisol subjected to zero-tillage, nitrogen application and stubble retention. *Aust. J. Soil Res.* 30:223-231.

Dalal, R.C. 1998. Soil microbial biomass-what do the numbers really mean? *Aust. J. Exp. Agric.* Vol. 38.

Dalal, R.C., P.A. Henderson, and J.M. Glasby. 1991a. Organic matter and microbial biomass in a Vertisol after 20 yr of zero-tillage. *Soil Biol. Biochem.* 23:435-441.

Dalal, R.C. and R.J. Henry. 1988. Cultivation effects on carbohydrate contents of soil and soil fractions. *Soil Sci. Soc. Am. J.* 52:1361-1365.

Dalal, R C. and A.J. King. 1994. Simulation of soil organic matter trends in Vertisols under continuous cultivation and cereal cropping using the CENTURY model. In: Organic Matter in Soils, Sediments and Waters. International Humic Substances Society, Adelaide, Dec. 1994.

Dalal, R.C. and R.J. Mayer. 1986a. Long-term trends in fertility of soils under continuous cultivation and cereal cropping in Southern Queensland. I. Overall changes in soil properties and trends in winter cereal yields. *Aust. J. Soil Res.* 24:265-279.

Dalal, R.C. and R.J. Mayer. 1986b. Long-term trends in fertility of soils under continuous cultivation and cereal cropping in Southern Queensland. II. Total organic carbon and its rate of loss from the soil profile. *Aust. J. Soil Res.* 24:281-292.

Dalal, R.C. and R.J. Mayer. 1986c. Long-term trends in fertility of soils under continuous cultivation and cereal cropping in Southern Queensland. III. Distribution and kinetics of soil organic carbon in particle-size fractions. *Aust. J. Soil Res.* 24:293-300.

Dalal, R.C. and R.J. Mayer. 1986d. Long-term trends in fertility of soils under continuous cultivation and cereal cropping in Southern Queensland. IV. Loss of organic carbon from different density fractions. *Aust. J. Soil Res.* 24:301-309.

Dalal, R.C. and R.J. Mayer. 1986e. Long-term trends in fertility of soils under continuous cultivation and cereal cropping in Southern Queensland. V. Rate of loss of total nitrogen from the soil profile and changes in carbon-nitrogen ratios. *Aust. J. Soil Res.* 24:493-504.

Dalal, R.C. and R.J. Mayer. 1987a. Long-term trends in fertility of soils under continuous cultivation and cereal cropping in Southern Queensland. VI. Loss of total N from different particle-size and density fractions. *Aust. J. Soil Res.* 25: 83-93.

Dalal, R.C. and R.J. Mayer. 1987b. Long-term trends in fertility of soils under continuous cultivation and cereal cropping in Southern Queensland. VII. Dynamics of nitrogen mineralisation potentials and microbial biomass. *Aust. J. Soil Res.* 25:461-472.

Dalal, R.C. and M.E. Probert. 1997. Nutrient depletion. p. 42-63. In: A.L. Clarke and P.J. Wylie (eds.), Sustainable agriculture for sub-tropical Australia. Department of Primary Industries, Brisbane, Queensland.

Dalal, R.C., W.M. Strong, E.J. Weston, and J. Gaffney. 1991b. Sustaining multiple production systems. 2. Soil fertility decline and restoration of cropping lands in subtropical Queensland. *Tropical Grasslands* 25:173-180.

Dalal R.C., W.M. Strong, E.J. Weston, J.E. Cooper, K.J. Lehane, A.J. King, and C.J. Chicken. 1995. Sustaining productivity of a vertisol at Warra, Queensland, with fertilisers, no-till or legumes .1. Organic matter status. *Aust J. Exp. Agric.* 37:903-913.

Detwiler, R.P. and C.A.S. Hall. 1988. Tropical forests and the global carbon cycle. *Science* 239:42-27.

Du Pont, G. 1997. Tree microclimate effects on understorey vegetation: Their impact on spatial and temporal modelling of pasture production in Queensland. Masters Thesis, University of Queensland, Brisbane.

Elliott, E.T. 1986. Aggregate structure and carbon, nitrogen and phosphorus in native and cultivated soils. *Soil Sci. Soc. Am. J.* 50: 627-633.

Eswaran, H., E. van den Berg, and P. Reich. 1993. Organic carbon in soils of the world. *Soil Sci. Soc. Am. J.* 57:192-194.

FAO. 1992. FAO Production Yearbook. Vol.46. Food and Agricultural Organisation, Rome.

Fisher, R.F. 1995. Amelioration of degraded rain forest soils by plantations of native trees. *Soil Sci. Soc. Am. J.* 59:544-549.

Fisher, M.J., I.M. Rao, M.A. Ayarza, C.E. Lascano, J.I. Sanz, R.J. Thomas, and R.R. Vera. 1994. Carbon storage by introduced deep-rooted grasses in the South American savannas. *Nature* (London). 371:236-238.

Ford, G.W., D.J. Greenland, and J.M. Oades. 1969. Separation of the light fraction from soils by ultrasonic dispersion in halogenated hydrocarbons containing a surfactant. *J. Soil Sci.* 20:291-296.

Gifford, M.R., N.P. Cheney, J.C. Noble, J.S. Russell, A.B. Wellinton, and C. Zammit. 1992. Australian land use, primary production of vegetation and carbon pools in relation to atmospheric carbon dioxide concentration. *Bureau Rural Resources Proceedings* 14:151-187.

Gillman, G.P.,D.F. Sinclair, R. Knowlton, and M.G. Keys. 1985. The effects on some soil chemical properties of the selective logging of a north Queensland rainforest. *For. Ecol. Manage.* 12:195-214.

Gokhale, N.G. 1959. Soil nitrogen status under continuous cropping and with manuring in the case of unshaded tea. *Soil Sci.* 87:331-333.

Grace, P.R., I.C. McRae, and R.J.K. Myers. 1992. Factors influencing the availability of mineral nitrogen in clay soils of the brigalow (*Acacia harpophylla*) region of central Queensland. *Aust. J. Agric. Res.* 43:1197-1215

Graham, T.W.G. 1987. The effect of renovation practices on nitrogen cycling and productivity on rundown of buffel grass pasture. PhD Thesis, University of Queensland.

Graham, T.W.G., A.A. Webb, and S.A. Waring. 1981. Soil nitrogen status and pasture productivity after clearing of brigalow (*Acacia harpophylla*). *Aust. J. Agric. Anim. Husb.* 21:109-118.

Greenland, D.J. and G.W. Ford. 1964. Separation of partially humified organic materials from soils by ultrasonic dispersion. *Trans. 8th Int. Congr. Soil Sci.* 3:137-148.

Gregorich, E.G. and H.H. Janzen. 1996. Storage of soil carbon in the light fraction and macroorganic matter. p. 167-190. In: M.R. Carter and B.A. Stewart (eds.), *Structure and Organic Matter Storage in Agricultural Soils*. Advances in Soil Science, Lewis Publishers, New York.

Haas, H.J., C.E. Evans, and E.F. Miles. 1957. Nitrogen and Carbon Changes in Great Plains Soils as Influenced by Cropping and Soil Treatments. U.S. Dep. Agric. Tech. Bull. No.1164.

Hobbs, J.A. and C.A. Thompson. 1971. Effect of cultivation on the nitrogen and organic carbon contents of a Kansas Argiustoll (Chernozem). *Agron. J.* 63: 66-68.

Holford, I.C.R. 1990. Effects of 8-year rotations of grain sorghum with lucerne, annual legume, wheat and long fallow on nitrogen and organic carbon on two contrasting soils. *Aust. J. Soil Res.* 28:277-291.

Hossain, S.A., R.C. Dalal, S.A. Waring, W.M. Strong, and E.J. Weston. 1996. Comparison of legume-based cropping systems at Warra, Queensland. I. Soil nitrogen and organic carbon accretion and potentially mineralisable nitrogen. *Aust. J. Soil Res.* 34:273-287.

Howden, S.M., G.M. McKeon, P.J. Reyenga, J.C. Scanaln, J.O. Carter, and D.H. White. 1995. Management options to reduce greenhouse gas emissions from tropical beef grazing systems. Report to Rural Industries Research and Development Corporation.

Insam, H., D. Parkinson, and K.H. Domsch. 1989. Influence of macro-climate on soil microbial biomass. *Soil Biol. Biochem.* 21:211-221.

Jenkinson, D.S., D.E. Adams, and A. Wild. 1991. Model estimates of CO_2 emissions from soil in response to global warming. *Nature*. (London). 351:304-306.

Jenkinson, D.S. and A. Ayanaba. 1977. Decomposition of carbon-14 labelled plant material under tropical conditions. *Soil Sci. Soc. Am. J.* 41:912-915.

Jenkinson, D.S. and J.H. Rayner. 1977. The turnover of soil organic matter in some of the Rothamsted classical experiments. *Soil Sci.* 123:298-305.

Jenkinson, D.S., D.D. Harkness, E.D. Vance, D.E. Adams, and A.F. Harrison. 1992. Calculating net primary production and annual input of organic matter to soil from the amount and radiocarbon content of soil organic matter. *Soil Biol. Biochem.* 24:295-308.

Kerr, D. 1993. Queensland Dairy Farm Study 1990-91. Department of Primary Industries Queensland Information Series Q193015.

Ladd, J.N., M. Amato, L. Jocteur Monrozier, and M. Van Gestel. 1990. Soil Microhabitats and carbon and nitrogen metabolism. *Proc. 14th Intern. Cong. Soil Sci.* 3: 82-87.

Lal, R. and B.T. Kang. 1982. Management of organic matter in soils of the tropics and subtropics. *12th ISSS Congress, New Delhi Symposia.* 1:152-178.

Lal, R. 1989. Conservation tillage for sustainable agriculture: tropics versus temperate environments. *Adv. Agron.* 42:85-191.

Lathwell, D.J. and D.R. Bouldin. 1981. Soil organic matter and soil nitrogen behaviour in cropped soils. *Trop. Agric.* (Trinidad). 58:341-348.

Lewis, T.J. and K. Wang. 1998. Geothermal evidence for deforestation induced warming: implications for the climatic impact of land development. *Geophys. Res. Letters* 25:535-538.

Long, S.P., E.G. Moya, S.K. Imbabma, A. Kamnalrut, M.T.F. Piedade, J.M.O. Scurlock, Y.K. Shen, and D.O. Hall. 1989. Primary productivity of natural grass ecosystems of the tropics: a reappraisal. *Plant Soil*. 115:155-166.

Maggs, J. and B. Hewett. 1993. Organic C and nutrients in surface soils from some primary rainforests, derived grasslands and secondary rainforests on the Atherton tableland in North East Queensland. *Aust. J. Soil Res.* 31:343-350.

Martin, A., A. Mariotti, J. Balesdent, P. Lavelle, and R. Vuattoux. 1990. Estimate of organic matter turnover rate in a savanna soil by ^{13}C natural abundance measurements. *Soil Biol. Biochem.* 22:517-523.

Miles, R.L. 1990. The land degradation situation of the mulga lands of south west Queensland. p. 78-86. In: Arid lands Administrators Conference, Charleville, April 1990.

Moody, P.W. 1994. Chemical fertility of Krasnozems: a review. *Aust. J. Soil Res.* 32:1015-1041.

Myers, R.J.K. and G.B. Robbins. 1991. Sustaining production pastures in tropics. 5. Maintaining productive sown grass pastures. *Tropical Grasslands* 25:104-110.

Neill, C., J.M. Melillo, P.A. Steudler, C.C. Cerri, J.F.L. De Moraes, M.C. Piccolo, and M. Brito. 1997. Soil carbon and nitrogen stocks following forest clearing for pasture in the southwestern Brazilian Amazon. *Ecological Applications*. 7:1216-1225.

NGGIC. 1997a. Australian Methodology for the Estimation of Greenhouse Gas Emissions and Sinks, Workbook 4.2, Carbon Dioxide from the Biosphere. Environment Australia.

NGGIC. 1997b. National Greenhouse Gas Inventory Committee. National Greenhouse Gas Inventory 1995 Australia. Environment Australia.

Oades, J.M. 1994. Krasnozems-Organic Matter. *Aust .J. Soil Res.* 33:43-57.

Obi, A.O. 1989. Long-term effects of the continuous cultivation of a tropical Ultisol in south-western Nigeria. *Exp. Agric.* 25:207-215.

Odell, R.T., S.W. Melsted, and W.M. Walker. 1984. Changes in organic carbon and nitrogen of morrow plot soils under different treatments, 1904–1974. *Soil Sci.* 137:160-171.

Parton, W.J., D.S. Ojima, and D.S. Schimel. 1996. Models to evaluate soil organic matter storage and dynamics. p. 421-448. In: M.R. Carter and B.A. Stewart (eds,), *Structure and Organic Matter Storage in Agricultural Soils*. Lewis Publishers, Boca Raton, FL.

Parton, W.J., R.L. Sanford, P.A. Sanchez, and J.W.B. Stewart. 1989. Modelling soil organic matter dynamics in tropical soils. p. 153-171. In: D.C. Coleman, J.M. Oades, and G. Uehara (eds.), *Dynamics of Soil Organic Matter in Tropical Ecosystems*. NifTAL Project, University of Hawaii, Hawaii.

Parton, W.J., D.S. Schimel, C.V. Cole, and D.S. Ojima. 1987. Analysis of factors controlling soil organic matter levels in Great Plains grasslands. *Soil Sci. Soc. Am. J.* 51:1173-1179.

Paul, E.A., R.F. Follett, S.W. Leavitt, A. Halvorson, G.A. Peterson, and D.J. Lyon. 1997. Radiocarbon dating for soil organic matter pool sizes and dynamics. *Soil Sci. Soc. Am. J.* 61:1058-1067.

Post, W.M., W.R. Emmanuel, P.J. Zinke, and A.G. Stangenberger. 1982. Soil carbon pools and world life zones. *Nature* (London) 298:156-159.

Post, W.M., A.W. King, and S.D. Wullschleger. 1996. Soil organic matter models and global estimates of soil organic carbon. p. 201-222. In: D.S. Powlson, P. Smith, and J.U. Smith (eds.), *Evaluation of Soil Organic Matter Models Using Existing Long-term Datasets*. Springer, New York.

Powlson, D.S., P.C. Brookes, and B.T. Christensen. 1987. Measurement of soil microbial biomass provides an early indication of changes in total soil organic matter due to straw incorporation. *Soil Biol. Biochem.* 19:159-164.

Probert, M.E., B.A. Keating, J.P. Thompson, and W.J. Parton. 1995. Modelling water, nitrogen, and crop yield for a long-term fallow management experiment. *Aust. J. Exp. Agric.* 35:941-950.

Rasmussen, P.E. and H.P. Collins. 1991. Long-term impacts of tillage, fertilisation, and crop residues on soil organic matter in temperate semi-arid regions. *Adv. Agron.* 45:93-134.

Rose, C.W. and Dalal, R.C. 1988. Erosion and runoff of nitrogen. p. 212- 235. In: J.R. Wilson (ed,), *Advances in Nitrogen Cycling in Agricultural Ecosystems*. CAB International, Wallingford, U.K.

Russell, J.S. 1981. Models of long term soil organic nitrogen change. p. 222-232. In: M.J. Frissel and J.A. van Veen (eds.), Simulation of Nitrogen Behaviour of Soil-plant Systems, Centre for Agricultural Publishing and Documentation, Wageningen.

Scurlock, J.M.O. and D.O. Hall. 1998. The global carbon sink: a grassland perspective. *Global Change Biol.* 4:229-233.

Skjemstad, J.O., V.R. Catchpoole, and R.P. Le Feuvre. 1994. Carbon dynamics in Vertisols under several crops as assessed by natural abundance ^{13}C. *Aust .J. Soil Res.* 32:311-321.

Skjemstad, J.O., P. Clarke, J.A. Taylor, J.M. Oades, and S.G. McClure. 1996. The chemistry and nature of protected carbon in soil. *Aust. J. Soil Res.* 34:251-71.

Skjemstad, J.O. and R.C. Dalal. 1987. Spectroscopic and chemical differences in organic matter of two Vertisols subjected to long periods of cultivation. *Aust. J. Soil Res.* 25:323-335.

Skjemstad, J.O., R.C. Dalal, and P.F. Barron. 1986. Spectroscopic investigations of cultivation effects on organic matter of Vertisols. *Soil Sci. Soc. Am. J.* 50: 354-359.

Skjemstad, J.O., R.P. Le Feuvre, and R.E. Prebble. 1990. Turnover of soil organic matter under pasture as determined by ^{13}C natural abundance. *Aust. J. Soil Res.* 28:267-277.

Spain, A.V. 1990. Influence of environmental conditions and some soil chemical properties on the carbon and nitrogen contents of some tropical Australian rainforest soils. *Aust. J. Soil Res.* 28:825-839.

Spain, A.V., G.L. Unwin, and D.F. Sinclair. 1989. Soils and vegetation of three rainforest sites in tropical north-eastern Queensland. CSIRO Aust. Div. Soils Rep. No. 105.

Sparling, G.P. 1992. Ratio of microbial biomass carbon to soil organic carbon as a sensitive indicator of changes in soil organic matter. *Aust. J. Soil Res.* 30:195-207.

Srivastava, S.C. and J.S. Singh. 1989. Effect of cultivation on microbial carbon and nitrogen in dry tropical forest soil. *Biol. Fert. Soils.* 8:343-348.

Standley, J., H.M. Hunter, G.A. Thomas, G.W. Blight, and A.A. Webb. 1990. Tillage and crop residue management affect Vertisol properties and grain sorghum growth over seven years in the semi-arid subtropics. 2. Changes in soil properties. *Soil Tillage Res.* 18:367-388.

Sutton, M.R., A.W. Wood, and P.G. Saffigna. 1996. Long term effects of green cane trash retention on Herbert river soils. p. 178-180. In: J.R. Wilson, D.M. Hogarth, and J.A. Campbell (eds.), Sugar 2000 Symposium: Sugarcane- Research Towards Efficient and Sustainable Production. CSIRO, Division of Tropical Crops and Pastures, Brisbane.

Tiessen, H. and J.W.B. Stewart. 1983. Particle size fractions and their use in studies of soil organic matter. II. Cultivation effects on organic matter composition in size fractions. *Soil Sci. Soc. Am. J.* 47:509-514.

Tothill, J.C. and C. Gillies. 1992. The Pasture Lands of Northern Australia: Their Condition, Productivity and Sustainability. Tropical Grasslands Society of Australia Occasional Publication No. 5.

Townsend, A.R. and P.M. Vitousek. 1995. Soil organic matter dynamics along gradients in temperature and land use on the island of Hawaii. *Ecology* 76:721-733.

Turton, S.M. and G.J. Sexton. 1996. Environmental gradients across four rainforest-open forest boundaries in northeastern Australia. *Aust. J. Ecol.* 21:245-254.

Van Veen, J.A. and E.A. Paul. 1981. Organic C dynamics in grassland soils. 1. Background information and computer simulation. *Can. J. Soil Sci.* 61:185-201.

Vitorello, V.A., C.C. Cerri, F. Andreux, C. Feller, and R.L. Victoria. 1989. Organic matter and natural carbon-13 distribution in forested and cultivated oxisols. *Soil Sci. Soc. Am. J.* 53:773-778.

Webb, A.A., B.J. Crack, and J.Y. Gill. 1977. Studies on the gilgaied clay soils (Ug5.2) of the Highworth land system in east-central Queensland. 1. Chemical characteristics. *Queensland J. Agric. Anim. Sci.* 34:53-65.

Whitehouse, M.J. and J.W. Littler. 1984. Effect of pasture on subsequent wheat crops on a black earth soil of the Darling Downs. II. Organic C, nitrogen and pH changes. *Queensland J. Agric. Anim. Sci.* 41:13-20.

Section V.

BASIC SOIL PROCESSES AND CARBON DYNAMICS

CHAPTER 16

Soil Aggregation and C Sequestration

R. Lal

I. Introduction

Realizing potential of soil as a sink for C requires understanding of processes involved in C sequestration. One of the processes is the formation of organo-mineral complexes or soil aggregation leading to development of heterogenous structural forms or architectural units. An aggregate is defined as "a naturally occurring cluster or group of soil particles in which the forces holding the particles together are much stronger than forces between adjacent aggregates" (Martin et al., 1955). The arrangement of particles to form aggregates is facilitated by the colloidal fraction comprising clay and humus contents. The humus fraction consists of large organic molecules that bind the clay particles together into secondary particles of varying size and shape. The binding process is also facilitated by polyvalent cations (e.g., Ca^{+2}, Mg^{+2}, Al^{+3}, Mn^{+3}, Fe^{+3}). Formation of stable secondary particles or aggregates influences C sequestration by physically protecting the organic matter from microbial enzymes. In addition, it leads to development of inter-connected voids or pores. Depending on the size and shape of the primary particles, pore size distribution varies widely. It is the size distribution and architecture of these pores that regulate plant root growth, gaseous exchange between the soil and the atmosphere, nutrient diffusion, and water retention and transmission characteristics. Soil's ability to function, both for plant growth and environmental regulation, depends on the size, stability, and shape of aggregates and the pores created by them. The environmental regulatory function of the soil involves water quality, gaseous exchange between soil and the atmosphere, and C sequestration within aggregates. Therefore, aggregation or formation of organo-mineral complexes is a process as important as but much less understood than photosynthesis.

There are two distinct scientific approaches to understanding the process of aggregation. These are, (i) physio-chemical approach based on electrokinetic processes involving flocculation and cementation, and (ii) pedological approach based on soil genesis and factors of soil formation. This manuscript reviews recent advances in the physio-chemical approach and the importance of soil organic carbon (SOC) in enhancing aggregation and sequestering C in soil.

II. Physico-Chemical Model of Aggregation

The process of aggregation involves flocculation of colloidal fraction followed by cementation of floccules. Floccules are cemented into micro-aggregates that in turn are clustered together into aggregates. The basic unit of micro-aggregates is clay particles bound together with polyvalent cations (M) and SOC (Tisdall, 1996). Different possible combinations include:

(clay - M - clay)	(Eq. 1)
(clay - SOC - clay)	(Eq. 2)
(clay - M - SOC - clay)	(Eq. 3)

Table 1. Model of aggregation and major stabilizing agents for an Alfisol

Stabilizing agent	Size of aggregation (μm)
Inorganic materials, organic polymers, electrostatic bonds, coagulation	< 0.2
Microbial and fungal debris	0.2 - 2 → 2 - 20
Plant and fungal debris	2 - 20 → 20 - 50
Roots and hyphae	20 - 250 → >2000
Polysaccharides	20 - 250 → >2000

(Modified from Tisdall and Oades, 1982.)

Table 2. Model of aggregation and major stabilizing agent for an Alfisol and Mollisol

Stabilizing agent	Size of aggregation (μm)
Microbial debris, inorganic materials	< 20
Plant debris	<20 → 20 - 90
Plant fragments	20 - 90 → 90 - 250
Roots and hyphae	20 - 250 → >2000

(Modified from Oades and Waters, 1991.)

These micro-aggregates are combined together to form aggregates (Eq. 4).

$$xy(\text{clay - M - SOC - clay}) \ll [(\text{clay - M - SOC - clay})_x]_y \qquad \text{(Eq. 4)}$$

There are different types of organic cementing materials that bind the micro-aggregates into aggregates, and the nature of the organic matter may differ among soils. Tisdall and Oades (1982) proposed that in Alfisols the binding material differs among different size fractions (Table 1). The binding materials for micro-aggregates <0.2 μm comprise inorganic compounds and organic polymers cemented together through coagulation and electrostatic bonds. The binding materials include microbial and fungal debris for 0.2 to 20 μm size, plant and fungal debris for 2 to 250 μm size, and roots and hyphae for 20 to 2000 μm. The latter are also bound together by polysaccharides (Tisdall and Oades, 1982). Similar observations were made for Alfisols and Mollisols by Oades and Waters (1991) (Table 2). In contrast, the primary binding agents are oxides of Fe and Al for Oxisols (Robert and Chenu, 1992), and allophones and amorphous aluminosilicates for Andosols (Robert and Chenu, 1992). The more stable and long-lasting the aggregates, the longer the duration for which the C is sequestered in them.

There are two distinct mechanisms of protection of SOC sequestered within an aggregate. One is the physical protection due to the small size of the pores in which the labile fraction (e.g., polysaccharides) may be located. A vast proportion of surfaces coated with SOC are physically protected because of inaccessibility to microorganisms and their enzymes (Adu and Oades, 1978). It is only due to the breakdown of aggregates (e.g., by tillage) that the C is quickly mineralized leading to a rapid flux of CO_2 (Reicosky, 1998). In addition to drying, physical protection may also be associated with the size distribution of primary particles. A large fraction of SOC content, ranging from 40 to 60% of the total amount, is often associated with the clay fraction (Christensen, 1996). The stability of adsorbed organic compounds depends upon physical and charge properties of soil separates, e.g., surface area, charge density, CEC, and nature of cations on the exchange complex.

The second mechanism involves chemical protection due to denaturing or polymerization of simple into complex compounds. The chemical transformation can be brought about by drying or water absorption by actively growing roots (Perfect et al., 1990; Dexter, 1991). The exact mechanism of the effect of drying on stabilization of SOC are not known (Dormaar and Foster, 1991). Formation of complex compounds or polymers with polyvalent cations and polyphenols is another mechanism of chemical protection (Tisdall, 1996). The relative magnitude of chemical protection depends on the degree of aggregation, composition of aggregates, nature of humic substances, and the nature of polyvalent cations involved.

III. Factors Affecting Aggregation

It is well established that aggregation increases with increase in SOC content (Tisdall and Oades, 1980; Tyagi et al., 1982). Over and above that, the process of aggregation is affected by physical, chemical, and biological factors or forces. The physical forces in tropical ecosystems originate due to wetting and drying, ultra-desiccation at high temperatures, water desorption due to actively growing plant roots, and pedoturbation due to the activity of soil fauna (Lal, 1987; 1988; 1991). The wetting and drying is an important process in Mollisols and Vertisols, ultra-desiccation in Alfisols and Aridisols, root and rhizosphere activity in Alfisols, Ultisols and Oxisols, and soil faunal activities or bioturbation in ecosystems that accentuate soil biodiversity (e.g., termite, earthworms, etc.). The relative importance of chemical, physical, and biological forces depends on several site-specific factors.

A. Climate

Soil temperature and moisture regimes affect aggregation. Bauer (1934) observed that soil aggregation decreased with increase in mean annual temperature in semi-arid regions but increased with increase in temperature in humid regions. Temperature and moisture regimes affect aggregation through their influence on (i) weathering leading to differences in texture (clay content) and clay minerals, and (ii) humus content and its quality. Soil moisture regime is governed by the ratio of precipitation:potential evapotranspiration (P/PET). Both clay and humus contents usually increase with increase in P/PET for 0.1 in Aridisols to about 0.5 in Vertisols. The humus content also increases with increase in P/PET because of the increase in C input (i.e., higher biomass production). Increase in temperature may decrease C input because of increase in decomposition rate. Taking into consideration these interactive effects, Dalal and Bridge (1996) observed that increase in P/PET ratio from 0 to 0.5 causes increase in percent of silt and clay aggregated in Aridisols, Entisols, Vertisols and Mollisols. Further increase in P/PET ratio from 0.5 to 1.0, however, causes decrease in percent silt and clay aggregated in Alfisols, Inceptisols and Ultisols. In contrast, Oxisols, formed in regions with high P/PET ratio of about 1, have high percent of silt and clay aggregated because of the strong cementing effect of hydroxides of Fe and Al.

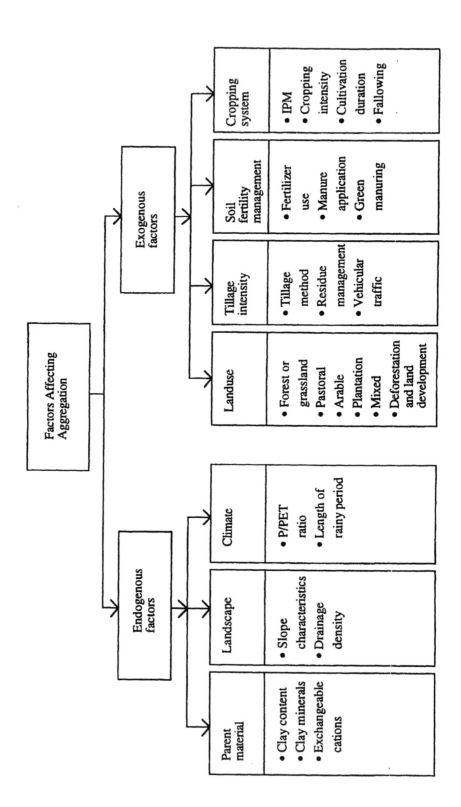

Figure 1. Factors affecting soil aggregates.

B. Textural Composition

The effect of climate on aggregation is compounded by differences in soil texture. Soils of different textural composition are formed from different parent materials even within the same climate. The SOC content is generally low in soils of coarse texture (i.e., <30% clay and < 20% silt), SOC is oriented on the outer surfaces of the clay, and the clay fraction is preferentially enriched in SOC content (Dalal and Mayer, 1986b). In this scenario, the SOC content is readily accessible to microorganisms and does not accumulate. In such cases, degree and stability of aggregation depend on continuous supply of organic material at high enough rate to compensate the rapid loss in SOC content. Plowing and clean cultivation of such soils would decrease SOC content and aggregation.

The SOC content relatively increases with change from coarse to medium texture (i.e., 30–50% clay). The interstices are filled with randomly oriented clay. In soils with fine matrix (i.e., > 50% clay), aggregation varies with clay minerals and the nature of the cations on the exchange complex.

C. Method of Deforestation and Land Development

Drastic mechanical soil disturbance with mechanical deforestation can have severe adverse impact on soil physical properties (Lal, 1996; 1997). In Nigeria, Lal and Cummings (1979) and Hulugalle et al. (1984) observed that soil structure was more adversely affected by mechanized than manual clearing methods. Similar observations were made for a soil in the humid region of Peru by Alegre and Cassel (1986). Hulugalle et al. (1984) observed that adverse effects of mechanized land clearing on aggregation can partly be alleviated by growing cover crops, e.g., *Mucuna utilis* or *Pueraria phaseoloides*.

D. Tillage Methods

All other factors remaining the same, aggregation decreases with increase in tillage intensity (Lal, 1989). In soils of humid and sub-humid regions, plowing increases the maximum soil temperature, decreases soil moisture content, increases the rate of decomposition by making SOC more accessible to microorganisms and their enzymes, decreases SOC content, and reduces degree and stability of aggregates. The data in Table 3 from Alfisols in western Nigeria show that SOC content increased with cultivation for the first 4 years, and then decreased for the next 3 years. The increase in SOC pool was due to addition of plant nutrients, high input of biomass, and enhancement of soil quality especially aggregation and earthworm activity. Soil quality improved more under no-till with mulch than plow-till or no mulch. Subsequent decline in SOC pool was due to severe degradation of soil structure, reduction in biomass production, and low return of biomass to soil. Strongly interacting with the tillage methods is the effect of crop residue management. Using crop residue as mulch increases aggregation more than when the residue is plowed under. Increase in rate of mulch application increases aggregation (Lal et al., 1980) primarily due to increase in activity and species diversity of soil fauna, e.g., earthworms. In comparison with incorporation, residue left on the soil surface as mulch improves soil moisture and temperature regimes, and increases the humification process (Aina, 1979; Lal, 1989). In a Nigerian savanna, Ike (1986) observed that disc plowing of an Ultisol decreased SOC content and aggregation. In Australia, Hamblin (1980) and Dalal (1989) observed higher SOC content and aggregation in no-till compared with plow-till methods of seedbed preparation. Use of compost, farm yard manure and other organic matter also have beneficial effects on SOC and aggregation similar to that of the no-till farming and residue mulching. The magnitude of the effect, however, may vary among soils, ecoregions, and land uses.

Table 3. Temporal changes in SOC pool of 0–10 cm depth for an Alfisol in western Nigeria managed by different tillage methods

Tillage	1980	1981	1983	1984	1985	1986	1987	Mean
					g kg⁻¹				
No-till + mulch	15.0	13.0	17.0	32.3	22.1	22.7	14.2	19.5
No-till + chisel	12.0	11.5	18.8	22.1	21.8	20.4	12.4	17.0
Plow	9.1	9.4	16.4	21.1	18.1	16.4	9.2	14.2
Disc	11.1	13.0	16.9	20.2	18.6	18.5	9.2	15.4
No-till - mulch	13.4	11.0	14.2	16.5	12.9	12.3	5.8	12.3
Summer plowing	12.1	10.3	18.4	16.2	13.9	13.7	6.7	13.0
Plowing + mulch	9.6	10.4	16.8	18.3	14.6	13.4	7.0	12.9
Ridge till	12.9	11.4	16.8	18.6	16.8	17.3	9.1	14.7
Mean	11.9	11.3	16.9	20.7	17.4	16.8	9.2	

(Recalculated from Lal, 1997.)

Interactive effects of tillage methods, residue management, and fertilizer use on temporal changes in SOC pool in an Alfisol are shown in Table 4. The SOC pool was maintained at about 24 Mg ha⁻¹ for a 5-year period with no-till, residue returned, and recommended rates of fertilizer application. In contrast, the SOC pool in the plow-till treatment without residue and fertilizer was only 12 Mg ha⁻¹, and declined at the rate of about 0.5 Mg ha⁻¹ yr⁻¹ for 0–10 cm depth. Reduction in aggregation and overall decline in soil structure may be the principal reasons for decline in SOC pool in systems where adequate amount of crop residue was returned. The high SOC pool in no-till treatment (with residue mulch and fertilizer) was associated with high activity of soil fauna and aggregation.

E. Cropping Duration

Duration and intensity of cultivation have significant impact on SOC content and aggregation through their effects on biomass input, soil aeration, soil temperature, and moisture regimes, and rate of decomposition. Experiments conducted on Alfisols and Vertisols in Australia showed that tillage and cropping of highly aggregated soils (formerly under pastures, virgin forest, or grasslands) caused strong reduction in aggregation (Dalal and Mayer, 1986a; Chan et al., 1988; Dalal et al., 1991). In west Africa, Siband (1974) reported drastic changes in SOC content with cultivation duration. The soil organic matter content of the top 0–10 cm layer decreased from 2.7% to 1.0% over a period of 90 years of cultivation. The decline in organic matter content over the 90-year period was from 1.5% to 0.8% for 10–20 cm depth and from 0.9 to 0.75% for 20–30 cm depth. Aggregate breakdown increases accessibility of SOC to micro-organisms.

F. Cropping Systems

Aggregation and SOC content also depend on land use and cropping system, which are generally more in pasture than in cropland. Within a cropping system, however, degree and stability of aggregation depend on the management, e.g., forage species, nutrient management, stocking rate, etc. In addition to altering the quantity, the land use and cropping systems can also alter the quality or type of SOC content and humic substances contained in it. Mendonca et al. (1991) studied the SOC characteristics

Table 4. Tillage, residue and fertilizer management effects on soil organic carbon pool of 0–10 cm depth of an Alfisol in western Nigeria

Tillage	Residue	Fertilizer	SOC pool (Mg ha^{-1})		
			1981	1984	1986
No-till	Removed	None	25.2	21.9	19.7
	Returned	None	22.9	22.9	23.4
	Removed	Applied	19.8	21.3	15.5
	Returned	Applied	23.6	23.3	24.0
Plow-till	Removed	None	14.9	15.2	12.4
	Returned	None	16.5	18.5	14.1
	Removed	Applied	24.9	19.1	13.6
	Returned	Applied	15.5	19.6	13.2
Treatment means					
No-till			22.9	22.4	20.7
Plow-till			17.8	18.1	13.3
No residue			21.2	19.4	15.3
With residue			19.6	21.1	18.7
No fertilizer			19.9	19.6	17.4
With fertilizer			21.0	20.8	16.6
Overall mean			20.4	20.2	17.0

(Recalculated from Lal, 1998.)

Table 5. Land use effects on SOC content for Red-Yellow Latosol in Mato Grosso, Brazil

Horizon	Natural forest	Rubber plantation	Grass
		% by weight	
A$_1$	2.50	2.04	2.12
A$_3$	1.55	1.29	–
B$_{11}$	0.90	0.76	1.29
B$_{12}$	0.58	0.54	0.94

(Adapted from Mendonca et al., 1991.)

of aggregates from a Red-Yellow Latosol for three land uses ,e.g., natural forest, rubber plantation, and grass in Mato Grosso, Brazil. Land use had a strong impact on total SOC content (Table 5). The SOC content of the A$_1$ horizon decreased by growing rubber and the pasture, but that of the B$_{11}$ and B$_{12}$ increased under pasture. The data in Table 6 also show that land use strongly impacted quality of SOC content. For the soil under forest, the SOC content in the micro-aggregates was more than in macro-aggregates without oxidation with H$_2$O$_2$. When treated with H$_2$O$_2$ to oxidize organic matter, the macro-aggregates contained more SOC content than micro-aggregates. However, the opposite was

Table 6. SOC content in A_1 and A_3 horizons of soil under natural forest and rubber plantation with and without treatment with H_2O_2

Land use	Aggregate	A_1 horizon		Fraction oxidized	A_3 horizon		Fraction oxidized
		Without H_2O_2	With H_2O_2		Without H_2O_2	With H_2O_2	
				— % —			
Forest	Macro[a]	2.63	0.49	81.4	20.5	0.56	72.7
	Micro[a]	2.79	0.36	87.1	2.13	0.34	84.0
Rubber	Macro	2.11	0.23	89.1	1.74	0.39	77.6
	Micro	1.63	0.23	85.9	1.33	0.47	64.7

[a]Macro = >2 mm; [a]Micro = <2 mm.

(Adapted from Mendonca et al., 1991.)

Table 7. Percent of SOC oxidized by H_2O_2 in A_3 horizon of soil from three land uses for different sized aggregates

Aggregate size (mm)	Forest	Rubber	Pasture
> 4	76.8a	84.6a	66.0a
2–4	69.9a	86.5a	66.4a
0.2–2	84.5a	81.5a	72.6a
0.05–0.2	83.2a	55.6b	74.0

(Adapted from Mendonca et al., 1991.)

the trend for the soil under rubber plantation. The SOC content in A horizon under forest was more strongly complexed with the mineral fraction than that in A_1 horizon. Therefore, SOC content after oxidation increased with depth in the forest system. The opposite trend in soil following conversion to rubber indicated the negative or adverse effect on the stability of the organo-mineral complexation in macro-aggregates (Table 6). Treatment with H_2O_2 lost more SOC content in the macro-aggregates of soil under rubber than under natural forest or grass. The least oxidation occurred under grass (Table 7). These differences in oxidation may be due to the differences in nature of humic substances and the degree of crystallization with Fe and Al compounds.

In India, Tyagi et al. (1982) observed that differences in aggregation among cropping systems were related to differences in SOC content ($R^2 = 0.87$). Soil aggregation is also improved by diversified farming system, e.g., agroforestry, mixed and relay cropping system. Lal (1989) observed that alley cropping systems increased aggregation especially in soil in close proximity to the hedgerow of shrubs and woody perennials. Fallowing increases aggregation, both natural fallowing as practiced in shifting cultivation (Nye and Greenland, 1960) and growing cover crops within a cropping cycle (Lal et al., 1978). Lal (1996) observed that soil compaction increased and aggregation decreased in pastures with a high stocking rate. A cropping system that produces a large quantity of biomass (especially the root biomass) is likely to cause more aggregation than the one that produces less biomass. The data in Table 8 show that regardless of the cropping system, the SOC pool of 0–10 cm depth declined with cultivation duration. The SOC pool declined even with reduction in cropping intensity and adoption

Table 8. Farming systems impact on temporal changes in SOC pool in 0–10 cm depth of Alfisols in western Nigeria

System	SOC Pool (Mg ha^{-1})			
	1982	1984	1985	1986
Alley cropping	22.7	18.5	19.0	19.3
Mucuna fallow (A)	23.7	19.8	20.9	20.4
Mucuna fallow (B)	21.4	19.6	15.9	14.5
Ley farming (A)	23.1	19.4	16.6	20.6
Ley farming (B)	21.9	16.6	17.9	15.3

A = severely eroded watersheds; B = moderately eroded watersheds;
(Recalculated from Lal, 1996.)

of recommended practices, e.g., alley cropping, Mucuna fallowing and pasture-based systems. Adopting these practices with mechanized farm operations may not have strong beneficial effects in enhancing SOC content because of soil compaction and the attendant increase in runoff and soil erosion. Decline in SOC pool in the cropping system reported in Table 8 may be attributed to many factors. Important among these are the reduction in aggregation and overall deterioration of soil structure. Decline in soil structure was accentuated by drastic reduction in activity and species diversity of soil fauna, notably earthworms. In addition, other factors that exacerbated decline in aggregation and reduction in SOC pool were low biomass produced and returned to the soil, depletion in soil fertility and nutrient pool (N, P, S, Zn, etc.), and soil degradation.

IV. Characterization of Structural Attributes in Relation to C Sequestration

Several studies on soils of the tropics have shown strong correlation between SOC content and aggregation. For soils from Mali and Burkina Faso, Dutartre et al. (1993) observed a very strong correlation between aggregation and SOC content. Therefore, several indices of evaluating soil aggregation are based on SOC content and its interaction with soil texture and chemical properties. Commonly used methods of evaluating aggregation and aggregate stability are based on wet sieving (Yoder, 1936) or dry sieving techniques (Kemper and Rosenau, 1986; Williams et al., 1966; Kemper and Chepil, 1965). Results may be expressed in terms of percent aggregation, mean weight diameter (MWD) or geometric mean diameter (GMD).

Based on the relative significance of a wide range of cementing agents, several indices of soil structure have been proposed for assessing structural attributes of soils of the tropics. Henin et al. (1958) proposed an Instability Index (s) for soils of tropical Africa (Eq. 5).

$$Is = \frac{(A + LF)\max\%}{1/3\,Ag\% - 0.9SG\%} \tag{Eq. 5}$$

Where (A+LF)max% is the maximum amount of dispersed 0–20 mm fraction obtained after three treatments of the initial soil sample: without any pretreatment (air dry), and following immersion in alcohol or benzene, Ag% refers to the >200 μm aggregates (air, alcohol, benzene) obtained after shaking (30 manual turnings and wet sieving of the 3 pretreated samples), and SG% represents the contents of coarse mineral same sand (> 200 μm), and (1/3 AG%– 0.9 SG%) represent mean % stable aggregates. This index was also found suitable for several soils from Nigeria (De Vleeschauwer et al., 1979).

FAO (1979) proposed an index of crusting (Ic) based on textural composition and soil organic matter content (Eq. 6).

$$Ic = \frac{1.55f + 0.75S_c}{Cl + (10 \times SOM)} \qquad \text{(Eq. 6)}$$

Where S_f is % fine silt, S_c is % coarse silt, Cl is % clay, and SOM is % soil organic matter content. The index of crusting is inversely related to the clay and SOC contents or the colloidal fraction that promotes aggregation. Pieri (1991) proposed the concept of critical level of SOM content needed to maintain a favorable level of soil structure (Eq. 7).

$$St = \frac{(SOM)\%}{(clay + silt)\%} \times 100 \qquad \text{(Eq. 7)}$$

Based on the analysis of about 500 samples from semi-arid regions of west Africa, Pieri proposed the following limits for characterizing soil structure (Eqs. 8 through 10).

St = <5%, loss of soil structure and high susceptibility to erosion   (Eq. 8)

St = 5% to 7%, unstable structure and risk of soil degradation   (Eq. 9)

St = >9%, stable soil structure   (Eq. 10)

Usefulness of these indices may be soil and ecoregion specific. Therefore, developing conceptual indices of soil structure for C sequestration is a research priority.

V. Conclusions

Soil aggregation or formation of organo-mineral complexes is an important process that governs soil quality and facilitates C sequestration in soil. Aggregation involves flocculation of clay into floccules or domains, and the clay domains are cemented together to form micro-aggregates through organic molecules and the complex polymers formed through association with polyvalent cations. Micro-aggregates are bound together into macro-aggregates by microbial byproducts, root hair, fungal hyphae and other particulate organic matter. Soil organic carbon sequestered within micro-aggregates is protected from microbial decomposition by physical and chemical protective mechanisms. Physical mechanisms involve inaccessibility of the organic matter due to small size of the pores. Chemical mechanisms comprise formation of complex polymers with polyvalent cations. These polymers have a long turnover time. Consequently, aggregation and aggregate stability are generally more in virgin forest or grasslands.

The degree and stability of aggregates thus formed depends on several exogenous and endogenous factors. Exogenous factors involve land use, cropping/farming systems, soil and crop management, tillage methods, and residue management systems. There exists a strong correlation between the ratio of precipitation: potential evapotranspiration and aggregation because of its influence on biomass input to soil, rate of decomposition, and soil biodiversity. Parent material and climate strongly influence textural composition, clay content, nature of clay minerals, and exchangeable cations. Soil aggregation is also influenced by land use and cropping/farming systems.

Methods of deforestation and land development affect aggregation through their impact on soil disturbance, biomass inputs, soil temperature, moisture regimes, and soil biodiversity. Cultural practices with beneficial effects on aggregation include no-till, residue mulching, cover crops and fallowing, use of manure and organic amendments, and diverse cropping systems, e.g., agroforestry.

Several indices used for characterizing soil structure are based on soil organic carbon content. Improving soil structure and aggregation is an important strategy of enhancing soil quality, improving agriculture productivity, and sequestering C in soil. Soil aggregation also impacts soil quality, agronomic productivity, and environmental moderating capacity through C sequestration in stable organo-mineral complexes.

Dynamics of SOC in tropical soils and processes affecting it are not widely understood. Therefore, there are several important research priorities:

(i) Structural properties of some soils are easily degraded when intensive cropping replaces traditional extensive cultivation. Factors responsible for decline of soil structure are not understood.

(ii) Decline in soil structure sets-in-motion soil degradative trends leading to crusting, compaction, decline in infiltration, high runoff rate, accelerated erosion, and desertification. Breakdown of aggregates exposes the C to microbial decomposition, and the process is exacerbated by accelerated soil erosion.

(iii) Restoration of soil structure is important to rehabilitation of degraded soils throughout the tropics and subtropics. An effective implementation of restorative measures involves understanding the resilience of soil structure, processes and factors governing it, and the critical limits of the key properties, e.g., SOC content.

(iv) Soil biodiversity, species composition, and activity of soil fauna and flora, play an important role in dynamics of soil structure. The cause-effect relationship between soil structure and soil biodiversity must be established. Soil and crop management practices should be developed that maintain high soil biodiversity.

References

Adu, J.K. and J.M. Oades. 1978. Physical factors influencing decomposition of organic materials in soil aggregates. *Soil Biol. Biochem.* 10:109-115.

Aina, P.O. 1979. Soil changes resulting from long-term management practices in western Nigeria. *Soil Sci. Soc. Am. Proc.* 43:173-177.

Alegre, J.C. and D.K. Cassel. 1986. Effect of land clearing methods and post-clearing management on aggregate stability and organic carbon content of a soil in the humid tropics. *Soil Sci.* 142: 289-295.

Bauer, L.D. 1934. A classification of soil structure and its relationship to the main soil groups. *Am. Soil Survey Assoc. Bull.* 15:107-109.

Chan, K.Y., W.D. Bellotti, and W.P. Roberts. 1988. Changes in surface soil properties of Vertisols under dryland cropping in a semi-arid environment. *Aust. J. Soil Res.* 26: 509-518.

Christensen, B.T. 1996. Carbon in primary and secondary organo-mineral complexes. p. 97-165. In: M.R. Carter and B.A. Stewart (eds.), *Structure and Organic Matter Storage in Agricultural Soils*, CRC Press, Boca Raton, FL.

Dalal, R.C. 1989. Long-term effects of no-tillage, crop residue and nitrogen application on properties of a Vertisol. *Soil Sci. Soc. Am. J.* 53:1511-1515.

Dalal, R.C. and R.J. Mayer. 1986a. Long-term trends in fertility of soils under continuous cultivation and cereal cropping in southern Queensland. II. Total organic carbon and its rate of loss from the soil profile. *Aust. J. Soil Res.* 24:281-292.

Dalal, R.C. and R.J. Mayer. 1986b. Long-term trends in fertility of soils under continuous cultivation and cereal cropping in southern Queensland. III. Distribution and kinetics of soil organic carbon in particle size fractions. *Aust. J. Soil Res.* 24:293-300.

Dalal, R.C., W.M. Strong, E.J. Weston, and J. Gaffney. 1991. Sustaining multiple production systems. 2. Soil fertility decline and restoration of cropping lands in subtropical Queensland. *Tropical Grasslands* 25:173-180.

Dalal, R.C. and B.J. Bridge. 1996. Aggregation and organic matter storage in sub-humid and semi-arid soils. p. 263-307. In: M.R. Carter and B.A. Stewart (eds.), *Structure and Organic Matter Storage in Agricultural Soils*, CRC Press, Boca Raton, FL.

De Vleeshauwer, D., R. Lal, and M. De Boodt. 1979. Comparison of detachability indices in relation to soil erodibility for some important Nigerian soils. *Pedology* 25:5-20.

Dexter, A.R. 1991. Amelioration of soil by natural processes. *Soil Tillage Res.* 20:87-100.

Dormaar, J.F. and R.C. Foster. 1991. Nascent aggregates in the rhizosphere of perennialryegrass (*Lolium perene*). *Can. J. Soil Sci.* 71:465-474.

Dutartre, P., F. Bartoli, F. Andreux, J.M. Portal, and A. Ange. 1993. Influence of content and nature of organic matter on the structure of some sandy soils from west Africa. *Geoderma* 56:459-478.

FAO 1979. Soil survey in irrigation investigations. FAO Soils Bulletin 42, FAO, Rome, Italy.

Hamblin, A.P. 1980. Changes in aggregate stability and associated organic matter properties after direct drilling and ploughing on some Australian soils. *Aust. J. Soil Res.* 18:27-36.

Henin, S., G. Monnier, and A. Combeau. 1958. Methode pour l'etude de la stabilite' structurale des sols. ann. *Agron.* 9:73-92.

Hulugalle, N., R. Lal, and C.H.H. ter Kuile. 1984. Soil physical changes and crop root growth following different methods of land clearing in western Nigeria. *Soil Sci.* 138:172-179.

Ike, I.F. 1986. Soil and crop responses to different tillage practices in a ferraginous soil in the Nigerian savanna. *Soil Tillage Res.* 6:261-272.

Kemper, W.D. and W.S. Chepil. 1965. Size distribution of aggregates. p. 499-519. In: C.A. Black, D.D. Evans, J.L. White, L.E. Ensminges, and F.E. Clark (eds.), *Methods of Soil Analysis. Part 1: Physical and Mineralogical Properties*, Agronomy Monogram 9, ASA, Madison, WI.

Kemper, W.D. and R.C. Rosenau. 1986. Aggregate stability and size distribution. p. 425-442. In: A. Klute (ed.), *Methods of Soil Analysis. Part I: Physical and Mineralogical Methods,* Agronomy Monograph 9, ASA, Madison, WI.

Lal, R. 1987. *Tropical Ecology and Physical Edaphology*. John Wiley & Sons, Chichester, U.K. 732 pp.

Lal, R. 1988. Effects of macrofauna on soil properties in tropical ecosystems. *Agric. Ecosyst. Environ.* 24:101-116.

Lal, R. 1989. Conservation tillage for sustainable agriculture: tropics vs. temperate environments. *Adv. Agron.* 42:85-91.

Lal, R. 1991. Soil conservation and biodiversity. p. 89-104. In: D.L. Hawkworths (ed.), *The Biodiversity of Micro-organisms and Invertebrates: Its Role in Sustainable Agriculture*. CAB International Wallingford, U.K.

Lal, R. 1996. Deforestation and land use effects on soil degradation and rehabilitation in western Nigeria. II. Soil chemical properties. *Land Degradation & Development* 7:87-98.

Lal, R. 1997. Long-term tillage and maize monoculture effects on a tropical Alfisol in western Nigeria. II. Soil chemical properties. *Soil & Tillage Res.* 42:161-174.

Lal, R. 1998. Soil quality changes under continuous cropping for seventeen seasons of an Alfisol in western Nigeria. *Land Degradation & Development* 9:259-274.

Lal, R., G.F. Wilson, and B.N. Okigbo. 1978. No tillage farming after various grasses and leguminous cover crops in tropical Alfisol. I. Crop Performance. *Field Crops Res.* 1:71-84.

Lal, R. and D.J. Cummings. 1979. Changes in soil and microclimate after clearing a tropical forest. *Field Crops Res.* 2:91-107.

Lal, R., D. De Vleeschauwer, and R.M. Nganje. 1980. Changes in properties of a newly cleared tropical Alfisol as affected by mulching. *Soil Sci. Soc. Am. J.* 44:827-833.

Martin, J.P., W.P. Martin, J.B. Page, W.A. Raney, and J.G. de Mont. 1955. Soil aggregation. *Adv. Agron.* 7:1-37.

Mendonca, E. des, W. Moura Filho, and L.M. Costa. 1991. Organic matter and chemical characteristics of aggregates from a Red-Yellow Latosol under natural forest, rubber plantation and grass in Brazil. p. 187-194. In: W.S. Wilson (ed.), Adv. in Soil Organic Matter Research: The Impact of Agriculture and the Environment. The Royal Society of Chemistry, London.

Nye, P.H. and D.J. Greenland. 1960. The soil under shifting cultivation. Tech. Comm 51, Commonwealth Bureau of Soils, U.K.

Oades, J.M. and A.G. Waters. 1991. Aggregate hierarchy in soils. *Aust. J. Soil Res.* 29: 815-828.

Perfect, E., B.D. Kay, W.K.P. Van Loon, R.W. Sheard, and T. Pojasok. 1990. Rates of change in structural stability under forages and corn. *Soil Sci. Soc. Am. J.* 54:179-186.

Pieri, C. 1991. *Fertility of Soils: A Future for Farming in the West African Savannah.* Springer-Verlag, Berlin.

Reicosky, D.C. 1998. Tillage methods and CO_2 loss: fall versus spring tillage. p. 99-111. In: R. Lal, J.M. Kimble, R.F. Follett, and B.A. Stewart (eds.), *Management of Carbon Sequestration in Soil*, CRC Press, Boca Raton, FL.

Robert, M. and C. Chenu. 1992. Interactions between soil minerals and microorganisms. *Soil Biochem.* 7:307-404.

Siband, P. 1974. Evolution des caractéres et de la fertilite' d'un sol rouge de Casamance. *L'Agron. Trop.* 29:1228-1248.

Tisdall, J.M. 1996. Formation of soil aggregates and accumulation of soil organic matter. p. 57-95. In: M.R. Carter and B.A. Stewart (eds.), *Structure and Organic Matter Storage in Agricultural Soils*, CRC Press, Boca Raton, FL.

Tisdall, J.M. and J.M. Oades. 1980. The effect of crop rotation on aggregation in a red-brown earth. *Aust. J. Soil Res.* 18:423-433.

Tisdall, J.M. and J.M. Oades. 1982. Organic matter and water stable aggregates in soils. *J. Soil Sci.* 33:141-163.

Tyagi, S.C., D.L. Sharma, and G.P. Nathani. 1982. Effect of different cropping patterns on the physical properties of medium black soils of Rajsthan. *Curr. Agric.* 6:172-176.

Williams, B.G., D.J. Greenland, G.R. Lindstrom, and J.P. Quirk. 1966. Techniques for the determination of the stability of soil aggregates. *Soil Sci.* 101:157-163.

Yoder, R.E. 1936. A direct method of aggregate analysis of soils and study of the physical nature of erosion losses. *J. Am. Soc. Agron.* 28:337-351.

Section VI.

MONITORING AND PREDICTION

Methods of Analysis for Soil Carbon:
An Overview

H.H. Cheng and J. Kimble

I. Introduction

Soil contains both inorganic and organic forms of carbon (C). Major inorganic forms are carbon dioxide (CO_2) and carbonates. Only in soils high in carbonates is there a significant amount of inorganic C present. Charcoal is mostly elemental C, which may be considered as an inorganic form, although it is derived from organic forms of C. Organic C ranges from the relatively undecomposed plant and animal residues, microbial biomass and dead tissues, decomposed and decomposable C-containing chemicals, naturally humified or recalcitrant carbonaceous materials, to manufactured wastes and sludges (biosolids). In most soils, C is organic in nature. Carbon constitutes approximately 50% of soil organic matter (SOM). It serves both a structural and a functional role in soil. Both the soil inorganic C (SIC) and soil organic C (SOC) play a critical role in many geochemical or biochemical processes. Dissolved CO_2 and soluble carbonates in ground or surface water can move into the ocean where they may precipitate as solid carbonates, making the ocean bottom the largest sink of inorganic C. Whereas biochemical oxidation of SOC can lead to emission of CO_2 into the atmosphere, making soil a source of atmospheric C, the humification process leads to sequestration of organic C. Whether soil is a source or sink of terrestrial C depends on the balance between the oxidation and humidification processes.

Interest in SOC has greatly increased in recent years because terrestrial organic C can be a key factor in understanding the effect of C emission on global climate change. The increase of CO_2 from anthropogenic sources has especially been the focus of public concern. Emission of CO_2 from oxidation of soil organic matter or from respiration of the above-ground biomass is still the largest source of CO_2 in the atmosphere (Schlesinger, 1991; Lal et al., 1998). Researchers are interested in knowing the factors influencing soil as a source or a sink of atmospheric CO_2 (Bouwman, 1990; Tans et al., 1990; Lal et al., 1995; Schimel, 1995). Schlesinger (1991) has reported on changing fluxes of greenhouse gases and their relationship to SOC. Paul et al. (1997) used the radiocarbon dating techniques to examine soil organic matter pool sizes and dynamics. Currently, research is being conducted to evaluate the below-ground accumulation of biomass as a technique to capture and store atmospheric CO_2 (Fisher et al., 1994). A number of studies have attempted to estimate the quantity of SOC sequestered in the terrestrial ecosystem. Their estimates have varied from 1220 Pg (Sombroek et al., 1993), to between 1462 and 1548 Pg (Batjes, 1996), to 1576 Pg (Eswaran et al., 1993; 1995). A valuable summary of current work on C sequestration can be found in Lal et al. (1998).

Much of the current database on terrestrial C contents has been gathered primarily from soil surveys, and the C contents were commonly determined by wet or dry combustion methods. If these estimates are to be accepted as accurate, the validity of the analytical methods used must be scrutinized. Moreover, many researchers are interested not only in the total soil C contents, but also in specific components of soil C, such as SOC, SIC, microbial biomass C, slow release C, and active or passive pools of soil C. If we are to model the soil carbon fluxes, the size of the various carbon pools must be properly established and the dynamics of their turnovers must be evaluated (Cole et al., 1996; Legros et al., 1994).

II. Analytical Methods

This overview has two objectives: (a) to summarize the various methods or approaches to determine the C contents of soil, and (b) to provide general guidelines for method selection. Numerous methods have been developed to analyze total soil C, SOC, or C in specific soil components following chemical fractionation or physical separation. Emphasis of this chapter will be on total C analysis, while a subsequent overview will be devoted to methods for estimating the various C fractions or pools. Furthermore, attention of this review will be on determination of organic C, even though the presence of SIC, such as carbonates and charcoal, should always be assessed.

A. Total Soil C Determination

Total soil C can be determined by dry or wet combustion methods. Dry combustion methods are based on thermal oxidation of soil C in the medium temperature range (~1000°C) in an oxygenated environment (purified O_2 and metal oxides), usually in the presence of a metal combustion accelerator to convert all C into CO_2 (Allison et al., 1965). Nelson and Sommers (1996) have recently updated the procedures for dry combustion by including information on analytical equipment currently available commercially.

Wet combustion or oxidation methods most commonly use potassium dichromate in concentrated sulfuric acid to oxidize SOC under various temperature regimes and heating time. The literature on wet combustion methods has been critically reviewed by Nelson and Sommers (1996). This reference is an excellent source of background information on soil C analysis.

The Walkley–Black method for SOC estimation (Walkley and Black, 1934) is probably the most widely used wet oxidation method for soil C characterization because of its ease of operation. However, its use is becoming more restricted because the procedure requires the use of toxic chromium compounds as oxidants. Many versions and variations of this method have been developed. Users of C data generated from the Walkley–Black method should be fully aware of the uncertainties of the data obtained by this method and should not over-extrapolate the data in their interpretations. For instance, because the Walkley–Black procedure depends only upon the heat generated from dilution of concentrated acid for the oxidation purpose, only the readily oxidizable C is estimated by this method.

In the 20 soils tested, Walkley and Black (1934) found that the organic C recovered by this method averaged 76% of the total SOC, but the actual recovery ranged from 60 to 86%. Subsequent studies, summarized by Nelson and Sommers (1996), showed recoveries ranged from 27 to 144%! In unpublished results obtained by the National Soil Survey Laboratory, the recovery factor was found to vary by soil order and with depth in the soil profile (Kimble, personal communications). In addition, the presence of other reducing substances in soil, such as manganous compounds, can give spurious results. Thus, using an average correction factor (commonly 1.3) for the Walkley–Black analysis to estimate total SOC in any individual soil can significantly under- or over-estimate the total SOC. Although it is common to carry out the wet oxidation procedure without applying external heat to obtain a quick estimate of soil C contents, methods involving extensive heating (e.g., Allison, 1960; Mebius, 1960) usually provide more consistent results.

B. Separation of Soil Inorganic and Organic C

In certain soils, the soil inorganic C (SIC) content can be significant (e.g., Eswaran et al., 1993). If no special attention is given to this component, errors in total soil C estimations can be significant. In a dry combustion procedure, SIC can be estimated by thermodecomposition of the organic C at 650°C before converting the carbonates to CO_2 at 1000°C. (Loeppert and Suarez, 1996). The CO_2 produced is absorbed in a dry absorbing bulb for gravimetric determination (Allison, 1960), or trapped in concentrated alkaline solutions and determined by acid titration (e.g., Cheng and Farrow, 1976). SIC can also be estimated from the difference between total C and SOC by the wet combustion procedure, as SOC is determined after treating soil with acid to convert the carbonates into CO_2.

If a soil contains a significant amount of elemental C, such as charcoal, precise determination of this form of C can be problematic, as it can be partially oxidized in the combustion procedures. The preferable way to deal with this fraction is to remove it from the soil sample by dry sieving or by wet floatation or other density gradient separation methods, since charcoal is lighter than mineral soil in weight. This approach becomes problematic if the soil contains a significant portion of fresh or partially decomposed plant residues.

C. Fractionation of SOC Components

Numerous approaches have been proposed to differentiate the various components in the soil organic matter (Cheng and Molina, 1995). Many fractionation schemes are variations of the classical approach based on solubilization of soil organic matter in strong alkali followed by separation of C components by acid precipitation. The original purpose of these fractionation efforts was to decipher the structural nature of the soil organic matter, although the exact chemical identity of these separated fractions is still mostly unknown. Other attempts have been aimed at isolating fractions of known chemical compositions from the soil. These methods, however, extract only a small amount of the total soil C.

In the past two decades, there has been a growing interest in developing methods using physical separation techniques to fractionate soil C components or using biochemical means to identify bioreactive C pools. Physical separation approaches include soil particle size partitioning using sedimentation or centrifugation, or particle density gradient separation techniques. Particulate organic matter may also be removed by simply picking the materials out of the soil sample. Biochemical approaches have been used to identify soil microbial biomass and readily decomposable fractions in the soil substrates. They tend to estimate the more labile or bioreactive components of the soil C rather than the total soil C, and they are more useful for characterizing and understanding soil C transformation processes than those used for chemical fractionation of soil C. These methods are also more relevant for characterizing soil carbon transformation dynamics in relation to global climate changes. Methodology for assessing soil carbon pools will be discussed in more detail in a subsequent overview.

III. Guidelines for Method Selection

The basic guidelines for selecting a specific analytical method should be based on two points of emphasis. One is the need for understanding the basic principles of an analytical procedure, and the other is the need for defining the purpose of analysis. Because C analysis is often considered to be routine, sometimes these basic guidelines are neglected. Soil C results have been reported in the literature without specifying how the C values were obtained or showing proof whether the method used to obtain the data was valid. Such C data have been used in extrapolation to other unrelated situations, making the reliability of interpretation of these data highly questionable.

Since all analytical procedures are operationally defined, it is essential that the method of analysis is specified in any research publication. Even when a specific C analysis method is referenced, there is often critical information missing. For instance, C was analyzed by a "modified" version of a certain method, without specifying which part of the method was modified. Or only partial information on the method used was given (e.g., only the method of detection) without specifying other critical aspects of a method (e.g., sample pretreatment). Thus, the need for defining the purpose and understanding all the components of an analytical method cannot be overemphasized. The following steps are not only critical components of methods for total C analysis, but are also important parts of practically all analytical methods.

A. Sampling and Sample Preparation

Sampling and sample preparation steps are often the most neglected aspect in many experimental studies. Considerations for any chemical analysis should start with how the sample should be obtained and handled prior to analysis. The results of an analysis is only as good as the sample used or how it is prepared for the analysis. If the sample is not representative of the soil in its natural setting, or the precautions required in its handling and preparation are not observed, or required pretreatment of the sample is not properly carried out, results obtained may be spurious or misleading.

Careful thoughts must be given to what constitutes a representative sample. The extent of spatial variability within a given field should determine how many soil samples are to be taken. Considerations are commonly given to how deep the sampling should be, and whether the depth should be divided by regular intervals or by horizons. Less attention is usually given to the spatial representation of the sample collected. The extent of data from a point sample can be scaled to represent a broad area needs to be addressed if a valid estimate of the terrestrial C concentration is to be obtained. If valid soil survey maps are available, they should be used as a basis for developing a sampling scheme.

Once the soil sample is obtained, decisions must be made as to whether it should be kept field moist, or air dried, or stored under refrigeration or frozen. While air drying of soil samples is the most common preparation step, significant loss of the mobile fractions of C could occur during the drying process, even though such a loss may be not obvious in the total C analysis. Other questions include how long a sample may be stored before analysis, and whether and how fine the sample should be ground. Samples to be used for biological studies should be refrigerated and used as soon as possible to prevent significant changes during storage. Similarly, samples used for total C analysis may be handled differently from those used for C component analysis.

B. Pretreatment, Conversion, and/or Separation and Purification

Before a sample is analyzed for its C contents, usually pretreatment steps are necessary, such as to remove the inorganic C prior to organic C analysis. Decisions are to be made on how best to treat the inorganic C component, charcoal C, and other specific fractions. Some methods require conversion of all forms of C into CO_2 or methane before analysis. Others require removal of interferences during the C conversion (e.g., oxidation) process. Methods of separation and purification of C-containing soil fractions from non-carbonaceous fractions include chemical solubilization or extraction, sonication or physical shaking, density gradient or chromatographic separations, and acid or alkaline hydrolysis. In each of the pretreatment or conversion steps, C may be lost or destroyed, resulting in incomplete recovery which would affect the accuracy of the final determination. Valid test results for recovery efficiencies should be included in all research reports to permit comparison of research findings. Some quick methods do not convert C quantitatively, but most include an average correction factor to estimate total C (e.g., the Walkley–Black method). Although the percentage of readily oxidizable C in different soils could differ significantly (see previous discussion), this variation has often not been reported.

C. Detection/Identification

Although many analytical methods for C analysis follow similar procedures for the sampling, pretreatment, and conversion steps, they often differ in the detection/identification step. Measurement of the amount of C present in a sample may be based upon the amount of CO_2 produced from conversion of all other forms of C by oxidation, or based upon the amount of oxidant used to convert all C into CO_2. Measuring CO_2 produced is considered a direct approach to total C estimation, but in reality, it is often not the CO_2 that is actually measured. For instance, the gravimetric method frequently used to determine total C actually measures the weight change of the absorption bulb before and after the CO_2 produced from the sample has passed through the bulb. It is assumed that only CO_2 is absorbed, while other gases or vapors are removed before the gas stream enters the absorbing bulb. Thus, extreme care is needed to reduce such interfering errors. In the case where the method of detection is by determining the amount of oxidant reduced, it is frequently assumed that all forms of C in the soil are oxidized to CO_2, and all soil C is at the most reduced oxidation state. This assumption is in error if there is a significant amount of carbonates present. If other interfering materials in the sample can also reduce the oxidant, they must be removed or error would result.

By converting all forms of C to CO_2 for total C determination, the nature of the original forms of soil C cannot be identified. Only by using nondestructive methods for extraction or solubilization to bring the various forms of C into solution can one hope to use infrared, nuclear magnetic resonance, or other methods of characterization to reveal the identity of the various forms of C in the soil. Because of the difficulty in discerning the total identity of SOC complexes, often only certain properties that are characteristic of a specific chemical is used in the identification step. For instance, chromatographic methods (liquid, gas, thin layer) are actually methods of separation and not true methods of identification. They are used for detection because each chemical has a specific, although not an exclusive, elution pattern through the chromatographic process. Care must be taken to use chromatographic methods for identification purposes.

D. Quantification and Data Interpretation/Extrapolation

Methods for estimation of C can vary from rapid color comparison for oxidant reduction to colorimetric, titrimetric, and chromatographic measurements. Decision on methods of quantifications depends upon the sensitivity of the detection limit and reproducibility of results required. A rough estimate of the soil organic matter may be sufficient as a guide for some soil management practices, but would be inadequate for predicting the dynamics of soil C transformations. Cheng and Molina (1995) showed that an analytical variation of 0.1% in SOC estimation could totally change the apparent role of soil as a source or a sink of atmospheric C. Yet some global C budget estimates are based on such imprecise data. When soil C measured in a sample of soil is extrapolated across a landscape, the field variation over space can be significant.

E. Method Selection

Selection of appropriate analytical methods to achieve the intended purpose must be the first decision in any experimental studies. Although one recognizes that there are always limitations facing any analytical laboratory, whether it is the availability of a specific analytical instrument or the ability of large numbers of samples to be analyzed in a timely fashion, it is still important to be reminded as a guiding principle that experimental data are as good as the analytical methods selected to obtain the results. Before any method is adopted for use, three important questions should be answered:

(a) What is the purpose of obtaining the specific C data?

(b) What analytical method is most appropriate for obtaining these C data?

(c) What are the precision and accuracy of the analytical method used?

Rather than defining the purpose of analysis before a meaningful experiment is designed and conducted, most studies tend to set up the experiment first and then make use of whatever method is available, but not necessarily the most appropriate, for analysis. This is especially common in many studies involving SOC.

When the purpose of analysis is identified, the mode of analytical approach can then be selected and analytical expectations can be defined. This would include such decisions as the number of samples to be handled, or the size of the sample to be collected to ensure its representativeness. All factors must be considered carefully before any experiments and analyses are performed. To compare the usefulness and appropriateness of various methods, an assessment of their procedures for sampling and sample preparation, separation, isolation, pretreatment, conversion, detection, and quantification will be necessary. Once an analytical method is selected, procedural requirements should be followed without arbitrary modifications, and results obtained should be validated, to ascertain that the data obtained are useful and applicable (Cheng and Mulla, 1988).

IV. Conclusion

As interest in characterizing soil carbon transformation dynamics in relation to global climate changes increases, demand for more precise C analysis will also increase. Selection of an appropriate method for analyzing soil C will become increasingly critical. This overview is not intended to reveal any new methodlogy for C analysis, but to offer a refresher course on the essential steps and components of analytical methods. Hopefully, an appreciation of analytical methodology will help to generate appropriate and valid data for evaluating the impact of global climate changes on terrestrial C dynamics.

References

Allison, L.E. 1960. Wet-combustion apparatus and procedures for organic and inorganic carbon in soil. *Soil Sci. Soc. Am. Proc.* 24:36-40.
Allison, L. E. 1965. Organic carbon. p. 1367-1378. In: C. A. Black, (ed.), *Methods of Soil Analysis*. Agronomy 9. American Society of Agronomy, Madison WI.
Batjes, N.H. 1996. Total carbon and nitrogen in the soils of the world. *European J. Soil Sci.* 47:151-163.
Bouwman, A.F. 1990. Exchange of greenhouse gases between terrestrial ecosystems and the atmosphere. In: A.F. Bouwman, (ed.), *Soils and the Greenhouse Effect*. John Wiley & Sons , Chichester, U.K. 575 pp.
Cheng, H. H. and F. O. Farrow. 1976. Determination of ^{14}C-labeled pesticides in soils by a dry combustion technique. *Soil Sci. Soc. Am. J.* 40:148-150.

Cheng, H. H. and J. A. E. Molina. 1995. In search of the bioreactive soil organic carbon: the fractionaation approaches. p. 343-350. In: R. Lal, J. Kimble, E. Levine, and B. A. Stewart (eds.), *Soils and Global Change*. Lewis Publishers, Boca Raton FL.

Cheng, H. H., and D. J. Mulla. 1988. Sample analyses for groundwater studies. p.90-96. In: D. W. Nelson and R. H. Dowdy, (eds.), Methods for Ground Water Quality Studies. University of Nebraska, Lincoln, NE.

Cole, C.V., C. Cerri, K. Minami, A. Mosier, N. Rosenberg, D. Sauerbeck, J. Dumanski, J. Duxbury, J. Freney, R. Gupta, O. Heinemeyer, T. Kolchugina, J. Lee, K. Paustian, D. Powlson, N. Sampson, H. Tiessen, M. van Noordwijk, and Q. Zhao. 1996. Agricultural options for mitigation of greenhouse gas emissions. p. 745-771. In: R.T. Watson, M. Zinyowera, and R.H. Moss, (eds.), Climate Change 1995: Impacts, Adaptations and Mitigation of Climate Change–Scientific-technical Analyses. IPCC Working Group II., Cambridge University Press, Cambridge, U.K.

Eswaran, H., E. Van den Berg, and P. Reich. 1993. Organic carbon in soils of the world. *Soil Sci. Soc. Am. J.* 57:192-194.

Eswaran, H., E. Van den Berg, P. Reich, and J. Kimble. 1995. Global soil carbon resources. p. 27-44. In: R. Lal, J. Kimble, E. Levine, and B.A. Stewart, (eds.), *Soils and Global Change*. Advances in Soil Science. Lewis Publishers, New York.

Fisher, M.J., I.M. Rao, M.A. Ayarza, C.E. Lascano, J.I. Sanz, R.J. Thomas, and R.R. Vera. 1994. Carbon storage by introduced deep rooted grasses in the South American savannas. *Nature* 371:236-238.

Lal, R., J. Kimble, and B.A. Stewart. 1995. World soils as a source or sink for radiatively-active gases. p. 1-8. In: R. Lal, J. Kimble, E. Levine, and B.A. Stewart (eds.), *Soil Management and Greenhouse Effect*. Lewis Publishers, Boca, Raton FL.

Lal, R., J.M. Kimble, R.F. Follett, and C.V. Cole. 1998. *The Potential of U.S. Cropland to Sequester Carbon and Mitigate the Greenhouse Effect*. Ann Arbor Press. Chelsea, MI. 128 pp.

Legros, J.P., P.J. Loveland, and M.D.A. Rounsevell. 1994. Soils and climate change - Where next? p. 258-266. In: M.D.A. Rounesvell and P.J. Loveland (eds.), *Soil Response to Climate Change*. NATO ASI Series. Vol. 23. Springer-Verlag, Berlin.

Loeppert, R. H. and D. L. Suarez. 1996. Carbonate and gypsum. p. 437-474. In: D. L. Sparks et al., (eds.), Methods of Soil Analysis, Part 3: Chemical Methods. SSSA Book Series No. 5. Soil Science Society of America, Madison WI.

Mebius, L. J. 1960. A rapid method for the determination of organic carbon in soil. *Anal. Chim. Acta* 22:120-121.

Nelson, D. W. and L. E. Sommers. 1996. Total carbon, organic carbon, and organic matter. p. 961-1010. In: D. L. Sparks et al. (eds.), Methods of Soil Analysis, Part 3: Chemical Methods. SSSA Book Series No. 5. Soil Science Society of America. Madison, WI.

Paul, E.A., R.F. Folett, S.W. Leavitt, A. Halvorson, G.A. Peterson, and D.J. Lyon. 1997. Radiocarbon dating for determination of soil organic matter pool sizes and dynamics. *Soil Sci. Soc. Am. J.* 61:1058-1067.

Schimel, D.S. 1995. Terrestrial ecosystems and the carbon cycle. *Global Change Biol.* 1:77-91.

Schlesinger, W.H. 1991. *Biogeochemistry: An Analysis of Global Change*. Academic Press, New York. 443 pp.

Sombroek, W.G., F.O. Nachtergale, and A. Hebel. 1993. Amounts, dynamics, and sequestration of carbon in tropical and subtropical soils. *Ambio* 22:417-426.

Tans, P.O., I.Y.Fung, and T. Takahashi. 1990. Observational constraints on the global atmosphere CO_2 budget. *Science* 247:1431-1438.

Walkley, A. and I.A. Black. 1934. An examination of the Degtjareffe method for determining soil organic matter and a proposed modification of the chromic acid titration method. *Soil Sci.* 37:29-38.

CHAPTER 18

Modeling Soil Carbon Dynamics in Tropical Ecosystems

P. Smith, P. Falloon, K. Coleman, J. Smith, M.C. Piccolo, C. Cerri,
M. Bernoux, D. Jenkinson, J. Ingram, J. Szabó and L. Pásztor

I. Introduction

Soil organic matter (SOM) plays a central role in nutrient availability, soil stability and in the flux of trace greenhouse gases between land surface and the atmosphere. It represents a major pool of carbon within the biosphere, estimated at between 1400 x 10^{15} g (Post et al., 1982) and 1500 x 10^{15} g (Batjes, 1996) globally, about twice that in atmospheric CO_2 and acts as both a source and a sink for carbon and nutrients.

The ability to predict the effects of climate, atmospheric composition and land-use change on SOM dynamics is essential in formulating environmental, agricultural and social / economic policies. Mathematical models encapsulate our best understanding of SOM dynamics and are essential tools for predicting the effects of environmental change, for testing specific scenarios and for developing strategies to mitigate the effects of environmental change. Changes in climate are likely to influence the rates of accumulation and decomposition of SOM through changes in temperature, moisture and the rate of return of plant residues to the soil. Other changes, especially in land use and management, may have even greater effects. Land-use change in the tropics is recognized to be of critical importance in the global carbon cycle (Schimel, 1995) since (a) SOM turnover is faster in tropical than in temperate ecosystems (Trumbore, 1993; Paustian et al., 1997b), (b) tropical ecosystems contain a large amount of carbon (Foley, 1994), and (c) land-use change is occurring rapidly in tropical regions (Dixon et al., 1994).

In this chapter, the soil carbon models participating in the Global Change and Terrestrial Ecosystems Soil Organic Matter Network (GCTE-SOMNET) are briefly reviewed. A discussion follows describing which of these have been applied to natural ecosystems and which have been evaluated for use in the tropics. The approaches used by these models are then described, with a description of the type of input data required and the type of output provided.

Many models have been used to describe and predict changes in soil carbon at the point, plot or field scale but few have been used at regional, national or continental scales. Modeling of carbon dynamics at the regional level and higher presents a number of problems, for example, how to simulate carbon turnover at depth, how to derive soil carbon inputs or estimates of net primary productivity (NPP), and how to acquire data on land-use / land-management history. These and other potential pitfalls are outlined briefly in this chapter.

Other problems arise when applying models in the tropics when they have been developed for use with nontropical climates, soils and ecosystems. These difficulties can be overcome by a full evaluation of the model using experiments from within the same region to check that the models are performing well and, if necessary, to recalibrate the models for the specific, local conditions.

Examples of the application of one soil carbon model, the Rothamsted Carbon Model (RothC), to a semi-arid agroecosystem in Syria and to a tropical Brazilian chronosequence where native forest has been cleared for pasture over the course of 23 years are provided.

In the last sections of this chapter, the application of RothC to a temperate dataset originating from Hungary is described. This application demonstrates the type of Carbon Model-Geographical Information System (GIS) linkage that will be required when modeling soil carbon dynamics at the regional level in tropical ecosystems, a strategy for which is described.

II. SOM Models Participating in SOMNET – Previous Evaluation in Tropical Ecosystems and Their Input Requirements

The Global Change and Terrestrial Ecosystems Soil Organic Matter Network (GCTE-SOMNET) is a network of SOM modellers and holders of data from long-term experiments that was established to facilitate scientific progress in predicting the effects on SOM of changes in land-use, agricultural practice and climate. Since 1995, SOMNET has attracted contributions from 29 SOM modellers and over 70 long-term experimentalists from around the world (Smith et al., 1996a, 1996b; Powlson et al., 1997; Smith et al., 1997a). Metadata (as well as a number of actual datasets) were collected from each participant; from these returns, an online database was constructed which is available for free global access at URL http://saffron.res.bbsrc.ac.uk/cgi-bin/somnet. The data is summarized in Smith et al. (1996a).

Twenty-four models registered with SOMNET in 1996 have recently been reviewed in detail (Molina and Smith, 1997; Smith et al., 1998a). In this chapter, the current 29 SOMNET models are examined for their input-data requirements, and to see which have previously been applied to tropical ecosystems. Table 1 lists the SOMNET models and shows the ecosystem-types and climatic regions for which each model has been evaluated.

Nine of the SOMNET models have been evaluated for use in tropical ecosystems (Table 1), i.e., CENTURY, DNDC, EPIC, McCaskill and Blair CNSP pasture model, MOTOR, RothC, SOCRATES, SOMM and Sundial. Of these, only CENTURY and RothC have so far been evaluated for grassland, forestry and arable ecosystems in the tropics.

A recent comparison of nine leading SOM models (Powlson et al., 1996; Smith et al., 1997d) was conducted in which they were evaluated against 12 long-term datasets from 7 long-term experiments in the temperate climatic region. A group of six models (RothC, CENTURY, DAISY, CANDY, NCSOIL and DNDC) performed significantly better than did three others (Smith et al., 1997e). Of these six, only two models were able to simulate all land-uses for the entire duration of each experiment (RothC and CENTURY).

Table 2 describes the data-input requirements for each model and the main soil outputs that each provide. Most models require meteorological and soil input data that are readily available, i.e., simple meteorological data on a daily or monthly timestep and simple descriptions of the soil. All models also require details of land-management and land-management history (Table 2) which are not always so easily available. The availability of good quality data in a form that can be made spatially explicit may often limit confidence in the use of SOM models at the regional level.

All of the models output changes in total carbon and many output a variety of other values (e.g., nitrogen dynamics, soil temperature and water dynamics, gaseous losses and other nutrients (Table 2). The information summarized in Tables 1 and 2 can be used to select a model for use in tropical ecosystems, depending upon the type of ecosystem to be modeled, and the breadth and quality of input data needed to run the models.

Table 1. Ecosystem type and climatic region for which each GCTE-SOMNET model has been evaluated (ecosystem separated from climatic region by hyphen)

Model	Ecosystem-climatic region	Reference
ANIMO	G-CT, A-CT	Rijtema and Kroes (1991)
CANDY	A-CT, A-WTST, G-CT, H-CT	Franko et al. (1995), Franko (1996), Franko et al. (1997)
CENTURY	N-CTB, N-CT, N-WTST, N-T, G-CT, G-WTST, G-T, A-CT, A-WTST, A-T, F-CT, F-WTST, F-T	Parton et al. (1988), Kelly et al. (1997)
Chenfang Lin Model	None - growth chamber only	Lin et al. (1987)
DAISY	A-CT, A-WTST, G-CT	Svendsen et al. (1995), Mueller et al. (1996)
DNDC	A-CT, A-WTST, A-T, G-CT	Li et al. (1994), Li et al. (1997)
DSSAT	None specified	Hoogenboom et al. (1994)
D3R	A-CTB, A-CT, A-WTST	Douglas and Rickman (1992)
Ecosys	A-CTB, A-CT, A-WTST	Grant et al. (1993a, 1993b), Grant (1995)
EPIC	A-CT, A-WTST, A-T	Williams (1990)
FERT	A-CT, A-WTST	Kan and Kan (1991)
GENDEC	N-PSP, G-WTST	Moorhead and Reynolds (1991)
Hurley Pasture ITE (Edinburgh) Forestry Model	N-CT, G-CT, F-CT	Thornley and Verberne (1989), Thornley and Cannell (1992), Arah et al. (1997)
ICBM	A-CTB, A-CT	Andrén and Kätterer (1997)
KLIMAT-SOIL-YIELD	A-CTB, A-CT	Sirotenko (1991)
McCaskill and Blair CNSP Pasture Model	N-CT, G-CT, G-WTST, G-T	McCaskill and Blair (1990a, 1990b)
Model of Humus Balance	A-CTB, A-CT	Schevtsova and Mikhailov (1992)

Table 1. continued --

Model	Ecosystem-climatic region	Reference
MOTOR	A-CT, A-T, G-CT, H-CT	Whitmore (1995), Whitmore et al., (1997)
NAM SOM	N-CTB, N-CT, A-CT	Ryzhova (1993)
NCSOIL	N-CTB, N-CT, N-WTST, A-CT, A-WTST, G-CT	Molina et al. (1983), Molina et al. (1997)
O'Leary Model	A-CT	O'Leary (1994)
Q-Soil	A-CTB, F-CTB, F-WTST	Ågren and Bosatta (1987)
RothC	N-CT, G-CTB, G-CT, G-WTST, G-T, A-CTB, A-CT, A-WTST, A-T, F-CT, F-WTST, A-T	Jenkinson and Rayner (1977), Coleman et al. (1997)
SOCRATES	G-WTST, A-CT, A-WTST, A-T	Grace and Ladd (1995)
SOMM	N-PSP, N-CTB, N-CT, N-WTST, N-T, G-CT, G-WTST, G-T, F-PSP, F-CTB, F-T	Chertov and Komarov (1996)
Sundial	A-CT, A-WTST, A-T, G-CT	Smith et al. (1996e), Bradbury et al. (1993)
Verberne	G-CT, A-CTB, A-CT, A-WTST, F-CT	Verberne et al. (1990)
VOYONS	G-CT, F-CT	André et al. (1994)
Wave	G-CT, A-CT, A-WTST	Vanclooster et al. (1995)

Key: <u>Ecosystem</u>: N = natural vegetation, G = grassland (managed and unmanaged), A = arable, H = horticulture, F = forestry/woodland; <u>Climatic region</u>: PSP = polar / sub-polar, CTB = cold temperate boreal, CT = cool temperate, WTST = warm temperate / sub-tropical, T = tropical.

Table 2. Broad categorization of input requirements for each GCTE-SOMNET model

Model	Timestep	Meteorology	INPUTS Soil and plant	Management	OUTPUTS Soil outputs	Notes
ANIMO	Day, week, month	P, AT, Ir, EvW	Des, Lay, Imp, Cl, OM, N, pH	Rot, Til, Fert, Man, Res, Irr	C, N, W, ST, AtN gas	
CANDY	Day	P, AT, Ir	D, Imp, W, N, C, Wi, PD, Nup	Rot, Til, Fert, Man, Res, Irr	C, N, W, ST, AtN gas	
CENTURY	Month	P, AT	W, Cl, OM, pH, C, N	Rot, Til, Fert, Man, Res, Irr	C, BioC, 13C, AtN 14C, N, W, ST, gas	
Chengang Lin	Day	ST	OM, BD, W	Man, Res	C, BioC, gas	
DAISY	Hour, day	P, AT, Ir, EvG	Lay, Cl, C, N, pH, PS	Rot, Til, Fert, Man, Res, Irr	C, BioC, N, AtN W, ST, gas	
DNDC	Hour, day, month	P, AT	Lay, Cl, Om, pH, BD	Rot, Til, Fert, Man, Res, Irr	C, BioC, N, AtN W, ST, gas	
DSSAT	Hour, day, month, year	P, AT, Ir	Des, Lay, Imp, W, Cl, PS, OM, ph, C, N	Rot, Til, Fert, Man, Res, Irr	C, BioC, N, W, ST	Decision support shell incorporating a range of models
D3R	Day	P, AT	Y, PS	Rot, Til, Res	Decomposition of surface and buried residue	Decomposition model

Table 2. continued --

Model	Timestep	Meteorology	INPUTS Soil and plant	Management	OUTPUTS Soil outputs	Notes
Ecosys	Minute, hour	P, AT, Ir, WS, RH	Lay, W, Cl, CEC, PS, OM, pH, N, BD, PG, PS	Rot, Til, Fert, Man, Res, Irr, AtN	C, BioC, N, W, ST, pH, Ph, EC, gas, ExCat	
EPIC	Day	P, AT	Lay, Imp, W, Cl, OM, pH, C, BD, Wi	Rot, Til, Fert, Man, Res, Irr	C, BioC, N, W, ST	
FERT	Day	P, AT, WS	Des, Lay, W, Cl, OM, pH, C, N, BD, W, Ph, K, Nup, Y, PS	Rot, Til, Fert, Man, Res, Irr	C, N, Ph, K	
GENDEC	Day, month	ST, W	W, InertC, LQ	Can be used - not essential	C, BioC, N, gas, LQ	Decomposition model
Hurley Pasture ITE (Edinburgh) Forestry Model	Day	P, AT, Ir, WS	W, Cl, PS	Rot, Fert, Irr, AtN	C, BioC, N, W, gas	
ICBM	Day, year	Combination of weather and climate	Many desirable: none essential	C inputs to soil	C	Simple two-component model solved analytically
KLIMAT-SOIL-YIELD	Day, year	P, AT, ST, Ir, EvG, EvS, VPD, SH	Des, Lay, Imp, W, Cl, PS, OM, pH, C, N	Fert, Man, Res, Irr	C, BioC, N, W, ST	

Table 2. continued --

Model	Timestep	Meteorology	INPUTS Soil and plant	Management	OUTPUTS Soil outputs	Notes
McCaskill and Blair CNSP Pasture Model	Day	P, AT, Ir	Lay, Imp, W, Cl, CEC, OM, pH, C, N, PS, AS	Fert	C, N, W, ST	
Model of Humus Balance	Year	Climate based on P and AT	Des, Lay, PS, OM, pH, C, N	Rot, Fert, Man	C, N	Statistical model
MOTOR	User specified	P, AT, EvG	Des, OM	Rot, Til, Fert, Man	C, BioC, 13C, 14C, gas	
NAM SOM	Year	P, AT	Des, PS, OM, Ero Man, RES		C, BioC	
NCSOIL	Day	ST (P, AT)	W, OM, C, N	Fert, Man, Res	C, BioC, 14C, N, 15N, gas	
O'Leary Model	Day	P, AT	Lay, W, Cl, pH, N	Til, Fert, Res	C, BioC, N, W, ST, gas, ResC, ResN	Wheat-fallow rotations only
Q-Soil	Year	Optional	C, N	Rot, Fert, Man, Res, AtN	C, BioC, 13C, N	
RothC	Month	P, AT, EvW	Cl, C, Inert C (can be estimated)	Man, Res, Irr	C, BioC, gas, 14C	
Socrates	Week	P, AT	CEC, Y	Rot, Fert, Res	C, BioC, gas	
SOMM	Day	P, ST	OM, N, AshL, NL	Man	C, N, gas	

Table 2. continued –

Model	Timestep	Meteorology	INPUTS Soil and plant	Management	OUTPUTS Soil outputs	Notes
Sundial	Week	P, AT, EvG	Imp, Cl, W, Y	Rot, Fert, Man, Res, Irr, AtN	C, BioC, N, 15N, W, gas	
Verberne	Day	P, AT, Ir, WS, EvS	Des, W, Cl, PS, OM, C, N	Man, AtN	C, BioC, N, W	
VOYONS	Day, week, month	P, ST	Cl, OM, C, N	Fert, Man, Res, Irr, AtN	C, BioC, 13C, 14C, N, gas	
Wave	Day	P, AT, Ir, EvG	Lay, OM, C, N, W, PG	Rot, Til, Fert, Man, Res, Irr, AtN	C, N, W, ST, gas	

Key: Meteorology: P = Precipitation, AT = Air Temperature, ST = Soil Temperature, Ir = Irradiation, EvW = Evaporation over water, EvG = Evaporation over grass, EvS = Evaporation over bare soil, WS = Wind Speed, RH = Relative Humidity, VPD = Vapour Pressure Deficit, SH = Sun Hours. Soil & Plant Inputs: Des = Soil description, Lay = Soil layers, Imp = Depth of impermeable layer, Cl = Clay content, OM = Organic matter content, N = Soil nitrogen content/dynamics, C = Soil Carbon content/dynamics, InertC = Soil inert carbon content, pH = pH, W = Soil water characteristics, Wi = Wilting point, PD = Soil particle size distribution, CEC = Cation Exchange Capacity, Ero = Annual erosion losses, BD = Soil bulk density, PG = Plant growth characteristics, PS = Plant species composition, AS = Animal species present, Y = Yield, Nup = Plant Nitrogen Uptake, LQ = Litter quality, AshL = Ash content of litter, NL = N content of litter. Management input details: Rot = rotation, Til = tillage practice, Fert = Inorganic fertilizer applications, Man = Organic manure applications, Res = Residue management, Irr = Irrigation, AtN = Atmospheric nitrogen inputs. Soil outputs: C, N, W, LQ and ST as above. BioC = Biomass carbon, 13C = ^{13}C dynamics, 14C = ^{14}C dynamics, 15N = ^{15}N dynamics, gas = gaseous losses (e.g., CO_2, N_2O, N_2), ResC = Surface residue carbon, ResN = Surface residue nitrogen, Ph = Phosphorus dynamics, K = Potassium dynamics, EC = Electrical conductivity, ExCat = Exchangeable cations. NB: N in the soil inputs and outputs section is used to denote all aspects of the N cycle.

III. Modeling SOC Dynamics at the Regional Level – Current Approaches

The modeling of SOC dynamics at the regional level spans a continuum between extremely simple approaches whereby empirical relationships are assumed to apply over large areas, to complex approaches whereby dynamic simulation models are linked to spatially explicit data, usually using Geographical Information Systems (GIS), in order to account for spatial differences in meteorology, soil and land-use. A theoretical discussion of issues involved in upscaling can be found elsewhere (Smith, 1996). Examples of simple approaches are given in such papers as Gupta and Rao (1994), Smith et al. (1996c), Smith et al. (1997b, 1997c) and Smith et al. (1998b). In the above-mentioned papers by Smith et al., the authors use simple statistical relationships to estimate changes in soil C pools over 100 years for the whole of Europe (to as far east as the Baltic States). The approach relies upon establishing relationships between a range of land management practices and changes in SOC over time, derived from 42 European long-term experiments (Smith et al., 1996d), and applying these changes to soil carbon pools calculated from soil organic matter maps of Europe. Yearly estimates of the carbon sequestration potential for six European scenarios based on this simple approach are shown in Figure 1.

Simple approaches such as those described above take no account of local variations in meteorology and soil. To account for such differences, spatially explicit data needs to be used, often through coupling a dynamic simulation model to a GIS database. Examples of this approach are given for the EPIC model in Lee et al. (1993), for CENTURY by Donigan et al. (1994), and for RothC by Parshotam et al. (1995) and Falloon et al. (1998). Details of the latter study are given in section VI below. Such approaches allow spatial variations in meteorology, soil and land-management to be accounted for explicitly, as well as allowing changes in SOC to be visualized spatially. Because of these advantages, the coupled simulation model-GIS system is the preferred option for regional SOC studies and is being used by European research consortia in the EU-funded modular modeling activities ETEMA (European Terrestrial Ecosystem Modeling Activity) and MAGEC (Modeling Agroecosystems under Global Environmental Change).

IV. Potential Pitfalls Associated with Large Scale Modeling of SOC Dynamics in Tropical Ecosystems

Despite model applications at large (global) spatial scales (CENTURY: Parton et al., 1987; Potter et al., 1993; Schimel et al., 1994; and RothC: Post et al., 1982; Jenkinson et al., 1991; Post et al., 1996; King et al., 1997), modeling of carbon dynamics at the regional level and higher presents a number of problems, the most important of which are outlined below.

The most obvious problem associated with large scale modeling in the tropics is the application of models to conditions other than those to which they were parameterized. Most models were originally developed for use in a specific climate (often cool temperate) in a specific ecosystem on a limited range of soil types. These difficulties can be overcome by evaluating the model on data from within the new region (see Table 1) to check that the models are performing well and, if necessary, to recalibrate the models for the specific, local conditions. Examples of the evaluation of one SOM model, RothC, in semi-arid and tropical ecosystems are presented in section V.

The simulation of carbon turnover at depth often presents a problem in that many of the models used for large-scale modeling are parameterized as single layer, topsoil models. RothC for example was originally designed to describe the top 23 cm of soil (Jenkinson and Rayner, 1977; Coleman and Jenkinson, 1996a), while CENTURY was designed to simulate the top 20 cm of soil (Parton et al.,

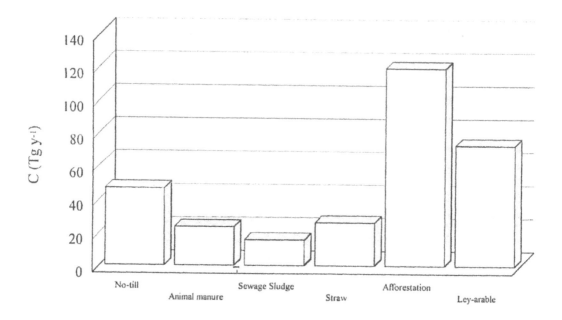

Figure 1. Figures are for the whole of Europe excluding most of the former Soviet Union but including Belarus and the Ukraine (total land area = 490 M ha; arable land area = 135 M ha; SOC content to 30 cm in arable land = 7.18 Pg). All values were estimated using (a) soil carbon contents as described in Smith et al. (1997b, 1997c), and (b) relationships between yearly changes in soil organic carbon content and management practices revealed by relevant European field experiments. No-till assumes conversion of all arable agriculture to no-till. Animal manure figures are for application of animal manure at 10 t ha^{-1} y^{-1} to all arable land (Smith et al., 1997b, 1997c). Sewage sludge figures are for application of sludge at 1 t ha^{-1} y^{-1} which would be sufficient to cover about 11% of arable land in Europe (Smith et al., 1997b, 1997c). Straw figures are for the incorporation of all cereal straws into the land on which the crops were grown. There is sufficient straw to provide an incorporation rate of about 5 t ha^{-1} y^{-1} (Smith et al. 1997b, 1997c). Afforestation is for natural woodland regeneration on 30% arable land which is the upper limit predicted to be surplus to arable requirements by 2010. It includes the carbon mitigation potential of the wood produced assuming 50:50 biofuel:bioproduct utilization of the wood (Smith et al., 1997b, 1997c). Ley-arable (or mixed cropping) figures are for extensification of arable agriculture (leaving current grassland at present level). The predicted 30% surplus of arable land by 2010 could be used to allow less intensive use of all land. The pasture or ley phases could then be used for less intensive animal production by raising pigs and poultry in outdoor units (Smith et al., 1997b, 1997c).

1987). Post et al. (1982), in their application of RothC, assumed that most of the carbon was in the upper layers of soil but applied the model with the same SOM turnover rate constants to a depth of 1 m. Differences in the rate of SOM turnover at depth may have profound implications for the overall SOM dynamics of an ecosystem. Figure 2 shows the predicted differences in average radiocarbon age of SOM at different depths for different rates of SOM turnover. The model (RothC) was fitted to radiocarbon dates (expressed as δ ^{14}C values) measured on soil samples taken in 1881 at three depths,

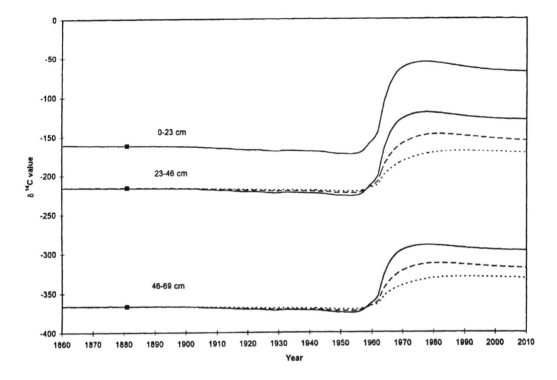

Figure 2. Difference in average radiocarbon content of SOM at different depths in the unfertilized control plot of the Broadbalk Continuous Wheat Experiment at Rothamsted assuming different rates for SOM turnover in deeper soil layers. Depths 0–23, 23–46 and 46–69 cm are labelled. Solid lines show predicted δ ^{14}C value using standard turnover rate constants, broken lines show values assuming 1/2 standard rate constant, and dotted lines show values assuming 1/4 standard rate constant. See text for further details.

0–23, 23–46 and 46–69 cm, on the Broadbalk Continuous Wheat Experiment at Rothamsted, UK. The model was then run using the standard turnover rate constants for each depth and using rate constants that were 1/2 and 1/4 the standard rates for depths 23–46 and 46–69 cm. A thermonuclear test ("bomb") effect can be seen at all depths during the 1960s (see Jenkinson et al., 1992).

Radiocarbon measurements clearly show (Trumbore, 1993) that SOC becomes older down the profile. SOC models should not be applied unmodified to subsoils when calculating SOC sequestration potential.

All SOM models are highly dependent upon the quality and quantity of carbon inputs to the soil system. Plant litter quality is frequently expressed as a C/N ratio, lignin/N ratio, or is fitted empirically to give a specified decomposable/resistant plant material ratio (see Molina and Smith, 1997; McGill, 1996). For the quantity of carbon returned to the soil, which is a critical factor in determining the accurate prediction of SOM dynamics by a model, some models (e.g., CENTURY; Parton, 1996) have simple NPP models which provide C inputs for a range of plant types, others (e.g., SUNDIAL; Smith et al., 1996e) have a series of default C inputs (which for agricultural crops is assumed to vary with yield) for various plant species derived from literature values, while others (e.g., RothC; Coleman and Jenkinson, 1996a) require that plant C inputs be entered. For any of these approaches, small errors in the amount of plant C returning to the soil can lead to large errors in predicted SOM dynamics. For this reason, plant C returns must be estimated effectively in any regional scale model application. In the GCTE approach to regional scale SOC modeling (see Section VII), it is intended that RothC and CENTURY will be given identical NPP estimates (either modeled or estimated from remotely-sensed data) so that this source of discrepancy between the models is removed.

The final potential problem addressed in this section is that of data adequacy. Models are often initially tested against data collected from detailed and well controlled experiments. Data of this quality cannot be collected for each portion of land for which a model will be run at large scales; the data that are available are necessarily more crude. While it is possible to collect good quality data on soil characteristics by soil survey, and good quality data on land-use and recent land-use change from remotely sensed data (aerial photographs and satellite imaging), it is often far more difficult to gain quality information on land-use history. Details of land-use history are essential for establishing initial SOM pool sizes—a site cleared from forest for arable use 10 years ago will have very different characteristics from a site cleared 200 years ago. This presents fewer problems when large areas of a region are still in a natural state, for example in uncleared portions of the Brazilian rainforest, but may cause significant problems in areas where land-use change has occurred over the past 200 or so years, e.g., many agricultural areas in North America. In many cases assumptions need to be made about land-use history. Where experimental sites exist within the region, the data can be used to "ground truth" the assumptions made regarding land-use history.

The potential pitfalls outlined in this section make it essential that models are well tested and that all shortcomings and uncertainties in the input data are considered when interpreting modeled regional results. Notwithstanding, the use of dynamic simulation models linked to GIS databases provides our best predictive tool for assessing the impacts on SOC of current and future environmental change.

V. Modeling SOC Dynamics in Semi-arid and Tropical Ecosystems – Some Examples

In this section examples of the application of one SOM model, RothC, to semi-arid and tropical ecosystems are presented beginning with a recent example of the application of RothC in a semi-arid agroecosystem in Syria and then the application of RothC to a tropical forest-pasture chronosequence in Brazil.

A. Modeling SOC Dynamics in a Semi-arid Region in Syria

Figure 3 shows the measured and predicted changes in SOC for the ICARDA site at Tel Hadya in Syria (Jenkinson et al., 1998) for two contrasting rotations: wheat (*Triticum aestivum*) -medic (*Medicago* spp.) and wheat-fallow. For calibration, the model was fitted to a measurement of SOC in 1993 on each rotation which was sampled independently of the main experiment.

Model performance can be evaluated quantitatively using statistics such as Root Mean Square Error (*RMSE*; Smith et al., 1996f). A full description of the range of statistics available for quantitative model evaluation is given in Smith et al. (1996f) and Smith et al. (1997e). As seen in Figure 3, the modeled and measured values are close for the wheat-fallow rotation (*RMSE* = 3.56) but less close for the wheat-medic rotation (*RMSE* = 6.19). This example demonstrates the general applicability of RothC in semi-arid agroecosystems but also shows that inaccurate site specific calibration can lead to discrepancies between modeled and measured values.

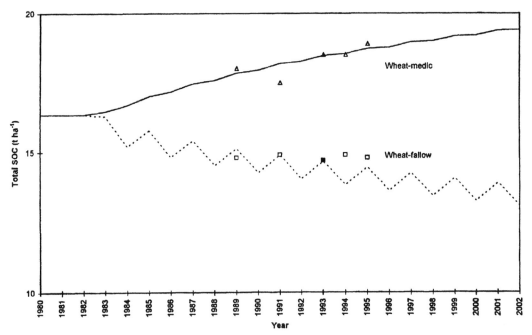

Figure 3. Measured and predicted (using RothC) changes in SOC for two different rotations from the ICARDA two-course rotation experiment at Tel Hadya, Syria: wheat-medic (soild line) and wheat-fallow (broken line). Measured data are shown as open symbols while independent calibration data (1993 only) are shown as solid symbols.

B. Modeling a Tropical Forest-pasture Chronosequence from Brazil

Data on C turnover were collected in 1992 from Fazenda Nova Vida, Rondônia, Brazil on two series of tropical pasture sites which were converted from natural forest between 1972 and 1992 (Moraes et al., 1995) forming two tropical forest-pasture chronosequences (hereafter refered to as CS1 and CS2). Native forest was cleared for pasture on these plots in 1972, 1979, 1983, 1987, 1989 and 1992 giving pastures aged 20, 13, 9, 5 and 3 years in 1992, as well as an area of native forest. The modeling of these two chronosequences using RothC are presented here. CS1 was used to fit the model to the data to estimate the yearly C inputs required to give observed SOC values (calibration). Using these C inputs, the model was then run independently for CS2.

   The initial C content of the soil (0–30 cm) under native forest in 1972 was estimated from measurements in 1992 as 40.8 t C ha[-1] for CS1 and 41.2 t C ha[-1] for CS2. The inert organic matter (IOM) content was estimated to be 1.9 t ha[-1] by fitting the model to the mean radiocarbon age value (δ [14]C = 109.8 for 0–30 cm) for the CS1 native forest site as described in Coleman and Jenkinson (1996b). A mean clay content (0–30 cm) across each of the different chronosequence sites was used for the model runs, though the clay contents differed quite widely (CS1: mean = 28.7%; range = 18.3–34.5%, CS2: mean = 22.9%; range = 10.3–39.86%).

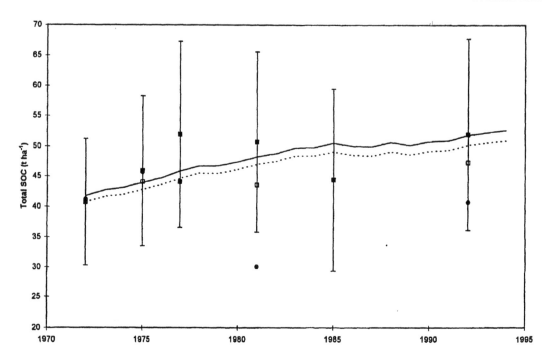

Figure 4. Modeled and measured C turnover (using RothC) on two tropical forest-pasture chronosequences (Moraes et al., 1995) at Fazenda Nova Vida, Rondônia, Brazil. Native forest was cleared for pasture on these plots in 1972, 1979, 1983, 1987, 1989 and 1992, giving pastures aged 20, 13, 9, 5 and 3 years in 1992 and an area of native forest. Chronosequence 1 (CS1) measured SOC points are shown by ■ (with standard error bars); modeled SOC for CS1 is shown by _____ ; measured SOC points for CS2 are shown by ● (standard error bars omitted for clarity); modeled SOC for CS2 (using mean clay%) is shown by --------; and modeled SOC points for CS2 (using actual clay%) are shown by □. See text for further details.

For CS1, the model was first run to equilibrium using a meteorological data file containing the 20-year local averages of monthly precipitation, air temperature and evaporation. The model needed an annual input of carbon under native forest of 7.6 t C ha^{-1} y^{-1} to account for the equilibrium SOC content of CS1. The model was then run for the pasture phase using local monthly meteorological records. The model predicted an annual input of carbon under pasture of 12.15 t C ha^{-1} y^{-1} to account for the SOC content of CS1. Using these C inputs for forestry and pasture, the model fit is reasonable (*RMSE* = 7.93; Figure 4). Variations in the measured values are attributable to differences in soil clay content at different chronosequence sampling sites (see below).

The model was then run again for CS2 using the same yearly C inputs (but using the CS2 mean clay content; 22.9%). Figure 4 shows the modeled SOC for CS2 (*RMSE* = 19.48) which is predicted to increase in a similar manner to CS1. The reason for the poor fit is the variation in clay content at each chronosequence site giving a scatter in the measured points; the lower than average clay content at sites corresponding to 1981 and 1992 in the chronosequence lead to a lower SOC content. A similar trend can be seen for CS1 (Figure 4) where the 1985 measured point is lower due to a clay content of 25.8% compared to the mean (28.7%).

To account for variations in clay content, the model was run again using the same annual C inputs, but with the actual clay contents. This provided a modeled point for comparison with each measured point, but the modeled points are derived from different model runs and therefore cannot be joined on the graph. The predicted points modeled in this manner are also shown in Figure 4 and show a much better fit (*RMSE* = 4.75); better in fact than the calibration run for CS1. This demonstrates that variations in clay content at different sites can have a profound affect on C accumulation and highlights the weakness of chronosequence data (where each sample is taken from a different plot) compared to long-term experimental data (where each sample is taken from the same plot). Had the clay content been the same at each site (as assumed in the first CS2 run), the model predicts a steady increase in SOC between 1972 and 1992.

It is interesting to note that plant C returns required to explain the observed increase in SOC after conversion of forest to pasture, are greater under pasture than under native forest. As well as greater inputs, there is a change in litter quality (i.e., the litter becomes more decomposable) under pasture, so that higher C inputs are required to maintain the same SOC level. Moraes et al. (1995) reported similar findings using the same data. Using δ ^{13}C values, they found that carbon derived from forest (Cdf) decreased sharply during the first 9 years until a stable value (about half the total SOC) was reached after 20 years, while carbon derived from pasture (Cdp) increased more rapidly than Cdf decreased. The total SOC (0–30 cm) after 20 years was about 17 to 20% higher under pasture than under native forest. The work of Veldkamp (1994) suggests that the major source of SOC under pasture is from below-ground net primary production. A similar increase in SOC has been observed elsewhere in the Amazon after deforestation (in areas that have not been subjected to severe burning or overgrazing; Moraes et al., 1995; Neill et al., 1997), though depletion of soil C has also been reported (e.g., Allen, 1985; Mann, 1986; Detwiller and Hall, 1988). Despite these increases in SOC, only about 10% of the total ecosystem C lost after deforestation (due to tree removal, burning, etc.) can be recovered (Fearnside, 1997; Neill et al., 1997).

VI. Linking Dynamic Simulation Models to Spatially Explicit Datasets – An Example

In section III the advantages of linking dynamic simulation models to GIS databases for regional simulation and prediction of SOC dynamics were outlined. In this section, it is shown how a C turnover model (RothC) and a GIS database (using the ArcView platform) can be linked. For demonstration purposes, the impacts on SOC after 100 years of the afforestation of all current arable land in a nontropical study area are presented. This scenario was chosen because it shows clear changes in SOC, not because it is regarded as a realistic land-use change option. Full details of this study are given in Falloon et al. (1998).

The study area was a central region in Hungary (an area of 24,804 km^2) bounded by coordinates (UTM 34) 5183816.64, 325048.59, 5340133.61, 472979.10. Soil data were taken from the HunSOTER database of Hungary (scale 1:500,000; Pasztor et al., 1996; Szabo et al., 1996) containing 275 representative soil profiles within the area. Variables for the uppermost horizon for SOC (%), clay % and soil bulk density were used. The profile data were linked to the 351 SOTER unit polygons (representing areas with unique soil, land form and lithology characteristics). The dataset was used to calculate the soil IOM content in the absence of radiocarbon data (see Falloon et al., 1997) and SOC (t C ha^{-1}) to 30 cm depth.

Land-use data were taken from the CORINE database for Hungary (scale 1:100,000; Büttner, 1997; Büttner et al., 1995a, 1995b) with 6470 polygons within the study area. The 44 original land-use codes were rationalized into 4 codes: arable, grassland, forestry and "not used" (includes marsh, water bodies and urban areas). Land use polygons were linked to meteorological data from the nearest of 17 meteorological stations, each providing long-term averages of mean monthly temperature, rainfall, and evaporation from 1931 to 1960 (Varga-Haszonits, 1977). This provided a linked land-use/meteorological layer.

The soil layer was overlaid upon the linked land-use/meteorological layer giving 12086 polygons representing a unique combination of land-use, soil, and meteorology. Excluding those land-uses not used, 9888 polygons were used in the modeling exercise.

The original source code of RothC was altered to take input from a fixed width ASCII file output from the GIS and write results to a new ASCII file, which could then be loaded into Excel, Access and GENSTAT for analysis and ArcView for visualisation. The model was run to equilibrium using default plant input values derived from a dataset of equilibrium land-use treatments at 60 SOMNET sites across the globe (default plant inputs were 3.55 t C ha^{-1} y^{-1} for arable, 3.72 t C ha^{-1} y^{-1} for grasslands, and 7.09 t C ha^{-1} y^{-1} for forestry). The modeled and measured SOC values were then matched by the model by analytically solving for the plant C inputs required to achieve the observed equilibrium SOC value. For each arable-land polygon, the model then ran for a further 100 years, using a hypothetical scenario of immediate afforestation (i.e., with default forestry plant C inputs and with the default forestry plant litter quality factor). The model run for all 9888 polygons was completed on a 66 MHz 486 personal computer in just 1.5 hours.

The total SOC stock for the whole area was calculated as 1.40 Tg, which increased to 1.89 Tg after 100 years. The change in SOC (t C ha^{-1}) 100 years after all arable land was afforested can be seen in Figure 5.

This is a change in SOC (0–30 cm) of 35% after 100 years over the whole area (i.e., 0.35% y^{-1} on average with more rapid changes earlier). For the same scenario, using the regression equation for the yearly percentage change in SOC of Smith et al., (1997b), the total SOC stock 100 years after afforestation of all arable land would be 1.97 Tg, slightly higher than this estimate of 1.89 Tg. However, we have more confidence in the present estimate since we have explicitly accounted for differences in soil type and climate. A more sophisticated approach to estimating NPP or plant C inputs (see section IV) could further improve the accuracy of the estimates presented here.

This simple example, in which a dynamic simulation model is linked to a GIS database, demonstrates a flexible and powerful methodology for assessing the impacts of different scenarios of land-use, management and climate change on SOC at the regional scale.

VII. Modeling the Impact of Land-use Change on Regional Soil Organic Carbon Stock in the Tropics

The spatial analysis method discussed in Section VI is equally applicable to tropical systems (subject to the potential problems outlined in Section IV) and is being developed as a major part of the GCTE soil organic matter research agenda (Ingram and Gregory, 1996). Building on the concept originally outlined by Elliott and Cole (1989) and refined by Paustian et al. (1997a), the strategy involves the integration of a series of approaches: process studies, simulation models and spatial extrapolation to build regional-scale projections of change (Figure 6).

The rationale for this approach is that, while methods based on summation of mapping unit values (e.g., Batjes, 1996) or interpolation (e.g., Moraes et al., 1995) may be suitable for estimating current soil carbon stocks at regional scale, only modeling approaches can give indications of future levels

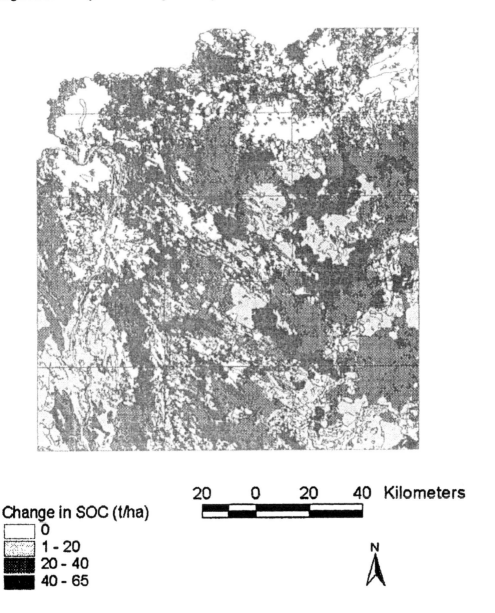

Figure 5. The change in SOC (t C ha⁻¹) 100 years after all arable land was hypothetically afforested in an area in Central Hungary. Areas showing no change are non-arable areas and were not changed in our scenario.

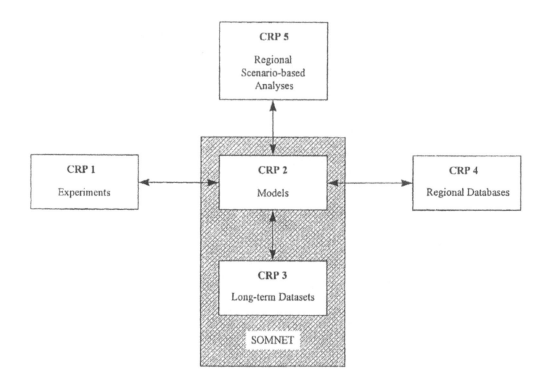

Figure 6. Conceptual framework of the GCTE soil organic matter research focus.

due to changed land-use, land-cover and/or climate. The modeling approach must however be evaluated against the other methods for the current situation. While it may not prove as satisfactory as mapping methods for estimating absolute SOC stocks, it is the only way of conducting scenario-based analyses, especially where the change in SOC stock is of primary interest (i.e., to improve the Climate Convention GHG inventory methodology). Nevertheless, a significant effort will be needed in gathering sufficient data to build adequate data layers to answer given scenario questions. The approach will therefore initially compare modeled estimates of current soil organic carbon stock at regional scale with estimates made by mapping unit methods. The next step is to estimate change over time in SOC stock (i.e., net CO_2 fluxes from soils) at national and sub-national scales as a consequence of change in driving variables, (i.e., climate, land cover, land management). A final step will be to estimate change in the soil organic carbon stock and CO_2 fluxes as a consequence of different policy options to assist in formulating policy options though improved land-use management.

Acknowledgement

We wish to thank Professor David Powlson of IACR-Rothamsted for helpful comments on this manuscript. We thank Professor Rattan Lal, Dr. John Kimble and Dr. E.A. Serrao, The Ohio State University and EMBRAPA for making it possible for P.Smith to attend the workshop at which this paper was presented. We would also like to thank Professor Tamás Németh of RISSAC, Budapest.

IACR receives grant-aided funding from the UK Biotechnology and Biological Sciences Research Council.

References

Ågren, G.I. and E. Bosatta. 1987. Theoretical analysis of the long-term dynamics of carbon and nitrogen in soils. *Ecology* 68:1181-1189.

Allen, J.C. 1985. Soil response to forest clearing in the United States and tropics: geological and biological factors. *Biotropica* 17:15-27.

André, M., J. Thiéry and L. Cournac. 1994. ECOSIMP2 model: prediction of CO_2 concentration changes and carbon status in closed ecosystems. *Adv. Space Res.* 14:323-326.

Andrén, O. and T. Kätterer. 1997. ICBM - the Introductory Carbon Balance Model for exploration of soil carbon balances. *Ecol. Applic.*

Arah, J.R.M., J.H.M. Thornley, P.R. Poulton and D.D. Richter. 1997. Simulating trends in soil organic carbon in long-term experiments using the ITE (Edinburgh) Forest and Hurley pasture ecosystem models. In: P. Smith, D.S. Powlson, J.U. Smith and E.T. Elliott (eds.) Evaluation and comparison of soil organic matter models using datasets from seven long-term experiments. *Geoderma* 81:61-74.

Batjes, N.H. 1996. Total carbon and nitrogen in the soils of the world. *Euro J. Soil Sci.* 47:151-163.

Bradbury, N.J., A.P. Whitmore, P.B.S. Hart and D.S. Jenkinson. 1993. Modelling the fate of nitrogen in crop and soil in the years following application of ^{15}N-labelled fertilizer to winter wheat. *J. Ag. Sci., Camb.* 121:363-379.

Büttner, G. 1997. Land Cover - Hungary; Final Technical Report to Phare, FOMI, Budapest (Internal Report).

Büttner, G., E. Csató and G. Maucha. 1995a. The CORINE Land Cover - Hungary project. p1813. In: 17th Int. Cartographic Conf., Barcelona, 1995.

Büttner, G., E. Csató and G. Maucha. 1995b. The CORINE Land Cover - Hungary project; p. 54-61. In: Proc. Int. Conf. on Environment and Informatics, Budapest, 1995.

Chertov, O.G. and A.S. Komarov. 1996. SOMM - a model of soil organic matter and nitrogen dynamics in terrestrial ecosystems. p. 231-236. In: D.S. Powlson, P. Smith and J.U. Smith (eds.) *Evaluation of Soil Organic Matter Models Using Existing, Long-Term Datasets.* NATO ASI I38. Springer-Verlag, Berlin.

Coleman, K. and D.S. Jenkinson. 1996a. RothC-26.3. A model for the turnover of carbon in soil. p. 237-246. In: D.S. Powlson, P. Smith and J.U. Smith (eds.) *Evaluation of Soil Organic Matter Models Using Existing, Long-Term Datasets.* NATO ASI I38. Springer-Verlag, Berlin.

Coleman, K. and D.S. Jenkinson. 1996b. RothC-26 3. A model for the turnover of carbon in soil. User Manual. Rothamsted Experimental Station.

Coleman, K., D.S. Jenkinson, G.J. Crocker, P.R. Grace, J. Klír, M. Körschens, P.R. Poulton, and D.D. Richter. 1997. Simulating trends in soil organic carbon in long-term experiments using RothC-23.6. In: P. Smith, D.S. Powlson, J.U. Smith and E.T. Elliott (eds.) Evaluation and comparison of soil organic matter models using datasets from seven long-term experiments. *Geoderma* 81:29-44.

Detwiller, R.P and A.S. Hall. 1988. Tropical forests and the global carbon cycle. *Science* 239:42-47.

Dixon, R.K., S.A. Brown, R.A. Houghton, A.M. Solomon, M.C. Trexler and J. Wisniewski. 1994. Carbon pools and flux of global forest ecosystems. *Science* 263:185-190.

Donigan, A.S. (Jr.), T.O. Barnwell (Jr.), R.B. Jackson (IV), A.S. Patwardhan, K.B. Weinrich, A.L. Rowell, R.V. Chinnaswamy and C.V. Cole. 1994. Assessment of Alternative Management Practices and Policies Affecting Soil Carbon in Agroecosystems of the Central United States. US EPA Report EPA/600/R-94/067, Athens.

Douglas, C.L. (Jr.) and R.W. Rickman. 1992. Estimating crop residue decomposition from air temperature, initial nitrogen content, and residue placement. *Soil Sci. Soc. Am. J.* 56:272-278.

Elliott, E.T. and C.V. Cole. 1989. A perspective on agroecosystem science. *Ecol.* 70:1597-1602.

Falloon, P., P. Smith, K. Coleman and S. Marshall. 1997. Estimating the size of the inert organic matter pool from total soil organic carbon content for use in the Rothamsted Carbon Model. *Soil Biol. Biochem.* 30:1207-1211.

Falloon, P., P. Smith, J.U. Smith, J. Szabó, K. Coleman and S. Marshall. 1998. Regional estimates of carbon sequestration potential: linking the Rothamsted carbon model to GIS databases. *Biol. Fert. Soil.*.

Fearnside, P.M. 1997. Greenhouse gases from deforestation in Brazilian Amazonia: Net committed emissions. *Clim. Change* 35:321-360.

Foley, J.A. 1994. Net primary productivity in the terrestrial biosphere: the application of a global model. *J. Geophys. Res.* 99(D-10):20773-20783.

Franko, U. 1996. Modelling approaches of soil organic matter turnover within the CANDY system. p. 247-254. In: D.S. Powlson, P. Smith and J.U. Smith (eds.) *Evaluation of Soil Organic Matter Models Using Existing, Long-Term Datasets.* NATO ASI I38. Springer-Verlag, Berlin.

Franko, U., B. Oelschlägel and S. Schenk. 1995. Simulation of temperature-, water- and nitrogen dynamics using the model CANDY. *Ecol. Mod.* 81: 213-222.

Franko, U., G.J. Crocker, P.R. Grace, J. Klír, M. Körschens, P.R. Poulton and D.D. Richter. 1997. Simulating trends in soil organic carbon in long-term experiments using the CANDY model. In: P. Smith, D.S. Powlson, J.U. Smith and E.T. Elliott (eds.) Evaluation and comparison of soil organic matter models using datasets from seven long-term experiments. *Geoderma* 81:109-120.

Grace, P.R. and J.N. Ladd. 1995. SOCRATES v2.00 User Manual. Cooperative Research Centre for Soil and Land Management. PMB 2, Glen Osmond 5064, South Australia.

Grant, R.F. 1995. Dynamics of energy, water, carbon and nitrogen in agricultural ecosystems: simulation and experimental validation. *Ecol. Mod.* 81:169-181.

Grant, R.F., N.G. Juma and W.B. McGill. 1993a. Simulation of carbon and nitrogen transformations in soil: mineralisation. *Soil Biol. Biochem.* 25:1317-1329.

Grant, R.F., N.G. Juma and W.B. McGill. 1993b. Simulation of carbon and nitrogen transformations in soil: microbial biomass and metabolic products. *Soil Biol. Biochem.* 25:1331-1338.

Gupta, R.K. and D.L.N. Rao. 1994. Potential of wastelands for sequestering carbon by reforestation. *Curr. Sci.* 66:378-380.

Hoogenboom, G., J.W. Jones, L.A. Hunt, P.K. Thornton and G.Y. Tsuji. 1994. An integrated decision support system for crop model applications. Paper 94-3025 presented at ASAE Meeting, Missouri, June, 1994.

Ingram, J.S.I. and P.J. Gregory (eds.). 1996. GCTE Activity 3.3 Effects of Global Change on Soils. GCTE Report 12. GCTE Focus 3 Office, Wallingford, U.K.

Jenkinson, D.S. and J.H. Rayner. 1977. The turnover of soil organic matter in some of the Rothamsted Classical Experiments. *Soil Sci.* 123:298-305.

Jenkinson, D.S., D.E. Adams and A. Wild. 1991. Global warming and soil organic matter. *Nature* 351:304-306.

Jenkinson, D.S., D.D. Harkness, E.D. Vance, D.E. Adams and A.F. Harrison. 1992. Calculating the net primary production and annual input of organic matter to the soil from the amount and radiocarbon content of soil organic matter. *Soil Biol. Biochem.* 24:295-308.

Jenkinson, D.S., H.C. Harris, J. Ryan, A.M. McNeill, C.J. Pilbeam and K. Coleman. 1998. Organic matter turnover in a calcereous clay soil from Syria under a two-course cereal rotation. *Soil Biol. Biochem.* (submitted).

Kan, N.A. and E.E. Kan. 1991. Simulation model of soil fertility. *Physiol. Biochem. Cult. Plant.* 23:3-16 (in Russian).

Kelly, R.H., W.J. Parton, G.J. Crocker, P.R. Grace, J. Klír, M. Körschens, P.R. Poulton and D.D. Richter. 1997. Simulating trends in soil organic carbon in long-term experiments using the CENTURY model. In: P. Smith, D.S. Powlson, J.U. Smith and E.T. Elliott (eds.) Evaluation and comparison of soil organic matter models using datasets from seven long-term experiments. *Geoderma* 81:75-90.

King, A.W., W.M. Post and S.D. Wullschleger. 1997. The potential response of terrestrial carbon storage to changes in climate and atmospheric CO_2. *Clim. Change* 35:199-227.

Lee, J.J., D.L. Phillips and R. Liu. 1993. The effect of trends in tillage practices on erosion and carbon content of soils in the US Corn Belt. *Wat. Air Soil Poll.* 70:389-401.

Li, C., S. Frolking and R. Harriss. 1994. Modelling carbon biogeochemistry in agricultural soils. *Glob. Biogeochem. Cyc.* 8:237-254.

Li, C., S. Frolking, G.J. Crocker, P.R. Grace, J. Klír, M. Körschens and P.R. Poulton. 1997. Simulating trends in soil organic carbon in long-term experiments using the DNDC model. In: P. Smith, D.S. Powlson, J.U. Smith and E.T. Elliott (eds.) Evaluation and comparison of soil organic matter models using datasets from seven long-term experiments. *Geoderma* 81:45-60.

Lin, C., T.-S Liu and T.-L. Hu. 1987. Assembling a model for organic residue transformation in soils. *Proc. Nat. Sci. Counc. (Taiwan) Part B* 11:175-186.

Mann, L.K. 1986. Changes in soil carbon storage after cultivation. *Soil Sci.* 142:279-288.

McCaskill, M. and G.J. Blair. 1990a. A model of S, P and N uptake by a perennial pasture. I. Model construction. *Fert. Res.* 22:161-172.

McCaskill, M. and G.J. Blair. 1990b. A model of S, P and N uptake by a perennial pasture. II. Calibration and prediction. *Fert. Res.* 22:173-179.

McGill, W.B. 1996. Review and classification of ten soil organic matter (SOM) models. p. 111-133. In: D.S. Powlson, P. Smith and J.U. Smith (eds.) Evaluation of soil organic matter models using existing, long-term datasets. NATO ASI I38. Springer-Verlag, Berlin.

Molina, J.A.E. and P. Smith. 1997. Modeling carbon and nitrogen processes in soils. *Adv. Agron.* 62:253-298.

Molina, J.A.E., C.E. Clapp, M.J. Shaffer, F.W. Chichester and W.E.Larson. 1983. NCSOIL, a model of nitrogen and carbon transformations in soil: description, calibration and behaviour. *Soil Sci. Soc. Am. J.* 47:85-91.

Molina, J.A.E., G.J. Crocker, P.R. Grace, J. Klír, M. Körschens, P.R. Poulton and D.D. Richter. 1997. Simulating trends in soil organic carbon in long-term experiments using the NCSOIL and NCSWAP models. In: P. Smith, D.S. Powlson, J.U. Smith and E.T. Elliott (eds.), Evaluation and comparison of soil organic matter models using datasets from seven long-term experiments. *Geoderma* 81:91-107.

Moorhead, D.L. and J.F. Reynolds. 1991. A general model of litter decomposition in the northern Chihuahuan Desert. *Ecol. Mod.* 56:197-219.

Moraes, J.F.L. de, B. Volkoff, C.C. Cerri and M. Bernoux. 1995. Soil properties under Amazon forest and changes due to pasture installation in Rondônia, Brazil. *Geoderma* 70:63-86.

Moraes, J.F.L. de, C.C. Cerri, J.M. Melillo, D. Kicklighter, C. Neill and P.A. Steudler. 1995. Soil carbon stocks of the Brazilian Amazon basin. *Soil Sci. Soc. Am. J.* 59:244-247.

Mueller, T., L.S. Jensen, S. Hansen and N.E. Nielsen. 1996. Simulating soil carbon and nitrogen dynamics with the soil-plant-atmosphere system model DAISY. p. 275-281. In: D.S. Powlson, P. Smith and J.U. Smith (eds.) *Evaluation of Soil Organic Matter Models Using Existing, Long-Term Datasets*. NATO ASI I38. Springer-Verlag, Berlin.

Neill, C., J.M. Melillo, P.A. Steudler, C.C. Cerri, J.F.L. de Moraes, M.C. Piccolo and M. Brito. 1997. Soil carbon and nitrogen stocks following forest clearing for pasture in the Southwestern Brazilian Amazon. *Ecol. Applic.* 7:1216-1225.

O'Leary, G.J. 1994. Soil water and nitrogen dynamics of dryland wheat in the Victorian Wimmera and Mallee. PhD Thesis, University of Melbourne.

Parshotam, A, K.R. Tate and D.J. Giltrap. 1995. Potential effects of climate and land-use change on soil carbon and CO_2 emissions from New Zealand's indigenous forests and unimproved grasslands. *Weath. Clim.* 15:47-56.

Parton, W.J., 1996. The CENTURY model. p. 283-293. In: D.S. Powlson, P. Smith and J.U. Smith (eds.) *Evaluation of Soil Organic Matter Models Using Existing, Long-Term Datasets*. NATO ASI I38. Springer-Verlag, Berlin.

Parton, W. J., D.S. Schimel, C.V. Cole and D. Ojima. 1987. Analysis of factors controlling soil organic levels of grasslands in the Great Plains. *Soil Sci. Soc. Am. J.* 51:1173-1179.

Parton, W.J., J.W.B. Stewart and C.V. Cole. 1988. Dynamics of C, N, P, and S in grassland soils: a model. *Biogeochem.* 5:109-131.

Pásztor, L., J. Szabó and G. Várallyay. 1996. Digging deep for global soil and terrain data. *GIS Europe* 5/8:32-34.

Paustian, K., E. Levine, W.M. Post and I.R. Ryzhova. 1997a. The use of models to integrate information and understanding of soil C at the regional scale. *Geoderma* 79:227-260.

Paustian, K., O. Andrén, H.H. Janzen, R. Lal, P. Smith, G. Tian, H. Tiessen, M. van Noordwijk and P.L. Woomer. 1997b. Agricultural soils as a sink to mitigate CO_2 emissions. *Soil Use Manage.* 13:230-244.

Post, W.M., W.R. Emanuel, P.J. Zinke and A.G. Stangenberger. 1982. Soil carbon pools and world life zones. *Nature* 298:156-159.

Post, W. M., A.W. King and S.D. Wullschleger. 1996. Soil organic matter models and global estimates of soil organic carbon. p. 201-222. In: D.S. Powlson, P. Smith and J.U. Smith (eds.) *Evaluation of Soil Organic Matter Models Using Existing, Long-Term Datasets*. NATO ASI I38. Springer-Verlag, Berlin.

Potter, C. S., J.T. Randerson, C.B. Field, P.A. Matson, P.M. Vitousek, H.A. Mooney and S.A. Klooster. 1993. Terrestrial ecosystem production: a process model based on satellite and surface data. *Glob. Biogeochem. Cyc.* 7:811-841.

Powlson, D.S., P. Smith and J.U. Smith (eds). 1996. *Evaluation of Soil Organic Matter Models Using Existing, Long-Term Datasets*. NATO ASI I38, Springer-Verlag, Berlin, 429 pp.

Powlson, D.S., P. Smith, K. Coleman, J.U. Smith, M.J. Glendining, M. Körschens and U. Franko. 1997. A European network of long-term sites for studies on soil organic matter. *Soil Till. Res.*

Rijtema, P.E. and J.G. Kroes. 1991. Some results of nitrogen simulations with the model ANIMO. *Fert. Res.* 27:189-198.

Ryzhova, I.M. 1993. Analysis of sensitivity of soil-vegetation systems to variations in carbon turnover parameters based on a mathematical model. *Euras. Soil Sci.* 25:43-50.

Schevtsova, L.K. and B.G. Mikhailov. 1992. Control of Soil Humus Balance Based on Statistical Analysis of Long-term Field Experiments Database. Moscow, 1992 (in Russian).

Schimel, D.S. 1995. Terrestrial ecosystems and the carbon cycle. *Glob. Change. Biol.* 1:77-91.

Schimel, D.S., B.H. Braswell (Jr)., E.A. Holland, R. McKeown, D.S. Ojima, T.H. Painter, W.J. Parton and J.R. Townsend. 1994. Climatic, edaphic, and biotic controls over storage and turnover of carbon in soils. *Glob. Biogeochem. Cyc.* 8:279-293.

Sirotenko, O.D. 1991. The USSR climate-soil-yield simulation system. *Meteorologia i Gidrologia* 4: 67-73 (in Russian).

Smith, J.U. 1996. Models and scale: up- and down-scaling. *Quant. App. Syst. Anal.* 6:25-42.

Smith, P., J.U. Smith and D.S. Powlson (eds.) 1996a. Soil Organic Matter Network (SOMNET): 1996 Model and Experimental Metadata. GCTE Report 7, GCTE Focus 3 Office, Wallingford, U.K.

Smith, P., D.S. Powlson, J.U. Smith and M.J. Glendining. 1996b. The GCTE SOMNET: a global network and database of soil organic matter models and long-term datasets. *Soil Use Manage.* 108: 57.

Smith, P., J.U. Smith and D.S. Powlson. 1996c. Moving the British cattle herd. *Nature,* 381:15.

Smith, P., D.S. Powlson and M.J. Glendining. 1996d. Establishing a European soil organic matter network (SOMNET). p. 81-98. In: D.S. Powlson, P. Smith and J.U. Smith (eds.) *Evaluation of Soil Organic Matter Models Using Existing, Long-Term Datasets.* NATO ASI I38. Springer-Verlag, Berlin.

Smith, J.U., N.J. Bradbury and T.M. Addiscott. 1996e. SUNDIAL: a user friendly PC-based system for simulating nitrogen dynamics in arable land. *Agron. J.* 88:38-43.

Smith, J. U., P. Smith and T.M. Addiscott. 1996f. Quantitative methods to evaluate and compare soil organic matter (SOM) models. p. 181-200. In: D.S. Powlson, P. Smith and J.U. Smith (eds.) *Evaluation of Soil Organic Matter Models Using Existing, Long-Term Datasets.* NATO ASI I38. Springer-Verlag, Berlin.

Smith, P., D.S. Powlson, J.U. Smith, and P. Falloon. 1997a. SOMNET. A global network and database of soil organic matter models and long-term experimental datasets. *The Globe* 38:4-5.

Smith, P., D.S. Powlson, M.J. Glendining and J.U. Smith. 1997b. Potential for carbon sequestration in European soils: preliminary estimates for five scenarios using results from long-term experiments. *Glob. Change Biol.* 3:67-80.

Smith, P., D.S. Powlson, M.J. Glendining and J.U. Smith. 1997c. Opportunities and limitations for C sequestration in European agricultural soils through changes in management. p. 143-152. In: R. Lal, J.M. Kimble, R.F. Follett and B.A. Stewart (eds.), *Management of Carbon Sequestration in Soil.* CRC Press, Boca Raton, FL.

Smith, P., D.S. Powlson, J.U. Smith and E.T. Elliott (eds.) 1997d. Evaluation and comparison of soil organic matter models using datasets from seven long-term experiments. *Geoderma* 81:1-225.

Smith, P., J.U. Smith, D.S. Powlson et al. 1997e. A comparison of the performance of nine soil organic matter models using datasets from seven long-term experiments. In: P. Smith, D.S. Powlson, J.U. Smith and E.T. Elliott (eds.) Evaluation and comparison of soil organic matter models using datasets from seven long-term experiments. *Geoderma* 81:153-225.

Smith, P., O. Andrén, L. Brussaard, M. Dangerfield, K. Ekschmitt, P. Lavelle, and K. Tate. 1998a. Soil biota and global change at the ecosystem level: describing soil biota in mathematical models. *Glob. Change Biol.* 4.

Smith, P., D.S. Powlson, M.J. Glendining and J.U. Smith. 1998b. Preliminary estimates of the potential for carbon mitigation in European soils through no-till farming. *Glob. Change Biol.* 4.

Svendsen, H., S. Hansen and H.E. Jensen. 1995. Simulation of crop production, water and nitrogen balances in two German agro-ecosystems using the DAISY model. *Ecol. Mod.* 81:197-212.

Szabó J, G. Várallyay and L. Pásztor. 1996. HunSOTER, a digitised database for monitoring changes in soil properties in Hungary. p. 150-156. In: Proc. of Soil monitoring in the Czech Republic, Brno.

Thornley, J.H.M. and M.G.R. Cannell. 1992. Nitrogen relations in a forest plantation - soil organic matter ecosystem model. *Ann. Bot.* 70:137-151.

Thornley, J.H.M. and E.L.J. Verberne. 1989. A model of nitrogen flows in grassland. *Plant Cell Env.* 12:863-886.

Trumbore, S.E. 1993. Comparison of carbon dynamics in tropical and temperate soils using radiocarbon measurements. *Glob. Biogeochem. Cyc.* 7:275-290.

Vanclooster, M., P. Viaene, J. Diels and J. Feyen. 1995. A deterministic evaluation analysis applied to an integrated soil-crop model. *Ecol. Mod.* 81:183-195.

Varga-Haszonits. Z. 1977. *Agrometeorologia.* Mezogazdasagi Kiado, Budapest.

Veldkamp, E. 1994. Organic carbon turnover in three tropical soils under pasture after deforestation. *Soil Sci. Soc. Am. J.* 58:175-180.

Verberne, E.L.J., J. Hassink, P. de Willigen, J.R.R. Groot and J.A. van Veen. 1990. Modelling soil organic matter dynamics in different soils. *Neth. J. Agric. Sci.* 38:221-238.

Whitmore, A.P. 1995. Modelling the mineralization and leaching of nitrogen from crop residues during three successive growing seasons. *Ecol. Mod.* 81:233-241.

Whitmore, A.P., H. Klein-Gunnewiek, G.J. Crocker, J. Klír, M. Körschens and P.R. Poulton. 1997. Simulating trends in soil organic carbon in long-term experiments using the Verberne/MOTOR model. In: P. Smith, D.S. Powlson, J.U. Smith and E.T. Elliott (eds.) Evaluation and comparison of soil organic matter models using datasets from seven long-term experiments. *Geoderma* 81:137-151.

Williams, J.R. 1990. The erosion-productivity impact calculator (EPIC) model: a case history. *Phil. Trans. R. Soc., Lond. B* 329:421-428.

CHAPTER 19

Evaluating Tropical Soil Properties with Pedon Data, Satellite Imagery, and Neural Networks

E. Levine, D. Kimes, S. Fifer and R. Nelson

I. Introduction

There are many uncertainties concerning the role of tropical forest soil carbon in the global carbon cycle (Delaney et al., 1997). Tropical soil carbon dynamics are difficult to assess at regional and global scales in this ecosystem due to the lack of multi-temporal, landscape level, ground reference observations. When working at regional or sub-continental scales, predictive models that integrate existing soil data and satellite imagery provide a tool that can be used to fill spatial and temporal gaps and to observe temporal trends that indicate change in an observable land attribute.

In order to develop predictive models of landscape parameters, comprehensive ground based data must be available. Soil pedon data represents the most comprehensive and detailed available set of laboratory measurements and descriptive properties of soils as they occur across the landscape. With proper analysis and interpretation of these data, researchers can obtain a "snapshot" of the biological, geological, climatic, and human history of an area. At the same time, these data describe the conditions at the time of sampling and provide information for predicting the properties of the soil at other locations. Modeling problems occur because the soil system is highly complex and its properties are related in a nonlinear fashion, so that it becomes very difficult to write simple equations or make inferences from this large amount of "raw" chemical, physical, and mineralogical soils data to practical information about general soil properties. "Pedo-transfer" functions have been developed using multiple linear regression methods to predict key soil parameters from other more readily available soil parameters. Because of the nonlinear complexity of soil properties, however, deriving adequate models can be difficult, and the accuracy of most pedo-transfer functions, as reflected by R^2 values, tends to be low.

Neural networks are a unique tool for modeling soil carbon from soil pedon data because of their ability to interpret complex nonlinear relationships in large data sets and either classify or predict information in a usable form. Their effectiveness with these data is a function of the way neural networks are organized which allows them to learn patterns or relationships in data from being shown a given set of inputs, generalize or abstract results from imperfect data (such as noise in the data, missing data, or a few incorrect values), and ignore minor variations in input (Anderson and Rosenfeld, 1988; Wasserman, 1989; Zornetzer et al., 1990).

With this method, input and output variables can be related without any knowledge or assumptions about the underlying mathematical representation. They attempt to find the best nonlinear function, based on the network's complexity, without the constraint of linearity or pre-specified nonlinearity use in regression analysis. In numerous studies, the neural network approach performs significantly better than traditional statistical methods because the neural network is free to learn functional relationships that could not always be envisioned by the researchers (Levine and Kimes, 1998; Pachepsky et al., 1996; Levine et al., 1996; Lam and Pupp, 1993).

The purpose of this study was to predict soil organic carbon content of tropical soils using soil pedon data and satellite imagery with a neural network approach. In this way, the best nonlinear model could be derived to assess soil carbon content across the tropical landscape.

II. Methods

A. Neural Network Training Data

The soil pedon data used to train the neural networks were from the United States Department of Agriculture Natural Resources Conservation Service (USDA NRCS) and the Projecto Radambrasil from areas in western and central Brazil. The USDA NRCS data was obtained from the National Soil Survey Center Soil Survey Laboratory Characterization Data Base (USDA NRCS, 1994). The Projecto Radambrasil data was digitized from the manuscripts of the Survey of Natural Resources (Projecto Radambrasil, 1973–1984). By combining these data sets, a total of 186 profiles were obtained for the western region of Brazil (Figure 1). Vegetation types for these profiles were classified as "dense tropical forest" according to the Projecto Radambrasil (1973–1984) report so that vegetation was not included as a variable.

Additional information was obtained for the location of each profile by overlaying satellite imagery and digital elevation data. The satellite imagery were from images from the Advanced Very High Resolution Radiometer (AVHRR) sensor at 8 km resolution from February and June of 1982 which was the earliest available date for imagery from the NOAA AVHRR sensor. Data for channels 1 (red) and 2 (near infrared) and normalized difference vegetation index (NDVI[1]) were obtained for each soil pedon location.   These data were composited over a 10 day period (to minimize effects of the atmosphere), and a mask for clouds was applied. Data from 1982 were assumed to be close enough to the original soil sampling dates, most of which occurred in the 1960s and 1970s. Finer resolution for this date would have been preferred to train the neural networks for each individual soil pedon location, but was not readily available for this study. It was assumed that cutting, burning, and other disturbance of the area at later dates may have affected the soil carbon content making it different from the amount originally sampled, and thus, invalid for training. Figure 1 shows the locations of the pedon sampling points overlaid on an image of NDVI for February 1982.

Koppen Climate Classification (Oliver, 1991) was also obtained for each pedon. All pedons fell within the A class which represents a tropical climate (A, coolest month > 18°C), with one of three subclasses: Af (constantly moist, driest month has at least 6 cm precipitation), Am (a dry season in an otherwise excessively moist climate), or Aw (dry in winter, wet in summer).

---

[1]NDVI is an index of greenness which has been correlated with known patterns of vegetation seasonality and variability across major biomes on a continental scale. It is calculated using the equation: [channel 2-channel 1]/[channel 2+channel] (Tucker, 1979). Studies using the Advanced Very High Resolution Radiometer (AVHRR), show that NDVI may be related to fractional absorbed photosynthetically active radiation (f_{APAR}), net primary productivity, vegetation vigor, or leaf area index (LAI) (Goward and Huemmrich, 1991; Lozano-García et al., 1991; Goward et al., 1987; Justice et al., 1985).

Figure 1. AVHRR NDVI Image for June 21–30, 1982 showing location of pedons used in this study.

B. Data Analysis

Table 1 lists the parameters that were provided to the neural network as a training set. These parameters were chosen because they were relevant to the analysis, and there were sufficient measurements of these within the pedon data to provide a complete set with no missing values. Two different approaches were taken to predict ground-measured soil carbon. First, soil characterization variables within the pedon data set were used to train the neural networks. In the second approach, spatial, non-soils information was used in the analysis.

1. Soil Property/Carbon Relationships

With the first approach, soil pedon data alone were used to train the neural networks to predict soil organic carbon. Horizons from each profile in the data set were regarded as individual points in the data. A subset of quantitative, descriptive, and taxonomic parameters were chosen from the large number available from the data base (Table 1). In this way, an evaluation of parameters that are important for carbon accumulation in soils could be made.

Table 1. Parameters used for the neural network training set

Latitude*	Silt (%)***
Longitude*	Clay (%)***
Elevation (m)*	Aluminum oxide (%)***
Koppen climate classification*	Iron oxide (%)***
AVHRR NDVI (1km) (February and June, 1995)**	pH ***
AVHRR Channel 1 and 2 (1 km) February and June, 1995)**	Available phosphorus (mg kg^{-1})***
Taxonomic class (US Soil Taxonomy)***	Exchangeable calcium (cmol kg^{-1})***
Soil order (US Soil Taxonomy)***	Exchangeable magnesium (cmol kg^{-1})***
Drainage class***	Exchangeable potassium (cmol kg^{-1})***
	Exchangeable sodium (cmol kg^{-1})***
Parent material***	Cation exchange capacity (cmol kg^{-1})***
Horizon nomenclature***	Sum of bases (cmol kg^{-1})***
	Base saturation (%)***
Horizon depth (cm)***	
	Total nitrogen (%)***
Sand (%)***	Organic carbon (%)*

*Used for both analyses.
**Used only for the soil carbon/imagery relationships analysis.
***Used only for the soil property/carbon relationships analysis.

2. Soil Carbon/Imagery Relationships

With the second approach, satellite imagery, latitude and longitude, elevation, and climate group (Koppen Classification; Oliver, 1991) were used to train neural networks to predict carbon and other soil parameters for the surface horizon (A) and for the total profile between 0 and 1 meter. Weighted values (thickness × % carbon) for the A horizons and for each horizon up to 1 m were calculated to be consistent with all profiles which may have been sampled to different depths.

In addition to the data set with continuous carbon variables, a data set was created in which % carbon was aggregated into 2 classes. For the surface horizon analysis, class 1 represented carbon ≤1.5%, and class 2 contained soils with >1.5% carbon. For the analysis of the top 1 m, class 1 represented carbon ≤0.7%, and class 2 contained soils with >0.7% carbon. These values were chosen because they fell at the point where about half of the total number of pedons could be grouped equally into each class. Data for bulk density were not available for the majority of the data points, so that the actual amount of carbon could not be calculated. However, a mean bulk density of 1.09 for the surface horizons and 1.22 for the 0–1 m profiles was obtained from the data that were available. Using these mean bulk density values, the cutoff values between class 1 and 2 for carbon in the soil surface represented approximately 16.4 kg/m^2, and about 8.5 kg/m^2 in the top 1 m.

C. Neural Network Training and Testing

From the subset of data (Table 1) used for each of the above analyses, an exhaustive search algorithm was used to identify those parameters within each horizon that would help predict organic carbon with the highest accuracy. The exhaustive search algorithm searches for the optimum subset of variables that behave synergistically to produce the best network performance. To construct the networks, 2/3 of the data were randomly selected and used for training, while the remaining 1/3 was used for testing the predictions of the trained network. A cascade method was used which adds a hidden node one at a time while testing network performance. As each hidden node is added, it is fully connected to all previous nodes. The transfer function used for the hidden nodes was the hyperbolic tangent function, and for the output node was the sigmoid function. An adaptive learning rule was also applied that uses a back-propagated gradient approach (Fahlmann and Labriere, 1990). Several methods were used to avoid overfitting the training data including reducing network structure, halting training when test performance began to decline, decreasing the number of input variables, and others. In all cases, network solutions were found where the network performance was very comparable between the training and testing data (e.g., root mean square values similar to within 4%).

The results of the neural network analyses were compared to traditional linear predictions and classification results to quantify differences. Continuous variables were predicted using standard multiple linear regression procedures. Multivariate linear models were developed using the same set of independent variables as used in the neural net analyses. Likewise, linear discriminant functions were calculated to predict class variables. Again, those independent variables identified and used by the comparable neural networks served as input for the discriminant functions.

D. Model Predictions

AVHRR imagery data were obtained for 1982 to include the Brazilian states of Amazonas, Rondônia, Roraima, Acre, and parts of Pará, Amapa, and Mato Grosso. These data were used with the most accurate neural network to predict the carbon class in the surface horizons and in the top 1 m of soil for these selected areas. Once the predictions were made, the results were mapped according to their geographic location.

III. Results And Discussion

A. Soil Property/Carbon Relationships

Using all horizons as inputs for training to predict % organic carbon, the most accurate neural network had a prediction accuracy of 0.82 R^2 (true value vs. predicted value), and root mean square error (rms) of 0.4%. The parameters that were used in the network development were horizon (A or B), depth, nitrogen, and base saturation. Using these same parameters, multiple linear regression was also run on the data. Results of the regression analysis produced an R^2 value of only 0.70, with an rms of 0.5%. When the A horizons for all profiles were run separately from the rest of the profile, the most accurate network used pH, nitrogen, calcium, and potassium to predict carbon, and had a prediction accuracy (R^2) of 0.74 and root mean square error of 0.6%. Multiple regression analysis applying only the same soil pedon parameters used in the neural network produced an R^2 value of 0.69 with rms of 0.5%. B horizon runs alone did not perform as well ($R^2 = 0.48$, rms = 0.5).

Table 2. Proportion of correct classification between True Classes (vertical) vs, Predicted Classes (horizontal)

a. Top 1 m

	Neural network		Discriminant analysis	
	Class 1	Class 2	Class 1	Class 2
Class 1	0.74	0.26	0.61	0.39
Class 2	0.24	0.76	0.44	0.56

b. Surface horizons

	Neural network		Discriminant analysis	
	Class 1	Class 2	Class 1	Class 2
Class 1	0.73	0.26	0.71	0.29
Class 2	0.24	0.76	0.43	0.57

These results indicate that when specific soil data (e.g., horizon, depth, nitrogen, and base saturation for total profile analysis) are available for similar soils, the amount of carbon in the horizon can be predicted using the neural network model. These parameters would have physical connections to carbon dynamics due to the association between carbon and nitrogen in organic matter and the probability of greater biomass producing carbon in soils of higher base saturation and pH (Jenny, 1980). In addition, the carbon distribution in soils of this region, as shown by Eswaran et al., (1995), tended to decrease below 50 cm of the profile.

B. Soil Carbon/Imagery Relationships

Neural networks were trained and tested on data for both the surface horizons and the top 1 m of soil for each profile. In both cases, the models predicting continuous values of carbon performed poorly. In order to improve the predicting ability of the neural networks, carbon in the top 1 m and the surface horizons were grouped into classes as described in Methods above.

The neural network model which best predicted soil carbon in the top 1 m used four variables from the total given in Table 1. These were June 1982 NDVI, Channel 1 and 2 for June 1982 , and Channel 2 for February 1982. The average accuracy of the neural network predicting the correct class was 0.75%. Table 2a shows the error matrix comparing the predictions of class 1 and class 2 for soil carbon using neural network and discriminant analysis. In this comparison, only the variables used by the neural network were applied in the discriminant analysis. Results of discriminant analysis had an average accuracy of only 58%. As shown on Table 2a, the neural network tended to classify both class 1 and class 2 incorrectly about 25% of the time, where the discriminant analysis misclassified class 2 as class 1 slightly more frequently than misclassifying class 1 as class 2.

The neural network that most accurately predicted soil carbon in the surface horizons used the same four parameters as did the network for the top 1 m prediction (June NDVI, Channel 1 and 2 for June, and Channel 2 for February) with the addition of elevation as a model parameter. With this model, predictions of class 1 and class 2 also had an average accuracy of about 75%, with about an equal misclassification of each class of about 25% (Table 2b). Discriminant analysis for the surface horizons performed better than for the top 1 m, but did not perform as well as the neural network. The average prediction accuracy from discriminant analysis was about 64%, and again, class 2 was misclassified more often than class 1.

The NDVI parameter has been used traditionally in studies assessing patterns of vegetation and thus indicates a relationship between soil carbon and vegetation "greenness". The use of channel 1 and 2 appears to add additional information that could not be obtained by the NDVI equation alone. The addition of elevation to the surface horizons estimate of carbon may be indicative of increases or decreases in organic decomposition that correlate with vegetation or temperature effects of elevation changes. These may not be as evident at lower depths where carbon may be in a more stable state and not as susceptible to climate conditions at the surface.

Using the most accurate neural network and satellite imagery for 1982, soil organic carbon in surface horizons and the top 1 m were predicted and mapped. Figure 2 shows the spatial distribution of predictions of class 1 (black) and class 2 (grey) values for organic carbon in the surface horizons (a) and in the top 1 m (b) over the selected regions of west central Brazil. According to Figure 2, the distribution of class 1 ($\leq$1.5% carbon) is about equal to the distribution of class 2 (>1.5% carbon) for the surface soil horizons in this region. Class 1 ($\leq$0.7% carbon) is greater than the distribution of class 2 (>0.7% carbon) in the map for representing the top 1 m of the soil profile. Areas of greater carbon content in the top 1 m tend to match those of high carbon content in the surface horizons for most areas except for the northeastern portion of the mapped area, where there is more carbon with depth predicted than in the surface horizons. Carbon in the surface horizons was predicted to be greater in the central and southeastern portion of the map than was determined for the top 1 m.

These predictions of soil organic carbon generally follow existing maps for the Amazon region of land form (Sombroek, 1990) and of global soil carbon produced by Eswaran et al., 1995. According to the map of Sombroek (1990), the predicted high carbon patterns (class 2) for both the surface and top 1 m match most of the Amazon area classified as "uplands". The upland areas are described as rolling to hilly dissected lands and rounded hills formed from primarily crystalline and some sedimentary rocks. Soil conditions vary from shallow to deep soils with weatherable minerals classified as Inceptisols (Cambisols), Fragiudults, or Hapludults (ferric or orthic Acrisols). Most of these soils sustain an open-canopy forest with a dense overgrowth. Closed-canopy high forest are also found in this area on Haplustox (orthic and rhodic Ferralsols) which contain an inactive clay mineralogy and little or no weatherable mineral reserve. An area of Paleudults (Nitosols) characterized as dusky red, deep, clayey soils of good structure with a high percentage of active clay-sized iron oxides, are also found in this region. These soils support a "luxuriant" forest cover and are sought after for growing cocoa because of their favorable physical and chemical properties (Sombroek, 1990). Thus, these regions would also tend to have a relatively high organic carbon content throughout the profile.

The southeastern portion of the predicted map, which shows carbon in the surface horizons being greater than the top 1 m, is mapped primarily as "relic valleys" by Sombroek (1990). This area is gently undulating, relatively high-lying land with concave slopes having primarily very deep, sandy loam to clayey acid yellowish to reddish soils of low to very low physico-chemical activity classified as Haplustox (xanthic or orthic Ferralsols) and Ustipsamments (ferralic Arenosols). They support high forest of large timber volume in most parts of this region (Sombroek, 1990). However, the discrepancy in carbon, predicted by the neural net model between the surface horizons and top 1 m may be due to the predominance of organic activity occurring closer to the area of litter fall at the soil surface, rather than in the sandy or highly weather subsoil.

The global organic carbon map of Eswaran et al., (1995) shows soil organic carbon in units of kg m^{-2} derived from the database from the World Soil Resources group of the Natural Resources

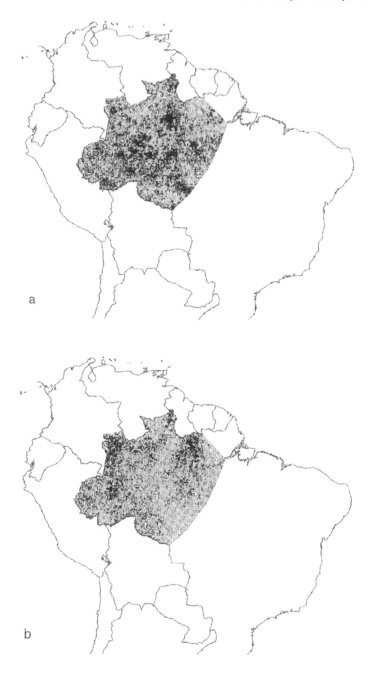

Figure 2. a: Predicted soil organic carbon classes for the surface horizons using the best fit neural network model; grey represents class 1 ($\leq 1.5\%$); black represents class 2 ($>1.5\%$); b: predicted soil organic carbon classes for the top 1 m using the best fit neural network model; grey represents class 1 ($\leq 0.7\%$); black represents class 2 ($>0.7\%$).

Conservation Service. On this map, each map unit was assigned a value for the organic carbon content to 1 m depth. If the map unit contained representative pedons, the unit was assigned a value based on an average carbon for that area. If there were no representative pedons for the mapping unit, a "best value" was assigned based on the soil classification and soil moisture and temperature regime of the area. The assigned value was multiplied by the area of the map unit to obtain the total carbon content for each map unit. Visual comparison of the soil organic carbon for the tropics in South America with the maps created by predictions of the neural network show similarities in patterns of carbon. In general, the map of predicted carbon for the top 1 m of soil matches the map of Eswaran et al., (1995) better than the predicted map of surface horizon carbon. However, actual comparisons are difficult to make since the predicted carbon map represents individual pixels, and the Eswaran et al., (1995) map uses a different scale to separate the classes of carbon.

IV. Conclusions

The approach to using neural networks for modeling soil organic carbon is a promising one. Prediction results give a high reliability that are better than those that could be obtained with traditional linear statistical approaches. In addition, the methods used in this study allow the creation of the most accurate model using a subset of many possible parameters. In this way, carbon can be predicted with reliability from other soil information that may be available or by using satellite imagery and elevation data alone.

The models generated by the neural network are specific for the soil data from which they were trained. Their applicability to soils in other regions would have to be verified before they were used. However, the general design and approach to data processing and to neural network testing and training could serve as a prototype for modeling soil carbon as well as other parameters in tropical regions and other ecosystems.

There are many drawbacks to this study that will require additional effort to remedy. Most importantly is the acquisition and use of finer resolution satellite data as close as possible to the dates of soil sampling. With this fine spectral data, additional site information will need to be collected including information on land use, vegetation type, climate, and atmospheric conditions. This will inevitably improve prediction ability of the neural networks as well as their ability to estimate continuous carbon content instead of classes of carbon. With these improved prediction tools, more accurate base maps of organic carbon can be derived. The improved neural network models can be applied to imagery from later years (corrected for sun angle and atmosphere effects), in combination with additional field data for model validation, to assess change in carbon content in soils due to land use practices such as deforestation or slash and burn agriculture in this region.

References

Anderson, J.A. and E. Rosenfeld (eds.). 1988. *Neurocomputing: Foundations of Research.* The MIT Press, Cambridge, MA.

Delaney, M., S. Brown, A.E. Lugo, A. Torre Lezama, and NB Quintero. 1997. The distribution of organic carbon in major components of forests located in five life zones of Venezuela. *J. Tropical Ecology* 13:697-708.

Eswaran, H., E. Van den Berg, P. Reich, and J. Kimble. 1995. Global soil carbon resources. p. 27-43. In: R. Lal, J. Kimble, E. Levine, and B.A. Stewart (eds.), *Soils and Global Change*. Advances in Soil Science, CRC Press. Boca Raton, FL.

Fahlmann, S. and C. Lebriere. 1990. The Cascade-Correlation Learning Architecture. In: D. Touretzky, (ed.), *Advances in Neural Information Processing Systems 2*. Morgan Kaufmann, San Mateo, CA.

Goward, S.N. and K.F. Huemmrich. 1991. Vegetation canopy PAR absorptance and the Normalized Difference Vegetation Index: an assessment using the SAIL model. *Remote Sensing of Environment* 39:119-140.

Goward, S.N., A. Kerber, D. Dye, and V. Kalb. 1987. Comparison of North and South American biomes from AVHRR observations. *Geocarto Int.* 2:27-39.

Jenny, H. 1980. The Soil Resource. Springer-Verlag, New York, NY.

Justice, C.O., J.R.G. Townshend, B.N. Holben, and C.J. Tucker. 1985. Analysis of the phenology of global vegetation using meteorological satellite data. *Int. J. Remote Sensing* 6:1271-1318.

Lam, D. and C. Pupp. 1993. Integration of GIS, expert system and modeling for state of environment reporting. Proc. of NCGIA 2nd Int. Conf. on Integrating GIS and Environ. Modeling. Breckenridge, CO.

Levine, E.R., D.S. Kimes, and V.G. Sigillito. 1996. Modeling soil structure using artificial neural networks. *Ecol. Modeling* 92:101-108.

Levine, E.R. and D.S. Kimes. 1998. Predicting soil carbon in Mollisols using Neural Networks. p. 473-484 In: R. Lal, J.M. Kimble, R.F. Follett, and B.A. Stewart (eds.), *Soil Processes and the Carbon Cycle*, CRC Press, Boca Raton, FL.

Lozano-García, D.F., R.N. Fernández, and C. Johannsen. 1991. Assessment of regional biomass-soil relationships using vegetation indexes. *IEEE Transactions on Geoscience and Remote Sensing* 29:331-339.

Oliver, J.E. 1991. The history, status and future of climatic classification. *Physical Geography* 12:231-251.

Pachepsky, Y.A., D. Timlin, and G. Varallyay. 1996. Artificial neural networks to estimate soil water retention from easily measurable data. *Soil Sci. Soc. Am. J.* 60:727-733.

Projecto Radambrasil. 1973–1984. Survey of Natural Resources. IGBE, Rio de Janeiro, Brazil.

Sombroek, W. 1990. Amazon landforms and soils in relation to biological diversity. International Workshop to Determine Priority Areas for Conservation in Amazonia, 10–20, January, 1990, Manaus, Brazil, Conservation International, New York, NY.

Tucker, C.J. 1979. Red and photographic infrared linear combinations for monitored vegetation. *Remote Sensing of Environment* 11:171-189.

USDA NRCS. 1994. National Soil Characterization Data. USDA National Soil Survey Center, Lincoln, NE.

Wasserman, P.D. 1989. *Neural Computing: Theory and Practice*. Van Nostrand Reinhold, New York, NY.

Zornetzer, S.F., J.L. Davis, and C. Lau. 1990. *An Introduction to Neural and Electronic Networks*. Academic Press, New York, NY.

CHAPTER 20

Geographic Assessment of Carbon Stored in Amazonian Terrestrial Ecosystems and Their Soils in Particular

W.G. Sombroek, P.M. Fearnside and M. Cravo

I. Introduction

The amount of carbon stored in various terrestrial ecosystems is an important factor in the cycling of atmospheric carbon dioxide, one of the greenhouse gases. Assessment of carbon in the aboveground standing biomass has received much attention, but only recently has awareness grown that the carbon pool in soils, in its quantity, its dynamics and its sustainability, is important as well.

Quantitative estimates of soil carbon have relied, by and large, on soil surveys that were carried out for planning of agricultural activities, in which the carbon content played a subordinate role only. Field and laboratory methods of soil characterization need to give more careful attention to the vertical transitions of content and type of roots, organic matter, carbonate carbon and charcoal, and to a much greater depth than the traditional 100 cm. Soil physical properties per horizon, including *in situ* bulk density and soil faunal activity, are required as well.

To better quantify terrestrial carbon pools, it is advocated that new soil mapping be carried out in a more holistic framework, taking natural landscape units as building blocks for digital data bases on landforms, soils, hydrology and vegetation or land use. In this way, small but carbon-significant inclusions in mapping units will be duly accounted for. An early example of such an approach is given for the Paragominas area in the southeast of Pará, Brazil. A contrasting one, for which the field studies are still in progress, deals with the situation in the Middle Madeira River area, south of Manaus, Brazil.

II. Assessment of the Aboveground Biomass and its Carbon Content

The amount of carbon stored in the various terrestrial ecosystems forms an important compartment in the cycle of atmospheric CO_2. Increasing concentration of CO_2 is considered to be a major driving force of worldwide climate change.

The assessment of the amounts and the fluxes of carbon in the standing biomass has received much attention since the 1970s. The storage of carbon in tropical forests, and its release upon deforestation, has prominently figured in international scientific and policy discussions on the hazards of a disastrous anthropogenic climate change (Houghton et al., 1988, 1992; Fearnside, 1996, 1997). The concerns that were raised in fact triggered a major international program — PPG7 — to support Brazil in its efforts to protect its tropical forests. More recently concerns gave rise to an international research program on the Amazon region as a whole: the "Large-Scale Biosphere Atmosphere Experiment in Amazonia" (LBA).

The dwindling acreage of forests worldwide is documented by FAO once per decennium, using remote sensing imagery and country reports. The latest inventory reflects the situation in 1990 (FAO, 1993a). The quantification of carbon stored in the forest's aboveground biomass is less easy. Mostly it is done by using conversion factors for data on timber volume, in m^3 ha^{-1}, of traditional forest inventories, as carried out for commercial purposes (Brown et al., 1989; Alves et al., 1997). These inventories consider only the merchantable trees with a diameter at breast height (DBH) of a minimum size — often 25 cm, but sometimes as small as 10 cm or as large as 50 cm. To arrive at the total aboveground biomass, dry-weight measuring is required at representative sites for calibration purposes. This can be further developed into linear regression models of tree biomass over structural tree parameters (Bruenig et al., 1979, on the MAB/IUFRO studies in San Carlos de Rio Negro). As outlined by Fearnside (1994), the sequence of timber volume measurements and its calibration with dry-weighing of carbon storage is fraught with pitfalls and uncertainties such as the degree of geographic representativeness of the forest inventories and the frequency of nonforest vegetation units or man-induced degradational phases (Sombroek, 1992); the density of the living wood of individual tree species; the volume occupied by palms, creepers, vines and shrubs per ha; the amount of dead wood on the ground; the thickness of the litter layer and the representativeness of the dry-weighing sites (Fearnside, 1994). On the latter, for instance, it is noteworthy that most sites for the Amazon are located in easily accessible parts, such as north of Manaus or in Rondônia. There is not a single weighing result for the vast area of dense but low forest on the Pleistocene "Içá" formation in the central-southern part of Amazonas State with its soils of plinthic character (unit Pp of Figure 1).

It is gradually becoming apparent that the early estimates of 360 t ha^{-1} (Woodwell, 1978) of the total aboveground carbon storage of tropical forest lands, as used for modeling of climate change in the framework of the UNEP/WHO International Scientific Panel on Climate Change (IPCC), are too high. The figure should be scaled down, but probably not as much as the 30% suggested by Brown and Lugo (1992). At one extreme, a "back of the envelope" estimate can be produced using average biomass and region-wide (Brazilian Legal Amazon) deforestation rate (e.g., Fearnside, 1985, 1987). At the next level of detail, state-level deforestation rates can be used (the smallest unit for which deforestation information is available throughout the region, as of 1997). These rates can be combined with state-level averages for carbon stocks, based on re-aggregation of the areas and biomass of the forest types within each state. This has been done at the National Institute for Research in Amazonia (INPA), producing estimates (Fearnside, 1990, 1991) based mainly on the forest type area estimates by Braga (1979) and biomass estimates from a variety of point estimates and anecdotal sources. These estimates were subsequently improved by using area estimates (Fearnside and Ferraz, 1995) from a geographic information system (GIS) analysis of the 1:5,000,000-scale vegetation map of the Brazilian Institute for Environment and Renewable Natural Resources (IBAMA) (IBGE, 1988), producing a series of estimates with successively better biomass information (Fearnside, 1992, 1997).

The success of the strategy of step-by-step improvement of the estimates adopted by the INPA group is in contrast with the results of parallel efforts at the National Institute of Space Research (INPE). There, efforts began in 1990 to digitize the RADAMBRASIL (1978) maps for GIS analysis, together with georeferenced deforestation data. The RADAMBRASIL (1978) maps, produced in the period 1973–1983, with 145 vegetation types in the Legal Amazon, are much more complex than the IBAMA map with 28 types in the same region. In addition, numerous inconsistencies exist among the RADAMBRASIL (1978) volumes in the map code adopted (see Fearnside, 1994).

An analysis at the level of RADAMBRASIL (1978) mapping units within each state is the logical next step in improving the level of detail of aboveground biomass assessment, combining the newly available information on vegetation/land use, geology and geomorphology, hydrology and soils.

It may be noted that recent measurements in Rondônia at the ABRACOS project (Grace et al., 1996) and north of Manaus at the BIONTE project (I. Biot and N. Higuchi, personal communications) have shown that the standing forest may at present be actively sequestering carbon, which would constitute

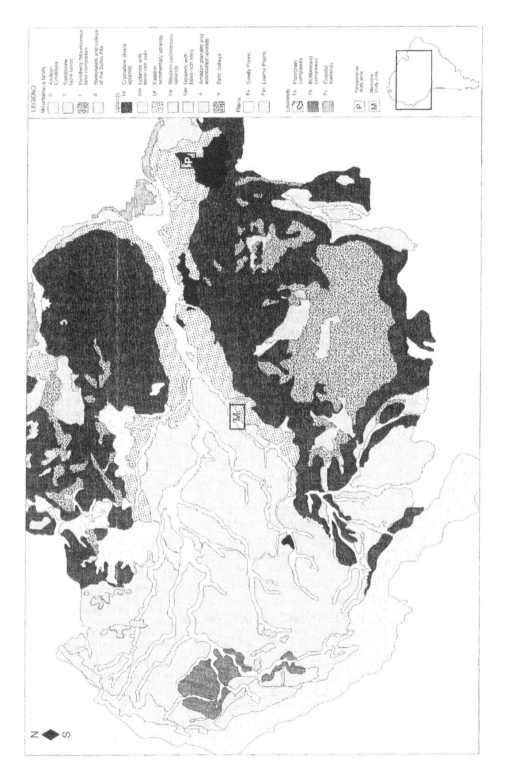

Figure 1. Main landforms of the Amazon region. (Adapted from ISRIC, July 1998.)

proof that the so-called CO_2 -fertilization process in agricultural crops is also active in natural woody vegetation.

III. Assessment of Soil Carbon Content

In determining the total potential for CO_2 flux into the atmosphere from large-scale deforestation, one should also look at the below-ground carbon pools. The amount of carbon stored in the living soil (i.e., those surface layers of the land into which roots, microbes and soil animals can penetrate) is very considerable (Delaney et al., 1997) For many temperate ecosystems, the amount is double or triple the amount stored in the aboveground natural or agricultural biomass, and a good part of it is an active component of the carbon cycle. Even the soils under tropical forest contain approximately the same amount of carbon as the lush vegetation above it. The amounts of carbon per main soil type — to 100 cm depth only — are summarized separately by Eswaran et al. (1993) and Sombroek et al. (1993). Klamt and Sombroek (1988) compared the forty-odd Ferralsol (Oxisol) profiles belonging to the collection of ISRIC and originating in all parts of the tropics and subtropics. They arrived at values of 0.64 to 10.19 g per 100 g soil (mean 2.73) for the surface horizons, and 0.07 to 1.37 (mean 0.48) for the subsurface horizons to 100 cm depth.

Recent quantifications of the soil carbon content for the Brazilian Amazon forests are given by Sombroek (1992), mainly on the basis of data gathered in the early 1960s, and more comprehensively by Moraes et al. (1995) on the basis of the RADAMBRASIL (1978) inventory data of the 1970s. However, there are many methodological flaws and uncertainties in determining the total amount and dynamics of the soil carbon and the part stored in the soil organic matter (SOM) in particular. These problems are of three kinds: the methods of sampling and laboratory analyses, the vertical transitions in the soil profile, and those horizontally across the landscape.

A. Sampling and Analytical Problems of Soil Carbon Assessment

The analytical problems are connected with the fact that traditional soil surveys and associated laboratory analyses of soil samples were carried out for planning or improving agricultural activities, in which carbon content per ha played a subordinate role. The density of roots in the various layers/horizons and the number of insects such as ants and termites is described in a qualitative way only, if at all. Sampling of roots for carbon analysis is almost never done.

Often, topsoils are sampled over a standard depth of 0–20 cm as being the future plow-layer. Forest soils, however, may contain two or three horizons within this distance. For instance, The A_{II} horizon of clayey *Latossolos Amarelos*[1] (Oxisols) under primary forest of the Brazilian Amazon may be only 1 to 3 cm thick; the sandy ones may have a 2 to 3 mm surface layer of loose bleached sand. The depth of detailed soil profile descriptions is normally not more than 100 cm when a pit is dug, or 200 cm when a subsequent auger-hole is made.

In extremely sandy or imperfectly drained soils, certain subsoil layers may contain relatively high amounts of carbon. Sampling of humus-rich layers in the subsoil is not always carefully done, in particular when only auger-hole examinations are made — as was the case by and large for the RADAMBRASIL (1978) mapping.

The assessment of carbon in the laboratory also faces many problems. First of all, all soil particles larger than 2 mm are removed by sieving, and are thus not accounted for, though this fraction may

---

[1] Brazilian nomenclature of EMBRAPA/CNPS; for the names in Soil Taxonomy, FAO-UNESCO legend and the World Reference Base for Soil Resources, see Table 1 (Beinroth, 1975; Camargo et al., 1986; EMBRAPA/SNLC, 1988; Soil Survey Staff, 1992; FAO, 1998).

contain a substantial amount of dead roots, pieces of charcoal (Saldarriaga et al., 1986), indurated SOM/iron complexes from Ortstein layers in *Podzólicos Hidromórficos*, or pedogenic solid calcium carbonates.

Secondly the method of determination of carbon content in the fine earth differs from laboratory to laboratory: Walkley–Black, Allison, or Tiurin methods. Differences may be up to 40% (Allison, 1965). With the claim that minor amounts cannot be measured with any degree of precision, carbon determinations for the deeper subsoil are often not carried out at all though their sum can be quite significant (see below).

Thirdly no conversion from g per 100 g soil (i.e., the fine earth) to g per m^3 or Mg ha^{-1} takes place. Such conversion is necessary to allow comparison with aboveground biomass carbon, which is always given in t ha^{-1}. The conversion would be easy if the bulk density of each soil layer were available routinely. But bulk-density values are seldom measured, and then often in a less-than-accurate manner. In the case of RADAMBRASIL (1978) soil samples, as studied by Moraes et al. (1995) for their carbon content, only about 8% of all horizons concerned had bulk-density figures. These figures, moreover, are not based on real bulk densities as occurring in the field, but rather on laboratory approximations: disturbed sample material is brought into the laboratory, its coarse fraction is removed by sieving, the fine earth is compacted in a metal cylinder of standard size (100 ml) and then weighed. The resulting data may be considerably different from field reality, especially when stony or dense soil layers are concerned. The correct alternative is to apply a coating of plastic sandwich-wrapping material to natural clods, or to collect the same volume of soil at field sampling, by means of a fixed number of metal cylinder fillings — for instance, the rings used for pF-curve determinations. This is easily done, but rarely practiced, and of course requires soil pits or machine-driven deep core probing rather than screw-auger profile examinations as was the practice in most of the exploratory soil inventories of RADAMBRASIL (1978).

In general, differences in bulk densities of soils within the same ecosystem can be up to 60%, and the ranking of soils by their carbon content in Mg ha^{-1} can therefore be substantially different from that in g per 100 g soil. For tropical soils the bulk density can vary from 0.90 to 1.65 g per cm^3. A number of rules of thumb apply: high-organic-matter topsoils have lower bulk density figures than low-content deeper layers; clayey soils are lower in bulk-density than sandy ones; low-iron *Latossolos* have lower bulk density than high-iron ones; subsoils (B-horizons) of *Latossolos* have lower bulk density than those of *Podzólico Vermelho-Amarelo* (Ultisols or Alfisols); imperfectly and poorly drained soils, such as *Plintossolos* and *Gleisolos* have higher bulk density than well-drained soils of the same overall texture.

The following figures are mostly based on bulk-density figures of disturbed samples, but sometimes calibrated with undisturbed samples of CNPS/EMBRAPA-ISRIC (dos Santos and Kauffman, 1995):

- The peaty soils of permanently submerged lowlands have bulk densities of 0.5 g cm^{-3} or less.
- The very heavy textured *Latossolos Amarelo* (LA) of the forested Amazon planalto ("Belterra clays"), have bulk-density values of only 0.90–1.10 down to 200 cm depth or more.
- The texture-differentiated *Podzólicos Vermelho-Amarelo* (PVA), baixa atividade, usually have bulk densities of 1.30–1.40.
- The iron-poor *Areias Quarzosas* under forest have bulk densities of 1.50–1.55 throughout.
- The *Plintossolos* of medium texture under natural savanna vegetation (*campo cerrado*) have bulk density values of 1.55–1.65 throughout.

After deforestation the bulk density of the topsoil increases, through rainfall impact and cattle trampling with 0.05 to 0.20 g cm^{-3} (Fearnside and Barbosa, 1998; Koutika et al., 1997), and this should be taken into account when trying to quantify soil carbon loss in degraded pasture lands or any gain in the few well-managed pastures (Serrão and Toledo, 1992; Cerri et al., 1996).

B. Vertical Differences in Soil Organic Carbon Content

As early as 1974, studies of organic carbon profiles in a number of representative *Latossolos* from all over Brazil, either under forest or under agriculture, were carried out by EMBRAPA, and some useful pedo-transfer functions were developed (Bennema, 1974).

Most soils have a decreasing amount of organic matter with increasing depth, while soil stability increases. There are notable exceptions also in tropical ecosystems (Sombroek et al., 1993). Tropical Podzols, when in a well-drained position, have a characteristic humus accumulation horizon in the subsoil (B_h horizon), which may be within the top 100 cm or at many meters depth. When in imperfectly or poorly drained positions, this accumulation layer is often cemented with iron or aluminum oxides (B_{hir} horizon, or "Ortstein"), which may be of a very irregular or broken nature. Planossolos or well-developed Plintossolos often have a small organic matter jump in the upper part of their pan-like B-horizon. Even well-drained *Podzólicos Vermelho-Amarelo* may have a degree of humus accumulation in the upper part of their B-horizon.

Noteworthy in the context of total carbon assessment in the soil is the irregular occurrence of SOM in the soils of floodplains or recent terraces where the sedimentary layering has not yet been erased by soil biological homogenization or pedogenic differentiation. Sedimentary layers rich in SOM may date from changes in flood regime, local bankfalls or the presence of buried humus-rich former surface horizons. Such a profile from the Lower Amazon area, used in the study of Sutmöller et al. (1966), shows 157 tons of SOM-carbon per ha, totaled over the various layers to 230 cm depth. On eastern Marajó Island, a similar profile showed 217 tons SOM-carbon per ha to 270 cm. As a consequence, the carbon stock of such floodplain soils gains significance in comparison to that of many terra firma soils. Mention should also be made of the scattered occurrences of Latossolo Roxo (an Oxisol) or Terra Roxa Estruturada (an Alfisol), developed from basic rock, with 50 to 100% higher amounts of soil carbon than the predominant soils. The same holds for the many patches of *Terra Preta* do Índio soils, a kind of Plaggensoil, occurring on and around dwelling sites of the pre-Columbian Amerindians, with approximately double the amount of SOM in comparison with neighboring nonanthropogenically enriched soils (Sombroek, 1966, 1992).

The total depth over which the SOM occurs varies highly. Obviously, shallow soils over hard rock (*Solos litólicos*) have limited depth, although rock fissures may contain significant amounts of SOM from former roots. Well-developed tropical soils in general are deep, often much more than 2 m, especially when the parent material consists of unconsolidated sediments (fluviatile, lacustrine, pyroclastic or aeolic). How deep exactly is partly a matter of conjecture, unless one has the opportunity to study deep, fresh road-cuts for the presence of living roots or active soil fauna, such as termites. Rare deep augerings — say to 400 cm depth — with sampling show that there is still some SOM present at this depth — admittedly in small amounts, but still appreciable when related to the thickness of the horizon involved. A striking example is the fresh-pit study of Nepstad et al. (1994) in the Paragominas area of eastern Amazonia. It shows that even at 8 m depth in the heavy textured *Latossolos Amarelo*, there are still living roots and SOM-carbon, with a degree of dynamics (see also Trumbore et al., 1995). Consequently, a full 60% of the SOM-carbon occurs below 100 cm depth. Such forest-growth related deep soil activity is likely to occur elsewhere too, especially when the soil has a low available-moisture storage capacity, a significant dry season (2 to 4 months with less than 100 mm rainfall each) and an absence of any mineral reserve in the upper few meters. This forces roots and termites to deep layers in search of moisture and micro-nutrients. The Nepstad results demonstrate that for the tropical forest zone, one should add a significant amount of soil carbon stock to the amounts mentioned by Eswaran et al. (1993) or Sombroek et al. (1993) for tropical soils in general, and by Moraes et al. (1995, 1996) for the Amazon soils, since these all refer to 100 cm depth only.

One can construct maplets of the total rootable depth that is required to allow forest to survive and grow in marginal rainfall areas (D. Nepstad, in preparation) and then compare these with the geographic reality for modeling the robustness of forest (re-)growth. The Nepstad data regarding total depth of "living soil" can be extrapolated to the deep and well-drained soils of the Cretaceous/Tertiary sediments of east-central Amazonia (land units A and Uf of Figure 1) (Latossolo Amarelo unit of EMBRAPA; see also Rodrigues, 1996). The well-drained soils of the crystalline basement zones of the Guyana and Brazilian shields, *Latossolos Vermelho-Amarelo* and *Podzólicos Vermelho-Amarelo* (PVA), would have a total rootable depth of 5 and 3 m, respectively, as an educated guess.

The imperfectly drained soils of Central Amazonia, such as the *Podzols Hidromórficos* of the upper Rio Negro area and the large extent of Plintossolos or plinthic PVA soils in the areas between the Rio Madeira and the Juruá, as well as those between the middle Rio Negro and the Rio Solimões, all developed on the Pleistocene silty sediments of the Içá formation (unit Pp of Figure 1), have a stop in the downward development of soil biological activity precisely because of the nature of the soil hydrological dynamics. In the well-developed Plintossolos, roots would not go deeper than 1 m or so, and about 2 m deep in the case of PVA plíntico. This is also the reason for the large extent of low forest formations in the above-mentioned areas, although there is no well-defined dry season, and of natural savannas where a degree of dry season is present (see Landsat image of the Middle Madeira area, Figure 2). The ranking of the different soils of the Brazilian Amazon by carbon content, as given by Moraes et al. (1995), is likely to change considerably when the above-suggested depths of the rootable soil are taken into account.

In conclusion, a new effort to quantify soil carbon stocks in the Brazilian Amazon region, and adjoining parts of Colombia, Peru and Bolivia, is called for. It should rely on real bulk-density data, on the absence or presence of deeper layers of living soil and, of course, should try to correlate these features with carbon data for the above-ground biomass of each major soil type. This brings us to the details of the horizontal pattern of tropical soils.

C. The Horizontal Geography of Tropical Soils

The early notion in temperate zone-originated textbooks, that soils of the humid tropics have very little diversity, is presently fully disproved (see FAO/UNESCO soil map of the world, 1994; FAO, 1993b). However, detailed patterns of the geography of soils under forest vegetation types are little studied, although they are basic to forest biodiversity and carbon storage.

In most soil surveys of the reconnaissance type, the mapping units are characterized in the legend by the pedologic classification of the main soil or soils, without precise information on their percentages or any mention of inclusions. At best, one can guess the relative importance from the sequence of soil names. Mention of "association to...", meaning a recognized geographic pattern, or "complex of...", when no pattern was apparent, makes the geographic relationship of soils less vague, and when it is stated "... with ..." or "... and inclusions of ..." one can make an educated guess of the percentage composition of the soil pattern in the landscape. Fully quantitative estimates of the percentage occurrence of all soil components in any mapping unit in reconnaissance surveys were a utopia in the old days, especially when forest- or bush-growth made access and horizontal overview very difficult. However, with the advance of remote sensing techniques such as aerial photography, radar and satellite imagery, one can delineate landscape units ("land system units" = "*unidades de paisagem*") with a high degree of accuracy and, after carefully planned field checks across the elements (or "facets") of such a landscape, arrive at good percentage estimates of the occurrence of these elements in the landscape unit concerned. Every experienced and ecologically / geomorphologically oriented soil surveyor knows that there is a system in the seemingly chaotic spatial distribution of the soils, that each constituent element of a landscape has not only its own topographic character, but also its own soil, its own microclimate, its own internal hydrology and therefore its own vegetation — or its own type of land use when agriculture is well established. Why then not use this knowledge to put soils information in such a landscape-ecological framework? Soil scientists should no longer be on the defensive by stressing a strictly pedologic focus and associated technical terminology to the exclusion of other elements of the landscape. Their soil geographic information will be much more useful to other disciplines if they make the effort to provide their legends in descrip-

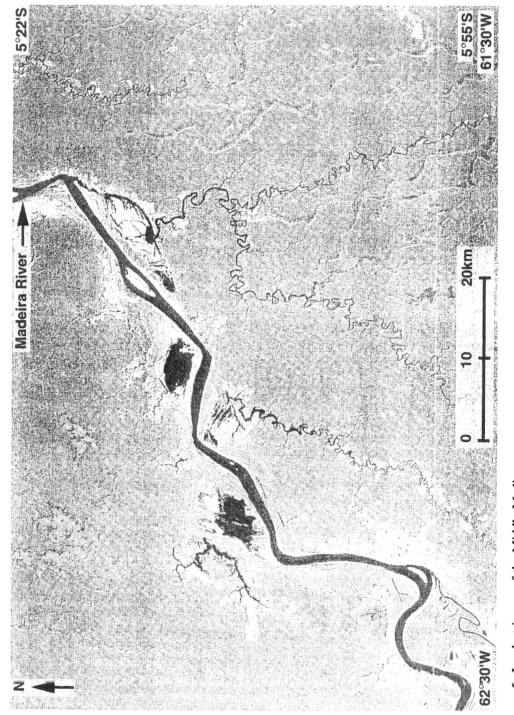

Figure 2. Landsat image of the Middle Madira.

tive terminology (drainage condition, depth, color, texture and texture sequence, mode of transition to subsoil, etc.), with the official pedologic classification only at the end and preferably between brackets. In these days of growing acknowledgment of the importance of soils to the diversity of terrestrial ecosystems, on the hydrological cycle and on the fluxes of most greenhouse gases, soil scientists should come out of their narrow agricultural trenches and join forces, on an equal level, with other disciplines in characterizing and evaluating landscapes in a multi-functional cadre.

The idea of using natural landscapes or "Land Systems" as basic building blocks for mapping of rural space is not new. It was first used in Australia, soon after the Second World War. It was subsequently applied in many English-language countries in Africa, by the British Land Resources Development Centre. The CIAT study on "Lands of Tropical America" has land systems as the starting point (Cochrane and Sanchez, 1982). The concept was also promoted by French geomorphologists such as Tricart (1977), which in turn influenced the recent work of IBGE geographers (viz. the PMACI study on the southwestern Brazilian Amazon region: IBGE, 1990), and of the Geography Department of the University of São Paulo (Ab'Saber, 1996). A generalized map of the main land systems in the whole Amazon region is presented in Figure 3.

IV. Examples of a Landscape-Ecological Approach to Carbon Assessment

The first author of the present article used the land-system approach in the early 1960s, for the reconnaissance soil survey cum forest inventory of the Guamá-Imperatriz area across the then-still-forested part of southeastern Pará (Sombroek, 1962, 1966). The area around the town of Paragominas, for instance, was denominated as the "Candiru" land-unit, with a total area of about 350,000 ha and characteristics as shown in Figure 1[2].

In this land unit, the forest cover was dense and high, with scattered emergents and relatively open undergrowth because of the near-absence of creepers and climbers. As reported by Glerum and Smit (1962), there was a high number of trees per hectare (124) and a substantially higher gross timber volume (191.6 m^3 ha^{-1}) than in the land units immediately north or south (Médio Guamá: 108 trees ha^{-1}, and 161.2 m^3 ha^{-1}; Alto Guamá: 94 trees ha^{-1} and 121.1 m^3 ha^{-1}). One tree species, Pau amarelo (*Euxilophora paraensis*) was strongly concentrated on the concretionary soils of the edges of the low-plateau and on its scarps (see also Appendix 3 of Sombroek, 1966 for a detailed sample area around 130 km south of Guamá). The central parts of the plateau (element *a*), had scattered Angelim pedra (*Hymenolobium excelsum*) as emergent while Pau roxo (*Peltogyne lecointei*) and Quaruba (*Vochysia maxima*) tended to be most frequent on the middle-terrace land. The latter facet had also a unique and exquisite fragrance, due to the presence of a "Vanilla" orchid.

Because of the high quality of the forest on this land unit, and at the specific request of the SPVEA — the regional development organization to which the FAO/UNESCO inventory team was attached — a good part of the land unit was recommended for sustainable timber production, to be protected by a buffer zone of controlled small-holder agricultural settlements that would also provide forestry labor. It is very disheartening to observe after 35 years that this early effort of ecologic-economic zoning in a tropical forest area failed miserably. Immediately after the technical survey data became available, they were used by unscrupulous individuals in a predatory way: indiscriminate clear-cutting of all valuable timber, followed by large-scale burning and establishment of extensive cattle ranches on most parts of which the soils soon became degraded because of lack of maintenance of the grass cover (Nepstad et al., 1991; Trumbore et al.October 13, 1999, 1995).

---

[2] Estimates on total rootable depth from unpublished field observations on the many fresh road-cuts in 1961; estimates on total SOM-carbon storage based on unpublished profile descriptions and analytical data of the EMBRAPA/CNPS central soil laboratory in Rio de Janeiro. For location of the Paragominas area, see site P of Figure 1.

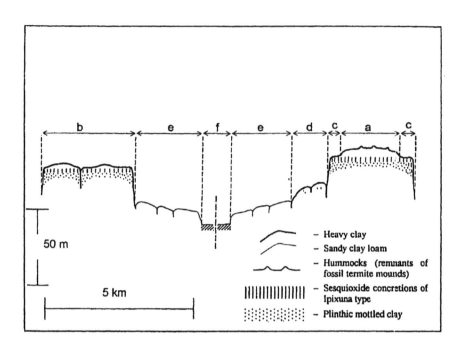

Figure 3. Sketch of the structure of the "Cardirú" land-unit in the Paragominas area of Pará, Brazil.

The above-described early example of a landscape-ecological approach was never applied in neighboring areas or even referred to in subsequent studies in the area itself. However, even in the present-day deforested situation, the basic physical elements of the local landscape remain intact, and their pattern forms a good starting point for the assessment and spatial differentiation of degradation features.

The multi-disciplinary physico-biotic database of landscape-system units can these days be harnessed, and manipulated for modeling purposes, much more easily than before because of the development of digital data storage systems and geographic information systems. An appropriate vehicle for geographic studies that contain a substantial soil element would be the SOTER system (Soil and Terrain Digital Data Base), developed by a working group of the International Society of Soil Science (ISSS). It was subsequently adopted by FAO and UNEP as an appropriate data storage and processing system, with an inbuilt capacity to be combined with relational databases on climate, hydrological resources, natural land cover or land-use and settlement patterns (UNEP/ISSS/ISRIC/-FAO, 1993).

The first Latin American application of the SOTER approach was through a joint project by national soil institutions of Brazil, Uruguay and Argentina in their frontier areas. Many more areas are now being covered with the same or a comparable approach. A number of software packages for practical uses of the database system have been developed, such as on erosion hazards or production capacity (FAO/UNEP/ISRIC, 1998). A similar package can be developed for geographic quantification of the SOM-carbon per landscape element or for the land-unit as a whole.

It may be mentioned that the Ecologic-Economic Zoning Program (*Zoneamento Ecológico-Econômico*), now starting in some Amazon priority areas of Integrated Environmental Management, takes the natural landscape units as a starting point for the physico-biotic part of its diagnostics. The soils element thereof will be taken care of by EMBRAPA specialists, this time cooperating very closely with geologists, geomorphologists, hydrologists and biologists in assembling the digital database information.

One of the first pilot areas of the zoning program is the Manicoré-Novo Aripuanã area along the Middle Madeira river in the state of Amazonas. Two strikingly different land systems are involved, on the left and right side of the river, respectively (see satellite image, Figure 2). In the center is the recent floodplain of the Madeira river, with its many pointbars "Restingas" and cut-off side lakes (oxbow lakes).

The land unit[3] on the western (left) bank of the river has silty sediments of the mid-Pleistocene Içá formation. Its flat interfluvial areas have a dense, low forest with many palms, or an open savanna vegetation — the blotchy parts — with imperfectly or poorly drained soils, moderately to strongly texture-differentiated and with a plinthic layer between 50 and 150 cm depth (*Plintossolos*; *Planossolos* when having an abrupt transition; Podzols when having a humus accumulation above the plinthite); the rootable depth is 100–150 cm, and the SOM-carbon content is about 60 t ha^{-1} (4 profiles). The sloping parts have a rather low, dense forest, of about 100 m^3 ha^{-1} timber volume (DBH 31.8 cm or more; 34 samples) and moderately well- to imperfectly-drained silty clay soils that easily slake and are mottled reddish at depth (*Podzólicos Vermelho Amarelo, plíntico*); their rootable depth is about 200 cm and the carbon content about 120 t ha^{-1} (6 profiles).

The land unit of the eastern (right) bank of the main river has mid-Pleistocene sandy to loamy sediments, probably derived from the Proterozoic Prainha formation farther to the south. Its flat terrace lands have high forest (about 130 m^3 ha^{-1}; 25 samples) and well-drained, very deep, friable and stable sandy clay loam soils (*Latossolos Amarelo*). The rootable depth is at least 5 m and the total SOM-carbon is the order of 160 t ha^{-1} (3 profiles). The terrace lands are interspersed with lower-level abandoned river-beds – the elongated narrow bands of vague grey – that are filled with white sands (presumably *Podzols hidromórficos* in the actual field observations) and are covered with a shrubby savanna vegetation. The higher river banks, along the present-day rivers and rivulets, are occupied by traditional small-holder farmers and forest-product gatherers (*ribeirinhos*); the mode of settlement is semi-sedentary rather than of shifting-cultivation type because of the nearly continuous presence of *Terra Preta do Índio* on these riverbank stretches (M. Von Roosmalen, personal communication).

V. Conclusion

A variety of strategies exists that might be adopted in estimating the stock of carbon presently contained in Amazonian landscapes and the emission of CO_2 that could be expected when these landscapes are subjected to land-use changes. The information that already exists has to be interpreted in the most effective way, and programs to acquire new information need to be designed, such that the uncertainty regarding carbon stocks and emissions is lowered as efficiently as possible — that is, obtaining the greatest reduction in uncertainty for a given amount of investment.

Existing information can be aggregated in a variety of ways to produce emission estimates. The degree of aggregation that is appropriate depends on the level of detail of the original information. Vegetation mapping units of the RADAMBRASIL (1978) type can be further subdivided on the basis of topography and soil differences to produce land units that make more sense as the basis both for development or conservation planning and for the current purpose of assessing carbon stocks and potential emissions.

---

[3] All descriptions and quantitative estimates are provisional, based on comparative analysis of relevant geological, geomorphological, soil and forestry survey data of the Purus volume of RADÁMBRASIL (1978), over an area of 60,000 km^2; more detailed sampling and analysis is ongoing for the 9000 km^2 of the area shown in Figure 2.

The gain to be expected from each successive increase in the level of detail of mapping can be estimated from comparing the results of carbon estimates made at different levels of aggregation within a defined study area that is small enough to make mapping at a detailed (i.e., land-units) level practical. The Middle Madeira area, as described above, is suggested for this purpose.

In summary, it would seem to be appropriate to use the landscape-ecological approach and the SOTER vehicle as a means of correlating carbon storage and dynamics in soils with that of the above-ground biomass and ecosystem carbon. Also, the many other functions of "land" in the holistic sense, as the term is used in UNCED's Agenda 21 (FAO, 1995), can best be studied in the landscape-ecological context. As recently stated by Turner et al. (1997), continued database development is required, including close attention to the methodologies used for quantifying carbon content and fluxes in the various compartments of ecosystems, if carbon budget assessments are to be sufficiently reliable for use by the international policy community in proposing climate change controls.

References

Ab'Saber, A.N. 1996. Zoneamento ecológico e econômico na Amazônia. p. 11-29. In: A.N. Ab'Saber, Amazônia: do Discurso à Praxis. Editora da Universidade de São Paulo, S.P. Brazil.

Allison, L.E. 1965. Organic Carbon. In: C.A. Black (ed.), *Methods of Soil Analysis, Part 2: Chemical and Microbiological Properties*. ASA, Madison, WI.

Alves, D.S., J.V. Soares, S. Amaral, E.M.K. Mello, S.A.S. Almeida, O.F. da Silva, and A.M. Silveira. 1997. Biomass of primary and secondary vegetation in Rondônia, Western Brazilian Amazon. *Global Change Biol.* 3:451-461.

Beinroth, F.H. 1975. Relationships between U.S. Soil Taxonomy, the Brazilian system, and FAO/UNESCO soil units. p. 97-108. In: E. Bornemisza and A. Alvarado (eds.), Soil Management in Tropical America: Proc. Sem. held at CIAT, Cali, Colombia, 10–14 February, 1974. North Carolina State University, Soil Science Department, Raleigh., NC.

Bennema, J. 1974. Organic carbon profiles in Oxisols. *Pedologie* 24:119-146.

Braga, P.I.S.. 1979. Subdivisão fitogeográfica, tipos de vegetação, conservação e inventário florístico da floresta Amazônica. *Acta Amazonica* 9(4), suplemento: 53-80.

Brown, S. and A.E. Lugo. 1992. Above-ground biomass estimates for tropical moist soils of the Brazilian Amazon. *Interciencia* 17:8-18.

Brown, S., A.J.R. Gillespie, and A.E. Lugo. 1989. Biomass estimation methods for tropical forests with application to forest inventory data. *Forest Science* 35:881-902.

Bruenig, E.F., D. Alder, and J. Smith. 1979. The International MAB Amazon Rainforest Ecosystem Pilot Project at San Carlos de Rio Negro: vegetation classification and structure. p. 67-100. In: Adisoemarto and E.F. Bruenig (eds.), Transactions of the Second International MAB-IUFRO Workshop on Tropical Rainforest Ecosystems Research, Jakarta, 21-25 October, 1978, Chair of World Forestry, Hamburg-Reinbek.

Camargo, M.N., E. Klamt, and J.H. Kauffman. 1986. Soil Classification as Used in Brazilian Soil Surveys. Annual Report 1986. ISRIC, Wageningen.

Cerri, C.C., M. Bernoux, and B.J. Feigl. 1996. Deforestation and use of soil as pasture: climatic impacts. In: R. Lieberei, C. Reisdorff, and A.D. Machado (eds.), Interdisciplinary Research on the Conservation and Sustainable Use of the Amazonian Rain Forest and its Information Requirements. Forschungszentrum Geesthacht (GKSS), Bremen.

Cochrane, T.T. and P.A. Sanchez. 1982. Land resources, soils and their management in the Amazon Region: a state of knowledge report. In: S.B. Hecht (ed.), Amazônia: Agriculture and Land Use Research. CIAT, Cali, Colombia.

Delaney, M., S. Brown, A.E. Lugo, A. Torres-Lezama, and N. Bello Quitero. 1997. The distribution of organic carbon in major components of forests located in five life zones of Venezuela. *J. Tropical Ecol.* 13:697-708.

dos Santos, H.J. and J.H. Kauffman (eds.). 1995. Soil Reference Profiles of Brazil. Field and Analytical Data. Country Report 9. ISRIC, Wageningen.

EMBRAPA/SNLCS. 1988. Critério para distinção de classes de solos e fases de unidades de mapeamento: normas em uso pelo SNLCS. Documento 1. SNLCS, Rio de Janeiro.

Eswaran, H.E., E. Van den Berg, and P. Reich. 1993. Organic carbon in soils of the world. *Soil Sci. Soc. Am. J.* 57:192-194.

FAO, 1993a. Forest resources assessment 1990: Tropical Countries. *Forestry Paper* 112. FAO, Rome.

FAO, 1993b. World Soil Resources; an explanatory note on the FAO World Resource Map at 1:25,000,000 scale. *World Soil Resources Reports* 66 (rev.1). FAO, Rome.

FAO, 1994. FAO-UNESCO Soil Map of the World 1:5,000,000, Vol. I, Legend. UNESCO, Paris.

FAO, 1995. Planning for sustainable use of land resources, towards a new approach (background paper to FAO's Task Managership for Chapter 10 of UNCED's Agenda 21. *Land and Water Bulletin* 3. FAO, Rome.

FAO, 1998. World Reference Base for Soil Resources. World Soil Resources Report 84. FAO, Rome.

FAO/UNEP/ISRIC. 1998. Soil and Terrain Database for Latin America and the Caribbean at 1:5 million scale. Land and Water Digital Media Series 5. (CD-ROM edition). FAO, Rome.

Fearnside, P.M. 1985. Brazil's Amazon forest and the global carbon problem. *Interciencia* 10: 179-186.

Fearnside, P.M. 1987. Summary of progress in quantifying the potential contribution of Amazonian deforestation to the global carbon problem. p. 75-82. In: D. Athié, T.E. Lovejoy, and P. de M. Oyens (eds.), Proc. Workshop on Biochemistry of Tropical Rain Forests: Problems for Research. Universidade de São Paulo, Centro de Energia Nuclear na Agricultura (CENA), Piracicaba, São Paulo.

Fearnside, P.M. 1990. Contribution to the greenhouse effect from deforestation in Brazilian Amazonia. p. 465-488. In: Intergovernmental Panel on Climate Change (IPCC), Response Strategy Working Group (RSWG), Subgroup on Agriculture, Forestry and other Human Activities (AFOS). Proceedings of the Conference on Tropical Forestry Response Options to Global Climate Change. U.S. Environmental Protection Agency, Office of Policy Assessment (USEPA-OPA, PM221), Washington, D.C.

Fearnside, P.M. 1991. Greenhouse gas contributions from deforestation in Brazilian Amazonia. p. 92-105. In: J.S. Levine (ed.), *Global Biomass Burning: Atmospheric, Climatic and Biospheric Implication.* MIT Press, Boston, MA.

Fearnside, P.M. 1992. Greenhouse Gas Contributions from Deforestation in Brazilian Amazonia. Carbon Emission and Sequestration in Forests: Case Studies from Developing Countries. Vol. 2 LBL-32758, UC-402. Climate Change Division, Environmental Protection Agency, Washington, D.C. and Energy and Environment Division, Lawrence Berkeley Laboratory (LBL), University of California (UC), Berkeley, CA.

Fearnside, P.M. 1994. Biomassa das florestas Amazônicas Brasileiras. p. 95-124. In: Anais do Seminário Emissão x Sequestro de CO_2. Companhia Vale do Rio Doce (CVRD), Rio de Janeiro.

Fearnside, P.M. 1996. Amazonia and global warming: annual balance of greenhouse gas emissions from land-use change in Brazil's Amazon region. p. 606-617. In: J.Levine (ed.), Biomass Burning and Global Change. Vol. 2: Biomass Burning in South America, Southeast Asia and Temperate and Boreal Ecosystems and the Oil Fires of Kuwait. MIT Press, Cambridge, MA.

Fearnside, P.M. 1997. Greenhouse gas from deforestation in Brazilian Amazonia: Net committed emissions. *Climatic Change* 35:321-360.

Fearnside, P.M. and J. Ferraz. 1995. A conservation gap analysis of Brazil's Amazonian vegetation. *Conservation Biol.* 9:1134-1147.

Fearnside, P.M. and R.I. Barbosa. 1998. Soil carbon changes from conversion of forest to pasture in Brazilian Amazonia. *For. Ecol. Manage.* 108:147-166.

Fernandes, E.C.M., P.P. Motavalli, C. Castilla, and L. Mukurumbira. 1997. Management control of soil organic matter dynamics in tropical land-use systems. *Geoderma* 79:49-67.

Glerum, B.B. and G.S. Smit. 1962. Combined forestry/soil survey along the road BR-14, from São Miguel to Imperatriz, FAO, EPTA Report 1483, Rome and SPVEA/SUDAM, Belém.

Grace, J., J. Lloyd, J. Mcintyre, A.C. Miranda, P. Meir, and H.S. Miranda. 1996. Carbon dioxide flux over Amazon rain forest in Rondônia. p. 307-318. In: J.H.C. Gash, C.A. Nobre, J.M. Roberts, and R.L. Victoria (eds.), *Amazonian Deforestation and Climate*. John Wiley & Sons, Chichester.

Houghton, J.T., B.A. Callander, and S.K. Varney (eds.). 1992. *Climate Change 1992*. The supplementary report to the IPCC Scientific Assessment. Cambridge University Press, Cambridge.

Houghton, R.A., G.M. Woodwell, R.A. Sedjo, R.P. Detwiler, C.A.S. Hall, and S. Brown. 1988. The global carbon cycle. *Science* 241:1736-1739.

IBGE. 1988. Mapa de Vegetação do Brasil. Map scale 1:5,000,000. Instituto Brasileiro do Meio Ambiente e dos Recursos Naturais Renováveis (IBAMA), Brasília.

IBGE. 1990. Diagnóstico geo-ambiental e sócio-econômico; área de influência da BR-364, trecho Porto Velho - Rio Branco - PMACI. Instituto Brasileiro de Geografia e Estatística, Rio de Janeiro.

Klamt, E. and W.G. Sombroek. 1988. Contribution of organic matter to exchange properties of Oxisols. p. 64-70. In: F.H. Beinroth, N.N. Camargo, and H. Eswaran. (eds.), Proc. Eighth International Soil Classification Workshop; classification, characterization and utilization of Oxisols. University of Puerto Rico, Mayaguez, PR.

Koutika, L.S., F. Bartoli, F. Andreau, C.C. Cerrie, G. Burtin, Th. Coné, and R. Philippy. 1997. Organic matter dynamic in soils under rainforest and pastures of increasing age in the eastern Amazon Basin. *Geoderma* 76:87-112.

Moraes, J.L., C.C. Cerri, J.M. Melillo, D. Kicklighter, C. Neil, D.L. Skole, and P.A. Steudler. 1995. Soil carbon stocks of the Brazilian Amazon Basin. *Soil Sci. Soc. Am. J.* 59:244-247.

Moraes, J.F.L. de, B. Volkoff, C.C. Cerri, and M. Bernoux. 1996. Soil properties under Amazon forest and changes due to pasture installation in Rondônia, Brazil. *Geoderma* 70:63-81.

Nepstad, D.C., C.R. de Carvalho, E.A. Davidson, P.H. Jipp, P.A. Lefebvre, J.H. Negreiros, E.D. da Silva, T.A. Stone, S.E. Trumbore, and S. Vieira. 1994. The role of deep roots in the hydrological and carbon cycles of Amazonian forests and pastures. *Nature* 372:666-669.

Nepstad, D.C., Chr. Uhl, and E.A.S. Serrão. 1991. Recuperation of a degraded Amazonian landscape: forest recovery and agricultural restauration. *Ambio* 20(8):248-255.

RADAMBRASIL, 1973-1983. Levantamento de Recursos Naturais (geologia, geomorfologia, pedologia, vegetação, uso potencial da terra). Folha SB20. Purus. Ministério das Minas e Energia, DNPM, Rio de Janeiro. Maps at scale of 1:1,000,000.

Rodrigues, T.E. 1996. Solos da Amazonia. In: V.H. Alvares, L.E.F. Fontes, and M.P.F. Fontes (eds.). O solo nos grandes domínios morfoclimáticos do Brasil e o desenvolvimento sustentado. Sociedade Brasileira da Ciência do Solo, UFV/DPS, Viçosa. MG.

Saldarriaga, J.G., D.C. West, and M.L. Tharp. 1986. Forest succession in the Upper Rio Negro of Colombia and Venezuela. Oak Ridge National Laboratory (ORNL), Environmental Sciences Division Publication No. 2694 (NTIS Pub. ORNL/TM-9712), National Technical Information Service (NTIS), US Dept. of Commerce, Springfield, VA.

Serrão, E.A.S. and J.M. Toledo. 1992. Sustainable pasture based production systems in the humid tropics. p. 257-280. in: T.E. Dowing, S.B. Hecht, H.A. Pearson, and C. Garcia-Dowing (eds.), *Development or Destruction: the Conversion of Tropical Forest into Pasture in Latin America*, Westview Press, Boulder, CO.

Soil Survey Staff. 1992. Keys to Soil Taxonomy, SMSS Technical Monograph 19, Pocahontas Press, Blacksburg, VA.

Sombroek, W.G. 1962. Reconnaissance Soil Survey of the area Guamá-Imperatriz (Area along the upper part of the Brazilian highway BR-4. Report, FAO/SPVEA, Belém.

Sombroek, W.G. 1966. Amazon Soils: A Reconnaissance of the Soils of the Brazilian Amazon Region, PUDOC, Wageningen.

Sombroek, W.G. 1992. Biomass and carbon storage in the Amazon ecosystems. *Interciência* 17(5):269-272.

Sombroek, W.G. 1996. Amazon landforms and soils in relation to biological diversity. In: G.T. Prance, T.E. Lovejoy, A.B. Rylands, A.A. dos Santos, and C. Miller (eds.), *Priorities for Conservation in the Amazonian Rainforests*. Smithonian Institution Press, Washington. DC. (in press). Advance publication in Annual Report 1990, ISRIC, Wageningen. (Map and text updated in 1996).

Sombroek, W.G., F. Nachtergaele, and A. Hebel. 1993. Amounts, dynamics and sequestering of carbon in tropical and subtropical soils. *Ambio* 22(7):417-426.

Sutmöller, P., A. Vahia de Abreu, J. van der Grift, and W.G. Sombroek. 1966. Mineral imbalances in cattle in the Amazon valley; the mineral supply of cattle in relation to landscape, vegetation and soils. Communication 53, Royal Tropical Institute, Amsterdam.

Tricart, J. 1977. Ecodinâmica. Supren. IBGE, Rio de Janeiro.

Trumbore, S.E., E.A. Davidson, P.B. de Camargo, D.C. Nepstad, and L.A. Martinelli. 1995. Belowground cycling of carbon in forests and pastures of Eastern Amazonia. *Global Biogeochemical Cycles* 9(4):515-528.

Turner, D.P., J.K. Winjum, T.P. Kolchugina, and M.A. Cairns. 1997. Accounting for biological and anthropogenic factors in national land-based carbon budgets. *Ambio* 26(4):220-226.

UNEP/ISSS/ISRIC/FAO. 1993. Global and national soils and terrain digital data bases (SOTER); procedures manual. World Soil Resources Report 74. FAO, Rome.

Woodwell, G.M., R.H. Whittaker, W.A. Reiners, G.E. Likens, C.C. Delwiche, and D.P. Botkin, 1978. The biota and the world carbon budget. *Science* 222:1081-1086.

CHAPTER 21

Carbon Dioxide Measurements in the Nocturnal Boundary Layer over Amazonian Tropical Forest

G. Fisch, A.D. Culf, Y. Malhi, C.A. Nobre and A.D. Nobre

I. Introduction

Tropical forest represents significant sources/sinks for trace gases (CO_2, O_3, CH_4), and the exchange of CO_2 between forest and the atmosphere is an important component of the global carbon cycle because of its greenhouse effect. The secular increase of CO_2 concentration is reasonably well defined (Keeling and Whorf, 1994) but a full comprehension of CO_2 cycles for different biomes and its impact on climate and ecology is not well understood. Baldocchi et al. (1996) have addressed this issue based on the conclusions of a workshop to discuss the strategies for monitoring and modeling CO_2 fluxes over terrestrial ecosystems.

The diurnal cycle of atmospheric CO_2 reflects the exchange of metabolic carbon between the atmosphere and the vegetation-soil system. The typical behavior of CO_2 concentration shows an increase during the night, when soil emissions and respiration combine as a source of CO_2 to the atmosphere and start to decrease during mid morning when atmospheric CO_2 is taken up by the vegetation. The nighttime CO_2 flux is positive, reversing its sign (to downward) in the early morning. Although the nighttime CO_2 flux above the canopy is a sum of all CO_2 released by the vegetation and the soil, the atmospheric stability near the canopy does not allow mixing between the rich CO_2 air inside canopy and that above it. Fitzjarrald and Moore (1990) have observed some events (waves) that occur during the night in the Amazon forest and associate these events with periods of higher turbulent mixing between the canopy and the atmosphere, leading to increasing of values of CO_2 exchange. They point out that these events occur on average 5 times during the night with a time interval ranging from 25 to 60 minutes . They also noted that these events happen when the windspeed above the canopy is typically around 1.8 m s^{-1}. Grace et al. (1995) noticed that in the early morning there was a higher flux of CO_2 out of the canopy due to the onset of turbulence. They pointed out that the onset of turbulence disrupted the stable layer at the crown, allowing mixing between air, which has been accumulating CO_2, and the atmosphere. They have called this phenomena the morning ventilation and it has high values of CO_2 flux: the maximum value observed in tropical forest (at Rondônia) is +20 μmol m^{-2}s^{-1}.

The CO_2 exchange between tropical forest and the atmosphere above it has been subject of a very few studies so far. Delmas et al. (1992) made measurements of the CO_2 concentration over the tropical forest in the Congo basin using gas chromatography. They found a higher concentration near the canopy due to the soil emissions. Maximum values of CO_2 concentration were observed at the end of night. The daily amplitudes of CO_2 concentration varied from 45 ppm during their first field campaign (1988) to 35 ppm next year (1989). This difference is due to the intensity of vertical exchanges between forest and the atmosphere during the two different years. The average baseline concentration was 373 ppm in 1988 and 344 ppm in 1989. Direct measurements of CO_2 exchange between vegetation and the atmosphere are becoming more common, although little data have been collected over tropical forest and for only limited periods (Fan et al., 1990; Grace et al., 1995; Malhi et al., 1998). Fan et al. (1990) made their measurements during the wet season in Central Amazonia (Manaus, AM), during the Amazon Boundary Layer Experiment - ABLE 2B campaign making measurements of the fluxes above the forest (41 m height). Their concentration data (using an infra red gas analyzer - IRGA) were collected above and inside the canopy (7 levels). The flux data (12 days composite data) shows an increase from the early evening until near sunrise with maximum values of above canopy flux around 1.26 μmol m^{-2} s^{-1}. The concentration of CO_2 at 41 m (10 m above canopy) also shows an increase during the night, with a value approximately 350 ppm at 18:00 Local Time rising to 380 ppm at 6:00 LT. Inside the canopy there were values up to 450 ppm near sunrise. Some estimates of the flux of CO_2 from the forest floor accounts for 1.40 μmol m^{-2} s^{-1}, without any diurnal cycle detectable. Grace et al. (1995) measured the CO_2 concentration and turbulent fluxes above a tropical forest at one of the Anglo-Brazilian Amazonian Climate Observational Study-ABRACOS experimental sites (Gash et al., 1996). Their data set (44 days during the end of wet season and beginning of dry season in 1993) shows a maximum averaged concentration of CO_2 of around 486 ppm at 7:00 LT and a minimum value of 360 ppm at 14:00 LT. They also has showed a higher efflux of CO_2 (+8.2 μmol m^{-2} s^{-1}) after sunrise, which came from the flushing of the in-canopy rich CO_2 air due to the onset of turbulence. Afterwards, the flux changes its sign (becomes negative, indicating an uptake of CO_2) with its maximum active at 13:00–14:00 LT. Culf et al. (1997) used this data set for some modeling studies, trying to represent the diurnal trend of surface CO_2 concentration.

The Manaus Atmospheric CO_2 Experiment (MACOE) was designed to collect data of CO_2 profiles and surface fluxes during the nighttime period in order to understand better the CO_2 budget of the tropical forest in the central Amazonian region. This experiment was a joint scientific cooperation between Brazil and the United Kingdom. The results can be used in improving the scaling-up micro meteorological measurements to the whole Amazon region. The aims of this chapter are (1) to present some of CO_2 profile data obtained during MACOE and (2) to examine the carbon budget from a mass balance inside the Nocturnal Boundary Layer-NBL between the derived profile and from turbulent fluxes. So far the MACOE data set is unique to characterize the CO_2 profile in the boundary layer as most of the previously observational data is taken from near the surface.

II. Site, Experimental Design and Data

The experimental site (2° 35' S; 60° 06' W and 90 m above sea level) is part of a continuous area of dense lowland evergreen tropical forest (Bacia Representativa do Rio Taruma Açu, hereafter referred as ZF-2), which is about 70 km from the city of Manaus (Figure 1). The access to the site is through a federal road (Manaus-Caracaraí - BR 174) for 50 km and then by a secondary road denominated ZF-2. The mean canopy height of the forest in this area is 30 m and the above ground biomass is 300 to 400 t ha^{-1}. The soil on the plateau is a yellow clay latosol and has a high (80%) clay content, low nutrient content, low pH (4.3) and porosity ranging from 50 to 80%. The climate is usually hot and humid, with average air temperatures higher than 25 °C, with a daily range between 20 and 35 °C. Windspeeds are generally low, although it can be gusty (up to 20 m s^{-1}) during convective showers. The rainfall distribution shows a wet season from December until May, when the total rainfall is around 250 mm month^{-1}. The climatological average for November is 211 mm but November 1995 was very high with 331 mm. During the field campaign, however, there was little rain (2.2 mm), all of which occurred during the daytime.

The data set used was collected during a field campaign held at Manaus (central Amazonia), during the period of 16 to 26 November 1995. The operations with a tethered balloon (5.25 m^3) required a small clearing for the winch and to store the balloon during the daytime. The measurements consist

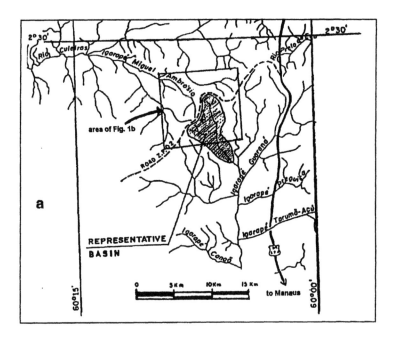

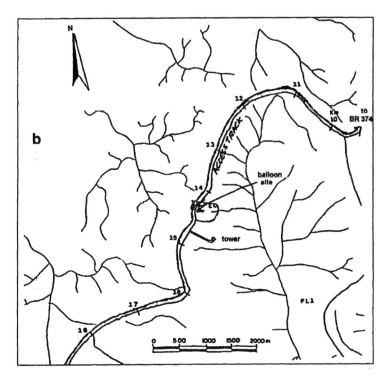

Figure 1. Map of ZF-2 and surroundings (a, top) and the balloon and tower site (b, bottom).

MACOE - 11/95   ———————   NBL
ZF-2 MANAUS

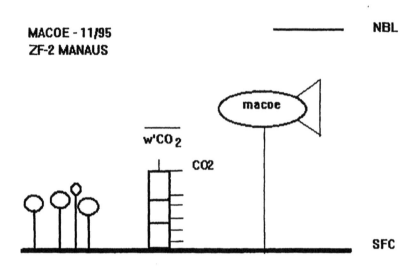

Figure 2. Schematic design of the MACOE experiment.

of a set of profiles of CO_2 concentration taken with the tethered balloon. A light and flexible special tube was attached to the tethered sonde and lifted by the balloon. The air at predetermined heights (each 50 m from the surface up to 300 m) was pumped down to a gas analyzer- IRGA (CIRAS 1, Hitchin, UK) from the early evening (at 18:00 Local Time) until sunrise (at 8:00 LT) at 3 hour-time intervals. The pump rate was 3 l min^{-1}. Simultaneous with this measurement, temperature, pressure, humidity and wind data were collected by the tethered sonde (A.I.R., Boulder, US) and transmitted to a receiver instrument (ADAS from A.I.R.). There were some occasions that the profile measurements were curtailed, when windspeeds were greater than 7.0 m s^{-1}, as there was a risk to the equipment (damage or even break the tethering line). The surface measurements from an automatic weather station were taken at the top of a micro meteorological tower (46 m height) about 500 m from the tethered base. Also at the top of the tower, the turbulent fluxes were measured with a system comprising a 3D Gill sonic anemometer (Solent Gill Instrument, Lymington, UK) and an IRGA (Li-Cor 6262, Lincoln, US) instrument. The software used to sample and compute the fluxes is the Edisol, developed by University of Edinburgh (Moncrieff et al., 1997). The sample rate of the sonic anemometer was set to 21 Hz with the IRGA samples at 5 Hz. A complete description of this software and the quality control analysis of the raw data is described by Moncrieff et al. (1997). Also mounted on the tower was a system to measure the concentration of CO_2 at the heights of 1.0, 9.0, 17.4, 25.3, 33.3 m high. These CO_2 measurements are reported in detail by Malhi et al. (1998) and they had been made using another IRGA (PP system CIRAS 1) with a slow sample rate of 1 observation every 5 minutes.

The experimental design is depicted in Figure 2. The tethered balloon activities were conducted only during MACOE. The climatic elements (rainfall, winds, temperature, solar radiation) and CO_2 measurements (concentration and fluxes) at the top of the tower as well as inside canopy are being taken as a collaborative study of long-term measurements of CO_2 gas exchange at tropical forest between Brazilian (Instituto Nacional de Pesquisas da Amazônia - INPA) and British (University of Edinburgh) organizations by the Joint Amazonian Carbon Experiment-JACAREX (Malhi et al., 1998).

III. Results and Discussion

A. Surface Measurements

In order to have a general view of the behavior of surface CO_2 fluxes and concentration at the top of the tower, an average ensemble with 11 days (from Nov. 16–27, 1995) have been made and are shown in Figure 3. The CO_2 flux (Figure 3b) has a peak (about +6.8 μmol m^{-2} s^{-1}) at sunrise due to the onset of turbulence which flushes out the rich CO_2 air from inside canopy. After this peak, the CO_2 flux reverts its exchange (from release to uptake) at 9:00 LT, intensifying this exchange until noon. At 18:00 LT the flux again becomes positive, increasing the flux during the night. The maximum uptake from vegetation was -15.6 μmol m^{-2} s^{-1} at 11:00 LT. The diurnal cycle of CO_2 concentration (Figure 3a) follows the behavior of the fluxes, showing a maximum of 423.0 ppm at 7:00 LT and then decreasing this value to a minimum of 353.5 ppm at 15:00 LT. During nighttime, the CO_2 concentration exhibits a steady increase due to the respiration from the vegetation and soil. Variations exist in both graphics as the days do not have the same synoptic conditions (cloudiness, rainfall, winds, etc.).

These surface measurements obtained during MACOE are similar to the figures collected by Fan et al. (1990) during ABLE. The CO_2 concentration also showed a good agreement, especially if we consider the secular trend of CO_2 increases. Some estimates of this trend have been made and we have found an increase of 1.5 ppm per year (from 1985–1992), which gives 12 ppm for the period from Fan et al.'s measurements (1987) and MACOE (1995). Grace et al. (1995) observed higher values than during MACOE, specifically in the CO_2 concentration: their maximum values are around 530 to 550 ppm. This is probably due to the dynamics involving the boundary layer at Ji-Paraná and Manaus and the differences in soil conditions and topography. The results from ABLE (Fan et al., 1990), ABRACOS (Culf et al., 1997) and this work have been summarized in Table 1.

B. Boundary Layer Measurements

During MACOE, we had 10 days of data that, after some consistency and quality control analysis, yielded 49 profiles. Although the maximum height of the profile is 300 m (due to the limitations of the flexible tube), it was very rare to reach this height as the windspeeds become too high for activities with the balloon.

As the final heights for each individual profile were not the same (due to the reasons explained above), a linear interpolation was made from the highest point in each individual profile to the height of 350 m at 25 m intervals. The height of 350 m has been chosen as the deep extension of the NBL (Fisch, 1995). No measurements of the CO_2 concentration value at 350 m height have been made so it was necessary to estimate it. The value of Wofsy et al. (1988) for the mixed boundary layer CO_2 concentration was 344.8 ppm and the average secular trend of annual CO_2 concentration for Mauna Loa records is 1.5 ppm per year (from 1985 until 1992). Assuming that the CO_2 difference between Mauna Loa and the Amazon region has not changed since 1985 and extrapolating from the last published record (November 1992) to the period of MACOE, the free atmosphere concentration was estimated to be 356.8 ppm. This value is thought to be the CO_2 concentration at the top of the boundary layer, which is not affected by the development and exchange process during NBL growth. These data have been extracted from Keeling and Whorf (1994).

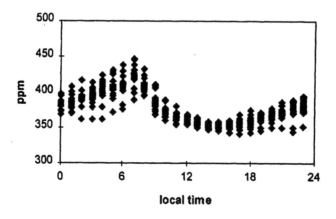

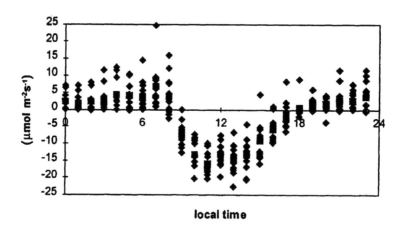

Figure 3. Diurnal cycle of CO_2 concentration (a, top) and flux (b, bottom) during MACOE.

A composite profile was made showing the time evolution of CO_2 profile in the NBL and it is presented in Figure 4. At 18:00 LT, the profile is still well mixed from the previous convective daytime conditions, showing a CO_2 concentration value of 360 ppm. As the night evolves, there is an increase of CO_2 concentration in the NBL, especially in the lower layers with the developing of sharp gradients. This is an indication of the small and shallow mixing between the layers in the NBL. At 21:00 LT, the profile shows a CO_2 concentration around 375 ppm at 50 m, increasing this value to 389 ppm at 24:00 LT, 409 ppm at 3:00 LT and 420 ppm at 5:30 LT; an overall increase of 60 ppm at 50 m from early evening until sunrise. At 100 m this variation is 36 ppm, decreasing to a value of 28 ppm at 150 m. It is interesting to notice that there are no large differences between the profiles at 3:00 and 5:30 LT. This small variation, mainly between 100–150 m high, is probably associated with an increase of windspeed during the night. Greco et al. (1992) have shown the existence of a nocturnal jet wind (at 300–500 m high) near Manaus, originating by the temperature difference between the forest and river (triggered by a thermal circulation—river breeze). The averaged profiles from 18:00 LT until 8:00 LT are presented at Table 2. The morning profile (8:00 LT) is not drawn in Figure 4.

Table 1. Typical measurements of CO_2 concentration and fluxes over tropical forest collected during ABLE (extracted from Figures 5 and 6 from Fan et al., 1992) and ABRACOS (extracted from Culf et al., 1997) and MACOE

	CO_2 (ppm)	Fluxes (μmol^{-2} s^{-1})
ABLE daytime	340 to 350	-11.8
nighttime	370 to 390	+2.4 to +4.8
ABRACOS daytime	350 to 380	-10.0
nighttime	480 to 530	+1.0 to +2.0
MACOE daytime	350	-15.0
nighttime	420	+2.0 to +4.0

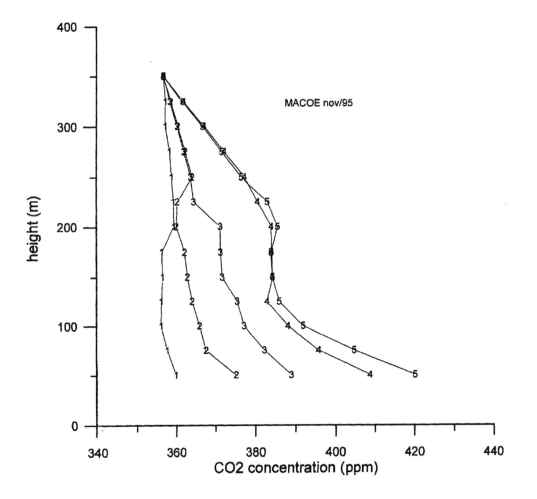

Figure 4. Composite profile of CO_2 concentration during MACOE; the numbers represent the time: 1 = 18:00 LT, 2 = 21:00 LT, 3 = 24:00 LT, 4 = 3:00 LT, 5 = 5:30 LT.

Table 2. Profile of CO_2 concentration (ppm) during MACOE (all times are local)

Height (m)	18:00	21:00	24:00	03:00	05:30	08:00
50	360.0	375.2	389.0	408.7	420.2	382.0
75	357.8	367.6	382.3	395.9	404.7	383.6
100	356.2	366.0	377.2	388.2	392.0	381.1
125	356.3	364.1	375.5	382.9	386.0	378.2
150	356.6	362.9	371.7	384.3	384.3	375.0
175	356.4	362.2	371.3	384.1	383.9	370.8
200	359.5	360.1	371.2	384.0	385.6	369.8
225	359.4	360.4	364.5	380.6	383.0	370.1
250	358.9*	364.1*	363.7	577.3	376.5	368.4
275	358.4*	362.3*	362.0*	372.2*	371.6*	366.8
300	357.3*	360.5*	360.3*	367.1*	366.7*	366.7
325	357.3*	358.7*	358.5*	361.9*	361.9*	
350	356.8**	356.8**	356.8**	356.8*	356.8**	

*Linear interpolated from an assumed (CO_2) height (**).

The time variations of CO_2 inside the NBL have been computed from the data at Table 2. The average CO_2 variation integrated from 50 m up to 350 m during all the night is 33.8 ppm h^{-1}. The maximum time variation occurred between 24:00 and 3:00 LT (54.1 ppm h^{-1}), with a very small variation at the subsequent period (13.8 ppm h^{-1} during 3:00–5:30 LT).

The data set from MACOE and from Grace et al. (1995) point out an increase of CO_2 flux and concentration just after sunrise, due to the breakdown of the thermal stability at canopy level and onset of the turbulence. An attempt has been made to quantify this effect on the boundary layer structure using the profile data collected at 8:00 LT. This profile (not shown in Figure 3 but presented in Table 2) gave a well-mixed CO_2 profile with an average of 373.9 ppm. Assuming that the CO_2 concentration in the free atmosphere or at the boundary layer is 356.8 ppm, there is an increase of 17.1 ppm, which is compatible with a decrease of the surface CO_2 concentration by 14 ppm from 7:00 to 8:00 LT.

As mentioned in Section 2, the operations with the tethered balloon require a clear space to raise and lower the balloon and this clearing is approximately 500 m from the tower. A simple comparison between CO_2 concentration at 50 m high (extracted from the profile) and the top of the tower was made and the dispersion pattern is shown in Figure 5. The linear regression between these data (55 pairs of values) had a slope of 1.01 and a correlation coefficient (r^2) of 0.70. This result gives confidence to match the CO_2 profile with the inside canopy data, and the average composite profile is depicted in Figure 6. The two independent data sets have matched extremely well. The build up of CO_2 during the night with high values of CO_2 concentration at sunrise can be clearly observed. Near to the surface there is maximum of 470 ppm due to the soil respiration. The air inside the canopy is CO_2 rich, and at crown level (35 m) it is 20 ppm higher than the lowest part of NBL. There is very little mixing inside the canopy, decoupling the upper part from the lower one. This configuration leads to a thermal stability inversion. This pattern of thermal stability was observed previously by Shuttleworth et al. (1985). In the early evening, the vertical gradient of CO_2 inside canopy is about 10 ppm, increasing to around 60 ppm at 6:00 LT. This same pattern occurs above the forest with less intensity: this vertical gradient is negligible at 18:00, being approximately 20 ppm at sunrise.

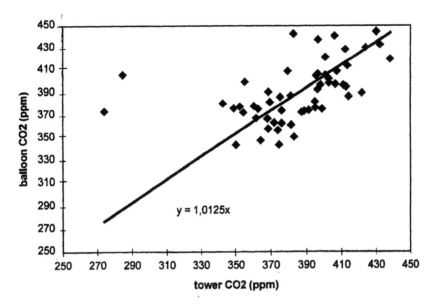

Figure 5. Comparisons between CO_2 concentration at the top of the tower and at 50 m (balloon measurement).

C. CO_2 Budget

The CO_2 balance has been computed using data from the tethered balloon and from the flux measurements at the tower. The CO_2 storage(SCO_2) in the NBL has been calculated as suggested by Denmead et al. (1996):

$$SCO_2 = \int_{50}^{h} \frac{\partial CO_2}{\partial t} dz \qquad (\mu mol\ m^{-2}\ s^{-1})$$

where the height (h) has been chosen as 350 m. The time variation of CO_2 ($\partial CO_2/\partial t$) was computed from the 18:00 and 5:30 LT profiles and it was integrated in the vertical variable (z). Table 3 presents the balance between this CO_2 storage and the nighttime integrated value from eddy-correlation equipment. There is good agreement during 3 nights (November 17,18 and 23) and disagreement on 3 nights (November 16, 19 and 22). In order to classify all the nights to see the influence of thermal stability, the Monin-Obhukov length (L) was calculated and is shown in Figure 7. It is clear that for small values of 1/L (which means weak stability) there was a tendency of the closure of CO_2 budget. On these nights (November 17,18 and 23) the closure balance is due to the high winds that mixed the NBL, bringing up air very rich in CO_2. During these nights the sensible heat flux is higher than normal, with less cooling. Also there was some cloudiness. On the other hand, the nights with poor agreement (November 16,19 and 22) showed remarkably low values of CO_2 and sensible heat fluxes. On those occasions, there were high values of the stability (typically between 0.5 and 1.0 m^{-1}) decoupling between the air inside the canopy and above it.

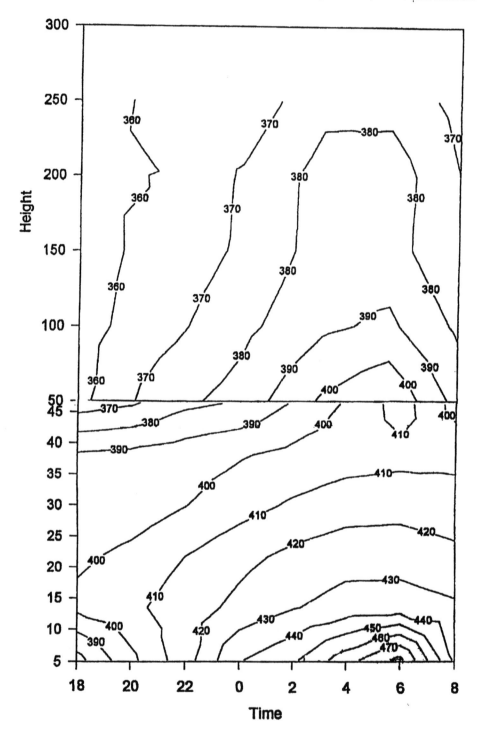

Figure 6. Nighttime evolution of CO_2 concentration inside and above canopy.

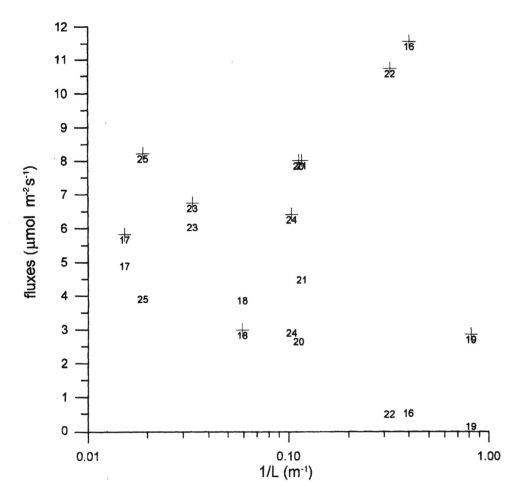

Figure 7. Stability class of the nights for both tower and balloon (crosses) estimates of integrated CO_2 fluxes; the labels represent the nights during MACOE.

IV. Concluding Remarks

The MACOE experiment was conducted in order to make regional CO_2 budgets over tropical forest in central Amazonia. The methodology used (a tethered balloon raising a flexible sample tube that brings air down for an IRGA) was shown to be practical and reliable, and the data set obtained is unique for a tropical forest.

The fluxes derived from the NBL profiles are estimates of the regional fluxes and they should not be expected to be exactly equal to the measured surface fluxes at a single point. There are some reasons that might explain this discrepancy, which is mostly positive (concentration profiles storage is higher than the turbulent fluxes). The most plausible explanation is that the surface CO_2 flux has

Table 3. CO_2 budget for individual nights during MACOE

Nights	CO_2 flux	Storage CO_2
	μmol $^{-2}$ s^{-1}	
16th*	0.48	11.53
17th*	4.84	5.79
18th	3.80	2.95
19th	0.08	2.83
20th	2.60	7.96
21st	4.43	7.97
22nd	0.45	10.69
23rd	5.98	6.71
24th	2.86	6.37
25th**	3.85	8.18

*Integrated from 21:00 until 5:30 LT; ** integrated from 18:00 until 3:00 LT.

a large spatial variability and there is an enhanced flux from river valleys and forest edges. The balloon profile measures the combined effect of this variability.

There are still unsolved questions related to the release and absorption of CO_2 by the vegetation, mainly relating these to topography (point measurements at plateau or valley) and spatial variations. Also, some higher altitude profiles are required to understand better the gas exchange between forest and vegetation during the nighttime growth, as some nights did not allow the balloon to reach the top of NBL. Also, measurements of the CO_2 concentration in the free atmosphere are needed. This will likely be provided in another field campaign to be held over tropical forest during the dry season of 1999.

Acknowledgements

The authors are pleased to thank all their Brazilian colleagues from INPE, INPA, UFAL who participated in the field campaign. We also acknowledge the financial support provide by the NERC through its Terrestrial Initiative in Global Environmental Research Programme (TIGER) awards number GST/02/605 and GST 91/III 1/5A. The Brazilian contribution was funded by Fundação de Amparo à Pesquisa do Estado de São Paulo -FAPESP (95/1596-4) and by the Projeto de Pesquisa Dirigida PPD-G7 "Balanço de Energia, Vapor D'água e CO_2 em áreas de floresta tropical na Amazônia Central (0966/95)." Special thanks go to Dr. John Roberts (Institute of Hydrology- United Kingdom) for his help in doing this manuscript.

References

Baldocchi, D., R. Valentini, S. Running, W. Oechel, and R. Dahlman. 1996. Strategies for measuring and modelling carbon dioxide and water vapour fluxes over terrestrial ecossystem. *Global Change Biol.* 2:159-168.

Culf, A.D., G. Fisch, Y. Malhi, and C.A. Nobre. 1997. The influence of the atmospheric boundary layer on carbon dioxide concentrations over a tropical forest. *Agric. For. Met.* 85:149-158.

Delmas, R.A., J. Servant, J.P. Tathy, B. Cros, and M. Labat. 1992. Sources and sinks of methane and carbon dioxide exchanges in mountain forest in equatorial Africa. *J. Geophys. Res.* 97:6169-6179.

Denmead, O.T., M.R. Raupach, F.X. Dunin, H.A. Cleugh, and R. Leuning. 1996. Boundary layer budgets for regional estimates of scalar fluxes. *Global Change Biol.* 2:255-264.

Fan, S.M., S. Wofsy, P.S. Bakwin, and D.J. Jacobs. 1990. Atmosphere-biosphere exchange of CO_2 and O_3 in the central Amazon forest. *J. Geophys. Res.* 95:16851-16864.

Fisch, G. 1995. Camada Limite Amazônica: aspectos observacionais e de modelagem. Instituto Nacional de Pesquisas Espaciais (PhD Thesis), Sao José dos Campos, Brazil.

Fiztjarrald, D.R. and K.E. Moore. 1990. Mechanisms of nocturnal exchange between the rain forest and the atmosphere. *J. Geophys. Res.* 95:16839-16850.

Gash, J.H.C., C.A. Nobre, J.M. Roberts, and R.L. Victoria. 1996. An overview of ABRACOS. p. 14-18. In: J.H.C. Gash, C.A. Nobre, J.M.Roberts, and R.L. Victoria (eds.), *Amazonian Deforestation and Climate*. John Wiley & Sons, Chichester, U.K.

Grace, J., J. Lloyd, J. McIntire, A. Miranda, P. Meir, H. Miranda, J. Moncrieff, J. Massheder, I. Wright, and J. Gash. 1995. Fluxes of carbon dioxide and water vapour over an undisturbed tropical forest in south-west Amazonia. *Global Change Biol.* 1:1-12.

Greco, S., S. Ulanski, M. Garstang, and S. Houston. 1992. Low-level nocturnal wind maximum over the central Amazon basin. *Boundary Layer Meteorol.* 58:91-115.

Keeling, C.D. and T.P. Whorf. 1994. Atmospheric CO_2 records from sites in the SIO air sampling network. In: T.A. Boden, D.P. Kaiser, R.J. Sepanski, and F.W. Stoss (eds.), Trends 93: A Compendium of Data on Global Change. ORNL/CDIAC-65, Carbon Dioxide Information Analysis Centre, Oak Ridge National Laboratoy, Oak Ridge, TN, USA.

Malhi, Y., A.D. Nobre, J. Grace, B. Kruijt, M.G.P. Pereira, A. Culf, and S. Scott. 1998. Carbon dioxide transfer over a central Amazonian rain forest. *J. Geophys. Res.*

Moncrieff, J.B., J.M. Massheder, H. de Bruin, J. Elber, T. Friborg, B. Heusinkveld, P. .Kabat, S. Scott, H. Soegaard, and A. Verhoef. 1997. A system to measure surface fluxes of momentum, sensible heat, water vapour and carbon dioxide. *J. Hydrol.* 188-189:589-611.

Shuttleworth, W.J., J.H.C. Gash, C.R. Lloyd, C.J. Moore, J.M. Roberts, A. de O. Marques Filho, G. Fisch, V.P. Silva Filho, M.N. Goes Ribeiro, L.C.B. Molion, L.D.A. Sa, C.A. Nobre, O.M.R. Cabral, S. Patel, and J.C. Moraes. 1984. Daily variations of temperature and humidity within and above Amazonian forest. *Weather* 40:102-108.

Wofsy, S.C., R.C. Harriss, and W.A. Kaplan. 1988. Carbon dioxide in the atmosphere over the Amazon basin. *J. Geophys. Res.* 93:1377-1387.

CHAPTER 22

Atmospheric CO_2 Fluxes and Soil Respiration Measurements over Sugarcane in Southeast Brazil

H.R. Da Rocha, O.M.R. Cabral, M.A.F. Da Silva Dias, M.A. Ligo, J.A. Elbers, H.C. Freitas, C. Von Randow and O. Brunini

I. Introduction

Sugarcane production has been a strong component of the Brazilian economy since the early 1970s as well as an attractive alternative to mitigate greenhouse gases emission. The associated agroindustry involves about 1 million jobs, using a cultivated area of 4.2 million ha (23.4% of the world area of cultivated sugarcane). The production has reached 273 million tons (harvested wet weight) in the 1996/97 season and is currently used for sugar (1/3, or 13.5 million tons) and ethanol (2/3, or 13.7 million m^3). The use of ethanol from sugarcane as car fuel (Goldenberg et al., 1993), either as anhydrous ethanol blended with gasoline or hydrated ethanol, has appeared to be a feasible alternative towards environmental sustainability. In fact, Macedo (1992; 1997) has estimated the Output/Input energy ratio of ethanol production equal to 9.2, comprising the whole energy consumption within sugarcane and ethanol production up to the entire produced ethanol and bagasse surplus used in energy cogeneration. The state of São Paulo (Figure 1), in the southeast region of Brazil, occupies 45% of the total cultivated area of the country, although its contribution to the national ethanol and sugar production is larger than 50%. This region was originally covered with tropical evergreen forest along the coast (Mata Atlântica) and tropical decidous forests and savanna (physiognomic types of Cerrado Restrito, Cerradão, Campo Cerrado and Campo) elsewhere (Ratter 1992; Eiten 1972). Today the land cover is pasture, sugarcane and crops.

The Department of Atmospheric Sciences of the University of São Paulo took the initiative in late 1996 to establish an environmental project over a sugarcane plantation in southeast Brazil, conceived to measure surface-atmosphere exchanges of heat, momentum, water vapor and carbon dioxide. The project is at present under development with the participation of other partners, namely the Empresa Brasileira de Pesquisa Agropecuária (EMBRAPA Meio Ambiente), Instituto Agronômico de Campinas (IAC) and the Winand Staring Centre (WSC/DLO/ Wageningen, Netherlands). The project pursues multi-disciplinary issues: (i) the long term measurement of the surface fluxes and its use to improve regional weather and climate numerical prediction; (ii) evaluation of CO_2 atmospheric fluxes including soil respiration, net ecosystem exchange, and soil carbon for preharvest burned sugarcane and green harvested sugarcane; (iii) comparability of the sugarcane agroecosystem with the original Cerrado (savanna) companion site (to be implemented in 1998). The sugarcane vegetation appeared relevant on two sides. It is a representative surface cover that influences the atmospheric state and the regional climatic variability, therefore it must be taken into account by the general circulation models (GCMs) used in the numerical simulations of weather and climatic studies, and second, it represents

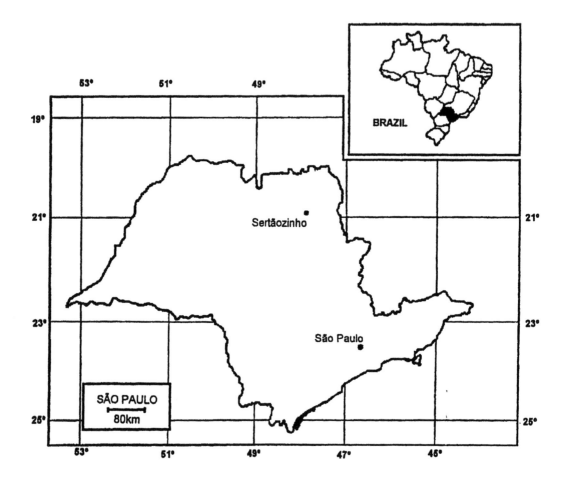

Figure 1. Geographical position of the state of São Paulo, Brazil (in black, top right corner), and the study site located in the city of Sertãozinho.

a feasible alternative to promote sustainability of environment, energy, climate change and social-economical aspects.

Sugarcane (*Saccharum* spp) is a C4 plant and in Brazil is a ratoon crop with an average life cycle of 4 to 5 years, usually harvested once a year during May to December. The productivity (wet harvest weight) ranges from 40 to 150 ton ha^{-1} year^{-1} depending on the variety, ratoon, climate, soil moisture, nutrients, soil type and management, pests, diseases and others. The harvest is usually preceeded by the burning dry leaves, then the culture is harvested manually or mechanically. On the other hand, it can be harvested green, mostly mechanically. In 1997 the mechanical harvest was about 20% in the state of São Paulo, a ratio which could possibly increase up to 50% over the next years (Macedo, 1997).

The N_2O emission from sugarcane has been estimated as 1.7 kg N_2O ha⁻¹ year⁻¹ (Macedo, 1997), what corresponds to an amount of 68 kg C equivalent ha⁻¹ year⁻¹, giving its potential greenhouse effect about 150 times that of CO_2. Adapted varieties in Brazil demand low input of nitrogen, about 60 kg N ha⁻¹ year⁻¹, a value usually much below those used in other countries (Urquiaga and Boddey, 1996). Therefore, compared to other countries, it means a reduced potential for N_2O greenhouse emissions.

The surface-atmosphere exchange of CO_2 for the sugarcane life cycle is the balance of several processes. During the growth period, carbon is fixed in the above and below ground biomass by photosynthesis, while loss occurs by maintenance and growth plant respiration. The balance of photosynthesis and total plant respiration is usually named net carbon assimilation. Soil respiration is the continuous loss of carbon by the respiration of roots, microbes, soil fauna and carbon oxidation. In the long term, once the soil organic carbon amount is kept constant, there must a balance, that is, soil respiration plus the carbon emission resulting from the harvest burning and sugarcane products (ethanol, sugar and other losses within the production process) must balance the net photosynthesis. How can carbon sequestration happen in these ecosystems, that is, the increase of soil organic carbon with time? The first constraint is a significant biomass input to the soil, a situation that already occurs in the regional circumstance. Favorable conditions are given by the tropical climate (warm temperature and plenty of rainfall) to stimulate photosynthesis, which on the other hand will affect soil respiration, increasing the microbial activity and carbon loss. Clayey soils, usually found in the tropics, are in principle favorable, as they are supposed to benefit soil aggregation and floculation (See Chapter 16). It seems that at the end, other factors such as tillage, mineralization and erosion, have to be reduced to create conditions to sequester carbon in the soil.

This chapter adresses the evaluation of the net ecosystem exchange of atmospheric CO_2 over sugarcane for specific periods during its life cycle, and the soil respiration, which will help to assess how the main mechanisms affect carbon balance. Measurements of above ground biomass and soil organic carbon are shown for the site.

II. Site and Methods

The study site is located in Sertãozinho (Figure 1), 330 km from São Paulo, 520 m height, 21°06'S, 48°04' W over an area owned by the sugar mill Cia Energética Santa Elisa. The site was implemented in October 1996, having the sugarcane variety SP 71-6180 in the 4th ratoon crop, planted in rows 1.35 m apart. The previous harvest was manual, preceded by the burning dry leaves in May 1996. The last harvest was in 10/11 Apr 1997, employing the same method. Prevailing soils are Red Latossols (Typic Eutortox, USA; Rhodic Ferralsol, FAO), with clay and sand fraction equal to 48% and 36%, respectively, and porosity equal to 55 to 58% at deeper levels.

The instrumental platform consists of a micrometeorological 10 m tower with sensors of air humidity and temperature (Campbell 107, assembled at the Institute of Hydrology, UK), soil temperature at 5, 10 and 20 cm (Campbell 107), precipitation (Texas 525), wind speed profile (RM Young), incoming and reflected solar radiation (Li-COR 200X), net radiation (REBS Fritschen) and soil heat flux (REBS HF7). A data logger (Campbell CR10X) sampled the data every 15 s and recorded the averages on a 10 min basis. Soil moisture was measured on a weekly basis using a neutron probe (CPN Boart Longyear, USA) in 5 access tubes near the tower, with measurements every 10 cm up to the 150 cm depth.

An eddy correlation device measured the atmospheric turbulent fluxes of sensible heat, evapotranspiration, momentum flux and total CO_2 flux. The system was set during intensive missions at the top of the tower, calculating the 30 min average fluxes. It comprised a three-dimensional sonic anemometer (A1012R Gill Solent), an infrared gas analyser for CO_2 and H_2O (Li-COR 6262) and a eletromechanical system to control the pumped air from the sampler tube near the anemometer, and record the 10 Hz frequency raw data (Moncrieff et al., 1997). The device was designed and assembled by the WSC/DLO (Wageningen, Netherlands). The eddy correlation method calculates the vertical flux of the atmospheric carbon dioxide, Fc, taking the correlation between the deviations of the 30 min average of the vertical wind speed (w') and the air CO_2 concentration (ρ_c'),

$$F_c = \overline{w' \rho_c'}.$$
Eq. 1

a term usually referred to as the net ecosystem exchange (NEE), which accounts for the sum of gross primary productivity, plant respiration and soil respiration.

Soil respiration and soil temperature at 1 cm were measured using a chamber (SRC-1, PP Systems, UK) sealed to the soil surface in closed circuit to a infrared gas analyser, with an encapsulated thermistor placed near the chamber, both connected to a data logger. Following Parkinson (1981), the soil respiration rate, Rs, is estimated taking the rate of increase of the CO_2 concentration inside the chamber, dc/dt, and having the area of soil covered, A, and the total volume system, V,

$$Rs = \left[\frac{dc}{dt}\right]\frac{V}{A}.$$
Eq. 2

Measurements with the chamber were carried out at specific intensive missions along the year, usually about 10 to 30 measurements a day for each mission. Estimates of the biomass and soil organic carbon were taken just before the 1997 harvest. The above ground biomass was sampled over two 2.7 m x 4 m plots before and after the harvest. The soil organic carbon was collected in 9 samples at the 0–20 cm depth, which was further analyzed in the laboratory using the Walkey–Black method (Black, 1965).

III. Results and Discussion

The harvest of the sugarcane usually begins in April or May, which coincides with the early dry season when the average precipitation and temperature begin to decrease, as shown by the climatological values in Figures 2 (c,e). Accordingly, the available solar radiation at the surface has a substantial seasonal pattern (Figure 2b) along the year. It is opportune to mention two anomalous events during the period of collected data. The first was the rainfall from 21 May to 16 June (days 141 to 167) (Figure 2e), that is also noticed on the secondary peaks of soil moisture, in the middle of the dry season. The second is the warmer and drier than average winter and spring periods, possibly caused by the effects of the El Niño-Southern Oscillation (ENOS) phenomenon. The ENOS 1997/1998 event has been ranked the strongest of the last 100 years. It induces intense air subsidence over the tropical latitudes in eastern South America, which supresses the rainfall, increases the low level air temperature and lowers the relative humidity. Therefore the absence of rainfall and the above normal temperatures on days 170 to 330 are mostly due to the effects of this large scale natural phenomenon. The 1997 harvest happened in day 100, as shown in Figure 2a, and is an important mark concerning the expected carbon fluxes of the ecosystem.

The above ground biomass measured before the harvest was 7.5 ton ha^{-1} (dry leaf weight) and 4.9 ton ha^{-1} (green leaf weight). After the harvest the sugarcane production was measured as 86 ton ha^{-1},

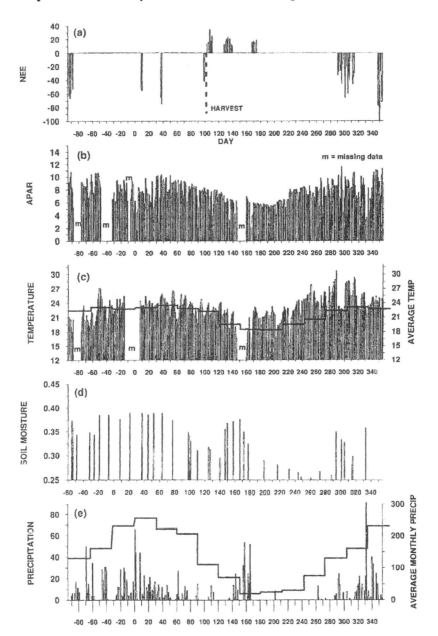

Figure 2. (a) Daily integral net ecosystem exchange (NEE), in kg C ha^{-1} day^{-1} (only specific days of intensive measurements are shown); (b) daily integral absorbed photosynthetic active radiation (APAR), in MJ m^{-2} day^{-1}; (c) daily average air temperature (left ordinate), and climatological monthly average temperature (right ordinate, shown as long horizontal bars), in °C; (d) volumetric soil moisture at 10 cm, in m^{-3} m^{-3} (only specific days of measurements are shown); (e) precipitation (left ordinate, in mm day^{-1}) and climatological monthly average precipitation (right ordinate, in mm, shown as long horizontal bars). Day 1 is 1st Jan 1997. Missing data periods are shown as m.

and the total oxidated biomass estimated as 9.8 ton ha^{-1}, which leads up to a ratio of burning efficiency equal to 79%, a result similar to that calculated by Macedo (1997). The average soil organic carbon percentage relative to the soil dry weight was calculated as (2.92 ± 0.21)%. Scaling up the sampling leads to an amount of SOC equal to 74.2 ton C ha^{-1} over the 0–20 cm depth. This is a significant amount when compared to the estimates of Cerri et al. (See Chapter 2) for the Brazilian Amazon (42 ton C ha^{-1} at 0–20 cm), and Woomer et al. (See Chapter 5) for the cropland conversed from forest (38 ton C ha^{-1} for the 0–30 cm depth).

A. Net Ecosystem Exchange

The measurements of the net ecosystem exchange were first made in October, 1996 (day -90 in Figure 2e), when the ratoon was 5 months old since the 1996 harvest. At that time the vegetation was vigorous and the canopy average height was 1.5 m. Other sparse measurements were carried out on days 9,10,40,41 and just before the 1997 harvest, on days 98 and 99. All these mentioned periods showed that the ecosystem was fixing CO_2, with rates ranging from -40 to -70 kg C ha^{-1} day^{-1} (negative values mean downward flux). After the harvest (day 100) the fluxes reversed the sign, which somehow could be anticipated since the vegetation was completely cut, although the absolute values were not expected. The observations on days 102 to day 175 revealed basically positive fluxes, with values ranging from 10 to 35 kg C ha^{-1} day^{-1}. Measurements of leaf photosynthesis were not made during this period, but it was possible to notice the very slow growth of the vegetation during the wintertime, which means that the soil respiration dominates the scene of the total carbon flux along this period. Following the chronology, later on days 290 to 310 (17 Oct to 06 Nov) and from days 345 to 352 (11 to 18 Dec), the total fluxes reversed the sign again, indicating the ecosystem was then fixing CO_2 from the atmosphere. Lack of data between days 180 to 290 made it impossible to know exactly when the ecosystem started fixing CO_2, or even if there was an abrupt or slow transition to detect such remarkable behavior.

B. Soil Respiration

Data of the soil respiration chamber and soil temperature at 1 cm were collected along eight intensive missions and used to fit a simple model. We have split the data in two sets, the first is typical of the wet season (28 Nov and 01 Dec 1996, Jan 22 and Feb 06 1997), and the second of the dry season (19 May, 28 Jun, 29 Jun and 03 Sep 1997). For a given soil temperature, the soil respiration was significantly higher in the wet period (Figure 3), with an average rate of 44 kg C ha^{-1} day^{-1}, than the dry period, with an average equal to 28 kg C ha^{-1} day^{-1}. Besides the different averages, the difference between the maximum and minimum observed in both periods was also significant: 53 kg C ha^{-1} day^{-1} (23 to 76) in the dry period and 34 kg C ha^{-1} day^{-1} (17 to 51) in the wet period. It seems that the larger variability of the soil temperature in the diurnal cycle of the winter influenced the larger amplitude of the soil efflux rates too.

Fitting these data resulted in an exponential relationship (Figure 3) as following:

$$Rs = exp(0.0704\ T_s - 0.2083),\ Q_{10} = 2.0;\ n = 50;\ R^2 = 0.45 \quad \text{(wet period)} \qquad \text{Eq. 3}$$

$$Rs = exp(0.0285\ T_s + 0.2269),\ Q_{10} = 1.3;\ n = 54;\ R^2 = 0.31 \quad \text{(dry period)} \qquad \text{Eq. 4}$$

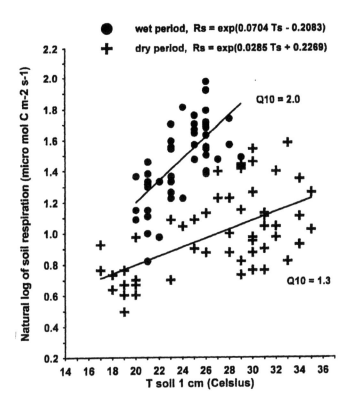

Figure 3. Natural logarithm of soil respiration rate over sugarcane, in μmol CO$_2$ m^{-2} s^{-1}, observed during the dry period (squares) and the wet period (circles), as a function of soil temperature at 1 cm depth (abscissa), in °C.

where Rs is in μmol CO$_2$ m^{-2} s^{-1} and T$_s$ the soil temperature at 1 cm, in °C, R^2 is the correlation coeficient calculated over n points, and Q$_{10}$ is the coefficient of the relative increase in soil respiration for a temperature increase of 10 °C.

Meir et al. (1996) has calculated the relationship between soil respiration and soil temperature at 1 cm using a similar approach, for a site at the southeast Amazonian rainforest and another at the Cerrado in central Brazil (Figure 4). Respiration at the sugarcane for the wet period was always lower than that in the rainforest. Nonetheless, the values of Q$_{10}$ in that case are not very far from that of the forest. On the other hand, for the dry period, the sugarcane site presents the lowest respiration rates compared to other cases. Even the sensitivity to the temperature is lower, as shown by Q$_{10}$ equal to 1.3. Meir et al. (1996) also found low correlation coefficients, equal to 0.21 for the forest and 0.33 for Cerrado. It suggests the need of further investigation about other physical process influencing soil respiration, for instance the problem of statistical spatial variability and the inclusion of soil moisture measured locally near the chamber.

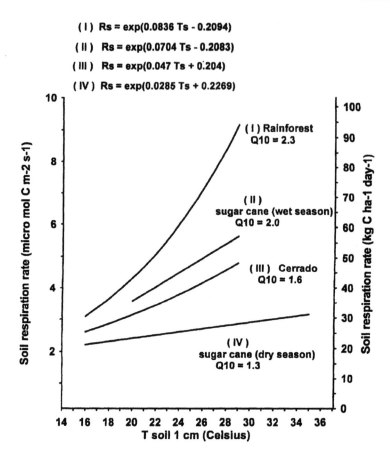

Figure 4. Fitted curves of the soil respiration rates, in μmol CO_2 m^{-2} s^{-1} (left ordinate) or in kg C ha^{-1} day^{-1} (right ordinate) as a function of soil temperature at 1 cm depth (abscissa), in °C, at the sugarcane site in the dry and wet period. Similar curves fitted for the amazonian rain forest (Rainforest) and for the cerrado (Cerrado) by Meir et al. (1996) are shown in the graph.

IV. Conclusion

The technique of the eddy correlation was used to measure the net ecosystem exchange over a sugarcane site in southeast Brazil collected during intensive missions within the approximate one year period from November 1996 to December 1997. We detected changes in the sign of the total CO_2 flux from the ecosystem before and after the harvest, as well as the average daily rates in these missions. The studies focused on the two main periods of the sugarcane life cycle. The first, just after the harvest, has shown positive CO_2 flux (carbon loss), with an average of 16 kg C ha^{-1} day^{-1}, persistent over at least the first 3 months after the harvest. Presumably the ecosystem lost carbon even longer than 3 months, but it was not possible to detect due to the lack of data. The second, representative of the vegetation about 5 to 7 months old, showed a negative CO_2 flux, with values varying from -15 to -100 kg C ha^{-1} day^{-1}.

The soil respiration had an important magnitude in the overall atmospheric carbon balance. The influence of the soil moisture was also a key factor determining the behavior of this process. We could possibly include the soil moisture as a second variable to fit the soil respiration in a mathematical model. We decided not to do it in this chapter for two reasons. The first concerns the validity of using the 10 cm measurement of the neutron probe and associating it directly with the chamber measurements. The spatial variability of soil moisture is in principle a limiting factor, a point far more critical for the top soil layer than deeper layers. The second is that keeping a simple model would enable us to know how well soil temperature controls soil respiration, as well as comparing the performance of this model with similar ones in different Brazilian ecosystems, that is, Rainforest and Cerrado (Meir et al., 1996).

The measurements reported here are site specific, and scaling up to other areas is still uncertain. For the sugarcane annual cycle, there exists a strong seasonal variation of soil temperature near the tropical latitudes. Also, the harvest is not carried out at the same time over all the cultivated area. It means that some areas will be under the vegetation cover for some periods of time while others will not, and this is a second type of influence on the yearly march of soil temperature.

Although the results shown here are not sufficient to calculate the net carbon flux at the sugarcane site on a yearly basis, it is reasonable to suggest the flux must be negative (sink of carbon). Looking strictly at our data, there is some evidence to support it: first, the magnitude of the negative fluxes is much higher, in absolute values, than that of the positive fluxes; second, the period when the positive fluxes prevail is relatively lower than that of the negative fluxes. Also, a question has been raised on the possibility of the sugarcane agroecosystem sequestering carbon in the soil. As mentioned previously, some points help to confirm this hypothesis in our study site, namely the climate, soil type and high productivity. The data provided here is useful as the beginning of a deeper investigation pursuing this major question. Presently, the introduction of the green harvested sugarcane management regionally will reinforce those chances, as the supression of the burning dry leaves allows a significant amount of carbon to return back to the soil instead of being released directly to the atmosphere. We intend to adress this question better in the coming years, when there will be an attempt to know the role of the soil respiration in the circumstances to compare green against burning dry leaves managements, and in parallel measuring the soil carbon evolution in the two scenarios.

Acknowledgements

We acknowledge the support of the Project 'Modelagem e Interação da Biosfera-Atmosfera em São Paulo' by the Fundação de Amparo à Pesquisa do Estado de São Paulo under grant 95/1816-4. We also thank the Usina Santa Elisa and the engineers Valmir Barbosa and Rodrigo Salgado for the strong support and help during the measurements and site maintenance.

References

Black, C.A. 1965. *Methods of Soil Analysis. Part 2. Chemical and Microbiological Properties*, Agronomy No. 9, American Society of Agronomy, Madison, WI, p. 417-429.

Eiten, G. 1972. The cerrado vegetation of Brazil. *The Botanical Review* 38:201-341.

Goldenberg, J., L.C. Monaco and I.C. Macedo. 1993. The Brazilian fuel-alcohol program. p. 841-862. In: T.B. Johansson (ed.), *Renewable Energy*.

Macedo, I.C. 1992. The sugar cane agro-industry - its contribution to reducing emissions in Brazil. *Biomass and Energy* 3:77-80.

Macedo, I.C. 1997. Greenhouse gas emissions and avoided emissions in the production and utilization of sugar cane, sugar and ethanol in Brazil: 1990-1994. Report for the Ministério da Ciência e Tecnologia (MCT). Coordenação de Pesquisa em Mudanças Globais. Piracicaba, Centro de Tecnologia Copersucar, (RT-CTC-002/97).

Meir, P., J. Grace, A. Miranda and J. Lloyd 1996. Soil respiration in a rainforest in Amazonia and in Cerrado in Central Brazil. p. 319-329. In: J.Gash, C. Nobre, J. Roberts and R. Victoria (eds.), *Amazonian Deforestation and Climate.*

Moncrieff, J.B., J.M. Massheder, H. de Bruin, J.Elbers, T. Friborg, B. Heusinkveld, P. Kabat, S. Scott, H. Soegaard and A. Verhof 1997. A system to measure surface fluxes of momentum, sensible heat, water vapour and carbon dioxide. *J. Hydrol.* 189:589-611.

Parkinson, K.J. 1981. An improved method for measuring soil respiration in the field. *J. Appl. Ecol.* 18:221-228.

Ratter, J.A. 1992. Transitions between cerrado and forest vegetation in Brazil. p. 417-429. In: P.A. Furley, J. Proctor and J.A. Ratter (eds.), *Nature and Dynamics of Forest-savanna Boundaries,* Chapman & Hall, London.

Urquiaga, S. and R.M. Boddey 1996. N2 fixation in sugar cane genotypes, the key to high energy balances in the brazilian Proalcool programme. p. B53-B54. In: Conference on Environmetrics, Universidade de São Paulo, São Paulo, June, 1996.

Section VII.

RESEARCH AND DEVELOPMENT PRIORITIES

What Do We Know and What Needs to Be Known and Implemented for C Sequestration in Tropical Ecosystems

R. Lal and J.M. Kimble

I. Introduction

Both tropical rainforest (TRF) and savanna ecosystems play major roles in ecologic processes that determine environment and economic opportunities that affect the well-being of human occupants of these regions. Total land area within 30° N and S of the equator, the region constituting tropics and subtropics, is about 5×10^9 ha (NRC, 1993). Prior to human intervention, the land area under forest cover was 3.2×10^9 ha. Thus the land area under TRF has declined by 600 million ha or by 20% of the original area. The present global rate of deforestation of the remaining TRF ecosystem is about 1.2% yr^{-1}. With modern techniques, including satellite imagery and GIS, the available statistics are improving. Total land area under tropical savanna (TS) prior to human settlement was as much as 1.7 billion ha. A large part of TS in Asia had long been converted to agricultural land use and is rapidly being converted in Central and South America and Africa.

Because of the large areas involved, changes in land use have resulted in depletion of the C pool from TRF and TS ecosystems. Estimates of the amount of biomass in the TRF range with an average of 460 Mg ha^{-1} comprising 350 Mg ha^{-1} above-ground and 110 Mg ha^{-1} below-ground biomass. The biomass in TS ecoregions may be less than 50% of that in TRF.

Both TRF and TS ecosystems are the last frontiers. Both TRF and TS ecosystems are also being rapidly converted to arable, pastoral and silvicultural land use in Central and South America, Africa, and Australia. Most TS ecoregions in Asia have long been converted to agricultural or forestry land uses. Consequently, large quantities of C (as CO_2 and CH_4) and N (as N_2O, NO_x) are emitted from these ecosystems to the atmosphere (Figure 1). Therefore land use and land cover changes in these ecoregions are directly related to the concerns about the "greenhouse effect."

II. Soil Organic Carbon in Tropical Ecosystems

The database on SOC pool in a tropical ecosystem is weak, with scanty information that is not readily available. The data in Table 1 show that soils under TRF ecosystem may contain 4 to 6 kg m^{-2} in 30 cm depth, 7 to 12 kg m^{-2} in 1 m depth, and 10 to 17 kg m^{-2} up to 2 m depth. In general, the SOC pool in TRF increases with an increase in the mean annual rainfalls. The SOC pool is lower in semi-deciduous forest in Nigeria than in humid regions of Indonesia and Brazil (Table 1). All other factors remaining the same, the SOC pool under grass savanna may be more than under TRF. The SOC pool

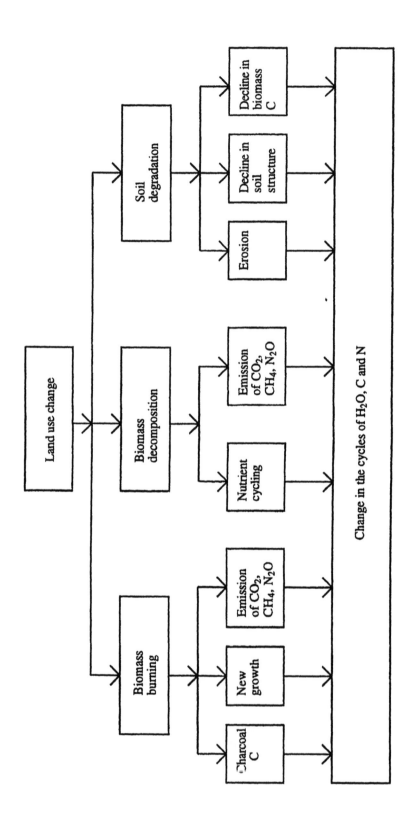

Figure 1. Land use change in TRF and TS and the emission of greenhouse gases.

Table 1. Land use effects on SOC pool in top 3 cm layer in the tropics

Land use	Soil	Moisture regime	SOC pool (kg m^{-2}) to different depths		
			30 cm	1 m	2 m
A. Indonesia (TRF)	Oxisol	Udic			
Forest (2)			6.38	12.22	16.93
Forest (4)			5.48	10.89	15.51
Manual clearing (1)			5.34	8.69	12.97
Manual clearing (7)			4.75	9.59	12.96
Rubber plantagion (8)			5.42	12.01	17.10
B. Brazil (TS)	Oxisol	Ustic			
Savanna (21)			9.35	19.46	28.63
Pasture (17)			7.52	13.61	18.22
Cultivated (22)			6.33	16.11	25.77
Cultivated (20)			6.10	18.45	28.16
C. Brazil (Subtropical forest)	Oxisol	Udic			
Forest (003)			13.04	28.99	29.92
Cultivated (001)			6.56	13.58	17.98
Cultivated (005)			7.90	16.25	23.92
D. Mali (TS)	Alfisol	Ustic			
Cultivated (1)			1.57	3.39	5.01
Cultivated (2) ·			1.70	4.33	6.07
E. Zimbabwe (TS)	Vertisol	Ustic			
Vertisol (1)			5.70	14.59	21.50
Cultivated (2)			4.52	13.04	17.76
Cultivated	Alfisol	Ustic	1.36	2.80	2.97
F. Malawi (TS)	Oxisol	Ustic			
Pasture			8.68	18.02	22.23
Cultivated			8.03	17.31	21.46
G. Nigeria	Alfisol	Ustic			
Semi-deciduous forest (Iwo)			5.61	9.60	12.40
Semi-deciduous forest (Egbeda)			4.24	7.62	10.17
Semi-deciduous forest (Ibadan)			5.17	8.03	10.40

(Unpublished data of J.M. Kimble from the NRCS data base, and recalculated from Moormann et al., 1975.)

under Brazilian savanna ranged from 8 to 10 kg m^{-2} in 30 cm depth, 18 to 20 kg m^{-2} in 1 m depth and 25 to 30 kg m^{-2} in 2 m depth (Table 1).

Conversion of TRF to plantation (e.g., rubber) may not cause drastic changes in SOC pool. Conversion to improved pastures (well-managed, deep-rooted grasses) may enhance SOC pool. Continuous cropping, especially subsistence agriculture based on minimal off-farm input, may lead to drastic reduction in SOC pool. Cultivated soils in Mali contain low level of SOC pool, especially in coarse-textured Alfisols.

The information of the type contained in Table 1 is not available for predominant soils in principal ecoregions and for major land uses in the tropics. Lack of reliable information of SOC pool to 2 m depth in undisturbed ecosystem and chronosequence data on temporal changes under different management system is a major constraint in planning appropriate land use. Availability of these data can facilitate identifying bright spots for enhancing SOC pool through adoption of appropriate land use and best management practices.

III. State of the Knowledge

Important activities that influence pools and fluxes of C with drastic impact on the atmospheric concentration of trace gases include the following:

A. Deforestation of TRF

Tropical deforestation accounts for about 15% of the total global warming potential, comprising 10% due to CO_2 and 5% due to other trace gases (e.g., N_2O, CH_4, and NO). The annual rate of deforestation is about 17 m ha or 1.2% of the total area (NRC, 1993). Deforestation leads to the emissions of CO_2, CH_4, N_2O and NO_x when the biomass is removed, burnt, or allowed to decompose. Estimates of CO_2 release by deforestation range from 400 to 2500 Tg C yr^{-1} (Tg = teragram = 1 x 10^{12} g). Depending on the subsequent land use and management, there also occurs a loss of the SOC pool. The principal reasons for deforestation include selective logging; clear cutting for timber, firewood and agricultural land use; expansion of human habitat and urbanization including infrastructure development, and mining.

B. Fire

Fire is a natural phenomenon that has shaped the TS ecosystems. It has also been used as a principal tool in removing the biomass in both TRF and TS ecosystems and leads to emissions of several radiatively-active gases, e.g., CO_2, CH_4, N_2O, NO_x and others. Biomass burning in all tropical regions may account for an annual release of about 3.4 Pg (Pg = petagram = 1 x 10^{15} g) of C.

C. Land Use and Management

Conversion to agricultural land use may have a very drastic impact on the pools and fluxes of C and other compounds that influence the greenhouse effect. The biomass production, an important determinant of the C pool and fluxes, depends on land use and management. The rate of wood and litter production in tropical plantations may be 10 to 12 Mg ha^{-1} yr^{-1}. Similarly, the rate of biomass production in improved pasture may be high, especially the below-ground biomass in deep-rooted grasses. Application of deficient nutrients and alleviation of other soil-related constraints are important to enhancing biomass production.

D. Carbon Sequestration Potential of the TRF Ecosystem

Research findings from an on-going experiment in the Amazon have shown that natural TRF ecosystems can sequester C at the rate of 2 to 3 Mg ha^{-1} yr^{-1}. The C sequestration by natural TRF may be attributed to, (i) CO_2 fertilization effect, (ii) tree mortality due to catastrophic events, and (iii) effect of El Niño in exacerbating tree mortality. Apparently, the rate of new growth exceeds the decomposition and mineralization of dead biomass and SOC.

E. Land Use Change

Conversion of natural (TRF and TS) ecosystems to plantation, pastoral and croplands has a drastic impact on carbon pool and emissions of greenhouse gases to the atmosphere. Adoption of best management practices in the converted landscape can compensate for, at least in part, the loss of C. The C sequestration potential of enriched/improved fallows and soil restorative measures for improving biomass productivity were emphasized.

F. Total System Carbon

It is important to assess the total system carbon rather than just the SOC content. The total system carbon includes that in the live and dead above- and below-ground biomass, soil fauna and flora, and all components of soil organic carbon and the soil inorganic carbon (SIC). Root biomass is an important component, especially in the TRF and plantation.

G. Soil Processes and C Pool

Important soil processes influencing C pool and dynamic include soil nutrient reserves and cycling, and soil aggregation and C sequestration. Availability of balanced soil nutrients is critical to enhancing biomass productivity of managed ecosystems, e.g., conversion of TS to croplands and pastures. Soil aggregation is a basic soil process that determines structural attributes with attendant impact on soil quality, biomass productivity, C sequestration within stable aggregates due to the formation of organo-mineral complexes, and mitigation of the potential greenhouse effect.

H. Methods of C Pool Assessment and Prediction

It is important to use appropriate methods of measuring different components of total soil carbon. The choice of appropriate methods(s) determines data credibility and reliability, and depends on several factors, e.g., the specific component to be monitored. There are also pros and cons of landscape approach to assessing the C pool. Potentials and constraints of different models need to be assessed in relation to the impact of land use and management on potential, attainable and actual SOC pools.

I. Soil Organic Carbon Pool

Estimates of the SOC pool are not available for all regions of TRF and TS. It is important to strengthen the database needed for obtaining these estimates.

IV. Knowledge Gaps And Researchable Priorities

Considerable progress has been made in assessing C pool and dynamics in tropical ecosystems. Yet, there are numerous knowledge gaps that need to be filled (Figure 2). While agricultural or managed ecosystems have been a source of emission of C and other greenhouse gases, they can also be a solution to the problems related to environmental and ecological issues. In this regard, the workshop identified several researchable priorities as outlined below.

A. Soil Carbon Pools and Fluxes

There are numerous researchable priorities that need to be addressed to enhance our knowledge of total system carbon pool and fluxes.

1. C Deep in the Soil Profile

The relative importance of C below 50 cm depth (Plate 1a,b) needs to be assessed. Carbon deep in the soil profile may comprise (i) SOC contents and (ii) live and dead biomass. The root biomass is particularly important in the forest and grassland ecosystems. The pool and dynamics of these components must be assessed especially in relation to land use and other anthropogenic activities. It is important to understand the magnitude and the fate of total system C deep in the soil profile.

2. Soil C Pool and Fluxes

In addition to improving the database on total system C pool, it is also important to understand the magnitude of fluxes and factors affecting them. Fluxes are proportional to the changes in pools, and both pools and fluxes are influenced by land use and management. Strategies for maintaining C pools (stocks) at antecedent levels are as important as those for enhancing pools (stocks). In most tropical ecosystems, enhancing C pools in soil may be difficult. Therefore, strategies for maintaining pools are extremely important.

3. Importance of Charcoal

Because fire is an important management tool (Plate 2), charcoal (Plate 3) may constitute a large component of total system carbon especially in the top 50 cm layer. The relative magnitude and dynamic of charcoal should be evaluated, and procedures to preserve need to be developed.

4. Losses of C From the Ecosystem

Principal mechanisms of C losses out of the ecosystem include soil erosion, volatilization and translocation of dissolved organic carbon (DOC) and inorganic carbon (DIC). The magnitude of these losses depend on land use, management and ecological conditions and need to be assessed. There is a lack of scientific data on the impact of erosional and depositional processes on pools of SOC and SIC.

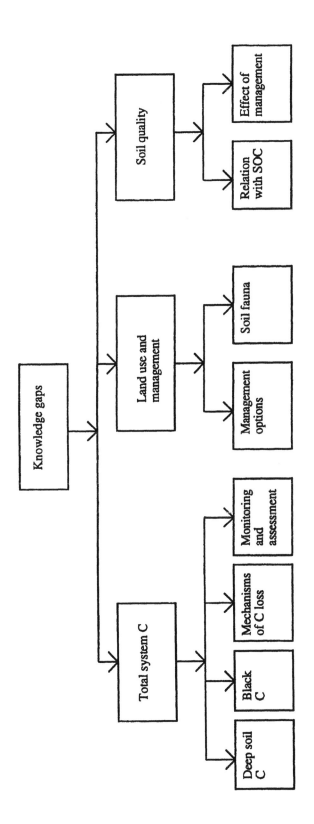

Figure 2. Principal knowledge gaps.

Plate 1a. A soil profile with C below 1 m depth.

Plate 1b. A typical profile with SOC concentration in the top 50 cm depth.

Plate 2. Fire in TRF is an important natural and anthropogenic factor.

Plate 3. Charcoal may comprise a considerable amount of the system C.

B. Land Use and Management

Land use and management play an important role in total system C pool and fluxes, and their impact needs to be quantified.

1. Land Use History

There is a strong need to quantify the impact of historical land use on C pools and fluxes. Systematic soil analyses and characterization need to be done for C pool and fluxes in relation to land use history for benchmark soils and ecoregions.

2. Climax Vegetation and C Fluxes

There is a need to develop a network of monitoring sites for representative soils in principal ecoregions to monitor the C flux in natural (TRF and TS including old and secondary growth) and managed ecosystems (plantation, pastures, enhanced fallows and cropland). The C sequestration potential of these systems must be quantified.

3. Management Options

Impact of soil and vegetation management on C pool and dynamics need to be assessed for diverse soils, land uses and cropping/farming systems. The effect of soil fertility management (N, P, K, and micronutrients), soil and water conservation and water management (e.g., irrigation, drainage), crop rotations, introduction of improved pastures, etc. on C pool and dynamics need to be evaluated.

4.Natural and Managed Fire

The impact of natural and managed fire on total system C pool and dynamics needs to be understood, especially in relation to seasonality in biomass production, nutrient cycling, gaseous emissions, and the charcoal component.

5. Soil Biota

Soil macro- and microfauna play an important role in carbon pool and dynamics. Activity and species diversity of soil fauna depend on land use and management and need to be quantified (Plate 4).

C. Processes and Mechanisms

There is a strong need to understand processes and mechanisms governing C pool and dynamic in tropical natural and managed ecosystems.

Plate 4. Termite mounds (white spots) in a TRF converted to pasture in the Amazon region of Brazil.

1. Net Primary Productivity (NPP) and Total System Carbon

The relationship between the aboveground biomass, NPP and the total system C depend on a multitude of interacting factors, e.g., soil, rainfall, water balance, vegetation, and land use and management. Quantification of these relationships is a research priority.

2. Potential, Attainable and Actual Total System Carbon

Ecoregional characteristics, land use and management systems determine the potential, attainable and actual total C and SOC pools. Conceptual and experimental relationships must be established between management scenarios and potential vs. attainable vs. actual total system C and SOC pool for different ecoregional characteristics.

3. Input vs. Fluxes

Fluxes of radiatively-active gases from terrestrial ecosystems to the atmosphere depend on both natural and anthropogenic inputs into the system. Establishing the cause-effect relationship between input and fluxes is very important. Such relationships would differ among ecoregions, vegetation, land use and soil types.

4. Soil Quality Dynamic

Total system C and SOC content are strong determinants of soil quality. Indices of soil quality should be determined in relation to SOC and SIC contents, aggregation and other soil attributes, and cause-effect relationship must be established between soil quality and fluxes of radiatively-active gases.

D. Soil Carbon Assessment and Prediction

There are numerous uncertainties about the research data and available information on C pool and its dynamic in tropical ecosystems (Figure 3).

1. Standardizing Methods

There is a conspicuous lack of standard methods for assessment of total system carbon and its components. The adequate assessment and monitoring of carbon in soils implies paying close attention to the standardization of soil sampling, sample preparation and storage, sample pretreatment and analytical procedures. Laboratory procedures should be developed for assessment of different components, and procedures involving isotopic methods should be simplified. There is a strong need to develop scaling procedures based on landscape units and watersheds involving remote sensing and GIS.

2. Soil Bulk Density Measurements and Soil Physical Quality Assessment

The usefulness of available data on SOC and SIC contents in relation to the global C cycle can be greatly enhanced by quantitative data on soil bulk density for corresponding horizons. There is a need to standardize methods for measurement of soil bulk density for sub-soil horizons of forested ecosystems containing large roots (Plate 5). There is a strong relationship between soil physical quality (e.g., aggregation, aeration, soil moisture regime) and C sequestration, and this relationship must be assessed. Some Andosols have low bulk density, are highly aggregated and less susceptible to erosion (Plate 6). The SOC dynamics of such soils need to be calculated under different cropping/farming systems.

3. Modeling

There is a need to model the impact of land use and different scenarios on C pool and dynamics. Soil and ecoregion-specific data are needed to run and validate various models. These data include information on (i) land use and cover history, (ii) soil type and vegetation characteristics, and (iii) canopy architecture and leaf physiology. Two specific questions that need to be addressed through modeling are (i) whether C sequestration in soil and terrestrial ecosystems is able to promote a significant reduction in the greenhouse radiative forcing at regional scales, and (ii) how potential climate change in ecologically sensitive ecoregions may alter the NPP and SOC pool.

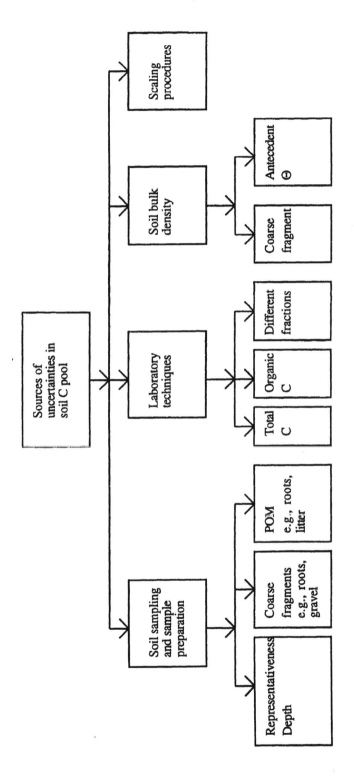

Figure 3. Sources of uncertainties in assessing soil carbon pool.

Plate 5. Measurement of soil bulk density around large roots is a challenge.

Plate 6. Highly aggregated Andosols in Rwanda have low bulk density and low erodibility.

E. Data Uncertainties, Reliability and Objectivity

Deforestation, global warming and biodiversity are important issues of the modern era. These issues also evoke emotions and are politically charged. It is of utmost importance, therefore, that scientists maintain objectivity and credibility. Maintaining objectivity and high scientific quality are necessary prerequisites to developing /identifying policy guidelines for sustainable management of natural resources. Rather than getting stuck in emotional rhetoric and regional political agendas, the scientific community needs to provide the strong/credible databases required to make objective decisions. Scientists should maintain dialogue and establish channels of communication with peers in other disciplines, policy makers, land managers and the public at large. Scientific objectivity and credibility must be maintained.

References

Moormann, F.R., R. Lal and A.S.R. Juo. 1975. Soils of IITA. Technical Bulletin 1. IITA, Ibadan, Nigeria.

National Research Council. 1993. *Soil and Water Quality: an Agenda for Agriculture*. National Academy Press, Washington, D.C. 516 pp.

Index

accelerated runoff 147, 157

active pool 266, 268

aggregates 85, 91,185-188, 190, 251,271, 272, 276, 294, 317-321, 323-327, 421

aggregation 17, 147, 148, 176, 190, 264, 265, 291, 293, 294, 299, 317-319, 321, 322, 324-327, 376, 385, 386, 407, 421, 428

agroforestry 17, 29, 90, 108, 110-112, 157, 270, 271, 274, 324, 327

Alfisols 3, 9, 10, 17, 26, 34, 35, 39, 40, 74, 77, 78, 91, 120, 148, 157, 283, 290, 292, 293, 297, 299, 302, 318, 319, 321, 322, 325, 379, 420

alley cropping 278

annual balance 231-233, 241-246

anthropogenic climate change 375

arable crops 278

arid ecosystems 76, 79, 83

Aridisols 3, 9, 40, 56, 61, 74, 76,77, 256, 319

aridity 89

atmosphere 30, 57, 59, 60, 89,91, 99, 157, 197, 208, 216, 219, 226, 231, 233, 251, 252, 254, 284, 295, 307, 317, 333, 341, 366, 373, 376, 378, 391, 392, 395, 398, 402, 405, 407, 410, 413, 417, 421, 427

balloon 392-395, 398, 399, 401, 402

biochemical approaches 335

biodiversity 17, 102, 113, 299, 319, 327, 381, 431

biomass 3, 9, 11, 14, 17, 19, 20, 25, 26, 30, 51, 53-55, 58, 59, 61-65, 90, 99, 102-105, 107, 109, 112, 119, 120, 127, 131, 136, 138-141, 147, 155, 160, 163, 173, 174, 177-182, 184, 185, 207, 213, 215-224, 226, 256, 266, 268, 275, 277, 278, 283, 287, 289, 291-294, 297, 298, 300, 303-307, 319, 321, 322, 324-327, 333-335, 348, 370, 375, 376, 378, 379, 381, 385, 392, 407, 408, 410, 417, 420-422, 426, 427

biotic activity 157

bulk density 29, 33, 36, 37, 41, 71, 72, 87, 102, 103, 148, 158, 160-162, 198, 254, 271, 272, 290, 301, 348, 355, 368, 375, 379, 428, 430

burn 15, 61, 63, 99-102, 104-111, 113, 118, 205-208, 215, 216, 219, 224-226, 231-236, 238, 239, 241, 243, 245, 246, 373

burning 11, 12, 14, 19, 20, 29, 51, 56, 57, 59-61, 63, 107, 117, 130, 185, 198, 205, 207, 216, 219, 222, 225, 226, 231-234, 237, 239, 241-246, 261, 293, 295, 298, 305, 306, 355, 366, 383, 406, 407, 410, 413, 420

burning efficiency 233, 234, 237, 239, 410

burning frequency 205, 207

burrowing 147

C:N ratio 25, 139, 140, 151, 152, 160, 161, 266, 268, 269

CANDY 342, 343, 345

canopy 4, 17, 52, 102, 285, 304, 371, 391, 392, 394, 395, 398-400, 410, 428

carbohydrate 225, 226

carbon dynamics 104, 107, 110, 111, 113, 193, 222, 224, 225, 239, 251, 261, 262, 264, 265, 268, 341, 342,346,349,370

cation exchange capacity 40, 122, 136, 169, 176, 186, 187, 251, 264, 298, 348, 368

cattle ranching 172, 213, 239

CENTURY model 111, 112, 294-296, 298, 304

cerrado 10, 174-176, 186, 187, 190, 232, 235, 236, 245, 379, 405, 411-413

charcoal 55, 56, 65, 102, 217, 218, 224, 232-234, 237, 241-245, 293, 295, 298, 305, 307, 333-336, 375, 379, 422, 425, 426

chronosequences 101, 103, 108, 113, 198, 206, 353, 354

clay content 37, 83, 122, 179, 188, 201, 208, 265, 284, 285, 287, 290-294, 305, 319, 327, 348, 353-355, 392

clear-cutting 33, 383, 420

climate 3, 10, 11, 16, 17, 41, 58, 71,78, 79, 83, 85, 87, 90, 91, 119, 147, 157, 172-174, 205, 215, 218, 231, 245, 256, 261, 262, 265, 269, 283, 287, 293, 294, 305-307, 319, 321, 327, 333, 335, 338, 341, 342, 346, 347, 349, 356, 358, 366, 368, 371, 373, 375, 376, 384, 386, 391, 392,

405-407, 413, 428
climate change 231, 245, 307, 333, 356, 375, 376, 386, 406, 428
CO_2 flux 99, 100, 239, 378, 391, 395, 398, 401, 402, 407, 412
collodial fraction 317, 326
compost 266, 270, 271, 273, 278, 321
contamination 157, 169
cover crops 12, 17, 26, 29, 148, 152, 253, 321, 324, 327
crop management practices 12, 179, 262, 290, 292, 299, 307, 327
crop production 3, 109, 117, 119, 127, 136, 205, 226, 251, 254, 256, 262, 273, 275
crop residue 17, 25, 117, 120, 123, 127-131, 133, 138-140, 147, 148, 160, 163, 252, 271, 299, 306, 321, 322
crop rotations 61, 278, 300, 301, 426
cropping period 215, 216, 219, 222, 224, 226, 261
crusting 123, 147, 155, 157, 163, 326, 327
cultivation 12, 16, 17, 21, 61, 76, 91, 99, 101, 111, 117, 119, 131, 136, 141, 150-152, 157, 159-162, 170, 172, 184, 185, 187, 189, 191, 193, 205, 206, 213-216, 227, 240, 251, 252, 254, 256, 262, 272, 273, 275, 276, 278, 289, 290, 292-297, 301, 305-307, 321, 322, 324, 327, 385
cultural practices 89, 150, 284, 327
cycling 17, 51, 63, 117, 124, 127, 140, 141, 155, 177, 178, 207, 375, 421, 426
DAISY 342, 343, 345
data interpretation 337
decay 57, 64, 139, 172, 187, 188, 231-233, 235-239, 241-243, 245, 246
decomposition 11, 12, 27-29, 51, 57, 60, 63, 64, 76, 78, 89, 91, 120, 131, 133, 136, 139, 167, 172, 173, 176, 177, 179, 184, 190, 191, 193, 205, 206, 218, 222, 225, 231, 232, 237, 238, 245, 251, 256, 283, 284, 287, 289-293, 295-297, 302, 305-307, 319, 321, 322, 326, 327, 341, 345, 346, 371, 421
deep-rooted trees 207
deforestation 9, 10, 12-15, 17-19, 21, 29, 99, 109-111, 113, 117, 123, 147, 170, 185, 197, 207, 208, 213, 231-233, 237, 239, 240, 242, 244-246, 275, 321, 327, 355,

373, 375, 376, 378, 379, 417, 420, 431
denuded soil surfaces 29
desert 56, 80,89, 117-120, 122, 123, 127, 129, 131, 134, 141, 264, 265, 285, 289
diffusion 317
disk harrow 174, 176, 184-187, 189, 191
disking 198, 205, 206, 208
DNDC 342, 343, 345
earthworm activity 148, 152, 155, 321
earthworm casts 148, 154, 155
earthworms 147, 148, 152, 154, 155, 319, 321, 325
ecosystems 3, 9-12, 26, 30, 33, 51, 54-56, 58, 76, 82, 87, 105, 147, 157, 158, 205, 217, 226, 251, 262, 264, 266, 268, 272, 273, 284, 285, 289, 290, 304, 319, 341, 342, 349, 352, 373, 375, 378, 380, 383, 386, 391, 407, 413, 417, 420-422, 426-428
Eddy correlation 238, 239, 407, 408, 412
enrichment plantings 220, 226
enrichment ratio 154, 155, 302
enzymes 268, 291-293, 317, 319, 321
EPIC 342, 343, 346, 349
evaporation 284, 298, 348, 354, 356
Entisols 10, 34, 35, 40, 72, 74, 120, 169
fallow 61, 101, 102, 104, 107, 109, 111, 117, 131, 133, 136, 141, 198, 205, 206, 213-227, 252-254, 261, 268-270, 290, 325, 347, 352, 353
fallow management 218, 219, 226, 227
faunal activity 375
fallow periods 215, 216, 218, 220, 224-226
fertility 12, 16, 17, 25, 61, 63, 100, 111, 117, 122, 127, 136, 141, 155, 157, 159, 174, 179, 186, 197, 215, 254, 261,266, 270, 271, 274, 275, 284, 290, 293, 294, 297, 302, 305, 307, 325, 426
fertilizer 25, 28, 29,60, 61, 64, 91, 117, 119, 127-131, 133, 136, 138-141, 147, 148, 155, 159-161, 214, 215, 219, 222, 226, 252, 254, 261, 266, 269-271, 273-275, 299, 301-303, 306, 307, 322, 323, 348
fire 55, 56, 59-61, 64, 65, 107, 219, 225, 226, 298, 302, 305-307, 420, 422, 425, 426
flocculation 317, 326
floodplains 380

fluxes 51, 52, 57, 123, 124, 158, 198, 231, 232, 238, 239, 245, 284, 289, 303-305, 307, 333, 334, 358, 375, 383, 386, 391, 392, 394, 395, 397, 399, 401, 405, 407, 408, 410, 413, 420, 422, 426-428

forest 8-11, 13-17, 28, 33, 38, 46, 48, 51-61, 63-65, 75, 79, 81, 91, 99-112, 177, 179, 185, 197-208, 213, 215, 217-219, 226, 231-246, 251, 262, 263, 265, 267, 268, 270, 274, 276, 284, 285, 287, 289, 292, 298, 322-324, 326, 342, 352-355, 365, 371, 376, 378-381, 383, 385, 391, 392, 394, 396-398, 401, 402, 405, 410-412, 417, 419, 422

gaseous exchange 317

geology 41, 376

geomorphology 376

global carbon content 71

global environmental markets 226

grain production 172, 302

graphitic particulate carbon 233, 245

grassland 60, 61, 64, 101, 104, 106, 107, 197, 283, 284, 287, 289, 297, 298, 307, 322, 326, 356

greenhouse effect 30, 71, 147, 157, 284, 299, 391, 407, 420, 421

harvest debris 59

human activity 208

human-induced land use change 206

humic substances 319, 322, 324

humus 11, 25, 28, 89, 254, 262, 266, 317, 319, 343, 347, 378, 380, 385

hydroelectric dams 239-241, 243

hydrometer method 148

Inceptisols 3, 34, 35, 39, 72, 74, 75, 77, 274, 290, 319, 371

indicators 51, 155, 197, 266

infiltration 123, 126, 129, 148, 251, 298-300, 327

infiltrometer 148

inherited emissions 231, 237, 242, 245

inorganic carbon 55, 65, 157, 333, 335, 336, 421, 422

intensive farming 273

isotopic methods 428

laboratory procedures 428

land clearing 12, 19, 20, 51, 56, 57, 60, 100, 206, 215, 226, 275, 284, 289, 290, 301, 306, 307, 321

land misuse 157

LANDSAT satellite imagery 305

leaching 56, 76, 122, 147, 157, 160, 163, 216, 269, 270, 298, 301

lime 141, 176, 184, 273, 276

litter 5, 9, 10, 55, 56, 58, 60, 62-65, 102, 103, 105, 107, 109, 217, 220, 221, 223, 238, 262, 266, 268, 278, 348, 3555, 356, 371, 376, 420

livestock population 172

logging 57, 59, 63, 64, 99, 105, 109, 232, 233, 239, 242, 244-246, 420

low-input agriculture 11, 12

Luvisols 34, 148

machinery 214, 216, 219, 225, 226

manure 138, 141, 176, 184, 252, 292, 297, 299, 302, 321, 327, 348, 350

mean annual temperature 157, 190, 215, 264, 265, 283, 319

mean weight diameter 186, 325

mechanized clearing 11

methane 235-239, 336

microbial activity 139, 176, 177, 179, 407

microbial biomass 136, 138-140, 173, 177, 179, 185, 266, 268, 277, 283, 292-294, 333-335

microorganisms 136, 139, 177, 185, 251, 264, 268, 291-293, 298, 319, 321

migrant farmers 110

migrants 101

mineralization 11, 51, 139, 140, 176, 179, 185, 190, 268, 293, 407, 421

modeling 307, 341, 342, 349, 352, 353, 356, 358, 365, 373, 376, 380, 384, 391, 392, 428

models 44, 111, 207, 222, 284, 295, 296, 306, 341, 342, 345, 349, 351, 352, 355, 356, 365, 369, 370, 373, 376, 405, 421, 428

moisture regimes 8, 9, 319, 322, 327

Mollisols 9, 35, 39, 73, 74, 76, 77, 83, 90, 318, 319

mulch 17, 29, 118, 119, 123, 124, 126, 127, 129, 131, 136, 141, 148, 149, 218, 219, 224-226, 252-254, 271, 321, 322

natural fallow 104, 107, 109, 111, 221

natural resources 29, 30, 366, 371, 376, 431

NCSOIL 342, 344, 347

nitrogen 58, 60, 120, 122, 131-133, 148, 151, 152, 160, 161, 184, 185, 215, 216, 220, 251-256, 262, 266, 268, 275, 298, 301, 303, 304, 342, 348, 368-370, 407

no till 12, 17, 29, 147, 157-164, 175, 252, 256, 296, 299, 300, 307, 321-323, 327, 350

nutrient cycling 17, 63, 117, 124, 141, 177, 207, 426

nutrient imbalance 147

nutrient losses 172, 216, 219, 226

nutrient management 147, 322

organic amendments 117, 129, 136, 141, 155, 327

organic carbon 11, 33, 37, 39, 48, 55, 58, 60-62, 64, 71, 74-82, 85-87, 91, 102, 105, 107, 111, 117, 120, 122, 124, 127, 129, 131, 133, 136, 138, 141, 147, 148, 151, 157, 161, 162, 172-179, 182, 184-186, 188-190, 193, 215, 218, 219, 251, 252, 255, 261-278, 284-307, 317, 323, 326, 327, 333-336 350, 356, 358, 366-369, 371-373, 380, 407, 408, 410, 417, 421, 422

organic matter 9-12, 17, 51, 54, 55, 57, 59-61, 63, 65, 71, 85, 89, 122, 126, 130, 133, 135, 136, 138-141, 152, 155, 169, 172-174, 176, 177, 184, 185, 187, 188, 190-192, 204-206, 208, 218, 224, 226, 251-254, 256, 264, 266, 269, 270, 283-285, 287, 289-291, 294-302, 307, 308, 317, 318, 321-323, 326, 333, 335, 337, 341, 342, 348, 349, 353, 356, 358, 370, 375, 378, 380

Oxisols 3, 9, 10, 34, 35, 39, 42, 74, 78, 79, 169, 172, 198, 201, 205, 207, 283, 293, 299, 300, 302, 318, 319, 378

ozone layer 71

pasture 17, 25, 28-30, 48, 51, 52, 58, 60, 61, 64, 100, 102, 104, 107, 109, 110, 147, 172, 174-176, 186, 187, 190, 191, 193, 197-208, 213, 215, 231, 232, 234-238, 240-246, 274, 289, 298-304, 306, 307, 322-325, 342, 343, 346, 347, 350, 352-355, 379, 405, 419, 420, 427

pasture age 198, 201-204, 208

pasture conservation 208

pasture productivity 303, 307

penetrometer 148

perhumid ecosystem 82

PET 119, 283, 284, 319

pH 37, 39, 43, 51, 83, 88, 122, 126, 127, 136, 137, 139, 147, 150-153, 162, 169, 174, 177, 179, 184, 215, 251, 273, 275-277, 285, 296, 298, 345-348, 365-370, 392

phosphorus 58, 60, 61, 117, 119, 122, 131, 132, 141, 160, 169, 172, 174, 184, 191, 205, 215, 216, 220, 253, 302, 348, 368

photosynthesis 317, 407, 410

plant nutrients 91, 122, 136, 141, 155, 176, 179, 251, 269, 270, 273, 298, 321

plant nutritional problems 185

plantation 13, 16, 22-24, 29, 51-65, 99, 100, 102, 110, 111, 174, 213, 215-218, 262, 271, 274-278, 323, 324, 405, 419-421, 426,

plastic mulch 124, 126

plowing 11, 12, 17, 159, 163, 172, 190, 219, 271, 321, 322

plow till 17, 29, 147, 149, 157-164, 321-323

pollution 157, 270

pores 317, 318, 326

porosity 123, 147, 392, 407

potential evapotranspiration 120, 283, 284, 286, 291, 319, 326

precipitation 89, 174, 197, 198, 201, 204, 205, 215, 253, 254, 256, , 261-264, 284, 290, 291, 319, 326, 319, 326, 335, 348, 354, 366, 407-409

primary forest 51,

primary forests 216, 232, 238

prompt emissions 231, 242

protein 225, 226

rain forest 55, 101, 262, 284, 289, 298, 412

rainfall 3, 8-10, 12, 58, 71, 79, 82, 85, 90, 91, 117-120, 122, 123, 126, 131, 133, 148, 169, 174, 205, 251, 254, 262, 264, 265, 267, 270, 283, 284, 286, 287, 290, 298, 302, 304, 305, 307, 356, 379, 380, 392, 394, 395, 407, 408, 427

rainfed agriculture 117

reforestation 65, 89

regeneration 59, 61, 109, 126, 208, 216, 350

regrowth 52, 54, 57, 59-61, 63, 65, 100,

109, 111, 159, 163, 164, 216, 219, 220, 235, 236, 245, 246, 302
retentive capacity 293
remote sensing 48, 376, 381, 428
residual slash 63
rhizosphere 138, 319
root biomass 17, 58, 62, 102, 107, 207, 217, 224, 268, 298, 300, 324, 421, 422
roots 9, 10, 17, 55-58, 60, 62, 64, 102, 104, 107, 140, 275, 278 141, 163, 177, 184, 197, 205, 206, 216, 217, 237, 240, 256, 290, 303, 307, 318, 319, 375, 378-381, 407, 428, 430
RothC 342, 344, 347, 349-356
row crop agriculture 208
runoff 122, 123, 126, 139, 147-149, 155, 157, 160, 163, 169, 172, 271, 325, 327
savanna 3, 5-8, 10-12, 16, 17, 25-27, 30, 57, 60, 64, 169, 177, 179, 184, 245, 285, 306, 307, 321, 379, 385, 405, 417, 419
secondary forest 102, 104, 109, 207, 208, 218, 219, 231-234, 237, 240-244, 246
secondary succession 214-216, 218
secondary vegetation 205, 214-216, 220, 240
semiarid ecosystem 79, 83
site preparation 59, 63, 64
slash-and-burn 15, 99-102, 104-111, 113, 206, 208, 215, 216, 219, 224-226, 238
small farmers 205, 213, 214, 226, 240
small holdings 214, 216, 220
soil acidification 124, 141
soil biological processes 136
soil chemical properties 129, 131, 150, 151, 162
soil compaction 17, 29, 123, 271, 324, 325
soil degradation 12, 30, 123, 147, 151, 158, 159, 161, 163, 262, 284, 307, 325, 326
soil fauna 147, 155, 208, 319, 321, 322, 325, 327, 380, 407, 421, 426
soil fertility 12, 16, 17, 25, 63, 117, 127, 136, 157, 159, 174, 186, 197, 215, 254, 261, 266, 270, 271, 275, 293, 294, 297, 305, 307, 325, 426
soil management practices 158, 293, 337
soil matrix 284, 290
soil moisture content 173, 179, 182, 321
soil nitrogen 151, 160, 161, 348

soil organic matter 51, 54, 55, 57, 59, 60, 65, 126, 135, 138, 141, 172, 173, 176, 177, 179, 185, 187, 188, 190, 206, 224, 226, 251, 253, 254, 266, 283, 284, 289, 295, 297-300, 307, 308, 322, 326, 333, 335, 337, 341, 342, 349, 356, 358, 378
soil physical properties 91, 124, 139, 193, 254, 272, 321, 375
soil productivity 91, 124, 127, 139, 251, 261, 298, 299
soil properties 3, 16, 51, 60, 126, 162, 187, 190, 218, 269, 277, 287, 306, 365
soil resilience 147
soil respiration 136, 138, 398, 405, 407, 408, 410-413
soil structure 51, 147, 155, 157, 163, 172, 176, 208, 251, 264, 321, 322, 325-327
soil survey 34, 35, 37, 71, 72, 76, 334, 336, 352, 366, 378, 383
soil temperature 12, 60, 89, 129, 173, 271, 289, 304, 319, 321, 322, 327, 342, 348, 407, 408, 410-413
SOMNET 341-343, 345, 356
SOTER system 384
spatial scale 34, 402
spatial variability 33, 41, 44, 48, 336, 349, 411, 413
stocking rate 17, 176, 191, 198, 205, 207, 298, 302, 304-307, 322, 324
stover 118, 119, 123-126, 128-131, 133
subhumid ecosystem 79
surface area 76, 85, 87, 89, 91, 319
surface fluxes 392, 401, 405
sustainable agriculture 176, 270, 275, 299
sustainable cropping practices 307
sustainable production 117
temperate climate 10, 11, 157, 261
temperate forests 9, 10, 12, 58, 197
termites 126, 235-238, 240-244, 298, 378, 380
terraces 264, 270, 380
texture 28, 100, 185, 198, 264, 295, 296, 319, 321, 325, 379, 383, 385
thinning 61-63
tillage intensity 321
tillage systems 160, 163, 174, 190-193, 252, 253, 256, 261, 272
topography 123, 385, 395, 402

trace gas 208, 231, 233, 235-237, 241-245

tree density 284

tropics 3, 7-12, 17, 29, 30, 89, 91, 99, 100, 109, 110, 120, 121, 127, 131, 147, 157, 158, 164, 213, 262, 266, 271, 285, 289, 295, 298, 301, 304-307, 325, 327, 341, 342, 349, 356, 373, 378, 381, 407, 417, 419, 420

turonover periods 289, 294, 295

Ultisols 3, 9, 10, 34, 39, 40, 74, 83, 169, 198, 201, 205, 207, 215, 290, 293, 299, 319, 379

vegetation 5-7, 9, 10, 14, 33, 41, 48, 51, 52, 54, 56-61, 71, 77, 79, 81, 82, 90, 99, 100, 102, 105, 107, 109, 117, 148, 157, 169, 170, 172, 174-176, 179, 184-186, 204, 205, 207, 214-223, 225, 227, 231-233, 237, 238, 240, 262, 264, 270, 283, 284, 286, 287, 289, 290, 293, 295, 298, 303-305, 308, 344, 366, 371, 373, 375, 376, 378, 379, 381, 385, 391, 392, 395, 402, 405, 410, 412, 413, 426-428

Vertisols 3, 9, 17, 27, 40, 60, 74, 76, 283, 290-292, 294, 299-303, 319, 322

volatilization 139, 160, 216, 219, 261, 269, 422

Walkley-Black method 334, 336, 379, 402

water balance 298, 427

water erosion 117, 123, 261, 296

watershed 158-164

wind erosion 123, 124

windrowing 63

windspeed 391, 396

winter cover crops 253

woodchip 59, 65

wood density 232

woodland 51, 52, 54-56, 60, 65, 173, 262, 285, 287, 289, 301, 304, 306, 344, 350